Lecture Notes in Electrical Engineering
Volume 35

For further volumes:
http://www.springer.com/series/7818

Sukhan Lee · Hanseok Ko · Hernsoo Hahn
Editors

Multisensor Fusion and Integration for Intelligent Systems

An Edition of the Selected Papers from the IEEE International Conference on Multisensor Fusion and Integration for Intelligent Systems 2008

Editors
Prof. Dr. Sukhan Lee
School of Information and
Communication Engineering
Sungkyunkwan University
300 Chunchun-Dong
Jangan-Ku, Kyunggi-Do 440-746
Korea
Lsh@ece.skku.ac.kr, Lsh1@ieee.org

Prof. Dr. Hanseok Ko
School of Electrical Engineering
Korea University
Seoul, 136-713
Korea
hsko@korea.ac.kr

Prof. Dr. Hernsoo Hahn
School of Electronic Engineering
Soongsil University
511 Sangdo-Dong, Dongjak-gu
Seoul 156-743
Korea
hahn@ssu.ac.kr

ISSN 1876-1100 e-ISSN 1876-1119
ISBN 978-3-540-89858-0 e-ISBN 978-3-540-89859-7
DOI 10.1007/978-3-540-89859-7
Springer Dordrecht Heidelberg London New York

Library of Congress Control Number: 2009920215

Cover design: eStudio Calamar S.L.

Printed on acid-free paper

Springer is part of Springer Science+Business Media (www.springer.com)

Preface

The field of multi-sensor fusion and integration is growing into significance as our society is in transition into ubiquitous computing environments with robotic services everywhere under ambient intelligence. What surround us are to be the networks of sensors and actuators that monitor our environment, health, security and safety, as well as the service robots, intelligent vehicles, and autonomous systems of ever heightened autonomy and dependability with integrated heterogeneous sensors and actuators. The field of multi-sensor fusion and integration plays key role for making the above transition possible by providing fundamental theories and tools for implementation.

This volume is an edition of the papers selected from the 7th IEEE International Conference on Multi-Sensor Integration and Fusion, IEEE MFI'08, held in Seoul, Korea, August 20–22, 2008. Only 32 papers out of the 122 papers accepted for IEEE MFI'08 were chosen and requested for revision and extension to be included in this volume. The 32 contributions to this volume are organized into three parts: Part I is dedicated to the Theories in Data and Information Fusion, Part II to the Multi-Sensor Fusion and Integration in Robotics and Vision, and Part III to the Applications to Sensor Networks and Ubiquitous Computing Environments. To help readers understand better, a part summary is included in each part as an introduction. The summaries of Parts I, II, and III are prepared respectively by Prof. Hanseok Ko, Prof. Sukhan Lee and Prof. Hernsoo Hahn.

It is the wish of the editors that readers find this volume informative and enjoyable. We would also like to thank Springer-Verlag for undertaking the publication of this volume.

Kyunggi-Do, Korea — *Sukhan Lee*
Seoul, Korea — *Hanseok Ko*
Seoul, Korea — *Hernsoo Hahn*

Contents

Contributors

Richard Altendorfer TRW Automotive, 56070 Koblenz, Germany, richard.altendorfer@trw.com

I. Amorim Department of Electrical and Computer Engineering, Institute of Systems and Robotics, University of Coimbra, Polo II, 3030 Coimbra, Portugal, ivonefamorim@isr.uc.pt

Elisabeth André Institute of Computer Science, University of Augsburg, Eichleitnerstr. 30, D-86159 Augsburg, Germany, andre@informatik.uni-augsburg.de

Hajime Asama Research into Artifacts, Center for Engineering, The University of Tokyo, Chiba, Japan

Joachim Bamberger Information and Communications, Corporate Technology, Siemens AG, Otto-Hahn-Ring 6, Munich, Germany, joachim.bamberger@siemens.com

Mohammed Benjelloun Université du Littoral Côte d'Opale, LASL 62228 Calais Cedex, France, Mohammed.Benjelloun@lasl-gw.univ-littoral.fr

Jeroen Bergmans TNO Defence, Security and Safety, Oude waalsdorperweg 63, The Hague, The Netherlands

Christophe Blanc Laboratoire des Sciences et Matériaux pour l'Electronique et d'Automatique, Université de Clermont-Ferrand, 24, avenue des Landais, 63177 Aubière cedex, France, Blanc.Christophe@lasmea.univ-bpclermont.fr

Rett Boehling Department of Mechanical Engineering, Virginia Tech, Blacksburg, VA 24060, USA

Herman Bruyninckx Department of Mechanical Engineering, Katholieke Universiteit Leuven, Celestijnenlaan 300B, B-3001 Leuven, Belgium, first.last@mech.kuleuven.be

Nina Camoriano Department of Mechanical Engineering, Virginia Tech, Blacksburg, VA 24060, USA

Alexander Carballo Intelligent Robot Laboratory, Graduate School of Systems and Information Engineering, University of Tsukuba, 1-1-1 Tennoudai, Tsukuba City, Ibaraki Pref., 305-8573, Japan, alexandr@roboken.esys.tsukuba.ac.jp

Wes Cardwell Department of Mechanical Engineering, Virginia Tech, Blacksburg, VA 24060, USA

François Chanier Laboratoire des Sciences et Matériaux pour l'Electronique et d'Automatique, Université de Clermont-Ferrand, 24, avenue des Landais, 63177 Aubière cedex, France, francois.chanier@lasmea.univ-bpclermont.fr

Paul Checchin Laboratoire des Sciences et Matériaux pour l'Electronique et d'Automatique, Université de Clermont-Ferrand, 24, avenue des Landais, 63177 Aubière cedex, France, Checchin.Paul@lasmea.univ-bpclermont.fr

Véronique Cherfaoui Heudiasyc UMR CNRS 6599, Université de Technologie de Compiègne, Centre de Recherche de Royallieu – BP 20529, 60205 Compiègne Cedex, veronique.cherfaoui@hds.utc.fr

Sung-Bae Cho Department of Computer Science, Yonsei University, 262 Seongsanno, Sudaemun-gu, Seoul 120-749, Korea, sbcho@cs.yonsei.ac.kr

Jun-Hyuk Choi Department of Mechanical Engineering, Chung-Ang University, 221 Huksuk-Dong, Dongjak-Ku, Seoul 156-756, South Korea, miyasaki@wm.cau.ac.kr

Kasper Claes Department of Mechanical Engineering, Katholieke Universiteit Leuven, Celestijnenlaan 300B, B-3001 Leuven, Belgium, kasper.claes@mech.kuleuven.be

Tinne De Laet Department of Mechanical Engineering, Katholieke Universiteit Leuven, Celestijnenlaan 300B, B-3001 Leuven, Belgium, tinne.delaet@mech.kuleuven.be

Joris De Schutter Department of Mechanical Engineering, Katholieke Universiteit Leuven, Celestijnenlaan 300B, B-3001 Leuven, Belgium, joris.deschutter@mech.kuleuven.be

J. Dias Department of Electrical and Computer Engineering, Institute of Systems and Robotics, University of Coimbra, Polo II, 3030 Coimbra, Portugal, jorge@isr.uc.pt

Fadi Fayad Heudiasyc UMR CNRS 6599, Université de Technologie de Compiègne, Centre de Recherche de Royallieu – BP 20529, 60205 Compiègne Cedex, fadi.fayad@hds.utc.fr

F. Ferreira Department of Electrical and Computer Engineering, Institute of Systems and Robotics, University of Coimbra, Polo II, 3030 Coimbra, Portugal, cfferreira@isr.uc.pt

Christian W. Frey Fraunhofer Institute for Information and Data Processing, Fraunhoferstrasse 1, 76131 Karlsruhe Germany, christian.frey@iitb.fraunhofer.de

Yasushi Hada National Institute of Information and Communications Technology, Ishikawa, Japan

David K. Han United States Naval Acedemy, USA, han@usna.edu

SunSin Han Intelligent Robot Laboratory, Pusan National University, Korea, ranger112@pusan.ac.kr

Uwe D. Hanebeck Intelligent Sensor-Actuator-Systems Laboratory (ISAS), Institute of Computer Science and Engineering, Universitaet Karlsruhe (TH), Karlsruhe, Germany, uwe.hanebeck@ieee.org

Thomas C. Henderson School of Computing, University of Utah, USA, tch@cs.utah.edu

Dennis W. Hong Director of the Robotics and Mechanisms Laboratory (RoMeLa), Department of Mechanical Engineering, Virginia Tech, Blacksburg, VA, USA, dhong@vt.edu

Jesse G. Hurdus TORC Technologies, LLC 2200 Kraft Dr. Suite 1325, Blacksburg, VA 24060, USA, hurdus@torctech.com

Seok-Woo Jang School of Information and Communication Engineering, Sungkyunkwan University, Korea, swjang@skku.edu

Greg Jannaman Department of Mechanical Engineering, Virginia Tech, Blacksburg, VA 24060, USA

Changwon Jeon School of Electrical Engineering, Korea University, Seoul, Korea, cwjeon@ispl.korea.ac.kr

Bonghyup Kang Samsung Techwin CO., LTD, Korea, bh47.kang@samsung.com

Leon Kester TNO Defence, Security and Safety, Oude waalsdorperweg 63, The Hague, The Netherlands

Dongjun Kim School of Electrical Engineering, Korea University, Seoul, Republic of Korea, djkim@ispl.korea.ac.kr

Jonghwa Kim Institute of Computer Science, University of Augsburg, Eichleitnerstr. 30, D-86159 Augsburg, Germany, kim@informatik.uni-augsburg.de

Kihyeon Kim School of Electrical Engineering, Korea University, Seoul, South Korea

Tae-Kuk Kim Department of Mechanical Engineering, Chung-Ang University, 221 Huksuk-Dong, Dongjak-Ku, Seoul 156-756, South Korea, kimtk@cau.ac.kr

Shawn Kimmel Department of Mechanical Engineering, Virginia Tech, Blacksburg, VA 24060, USA

Hanseok Ko School of Electrical Engineering, Korea University, Seoul, Korea, hsko@korea.ac.kr

Christian Laugier INRIA RHONE-ALPES, Grenoble, France, yong.mao@inrialpes.fr

Sukhan Lee Intelligent Systems Research Center, Sungkyunkwan University, 300 Cheoncheon-dong, Jangan-gu, Suwon, Gyeonggi-do, 440-746, South Korea, lsh@ece.skku.ac.kr

JangMyung Lee Pusan National University, KeumJung Gu, JangJeon Dong, BUSAN, 609-735, South Korea, jmlee@pusan.ac.kr

Yong-Ju Lee Department of Mechanical Engineering, Korea University, Seoul, Republic of Korea, yongju_lee@korea.ac.kr

Soojung Lim Department of Computer Science, Yonsei University, 262 Seongsanno, Sudaemun-gu, Seoul 120-749, Korea, soojung@sclab.yonsei.ac.kr

Flavien Maingreaud Institut des Systèmes Intelligents et de Robotique (ISIR), Université Pierre et Marie Curie, France, maingreaud@robot.jussieu.fr

Yong Mao INRIA RHONE-ALPES, Grenoble, France, yong.mao@inrialpes.fr

Jan Willem Marck TNO Defence, Security and Safety, Oude waalsdorperweg 63, The Hague, The Netherlands, jan willem.marck@tno.nl

Philippe Martinet LASMEA, Blaise Pascal University, Clermont-Ferrand, France, Philippe.Martinet@lasmea.univ-bpclermont.fr

Yoshiteru Matsushita Department of Mechanical Engineering, Osaka University, matsushita@cv.mech.eng.osaka-u.ac.jp

Stephan Matzka Heriot-Watt University, Edinburgh, EH14 4AS, UK, sm217@hw.ac.uk

U. Mayer Institute for Data Processing and Electronics (IPE), Forschungszentrum Karlsruhe, Germany, mayer@ipe.fzk.de

Kamel Mekhnacha Probayes SAS, 38330 Montbonnot, France, kamel.mekhnacha@probayes.com

Jun-Ki Min Department of Computer Science, Yonsei University, 262 Seongsanno, Sudaemun-gu, Seoul 120-749, Korea, loomlike@sclab.yonsei.ac.kr

Jun Miura Department of Information and Computer Sciences, Toyohashi University of Technology, jun@ics.tut.ac.jp

Chongning Na Siemens AG, Corporate Technology, Information and Communications, Otto-Hahn-Ring 6, 81739, Munich, Germany, na.chongning.ext@siemens.com

Takuji Narumi Graduate School of Engineering, The University of Tokyo/JSPS, 7-3-1 Bunkyo-Ku Hongo, Japan, narumi@cyber.t.u-tokyo.ac.jp

Dragan Obradovic Siemens AG, Corporate Technology, Information and Communications, Munich, Germany, dragan.obradovic@siemens.com

Akihisa Ohya Intelligent Robot Laboratory, Graduate School of Systems and Information Engineering, University of Tsukuba, 1-1-1 Tennoudai, Tsukuba City, Ibaraki Pref., 305-8573, Japan, ohya@roboken.esys.tsukuba.ac.jp

Wei Chuan Ooi School of Electrical Engineering, Korea University, Seoul, Republic of Korea, wcooi@ispl.korea.ac.kr

Han-Saem Park Department of Computer Science, Yonsei University, 262 Seongsanno, Sudaemun-gu, Seoul 120-749, Republic of Korea, sammy@sclab.yonsei.ac.kr

Edwige Pissaloux Institut des Systèmes Intelligents et de Robotique (ISIR), Université Pierre et Marie Curie, France, pissaloux@robot.jussieu.fr

Mario Prats Robotic Intelligence Lab, Jaume-I University, Castellón, Spain, mprats@icc.uji.es

Alex Purcell Department of Mechanical Engineering, Virginia Tech, Blacksburg, VA 24060, USA

David Raulo Probayes SAS, Grenoble, France, kamel.mekhnacha@probayes.com

R. Rocha Department of Electrical and Computer Engineering, Institute of Systems and Robotics, University of Coimbra, Polo II, 3030 Coimbra, Portugal, rprocha@isr.uc.pt

Dan Ross Department of Mechanical Engineering, Virginia Tech, Blacksburg, VA 24060, USA

N. V. Ruiter Institute for Data Processing and Electronics (IPE), Forschungszentrum Karlsruhe, Germany, ruiter@ipe.fzk.de

Eric Russel Department of Mechanical Engineering, Virginia Tech, Blacksburg, VA 24060, USA

Pedro J. Sanz Robotic Intelligence Lab, Jaume-I University, Castellón, Spain, sanzp@icc.uji.es

Felix Sawo Intelligent Sensor-Actuator-Systems Laboratory (ISAS), Institute of Computer Science and Engineering, University of Karlsruhe, Germany, sawo@ira.uka.de

G. F. Schwarzenberg Institute for Data Processing and Electronics (IPE), Forschungszentrum Karlsruhe, Germany, schwarzenberg@ipe.fzk.de

Christopher Sikorski School of Computing, University of Utah, USA, sikorski@cs.utah.edu

Ruben Smits Department of Mechanical Engineering, Katholieke Universiteit Leuven, Celestijnenlaan 300B, B-3001 Leuven, Belgium, ruben.smits@mech.kuleuven.be

Jae-Bok Song Department of Mechanical Engineering, Korea University, Seoul, Korea, jbsong@korea.ac.kr

Taek Lyul Song Department of Control and Instrumentation Engineering, Hanyang University, Sa 1 Dong 1271, Sangnok-Gu, Ansan-Si, Gyeonggi-Do, 426-791, Republic of Korea, tsong@hanyang.ac.kr

Andrei Szabo Information and Communications, Corporate Technology, Siemens AG, Otto-Hahn-Ring 6, Munich, Germany, andrei.szabo@siemens.com

Clark N. Taylor Department of Electrical and Computer Engineering, Brigham Young University, Provo, UT 459 CB, 84602, clark.taylor@byu.edu

Khalid Touil Université du Littoral Côte d'Opale, Laboratoire d'Analyse des Systèmes du Littoral, (LASL-EA 2600), 62228 Calais Cedex, France, Khalid.Touil@lasl.univ-littoral.fr

Laurent Trassoudaine Laboratoire des Sciences et Matériaux pour l'Electronique et d'Automatique, Université de Clermont-Ferrand, 24, avenue des Landais, 63177 Aubière cedex, France, Trassoudaine.Laurent@lasmea.univ-bpclermont.fr

Hung Q. Truong School of Information and Communication Engineering, Sungkyunkwan University, Suwon, Republic of Korea, hungtruong@skku.edu

Kunihiro Tsuji Kunihiro Tsuji Design, Japan

Eelke van Foeken TNO Defence, Security and Safety, Oude waalsdorperweg 63, The Hague, The Netherlands

Miranda van Iersel TNO Defence, Security and Safety, Oude waalsdorperweg 63, The Hague, The Netherlands

Ramiro Velázquez Mecatrónica y Control de Sistemas (MCS), Universidad Panamericana, Mexico, rvelazquez@ags.up.mx

Hui Wang Information and Communications, Corporate Technology, Siemens AG, Otto-Hahn-Ring 6, Munich, Germany, hui.wang@siemens.com

Sung-Ihk Yang Department of Computer Science, Yonsei University, 262 Seongsan-ro, Sudaemoon-gu, Seoul 120-749, Republic of Korea, unikys@sclab.yonsei.ac.kr

Seokwon Yeom Department of Computer and Communication Engineering, Daegu University, Gyeongsan, Gyeongbuk, 712-714, South Korea, yeom@daegu.ac.kr

Shin'ichi Yuta Intelligent Robot Laboratory, Graduate School of Systems and Information Engineering, University of Tsukuba, 1-1-1 Tennoudai, Tsukuba City, Ibaraki Pref., 305-8573, Japan, yuta@roboken.esys.tsukuba.ac.jp

Mourad Zribi Université du Littoral Côte d'Opale, LASL 62228 Calais Cedex, France, Mourad.Zribi@lasl.univ-littoral.fr

Part I
Theories in Data and Information Fusion

Hanseok Ko

Information fusion is a concept describing a process that exploits the synergy offered by the information originating from various sources. By combining additional independent and/or redundant data, an improvement of the results can be obtained in the form of robustness and reliability, extended coverage and dimensionality in space and time, and reduced ambiguity, to name a few. Research in information fusion expresses the properties of the information to be fused, of the methods for fusion, of the architectures, thus permitting better design, implementation and analysis of fusion processes. This chapter reports on some recent efforts on information fusion with 11 contributions, addressing various forms of information fusion and their performance results by combining multiple information sources, all aimed at capturing the best possible synergy effects.

1. "Performance Analysis of GPS/INS Integrated System by Using a Nonlinear Mathematical Model": This paper develops an inertial error model in vehicle navigation system. The authors' proposed model, which can be used for the GPS/INS integration, employs a combination of Bayesian state estimation and Monte Carlo method. The proposed method is demonstrated to show superior performance in terms of position estimates, compared to the classical models.
2. "Object-level Fusion and Confidence Management in a Multi-sensor Pedestrian Tracking System": Fayad and Cherfaoui propose to fuse asynchronous data provided by different sensors with complementary and supplementary fields of view, to detect, recognize, and track pedestrians. The confidence in detection and recognition is measured based on geometric features and updated using the Transferable Belief Model Framework.
3. "Effective Lip Localization and Tracking for Achieving Multimodal Speech Recogntion": The authors design an effective lip motion analysis system aimed at deploying it to a multimodal speech recognition system. Several levels of motion extraction sequence under various environmental conditions are presented and challenges are discussed.
4. "Optimal View Selection and Event Retrieval in Multi-Camera Office Environment": This paper describes multiple cameras used to provide multiple views of an object for optimal view selection and event retrieval.

5. "Emotion-specific Dichotomous Classification and Feature-level Fusion of Multichannel Bio-signals for Automatic Emotion Recognition": This paper describes all the the essential stages of an automatic emotion recognition system using multichannel physiological measures, from data collection to the classification process. A wide range of physiological features from various analysis domains is presented to correlate them with emotional states.
6. "A Comparison of Track-to-Track Fusion Algorithms for Automotive Sensor Fusion": The authors compare the performance of a standard asynchronous Kalman filter applied to tracked sensor data to several algorithms for the track-to-track fusion of sensor objects of unknown parameters. Use of covariance intersection and cross-covariance are found to yield significantly lower errors than a Kalman filter at a comparable computational load.
7. "Effective and Efficient Communication of Information": As a part of distributed sensor network issue, Mack et al propose an information fusion procedure by effective communication which then makes the interaction between different fusion processes more efficient and effective.
8. "Most Probable Data Association with Distance and Amplitude Information for Target Tracking in Clutter": Song develops a new filter structure accommodating the probabilistic nature of the data association and builds a target tracking. The performance is evaluated by a series of Monte Carlo tests and demonstrates to be effective in terms of error.
9. "Simultaneous Multi-information Fusion and Parameter Estimation for Robust 3-D Indoor Positioning Systems": This paper develops a theoretical framework fusing pressure measurements and a topological building map with received signal strength of a typical WLAN based indoor positioning systems, to render a robust 3-D indoor positioning system.
10. "Efficient Multi-target Tracking with Sub-event IMM-JPDA and One-point Prime Initialization": The author addresses an interacting multi-model joint probabilistic data association tracker with sub-event decomposition and one-point prime initialization, which demonstrates to significantly reduce the number of hypotheses without reducing tracking performance.
11. "Fusion of Inertial, Vision, and Air Pressure Sensors for MAV (miniature unmanned aerial vehicles) Navigation": This paper introduces a system for fusing information from two additional sensors with the IMU (Inertial Measurement Unit) to improve the navigation performance of the MAV.

Performance Analysis of GPS/INS Integrated System by Using a Non-Linear Mathematical Model

Khalid Touil, Mourad Zribi and Mohammed Benjelloun

Abstract Inertial navigation system (INS) and global position system (GPS) technologies have been widely utilized in many positioning and navigation applications. Each system has its own unique characteristics and limitations. In recent years, the integration of the GPS with an INS has become a standard component of high-precision kinematics systems. The integration of the two systems offers a number of advantages and overcomes each system's inadequacies. In this paper an inertial error model is developed which can be used for the GPS/INS integration. This model is derived by employing the Stirling's interpolation formula. The Bayesian Bootstrap Filter (BBF) is used for GPS/INS integration. Bootstrap Filter is a filtering method based on Bayesian state estimation and Monte Carlo method, which has the great advantage of being able to handle any functional non-linearity and system and/or measurement noise of any distribution. Experimental result demonstrates that the proposed model gives better positions estimate than the classical model.

Keywords Navigation · GPS/INS integration · Stirling's interpolation Bayesian Bootstrap Filter

1 Introduction

Most vehicle navigation systems estimate the vehicle position from Inertial Navigation System (INS) [1] and Global Positioning System (GPS) [2]. GPS is the most attractive one for the vehicle navigation system. This is because the position can be calculated on the globe if more than four satellites are detected. GPS can provide positioning and navigation information quickly and accurately at relatively low cost. GPS has made a significant impact on almost all positioning and navigation applications. However, GPS alone is insufficient to maintain continuous positioning because of inevitable obstructions caused by buildings and other natural features.

K. Touil (✉)
Laboratoire d'Analyse des Systèmes du Littoral, (LASL-EA 2600), Université du Littoral Côte d'Opale, 62228 Calais Cedex, France
e-mail: Khalid.Touil@lasl.univ-littoral.fr

S. Lee et al. (eds.), *Multisensor Fusion and Integration for Intelligent Systems*, Lecture Notes in Electrical Engineering 35, DOI 10.1007/978-3-540-89859-7_1,

GPS appears then as an intermittent positioning system that demands the help of an INS. INS is one of the most widely used dead reckoning systems [3]. It can provide continuous position, velocity, and also orientation estimates, which are accurate for a short term, but are subject to drift due to sensors drifts. Unfortunately, most of the available positioning technologies have limitations either in accuracy of the absolute position GPS accumulated error INS. In general, GPS/INS integration provides reliable navigation solutions by overcoming each of their shortcomings, including signal blockage for GPS and growth of position errors with time for INS. Integration can also exploit advantages of the two systems, such as the uniform high accuracy trajectory information of GPS and the short term stability of INS. In this paper an inertial error model is developed which can be used for the GPS/INS integration. This model is derived by employing the Stirling's interpolation formula [4]. It is independent on the initial state and can avoid the divergence problem, but its drawback is the heavy computational load in the update stage. The Bayesian Bootstrap Filter (BBF) is used for GPS/INS integration [5]. Bootstrap Filter is a filtering method based on Bayesian state estimation and Monte Carlo method, which has the great advantage of being able to handle any functional non-linearity and system and/or measurement noise of any distribution. The paper is organized as follows. In Sect. 2, we introduce an overview of approximation techniques. Section 3 presents a dynamic and measurement models. In Sect. 4, a Bayesian Bootstrap Filter algorithm is described. Experimental results are presented to demonstrate the accuracy of the proposed model in Sect. 5. Finally, Conclusions are made in Sect. 6.

2 Overview of Approximation Techniques

Numerous approximation techniques for point estimation on nonlinear systems have been proposed. This section deals with polynomial approximations of arbitrary functions. In particular we will compare approximations obtained with Taylor's formula and Stirling's interpolation formula.

2.1 Taylor-Series Approximation

Taylor-Series expansion (TSE) is a fundamental tool for handling nonlinearity [6]. We denote the nth-order TSE approximation of a function f at $\bar{x}$ by: $f(x) \approx TSE(x, n, \bar{x})$. For an analytic function $f(x)$, it is given by:

$$
\begin{aligned}
TSE(x, n, \bar{x}) = f(\bar{x}) + f'(\bar{x})(x - \bar{x}) + \frac{f''(\bar{x})}{2!}(x - \bar{x})^2 \\
+ \frac{f^{(3)}(\bar{x})}{3!}(x - \bar{x})^3 + \ldots\ldots + \frac{f^{(n)}(\bar{x})}{n!}(x - \bar{x})^n
\end{aligned}
\tag{1}
$$

The principle of the TSE is that the approximation inherits still more characteristics of the true function in one particular point as the number of terms increases. Although the assumption that f is analytic implies that any desired accuracy can be achieved provided that a sufficient number of terms are retained, it is in general advised to use a truncated series only in the proximity of the expansion point unless the remainder terms has been properly analyzed. Several interpolation formulas are available for deriving polynomial approximations that are to be used over an interval. In the following we will consider one particular formula, namely Stirling's interpolation formula.

2.2 Stirling's Interpolation Formula

Let the operators δ and μ perform the following operations (h denotes a selected interval length):

$$\delta f(x) = f\left(x + \frac{h}{2}\right) - f\left(x - \frac{h}{2}\right) \tag{2}$$

$$\mu f(x) = \frac{1}{2}\left(f\left(x + \frac{h}{2}\right) - f\left(x - \frac{h}{2}\right)\right) \tag{3}$$

With these operators Stirling's interpolation formula used around the point $x = \bar{x}$ can be expressed as [4]:

$$\begin{aligned} f(x) = f(\bar{x} + ph) &= f(\bar{x}) + p\mu\delta f(\bar{x}) \\ &+ \frac{p^2}{2!}\delta^2 f(\bar{x}) + \binom{p+1}{3}\mu\delta^3 f(\bar{x}) \\ &+ \frac{p^2(p^2-1)}{4!}\delta^4 f(\bar{x}) + \binom{p+2}{5}\mu\delta^5 f(\bar{x}) + \ldots\ldots \end{aligned} \tag{4}$$

In the case of first and second-order polynomial approximations, the formula (4) is given by:

$$f(x) \approx f(\bar{x}) + f'_{DD}(\bar{x})(x - \bar{x}) + \frac{f''_{DD}(\bar{x})}{2!}(x - \bar{x})^2 \tag{5}$$

where

$$f'_{DD}(\bar{x}) = \frac{f(\bar{x} + h) - f(\bar{x} - h)}{2h} \tag{6.a}$$

and

$$f''_{DD}(\bar{x}) = \frac{f(\bar{x} + h) - f(\bar{x} - h) - 2f(\bar{x})}{h^2} \tag{6.b}$$

In Fig. 1, the expansion point is $\bar{x} = 0$ and for the interpolation formula the interval length was selected to $h = 3.5$. The solid line shows the true function, the dot-dashed line is the second-order Taylor approximation while the dashed line is

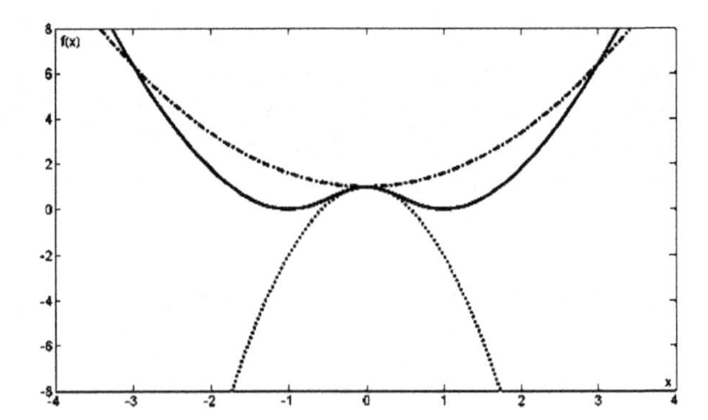

Fig. 1 Comparison of a second-order polynomial approximation obtained with Taylor's formula and Stirling's formula

the approximation obtained with the Stirling's interpolation formula. Obviously, the Taylor polynomial is a better approximation near the expansion point while further away the error is much higher than for the approximation obtained with the Stirling's interpolation formula. In the following two sections we are going to use the Stirling's interpolation formula in a dynamic model of the INS and in a Bayesian Bootstrap filtering.

3 Dynamic and Measurement Models

In this paper, we propose to use two kinds of sensors:

- Absolute position measurements issued from a GPS. The design of the GPS requires four satellites to be tracked in order to solve for three dimensional positions.
- The INS provides us with rotation rate (gyroscopes) and acceleration (accelerometers).

3.1 The Proposed Dynamic Model

The navigation frame inertial navigation equations can be described as:

$$\begin{pmatrix} \dot{r}^n \\ \dot{v}^n \\ \dot{C}_b^n \end{pmatrix} = \begin{pmatrix} D^{-1} v^n \\ C_b^n f^b - (\Omega_{ie}^n + \Omega_{in}^n) v^n + g^n \\ C_b^n (\Omega_{ib}^b - \Omega_{in}^b) \end{pmatrix} \tag{7}$$

where

$$D^{-1} = \begin{pmatrix} 1/(M+h) & 0 & 0 \\ 0 & 1/(N+h)\cos\varphi & 0 \\ 0 & 0 & -1 \end{pmatrix} \tag{8}$$

where Ω represents the skew symmetric matrix form of the rotation rate vector ω, Ω_{ie}^{n} is the rotation rate of the e-frame with respect to the i-frame projected to the n-frame, Ω_{ib}^{b} is the outputs of the strapdown gyroscopes, Ω_{in}^{b} is the rotation rate of the n-frame with respect to the i-frame projected to the b-frame, C_b^n is the direction cosine matrix (DCM) from the n-frame to the b-frame [7], f^b is the specific force vector defined as the difference between the true acceleration in space and the acceleration due to gravity, g^n is the gravity vector and M, N are radii of curvature in the meridian and prime vertical given by Schwarz and Wei [8].

The state vector is composed of the INS error that is defined as the deviation between the actual dynamic quantities and the INS computed values $\delta X = X - X_{INS}$. The state model describes the INS error dynamic behaviour depending on the instrumentation and initialization errors. It is obtained by using the second-order polynomial approximations of Stirling's interpolation formula (5) to the ideal equation (7) around the INS estimates as follow:

$$\delta\dot{X} = f(X, U) - f(X_{INS}, U_{INS}) \tag{9}$$

$$\delta\dot{X} = f'_{DD}(X_{INS}, U_{INS})\ \delta X + \delta X^t \frac{f''_{DD}(X_{INS}, U_{INS})}{2!}\delta X \tag{10}$$

The state vector is usually augmented with systematic sensor errors:

$$\delta X = (\delta r, \delta v^n, \delta\rho, b_a, b_g, b, d) \tag{11}$$

where all the variables are expressed in the navigation frame NED (North, East, Down).

- $\delta r = (\delta\varphi, \delta\lambda, \delta h)$ is the geodetic position error in latitude, longitude and altitude,
- $\delta v^n = (\delta v_N, , \delta v_E, \delta v_D)$ is the velocity error vector,
- $\delta\rho$ is the attitude (roll, pitch, and yaw) error vector,
- b_a and b_g represent the accelerometers and gyrometers biases,
- $b = c\tau_r$ and d are respectively the GPS clock offset and its drift. τ_r is the receiver clock offset from the GPS time and c is the speed of light (3×10^8m/s).

For short-term applications, the accelerometers and gyrometers can be properly defined as random walk constants $\dot{b}_a = \omega_a$ and $\dot{b}_g = \omega_g$. Note that the standard deviations of the white noises ω_a and ω_g are related to the sensor quality. The navigation solution also depends on the receiver clock parameters b and d models as $\dot{b} = d + \omega_b$ and $\dot{d} = \omega_d$, where ω_b and ω_d are mutually independent zero-mean Gaussian random variables [9]. For simplicity, denote as X (instead of δX) the state vector.

For each $j \in [1,$ length of $X_{INS,k}]$, the discrete-time state model takes the following form:

$$X_{INS,k+1}^{j} = A_k^j X_{INS,k} + \frac{1}{2} X_{INS,k}^{t} B_j^k X_{INS,k} + v_k^j \tag{12}$$

where v_k denote the dynamical zero-mean Gaussian noise vector with associated covariance matrix Σ_v, A_k^j and B_k^j are obtained by using respectively (6.a) and (6.b) to the j-th element of the second member of (7).

In the classical dynamic model, the state equation is given by the following form:

$$X_{INS,k+1} = A_k X_{INS,k} + v_k \tag{13}$$

where A_k is a block diagonal matrix which elements are detailed in many standard textbooks such as [9].

3.2 Measurement Equation

The standard measurement of the GPS system is the pseudo-range. This defines the approximate range from the user GPS receiver antenna to a particular satellite. Consequently, the observation equation associated to the ith satellite can be defined as:

$$\rho_i = \sqrt{(X_i - x)^2 + (Y_i - y)^2 + (Z_i - z)^2} + b + \omega_i \tag{14}$$

where $i = 1, \ldots\ldots, n_s$ (recall that n_s is the number of visible satellites). The vectors $(x, y, z)^T$ and $(X_i, Y_i, Z_i)^T$ are respectively the positions of the vehicle and the ith satellite expressed in the rectangular coordinate system WGS-84 [9].

The position of the vehicle is transformed from the geodetic coordinate to the rectangular coordinate system as follows:

$$\begin{cases} x = (N + h_{INS} + \delta h)\cos(\lambda_{INS} + \delta\lambda)\cos(\varphi_{INS} + \delta\varphi) \\ y = (N + h_{INS} + \delta h)\sin(\lambda_{INS} + \delta\lambda)\cos(\varphi_{INS} + \delta\varphi) \\ z = (N(1 - e^2) + h_{INS} + \delta h)\sin(\varphi_{INS} + \delta\varphi) \end{cases} \tag{15}$$

where $N = \frac{a}{\sqrt{1-e^2\sin^2\varphi}}$. The parameters a and e denote respectively the semi-major axis length and the eccentricity of the earth's ellipsoid. The expression (15) has to be substituted in (14) to obtain the highly nonlinear measurement equation:

$$Y_{GPS,k} = g(X_{INS,k}) + \beta_k \tag{16}$$

where $\beta_k \sim N(0, \Sigma_\beta)$ and $Y_{GPS} = (\rho_1, \ldots\ldots, \rho_n)$.

In the Bayesian bootstrap filter, it is not necessary that the measurement noise β_k must be the white Gaussian. Now we can apply the bootstrap filter using the above system dynamic and measurement models.

4 Bayesian Bootstrap Filter

The Bayesian bootstrap filtering approach is to construct the conditional probability density function (PDF) of the state based on measurement information [10]. The conditional PDF can be regarded as the solution to the estimation problem. We shall briefly explain the recursive Bayesian estimation theory and the Bayesian Bootstrap filter.

4.1 Recursive Bayesian Estimation

The system model is assumed to have the discrete form:

$$x_{k+1} = f(x_{k-1}, w_k) \tag{17}$$

where $f : IR^n \times IR^m \rightarrow IR^n$ is the system transition function and $w_k \in IR^m$ is a zero-mean noise process independent of the system states. The PDF of w_k is assumed to be known as $p_w(w_k)$. At discrete time, measurements are denoted by $y_k \in IR^p$, which are related to the state vector via the observation equation:

$$y_k = h(x_k, v_k) \tag{18}$$

where $h : IR^n \times IR^r \rightarrow IR^p$ is the measurement function, and $v_k \in IR^r$ is the observation noise, assumed to be another zero mean random sequence independent of both state variable x_k and the system noise w_k. The PDF of v_k is assumed to be known as $p_v(v_k)$. The set of measurements from initial time step to step k is expressed as $Y_k = \{y_i\}_{i=1}^k$. The distribution of the initial condition x_0 is assumed to be given by $p(x_0/Y_0) = p(x_0)$.

The recursive Bayesian filter based on the Bayes' rule can be organized into the time update state and the measurement update stage [11]. The time update state can be constructed as:

$$p(x_k/Y_{k-1}) = \int p(x_k/x_{k-1}) \times p(x_{k-1}/Y_{k-1})dx_{k-1} \tag{19}$$

where $p(x_k/x_{k-1})$ is determined by $f(x_{k-1}, w_{k-1})$ and the known PDF $p_w(w_{k-1})$. Then at time step k, a measurement y_k becomes available and may be used to update the prior according to the Bayes' rule:

$$p(x_k/Y_k) = \frac{p(y_k/x_k) \times p(x_k/Y_{k-1})}{\int p(y_k/x_k) \times p(x_k/Y_{k-1})dx_k} \tag{20}$$

where the conditional PDF $p(y_k/x_k)$ is determined from the measurement model and the known PDF, $p_v(v_k)$.

4.2 Bayesian Bootstrap Filter

The Bayesian Bootstrap Filter (BBF) is a recursive algorithm to estimate the posterior $p(x_k/Y_k)$ from a set of samples [10]. Suppose we have a set of random samples $\{x_{k-1}(i) : i = 1, ..., N\}$ from the PDF $p(x_{k-1}/Y_{k-1})$. Here, N is the number of Bootstrap samples. The filter procedure is as follows:

- *Prediction*: Each sample from PDF $p(x_{k-1}/Y_{k-1})$ is passed through the system model to obtain samples from the prior at time step k:

$$x_k^*(i) = f(x_{k-1}(i), w_k(i)) \tag{21}$$

  where $w_k(i)$ is a sample draw from PDF of the system noise $p_w(w_k)$.
- *Update*: on receipt of the measurement y_k, evaluate the likelihood of each prior sample and obtain the normalized weight for each sample:

$$q_i = \frac{p(y_k/x_k^*(i))}{\sum_{j=1}^{N} p(y_k/x_k^*(j))} \tag{22}$$

Define a discrete distribution over $\left\{x_k^*(i) : i = 1, ..., N\right\}$, with probability mass q_i associated with element i. Now resample N times from the discrete distribution to generate samples $\{x_k(i) : i = 1, ..., N\}$, so that for any j, Prob $\left\{x_k(j) = x_k^*(i)\right\} = q_i$. It can be contented that the samples $x_k(i)$ are approximately distributed as the required PDF $p(x_k/Y_k)$ [10]. Repeat this procedure until the desired number of time samples has been processed. The resampling update stage is performed by drawing a random sample u_i from the uniform distribution over [0, 1]. The value $x_k^*(M)$ corresponding to:

$$\sum_{j=0}^{M-1} q_j < u_i \leq \sum_{j=0}^{M} q_j \tag{23}$$

where $q_0 = 0$, is selected as a sample for the posterior. This procedure is repeated for $i = 1, \ldots, N$. It would also be straightforward to implement this algorithm on massively parallel computers, raising the possibility of real time operation with very large sample sets.

5 Simulation

The analysis of some simulations will enable us to evaluate the performance of the proposed model by using the BBF. The kinematics data used were generated by SatNav Toolbox for Matlab created by GPSoft. A GPS-INS simulation can be divided into three parts:

- *Trajectory*: the vehicle dynamics is simulated according to position-velocity-acceleration model.
- *INS data*: the INS estimates of the vehicle dynamics are then computed for low cost inertial sensors. The accelerometer bias instability and random walk are given by $10^{-3}\mathrm{m/s^2}$ and $4.5\,\mathrm{m/s/\sqrt{hour}}$. The gyrometer bias instability and random walk are given by $5\,\mathrm{deg/hour}$ and $16\mathrm{deg/\sqrt{hour}}$.
- *GPS data*: the pseudo-ranges corresponding to the visible satellites from the vehicle is evaluated (the standard deviations of the GPS measurements noises are chosen as $\sigma_{\beta_k} = 10\,\mathrm{m}$ for pseudo-range).

We have fixed the number of particles of the BBF to 2000 and the time of simulation is equal to 1200 s. In our work, we compare the proposed method with the classical method [5] for the following two cases:

- Vehicle Dynamics is moderate (i.e., the speed of vehicle is between 10 and 50 m/s).
- Vehicle Dynamics is fast (i.e., the speed of vehicle is between 100 and 200 m/s).

5.1 First Case

The obtained results by using the proposed method and the classical method are compared for simulation duration of 1200 s. For each method, the horizontal positioning root mean square error and the horizontal velocity root mean square error are competed from 100 Monte Carlo runs. Figures 2 and 3 shows the results obtained with the classical method (solid-line) and proposed method (dashed-line).

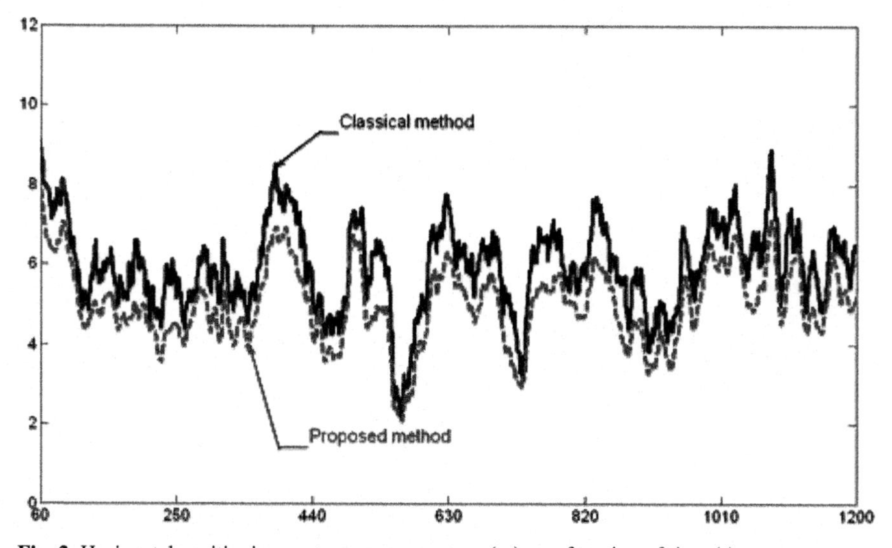

Fig. 2 Horizontal positioning root mean square error (m) as a function of time (s)

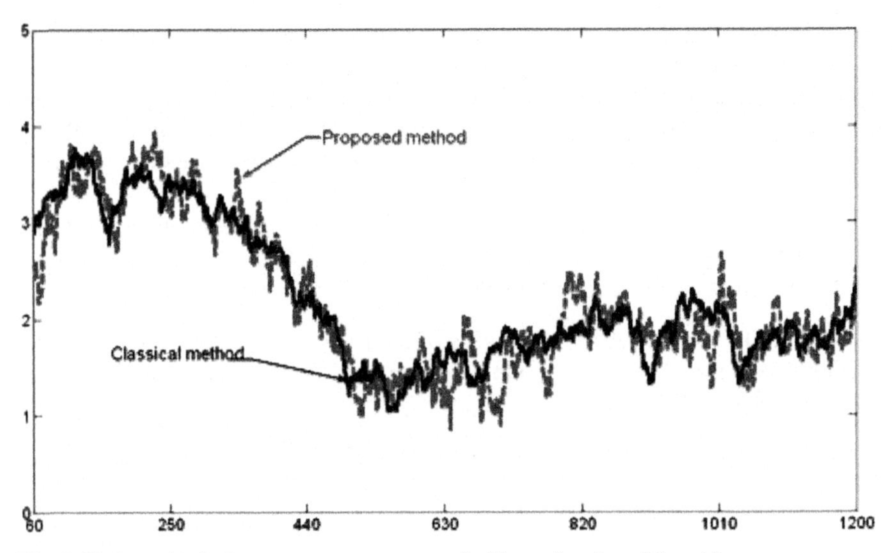

Fig. 3 Horizontal velocity root mean square error (m/s) as a function of time (s)

In Figs. 2 and 3, we note that the both method have the same precision. It is due to the dynamics errors which are reasonable (where the local linearization is possible).

5.2 Second Case

The Figs. 4 and 5 represent respectively the horizontal positioning root mean square error and the horizontal velocity root mean square error, which are competed from

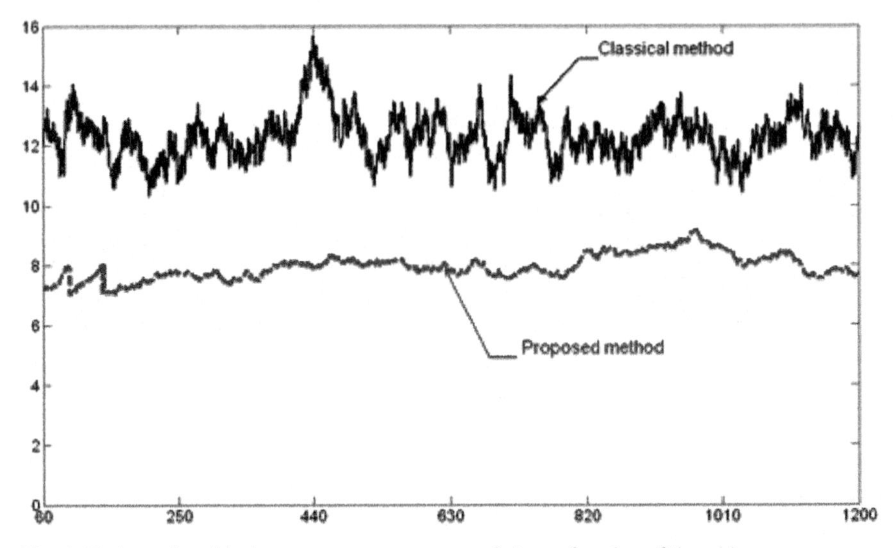

Fig. 4 Horizontal positioning root mean square error (m) as a function of time (s)

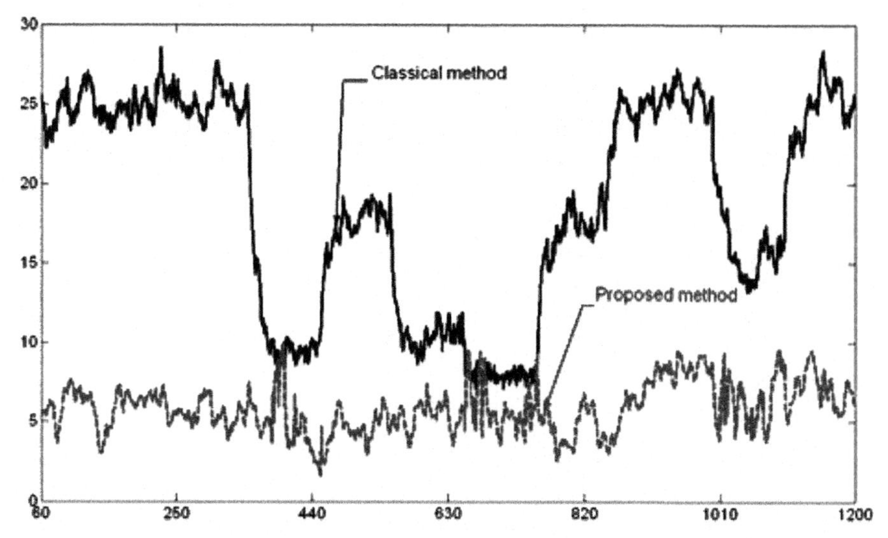

Fig. 5 Horizontal velocity root mean square error (m/s) as a function of time (s)

Table 1 Root mean errors during satellites outage for first case

	Horizontal positioning (m)	Horizontal velocity (m/s)
Proposed method	6.9	3.65
Classical method	7.43	3.81

Table 2 Root mean errors during satellites outage for second case

	Horizontal positioning (m)	Horizontal velocity (m/s)
Proposed method	10.32	7.11
Classical method	15	14.73

100 Monte Carlo runs. In Figs. 4 and 5 the solid-line and dashed-line represent respectively the results obtained by classical method and proposed method. In Figs. 4 and 5, we note that the proposed method give better precision that the classical method. It is due to the dynamics errors which are fast (where the local linearization is not possible).

Finally a complete GPS signal outage of 50 s was introduced within the GPS data and both methods were used to predict the INS dynamic, during this period. The root mean square errors of the two methods in this period are compared respectively in Tables 1 and 2 for first case and second case.

6 Conclusion

This paper has studied the performance analysis of GPS/INS integrated system by using a non-linear mathematical model. This model is derived by employing the second order polynomial approximation obtained with Stirling's formula. The

Bayesian Bootstrap Filter (BBF) is used for GPS/INS integration. In the simulation results, we showed that the proposed model gives better positions estimate than the classical model where the vehicle dynamics is fast. But, the disadvantage of the proposed model is the computing time compared to the classic model. The next step of this work is to validate the performance of the proposed model on the real data.

Acknowledgments This work was sponsored by FEDER (Fonds Européen de Développement Régional), DRESSTIC (Direction Recherche, Enseignement Supérieur, Santé et Technologies de l'Information et de la Communication) and CRNC (Conseil Régional Nord-Pas de Calais).

References

1. X. Yun et al., An inertial navigation system for small autonomous underwater vehicles, *ICRA, IEEE International Conference*, 2, 2000, 1781–1786.
2. A. Leick, *GPS Satellite Surveying*, New York: John Wiley, 1995.
3. Q. Li, The application of integrated GPS and dead reckoning positioning in automotive intelligent navigation system, *Journal of Global Positioning Systems*, 3(1–2), 2004, 183–190.
4. E. T. Whittaker and G. Robinson, Stirling's approximation to the factorial, §70 in *The Calculus of Observations: A Treatise on Numerical Mathematics*, 4th ed. New York: Dover, 1967, pp. 138–140.
5. K. Touil et al., Bayesian Bootstrap Filter for integrated GPS and dead reckoning positioning, *IEEE, International Symposium on Industrial Electronics*, ISIE07, Spain, 2007, pp. 1520 –1524.
6. R. Hedjar et al., Finite horizon nonlinear predictive control by the Taylor approximation: application to Robot tracking trajectory, *International Journal of Applied Mathematics and Computer Science*, 15(4), 2005, 527–540.
7. D. H. Titterton and J. L. Weston, *Strapdown Inertial Navigation Technology*, London: Peter Peregrinus Ltd, 1997.
8. K. P. Schwarz and M. Wei, INS/GPS integration for geodetic applications, *Lectures Notes ENGO 623.* Department of Geomatics Engineering, The University of Calgary, Calgary, Canada, 2000.
9. M. R. Robert, *Applied Mathematics in Integrated Navigation Systems*, 2nd ed., *AIAA Education Series*, 2003.
10. N. J. Gordon et al., Novel approach to nonlinear/non-Gaussian Bayesian state estimation, *IEEE Proceedings-F*, 140, 1993, 107–113.
11. A. Doucet et al., On sequential Monte Carlo sampling methods for Bayesian filtering, *Statistics and Computing*, 10, 2000, 197–208.

Object-Level Fusion and Confidence Management in a Multi-Sensor Pedestrian Tracking System

Fadi Fayad and Véronique Cherfaoui

Abstract This paper[1] describes a multi-sensor fusion system dedicated to detect, recognize and track pedestrians. The fusion by tracking method is used to fuse asynchronous data provided by different sensors with complementary and supplementary fields of view. Having the performance of the sensors, we propose to differentiate between the two problems: object detection and pedestrian recognition, and to quantify the confidence in the detection and recognition processes. This confidence is calculated based in geometric features and it is updated under the Transferable Belief Model framework. The vehicle proprioceptive data are filtered by a separate Kalman filter and are used in the estimation of the relative and absolute state of detected pedestrians. Results are shown with simulated data and with real experimental data acquired in urban environment.

Keywords Multi-sensor data fusion · Pedestrians' detection and recognition · Transferable belief Model · Confidence management

1 Introduction

Recent projects on pedestrian detection [1] or obstacle detection [2] have highlighted the use of multi-sensor data fusion and more generally the multiplication of data sources in order to obtain more reliable, complete and precise data. The Vehicle to Vehicle communication is an example of enlarging the field of view of one vehicle by the information coming from other vehicles [3, 4].

The work presented in this paper is a contribution to the development of an "Advances Driver Assistance Systems" (ADAS). A generic multi-sensor pedestrian

F. Fayad (✉)
Heudiasyc UMR CNRS 6599, Université de Technologie de Compiègne, Centre de Recherche de Royallieu – BP 20529, 60205 Compiègne Cedex, France
e-mail: fadi.fayad@hds.utc.fr

[1] This work is a part of the project LOVe (Logiciel d'Observation des Vulnérables http://love.univ-bpclermont.fr/) and supported by the French ANR PREDIT.

S. Lee et al. (eds.), *Multisensor Fusion and Integration for Intelligent Systems*, Lecture Notes in Electrical Engineering 35, DOI 10.1007/978-3-540-89859-7_2,

detection, recognition and tracking system, is introduced. However, sensors are not synchronized and have not the same performance and field of view. Thus to explore the whole capability of sensors in order to benefit of all available data and to solve the problem of asynchronous sensors, we present a generic method to fuse data provided by different sensor, with complementary and/or supplementary fields of view, by tracking detected objects in a commune space and by combining the detection and/or the recognition information provided by each sensor taking into consideration its performance.

This paper is organized as follows: Sect. 2 presents the proposed multi-sensor fusion system architecture and describes the Object-Level Fusion by Tracking method. Section 3 described the state models used to filter and estimate vehicle and pedestrians' kinematical state. Section 4 presents the detection and recognition confidences calculation and update. Experimental results are shown in Sect. 5 illustrating the effect of sensors performance. Conclusion and perspectives will be proposed in the last section.

2 Object-Level Fusion by Tracking

2.1 Overview of the System

The described multi-sensor pedestrian tracking system is an in-vehicle embedded real-time system. This generic fusion system (Fig. 1) has as input the unsynchronized data provided by independent unsynchronized sensors with complementary and supplementary fields of view (Fig. 2).

The system is composed of one "Object-Level Fusion Module" and one "Sensor Module" per sensor. Each Sensor Module analyzes data provided by the corresponding sensor to supply the Object Level Fusion Module by a list of objects supposed present in the scene of its field of view. A lot of works in ADAS and robotics applications are dedicated to the object detection capabilities. For example for pedestrian

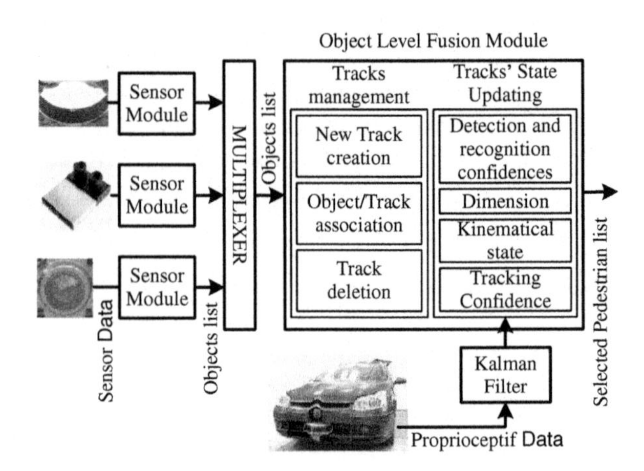

Fig. 1 System architecture

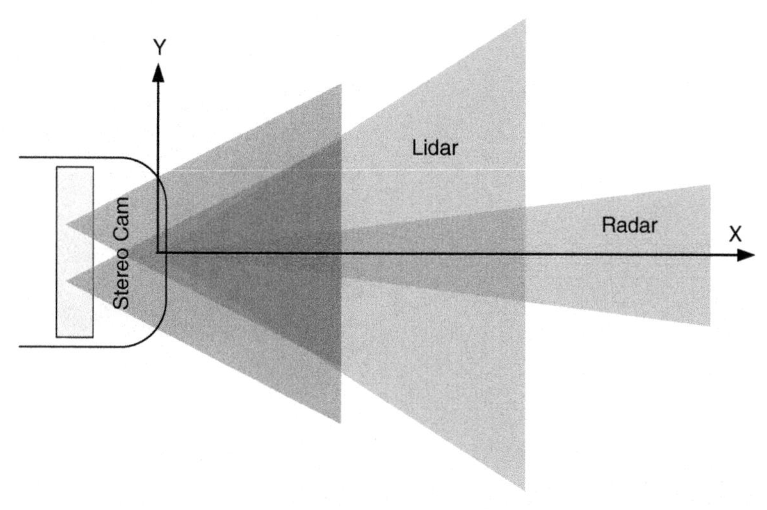

Fig. 2 Complementary and supplementary fields of view of the different sensors: Stereo-camera, Lidar and Radar

detection, [5] proposes obstacle detection and identification with Lidar sensor; [6] proposes stereo-vision obstacle detection with disparity analysis and SVM based on pedestrian classification, [7] gives pedestrian classification resulting from monocular vision with AdaBoost algorithm. The Object Level Fusion Module takes any ready object list and combine it with the existing track list, tacking into consideration the vehicle proprioceptive data (filtered by a separate Kalman filter) and the performance of each detection module (stored in a configuration file with other tuning parameters). Latency problem can be solved by a time indexed buffer of observations and state vectors as in [8]. The buffer size depends on the maximum acceptable observation delay.

2.2 *Object Level Input/Output*

The Sensor Module works at the frequency of the corresponding sensor, it provides at each detection cycle a list of objects supposed present in the scene of its field of view. Objects are described by their position (relative to the sensor), position error, dimension (if detected), dimension error and two indicators quantifying the confidence in detection and the confidence in recognition if the sensor is capable to recognize pedestrians or any type of obstacles. The performance of each sensor module is quantified by two probability values: P_{FR} representing the probability of false pedestrian recognition and P_{FA} the probability of false alarm or false detection. Sensor performance is propagated to the object's detection and recognition confidences.

The Object Level Fusion Module has to run at the frequency of the incoming object lists. It has to combine any ready object list with the existing track list to

Fig. 3 Commune relative Lidar coordinate system

provide a list of estimated tracks at the current time. Tracks are characterized by their relative position, position error, speed, speed error, dimension, dimension error and three indicators quantifying the confidences in detection, recognition and tracking.

To fuse data, all information is represented in the same 3D coordinate system $(X_L Y_L Z_L)$ showed in Fig. 3: the origin is the center of the Lidar reflection mirror and the plan $(X_L Y_L)$ is parallel to the ground. A special calibration procedure is developed to project vision data into 3D coordinate system and vice versa.

2.3 Fusion by Tracking

The Track's state is divided into four parts updated by four different processes (Fig. 1) with the same update stages and models for all tracks:

1. Kinematical state (track's position and velocity) loosely-coupled with the vehicle state and updated by a classical Kalman filter detailed in Sect. 3.
2. Tracking confidence: calculated and updated based on the score of Sittler using likelihood ratio [9].
3. Dimension: updated by a fixed gain filter taking into consideration the objects partial occultation problem [10].
4. Detection and recognition confidences: updated by a credibilistic model based on the belief functions and presented in Sect. 4.

When the fusion and state updating module receive a list of object at its input, it predicts the last list of tracks' state to the current object list time, and then it runs an object to track association procedure based on a modified version of the nearest neighborhood association method. This modified method takes into consideration the occultation problem by geometrically detecting the occultation areas and allowing multi-object to track association to associate all parts of a partially hidden object to the corresponding track.

3 Pedestrian Model for In-Vehicle Tracking

3.1 Coordinate Systems Transformation

Let $(o, \underline{i}, \underline{j})$ be an absolute fixed coordinate system and $(O, \underline{I}, \underline{J})$ and $(O_L, \overrightarrow{I_L}, \overrightarrow{J_L})$ be two relative coordinate systems attached respectively to the center of the rear wheel axle and the center of Lidar rotating mirror (Fig. 4). The x-axis is aligned with the longitudinal axis of the car. Let M be a point of the space and let (x, y), (X, Y) and (X_L, Y_L) be its respective Cartesian coordinate in the three systems. (X, Y) and (X_L, Y_L) are related by the equations:

$$\begin{cases} X = X_L + L \\ Y = Y_L \end{cases} \tag{1}$$

The geometry of Fig. 4 shows that:

$$\begin{cases} \overrightarrow{oM} = \overrightarrow{oO} + \overrightarrow{OM} \\ \overrightarrow{oO} = x_o \underline{i} + y_o \underline{j} \\ \overrightarrow{OM} = X\underline{I} + Y\underline{J} \\ \overrightarrow{I} = \underline{i} \cos\theta + \underline{j} \sin\theta \\ \overrightarrow{J} = -\underline{i} \sin\theta + \underline{j} \cos\theta \end{cases} \tag{2}$$

Therefore:

$$\begin{cases} x = x_o + X\cos\theta - Y\sin\theta \\ y = y_o + X\sin\theta + Y\cos\theta \end{cases} \tag{3}$$

Then:

$$\begin{cases} X = (x - x_o)\cos\theta + (y - y_o)\sin\theta \\ Y = -(x - x_o)\sin\theta + (y - y_o)\cos\theta \end{cases} \tag{4}$$

The absolute speed vector of the point M is the derivative of its position vector:

$$\vec{v} = \frac{d\overrightarrow{oM}}{dt} = \frac{d\overrightarrow{oO}}{dt} + X\frac{d\overrightarrow{I}}{dt} + Y\frac{d\overrightarrow{J}}{dt} + \frac{dX}{dt}\overrightarrow{I} + \frac{dY}{dt}\overrightarrow{J} \tag{5}$$

Let $\overrightarrow{V} = (dX/dt)\overrightarrow{I} + (dY/dt)\overrightarrow{J}$ be the relative speed of M with respect to the vehicle coordinate system $(O, \underline{I}, \underline{J})$ and $\overrightarrow{v_o} = d\overrightarrow{oO}/dt$ be the absolute speed of O. The derivatives of the vectors $\underline{I}$ and $\underline{J}$ are:

$$\begin{cases} d\overrightarrow{I}/dt = \left[(-\sin\theta)\underline{i} + (\cos\theta)\underline{j}\right] d\theta/dt \\ d\overrightarrow{J}/dt = \left[(-\cos\theta)\underline{i} + (-\sin\theta)\underline{j}\right] d\theta/dt \end{cases} \tag{6}$$

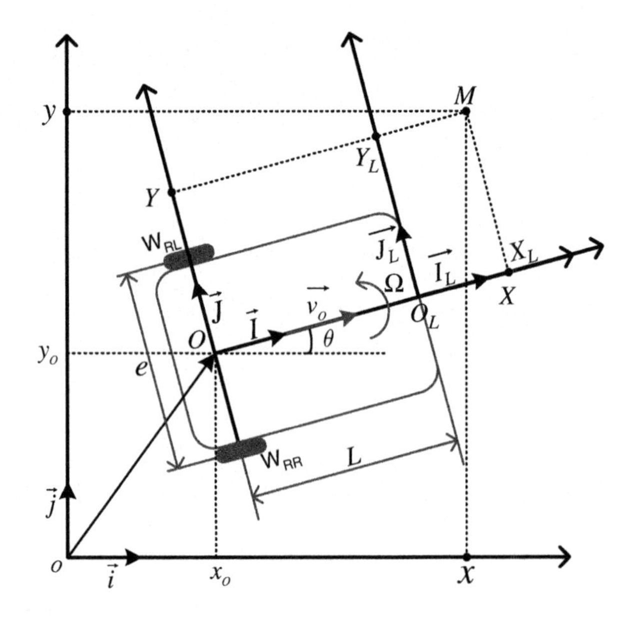

Fig. 4 Absolute and relative coordinate system

Let $\Omega = d\theta/dt$ be the absolute rotation speed of the vehicle around the point O, then (4) can be written as:

$$\vec{v} = \vec{V} + \vec{v_o} + \Omega \begin{bmatrix} -\sin\theta & -\cos\theta \\ \cos\theta & -\sin\theta \end{bmatrix} \begin{bmatrix} X \\ Y \end{bmatrix} \tag{7}$$

3.2 Vehicle Model

In modern cars, braking is assisted by ABS systems that use angular encoders attached to the wheels. In such a case, the sensors basically measure the wheel speeds. We propose in this paper to use this data to model vehicle movement and to estimate its kinematical state.

Figure 5 shows the elementary displacement of the vehicle between two samples at t_k and t_{k+1}.The presented car-like vehicle model is a real odometric model [11] and not the discretized kinematics model used in [12]. Assumptions are made on the elementary motions and geometric relationships are expressed to provide relations between the rotations of the wheels and the displacements. The rear wheels' speeds are read from the CAN bus of the experimental vehicle. They are supposed constants between two readings. On the assumption that the road is perfectly planar and the motion is locally circular, the vehicle's linear speed v_o and angular speed Ω can be calculated from the rear wheels speed as follow:

$$\begin{cases} v_o = (V_{RR} + V_{RL})/2 \\ \Omega = (V_{RR} - V_{RL})/e \end{cases} \tag{8}$$

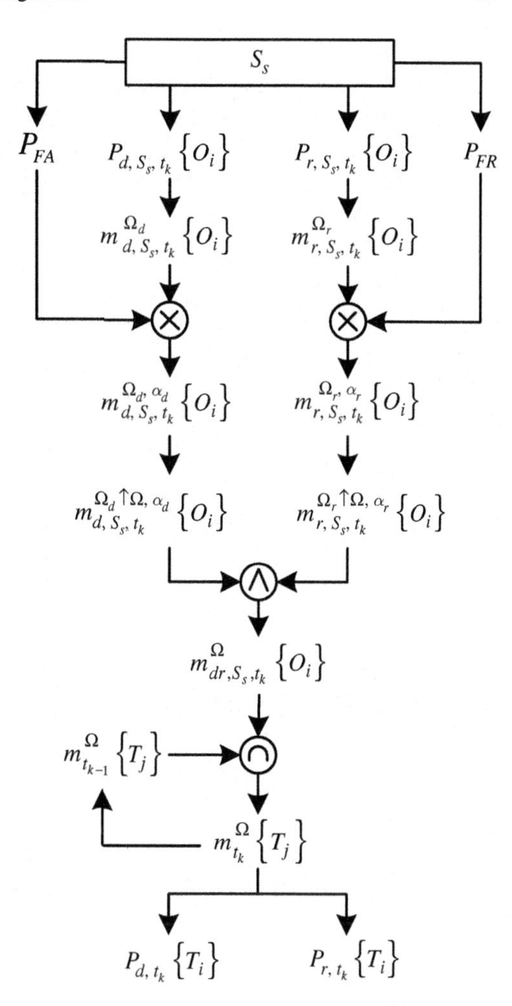

Fig. 5 Confidence updating algorithm

Where V_{RR} and V_{RL} represent respectively the rear right and left wheel speeds, and e is the distance between their points of contact with the road plane.

With the assumption of constant wheels speed between two CAN readings (with sampling time of T_e), the Eq. (7) prove that the linear and angular speeds are also constant; the vehicle state evaluation between the time t_k and $t_{k+1} = t_k + T_e$ can be written as:

$$\begin{cases} v_{o,k+1} = v_{o,k} \\ \theta_{k+1} = \theta_k + \Omega_k T_e \\ \Omega_{k+1} = \Omega_k \end{cases} \tag{9}$$

Where θ_k represents the absolute heading angle of the vehicle.

The vehicle state is filtered and estimated with a traditional Kalman filter having the state vector: $\begin{bmatrix} v_o & \theta & \Omega \end{bmatrix}^T$ and the measurement vector: $\begin{bmatrix} V_{RR} & V_{RL} \end{bmatrix}^T$

The model error covariance matrix is experimentally approximated based on the maximum error provided by the assumption of constant angular and linear speed model. The measurement error covariance matrix is calculated based on the ABS angular encoders' error.

3.3 Pedestrian Model

Pedestrians are supposed moving linearly at constant speed. The evaluation of the absolute position (x, y) and speed (v_x, v_y) of a pedestrian, with respect to the coordinate system $(o, \underline{i}, \underline{j})$, between the time t_k and t_{k+1} is:

$$\begin{cases} x_{k+1} = x_k + v_{x,k} T_e \\ y_{k+1} = y_k + v_{y,k} T_e \\ v_{x,k+1} = v_{x,k} \\ v_{y,k+1} = v_{y,k} \end{cases} \tag{10}$$

From the Eqs. (2), (4), (5), (7) and (10) we calculated the relative position and velocity of a pedestrian with respect to the coordinate system $(O, \underline{I}, \underline{J})$:

$$\begin{aligned} X_{k+1} &= X_k \cos(\Omega_k T_e) + Y_k \sin(\Omega_k T_e) - v_{o,k} T_e \cos(\Omega_k T_e/2) \\ &\quad + v_{x,k} T_e \cos(\theta_k + \Omega_k T_e) + v_{y,k} T_e \sin(\theta_k + \Omega_k T_e) \\ Y_{k+1} &= Y_k \cos(\Omega_k T_e) - X_k \sin(\Omega_k T_e) + v_{o,k} T_e \sin(\Omega_k T_e/2) \\ &\quad - v_{x,k} T_e \sin(\theta_k + \Omega_k T_e) + v_{y,k} T_e \cos(\theta_k + \Omega_k T_e) \end{aligned} \tag{11}$$

$$\begin{aligned} V_{x,k+1} &= V_{x,k} + \Omega_k T_e \left(v_{y,k} + v_{o,k} \sin(\theta_k - \Omega_k T_e/2) \right) \\ V_{y,k+1} &= V_{y,k} + \Omega_k T_e \left(-v_{x,k} + v_{o,k} \cos(\theta_k + \Omega_k T_e/2) \right) \end{aligned}$$

Pedestrians state is filtered and estimated with a traditional Kalman filter (one filter per pedestrian) having the state vector:

$$\begin{bmatrix} v_x & v_y & X & Y & V_x & V_y \end{bmatrix}^T$$

and the measurement vector:

$$\begin{bmatrix} X \\ Y \end{bmatrix} = \begin{bmatrix} X_L + L \\ Y \end{bmatrix} \tag{12}$$

The model error covariance matrix is experimentally approximated based on the maximum error provided by the assumption of pedestrian constant speed model. The measurement error covariance matrix is calculated based on the sensor's resolution saved in a configuration file with other tuning parameters.

After updating the kinematical state by the Kalman filter, the next section will describe the update method used for the detection and the recognition confidences by a credibilistic model based on the belief functions.

4 Confidence Indicators

4.1 Definition of Pedestrian's Confidence Indicators

The objective of the system is the detection and the recognition of pedestrians. To quantify this goal, we defined two numerical indicators representing respectively the confidence in detection and in recognition. These indicators can be calculated, for example, based on statistical approaches or on geometrical features analysis. As an example, a calculation method of theses indicators is described in [13] for the case of 4-planes Lidar.

4.2 Confidence Indicators Updating

4.2.1 TBM Principle and Notation

The transferable belief model TBM is a model to represent quantified beliefs based on belief functions [14]. It has the advantage of being able to explicitly represent uncertainty on an event. It takes into account what remains unknown and represents perfectly what is already known.

(a) *Knowledge representation*

Let Ω be a finite set of all possible solution of a problem. Ω is called the frame of discernment (also called state space); it's composed of mutually exclusive elements. The knowledge held by a rational agent can be quantified by a belief function defined from the power set 2^{Ω} to [0,1]. Belief functions can be expressed in several forms: the basic belief assignment (BBA) m, the credibility function bel, the plausibility function pl, and the commonality function q which are in one-to-one correspondence. We recall that $m(A)$ quantifies the part of belief that is restricted to the proposition "the solution is in $A \subseteq \Omega$" and satisfies:

$$\sum_{A \subseteq \Omega} m(A) = 1 \tag{13}$$

Thus, a BBA can support a set $A \subseteq \Omega$ without supporting any sub-proposition of A, which allows to account for partial knowledge. Smets introduced the notion of open world where Ω is not exhaustive; this is quantified by a non zero value of $m(\emptyset)$.

(b) *Information fusion*
n distinct pieces of evidence defined over a common frame of discernment and quantified by BBAs $m_1^\Omega \cdots m_n^\Omega$, may be combined, using a suitable operator. The most common are the conjunctive and the disjunctive rules of combination denoted respectively $\textcircled{\cap}$ and $\textcircled{\cup}$, and defined by the Eqs. (14) and (15).

$$\begin{aligned} m^\Omega_{\textcircled{\cap}}(A) &= m_1^\Omega(A) \textcircled{\cap} \cdots \textcircled{\cap}\, m_n^\Omega(A) \\ &= \sum_{A_1 \cap \cdots \cap A_n = A} m_1^\Omega(A_1) \times \cdots \times m_n^\Omega(A_n) \qquad \forall A \subset \Omega \end{aligned} \tag{14}$$

$$\begin{aligned} m^\Omega_{\textcircled{\cup}}(A) &= m_1^\Omega(A) \textcircled{\cup} \cdots \textcircled{\cap}\, m_n^\Omega(A) \\ &= \sum_{A_1 \cup \cdots \cup A_n = A} m_1^\Omega(A_1) \times \cdots \times m_n^\Omega(A_n) \qquad \forall A \subset \Omega \end{aligned} \tag{15}$$

Obtained BBAs should be normalized in a closed world assumption.
The conjunctive and disjunctive rules of combination assume the independence of the data sources. In [15] and [16] Denoeux introduced the cautious rule of combination (denoted by $\textcircled{\wedge}$) to combine dependent data. This rule has the advantage of combining dependent BBAs without increasing total belief: the combination of a BBA with itself will give the same BBA: $m = m \textcircled{\wedge} m$ (idempotence property). The cautious rule of combination is based on combining conjunctively the minimum of the weighted function representing dependent BBAs.

(c) *Reliability and discounting factor*
The reliability is the user opinion about the source [17]. The idea is to weight most heavily the opinions of the best source and conversely for the less reliable ones. The result is a discounting of the BBA m^Ω produced by the source into the new BBA $m^{\Omega,\alpha}$ where:

$$\begin{cases} m^{\Omega,\alpha}(A) = (1-\alpha)m^\Omega(A), & \forall A \subset \Omega,\ A \neq \Omega \\ m^{\Omega,\alpha}(\Omega) = \alpha + (1-\alpha)m^\Omega(\Omega) \end{cases} \tag{16}$$

The discounting factor $(1-\alpha)$ can be regarded as the degree of trust assigned to the sensor.

(d) *Decision making*
The couple (credibility, plausibility) is approximated by a measurement of probability by redistribute the mass assigned to each element of 2^Ω, different from singleton, to the elements which compose it. The probability resulting from this approximation is called pignistic probability $BetP$; it's used for decision making:

$$\forall \omega_i \in \Omega \quad \Rightarrow \quad BetP^\Omega(\omega_i) = \sum_{\omega_i \in A \subseteq \Omega} \frac{m^\Omega(A)}{|A|\,(1 - m^\Omega(\varnothing))} \tag{17}$$

4.2.2 Confidence Calculation

(a) *Defining the frames of discernment*
Before defining any quantified description of belief with respect to the objects' detection and/or pedestrians' recognition, we must define a frame of discernment Ω on which beliefs will be allocated and updated.

For the objects detection problem, we can associate two general cases: object O and non object NO. The object can be a pedestrian or a non pedestrian object, but with no object identification, the frame of discernment of the object detection process is limited to: $\Omega_d = \{O,NO\}$. As an example, a disparity image analyzer of a stereo-vision system can have Ω_d as its frame of discernment.

A mono-vision pedestrian recognition process based on an AdaBoost algorithm for example, gives the probability of detecting a pedestrian P or non pedestrian NP. The non pedestrian can be a non pedestrian object or a false alarm. Let $\Omega_r = \{P, NP\}$ be the frame of discernment of this type of recognition processes.

The update stage requires a commune frame of discernment, let $\Omega = \{PO, NPO, FA\}$ be the frame containing all possible solutions of the detection and the recognition problem. The relation between the three frames Ω, Ω_d and Ω_r is represented in Fig. 6.

Transforming BBAs from Ω_d and Ω_r to a more detailed frame, such as Ω, is called the refinement process and it's denoted $\uparrow$. The inverse transformation is called coarsening and it's denoted $\downarrow$ [14].

(b) *Basic belief assignment calculation*
The outputs of the detection and the recognition processes are Bayesian probability functions. With no additional information, we have to build, based on these probabilities, the basic belief assignments $m^{\Omega_d}_{d,S_s,t_k}\{O_i\}$ (BBA of the detection module of the object O_i detected by the source S_s and defined over the frame of discernment Ω_d at time t_k) and/or $m^{\Omega_r}_{r,S_s,t_k}\{O_i\}$ (BBA of the recognition module of O_i detected by S_s and defined over Ω_r at time t_k).
We are using the inverse pignistic probability transform proposed by Sudano [18] to calculate belief functions from Bayesian probability functions. So, to build the BBAs, we calculate from the probability functions the less informative BBAs who regenerate the same probability as its pignistic probability [19].

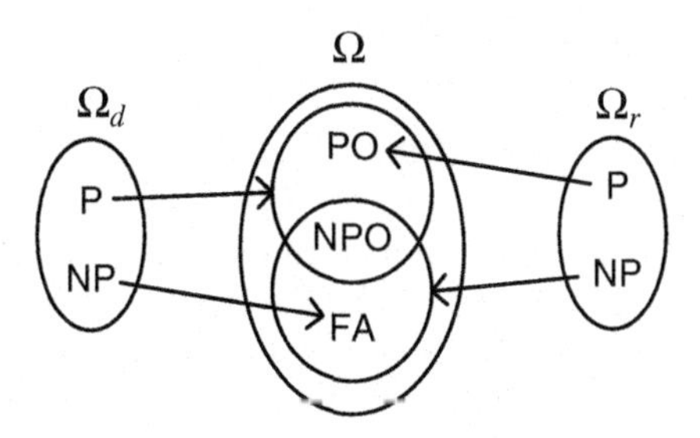

Fig. 6 Relation between the state spaces of the detection and the recognition processes

4.2.3 Confidence Updating Algorithm

The fusion and tracking module updates all tracks information such as track's state and track's detection and recognition confidences. The algorithm of track detection and recognition confidence update with object detection and recognition confidence consists in: (Fig. 5)

- Transform the probabilities $P_{d,S_s,t_k}\{O_i\}$ and $P_{r,S_s,t_k}\{O_i\}$ into basic belief assignment BBAs: $m^{\Omega_d}_{d,S_s,t_k}\{O_i\}$ and $m^{\Omega_r}_{r,S_s,t_k}\{O_i\}$ (see Sect. 4.2.2 b)
- Transform the performance of the sensor module into discounting values: the probability of false alarm P_{FA} and the probability of false recognition P_{FR} of the sensor module transform the last BBAs into $m^{\Omega_d,\alpha_d}_{d,S_s,t_k}\{O_i\}$ and $m^{\Omega_r,\alpha_r}_{r,S_s,t_k}\{O_i\}$ where $\alpha_d = 1 - P_{FA}$ and $\alpha_r = 1 - P_{FR}$ are respectively the discounting factors of the detection and the recognition processes (see Eq. 16).
- Transform beliefs from Ω_d and Ω_r to the commune frame of discernment Ω by doing the refinement process, that is, moving the belief on a subset of Ω_d (respectively Ω_r) to the corresponding subset of Ω. We get: $m^{\Omega_d\uparrow\Omega,\alpha_d}_{d,S_s,t_k}\{O_i\}$ and $m^{\Omega_r\uparrow\Omega,\alpha_r}_{r,S_s,t_k}\{O_i\}$ (see Sect. 4.2.2 a).
- The two obtained BBAs are not really independent because they are calculated in a sensor module based on the same data set. We decided to combine them using the cautious rule of combination, which take into consideration the dependency in data (see Sect. 4.2.1 b):

$$m^{\Omega}_{dr,S_s t_k}\{O_i\} = m^{\Omega_d\uparrow\Omega,\alpha_d}_{d,S_s,t_k}\{O_i\} \owedge m^{\Omega_r\uparrow\Omega,\alpha_r}_{r,S_s,t_k}\{O_i\}.$$

- We suppose the temporal independency of data provided by a sensor, thus the obtained BBA is combined conjunctively (using Eq. 14) with the associated track belief function $m^{\Omega}_{t_{k-1}}\{T_j\}$ to get $m^{\Omega}_{t_k}\{T_j\}$ as result of the combination and update process:

$$m^{\Omega}_{t_k}\{T_j\} = m^{\Omega}_{dr,S_s,t_k}\{O_i\} \textcircled{\cap} m^{\Omega}_{t_{k-1}}\{T_j\}$$

- Finally, the track's detection and recognition confidence $P_{d,t_k}\{T_i\}$ and $P_{r,t_k}\{T_i\}$ are the pignistic probability $BetP$ calculated from $m^{\Omega}_{t_k}\{T_j\}$ (see Sect. 4.2.1 d).

5 Results

5.1 Simulation Results

To show the effect of the proposed confidence updating method with two sensors with different reliabilities, we simulated the data coming from two unsynchronized sensors. Results shows that with constant object detection and recognition confidences, the more reliable sensor dominates.

In (Fig. 7), the sensor 1 is reliable in detection with a probability of false detection of 20%, but not in recognition (80% false recognition) and the sensor 2 is reliable in recognition (20% false recognition) but not in detection (80% false detection): fusion results shows that the corresponding track is well detected and recognized. Figure (Fig. 8) shows the inverse case: the track detection and recognition confidences go down to zero with the low confidences in detection and recognition given by the reliable sensors.

5.2 Experimentations

The algorithms are tested as a real time embedded system implemented in the experimental vehicle CARMEN (Fig. 3) of the laboratory Heudiasyc. CARMEN is equipped with different sensors such as 4-plans Lidar, stereo and mono cameras

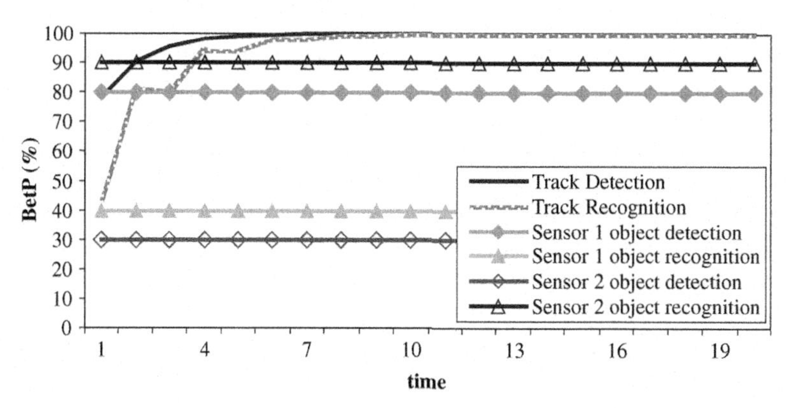

Fig. 7 Track detection and recognition with two sensors: confidences follow the powerful sensor up to 100%

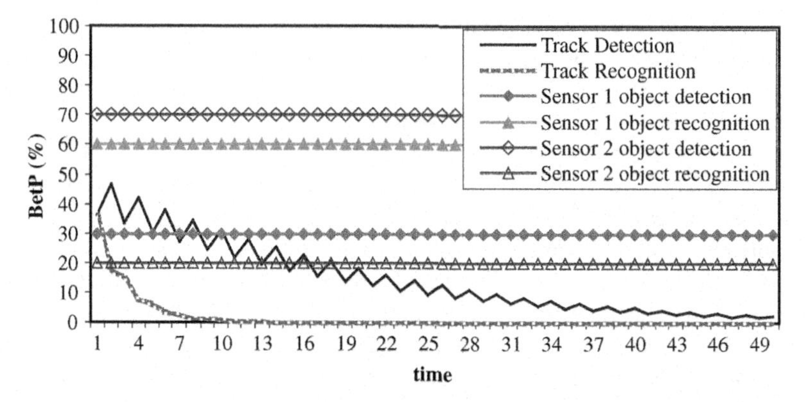

Fig. 8 Track detection and recognition with two sensors: confidences follow the powerful sensor down to 0%

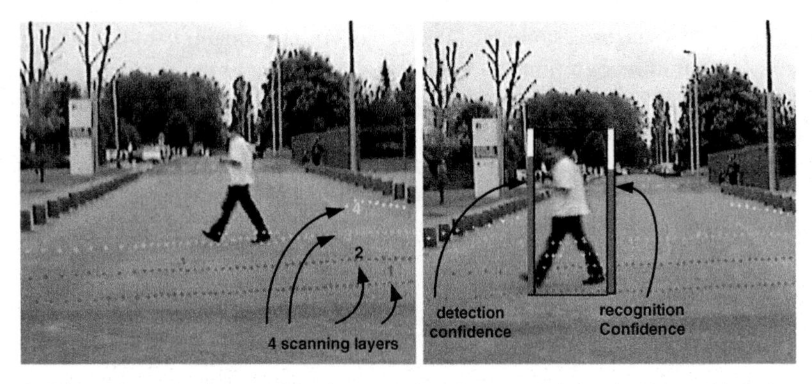

Fig. 9 Projection on the image of the four scanning layers and the pedestrians with their corresponding detection and recognition confidences

and radar. Proprioceptive data, such as wheels speed, is read from the vehicle CAN bus. Only Lidar and proprioceptive data are used in this experiment while image data provided by cameras is used to validate results by projecting laser data, tracks and confidences on the corresponding image (Fig. 9). Experimentations are done in an urban environment.

To simulate two unsynchronized sensors with different performance, Lidar data are assigned at each scanning period to one of two virtual Lidars having different detection and recognition confidences but the same measurement precision as the real Lidar.

5.3 Experimental Results

Results show the efficiency of the described method in unsynchronized data fusion especially when the frequency of the incoming data is unknown or variable. As an example, we will show the detection and recognition confidence result of tracking one pedestrian detected by the laser scanner.

The probability of false alarm P_{FA} and false recognition P_{FR} of the first virtual Lidar are fixed respectively to 10% and 40%, while the second virtual Lidar has more false alarms with $P_{FA} = 40\%$ and less false recognition with $P_{FR} = 10\%$.

Figures 10 and 11 show the results of tracking the same pedestrian during 90 Lidar scans. The 90 scans are distributed between the two virtual Lidar sensors. Figure 10 shows that the track detection confidence follows the confidence variation of the object detected by the first sensor having less false alarm probability then the second sensor. While Fig. 11 shows the variation of the tracked pedestrian's recognition confidence with the variation of the objects confidence detected by the two sensors.

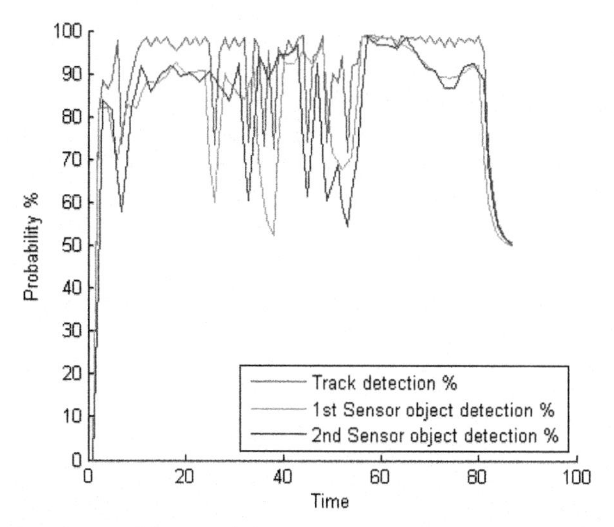

Fig. 10 Track detection confidence variation of one pedestrian detected by two sensors having different detection performance

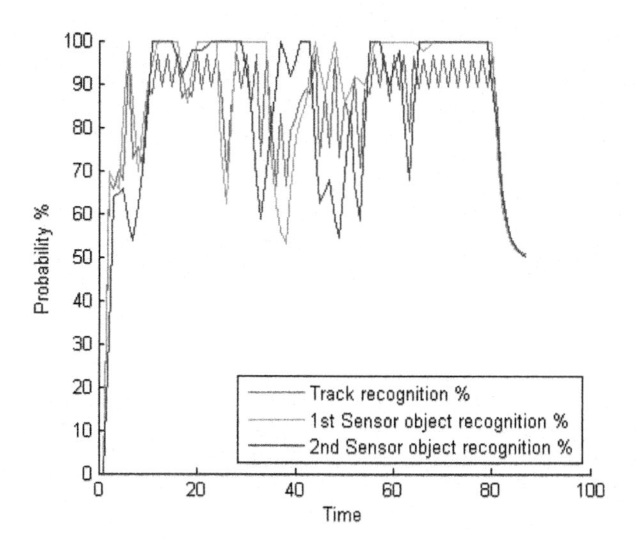

Fig. 11 Track recognition confidence variation of one pedestrian recognized by two sensors having different recognition performance

6 Conclusion and Future Works

In this paper we presented a multi-sensor fusion system dedicated to detect, recognize and track pedestrians. We differentiated between the two problems: object detection and pedestrian recognition, and quantified the confidence in the detection and recognition processes. The fusion by tracking method is used to solve the problem of asynchronous data provided by different sensors. The tracks state is divided into four parts and updated with different filters. Two of them are presented in this article: Kalman filter used for the kinematical state, and the detection and recognition confidences updated under the transferable belief framework. Results are shown with simulated and experimental data acquired in urban environment. Future works will concentrate on the validation of the method with multi-sensor data such as image and radar that have different performance in the detection and the recognition processes.

References

1. Sensors and system architecture for vulnerable road users protection, http://www.save-u.org/
2. Integrated solutions for preventive and active safety systems, http://www.prevent-ip.org/
3. Cooperative vehicles and road infrastructure for road safety, http://www.safespot-eu.org/
4. Cherfaoui V., Denoeux T., and Cherfi Z-L. Distributed data fusion: application to the confidence management in vehicular networks. Proceedings of the 11th International Conference on Information Fusion, Cologne, Germany, pp. 846–853 , July 2008.
5. Fuerstenberg K.C. and Lages U. Pedestrian detection and classification by Laser scanners. 9th EAEC International Congress, Paris, France, June 2003.
6. Suard F. et al. Pedestrian detection using stereo-vision and graph kernels. Proceedings of the IEEE Intelligent Vehicles Symposium IV'2005, pp. 267–272, June 2005.
7. Viola P., Jones M.J., and Snow D. Detecting pedestrians using patterns of motion and appearance. Proceedings of 9th IEEE International Conference on the Computer Vision, Vol. 63, pp. 153–161, Nice, France, October 2003.
8. Tessier C. et al. A real-time, multi-sensor architecture for fusion of delayed observations: application to vehicle localization. IEEE ITSC'2006, Toronto, Canada, September 2006.
9. Sittler R.W. An optimal data association problem in surveillance theory. *IEEE Transactions on Military Electronics*, 8(2), 125–139, 1964.
10. Fayad F. and Cherfaoui V. Tracking objects using a laser scanner in driving situation based on modeling target shape, Proceedings of the IEEE Intelligent Vehicles Symposium IV'2007, pp. 44–49 Istanbul, Turkey, June 2007.
11. Bonnifait Ph. et al. Dynamic localization of car-like vehicle using data fusion of redundant ABS sensors. *The Journal of Navigation*, 56, 429–441, 2003.
12. Julier S. and Durrant-Whyte H. Process models for the high-speed navigation of road vehicles. Proceedings of the IEEE International Conference on Robotics and Automation, Vol. 1, pp. 101-105, Nagoya, Japan, May 1995.
13. Fayad F., Cherfaoui V., and Dherbomez G. Updating confidence indicators in a multi-sensor pedestrian tracking system. Proceedings of the IEEE Intelligent Vehicles Symposium IV'2008, Eindhoven, Holland, June 2008.
14. Smets Ph. and Kennes R. The transferable belief model. *Artificial Intelligence*, 66(2), 191–234, 1994.
15. Denoeux T. The cautious rule of combination for belief functions and some extensions. Proceedings of the 9th International Conference on Information Fusion, Florence, Italy, pp. 1–8, July 2006.
16. Denoeux T. Conjunctive and disjunctive combination of belief functions induced by non distinct bodies of evidence. *Artificial Intelligence*, 172, 234–264, 2008.
17. Elouedi Z. et al. Assessing sensor reliability for multisensor data fusion within the transferable belief model. *IEEE Transactions on Systems, Man and Cybernetics*, 34, 782, 2004.

18. Sudano J.J. Inverse pignistic probability transforms. Proceedings of the 5th International Conference on Information Fusion, Vol. 2, pp. 763–768, 2002.
19. Fayad F. and Cherfaoui V. Detection and recognition confidences update in a multi-sensor pedestrian tracking system. Proceedings of the 12th International Conference on Information Processing and Management of Uncertainty in Knowledge-Based Systems IPMU2008, Malaga, Spain, June 2008.

Effective Lip Localization and Tracking for Achieving Multimodal Speech Recognition

Wei Chuan Ooi, Changwon Jeon, Kihyeon Kim, Hanseok Ko and David K. Han

Abstract Effective fusion of acoustic and visual modalities in speech recognition has been an important issue in Human Computer Interfaces, warranting further improvements in intelligibility and robustness. Speaker lip motion stands out as the most linguistically relevant visual feature for speech recognition. In this paper, we present a new hybrid approach to improve lip localization and tracking, aimed at improving speech recognition in noisy environments. This hybrid approach begins with a new color space transformation for enhancing lip segmentation. In the color space transformation, a PCA method is employed to derive a new one dimensional color space which maximizes discrimination between lip and non-lip colors. Intensity information is also incorporated in the process to improve contrast of upper and corner lip segments. In the subsequent step, a constrained deformable lip model with high flexibility is constructed to accurately capture and track lip shapes. The model requires only six degrees of freedom, yet provides a precise description of lip shapes using a simple least square fitting method. Experimental results indicate that the proposed hybrid approach delivers reliable and accurate localization and tracking of lip motions under various measurement conditions.

1 Introduction

A multimodal speech recognition system is typically based on a fusion of acoustic and visual modalities to improve its reliability and accuracy in noisy environments. Previously, it has been shown that speaker lip movement is a significant visual component that yields linguistically relevant information of spoken utterances. However, there have been very few lip feature extraction methods that work robustly under various conditions. The difficulty is caused by variation of speakers, visual capture devices, lighting conditions, and low discriminability in lip and skin color.

W.C. Ooi (✉)
School of Electrical Engineering, Korea University, Seoul, Korea
e-mail: wcooi@ispl.korea.ac.kr

S. Lee et al. (eds.), *Multisensor Fusion and Integration for Intelligent Systems*, Lecture Notes in Electrical Engineering 35, DOI 10.1007/978-3-540-89859-7_3,

Historically, there have been two main approaches [1] in extracting lip features from image sequences. The first method is called the Image-based approach. In this approach, image pixels (e.g. intensity values) around the lip region are used as features for recognition. For instance, these approaches are based on a DCT or a PCA method. The projected low dimensional features are used for speech recognition. The extracted features not only consist of lip features but also of other facial features such as tongue and jaw movement depending on ROI size. The drawback is that it is sensitive to rotation, translation scaling, and illumination variation.

The second type is known as the model-based method. A lip model is described by a set of parameters (e.g. height and width of lips). These parameters are calculated from a cost function minimization process of fitting the model onto a captured image of the lip. The active contour model, the deformable geometry model, and the active shape model are examples of such methods widely used in lip tracking and feature extraction. The advantage of this approach is that lip shapes can be easily described by low order dimensions and it is invariant under rotation, translation, or scaling. However, this method requires an accurate model initialization to ensure that the model updating process converges.

In this paper, we propose a model-based method designed primarily to improve accuracy and to reduce the processing time. The model-based method requires good initialization to reduce the processing time. By integrating color and intensity information, our algorithm maximizes contrast between lip and non-lip regions, thus resulting accurate segmentation of lip. The segmented lip image provides initial position for our point based deformable lip model which has built in flexibility for precise description of symmetric and asymmetric lip shapes.

We describe in more detail of the PCA based color transformation method in Sect. 2. In Sect. 3, we present a new deformable model for Lip Contour Tracking. We also describe cost function formulation and model parameter optimization in the same section. Experimental Results and comparison with other color transformation are presented in Sect. 4. The conclusion is presented in Sect. 5.

2 Lip Color and Intensity Mapping

Many methods have been proposed for segmenting the lip region that based on image intensity or color. We propose a new color mapping of the lips by integrating color and intensity information. Among color based lip segmentation methods are Red Exclusion [2], Mouth-Map [3], R–G ratio [4], and Pseudo hue [5]. Theoretically, pseudo hue method gives better color contrast, but we found that it is only useful in performing coarse segmentation which is not adequate for our purpose. Thus, we perform a linear transformation of RGB components in order to gain maximum discrimination of lip and non-lip colors. We employ a PCA to estimate the optimum coefficients of transformation. From a set of training images, N pixels of lip and non-lip are sampled and its distribution shown in Fig. 1(a). Each pixel is regarded as three dimensional vector $\mathrm{x}_i = (R_i, G_i, B_i)$. The covariance matrix

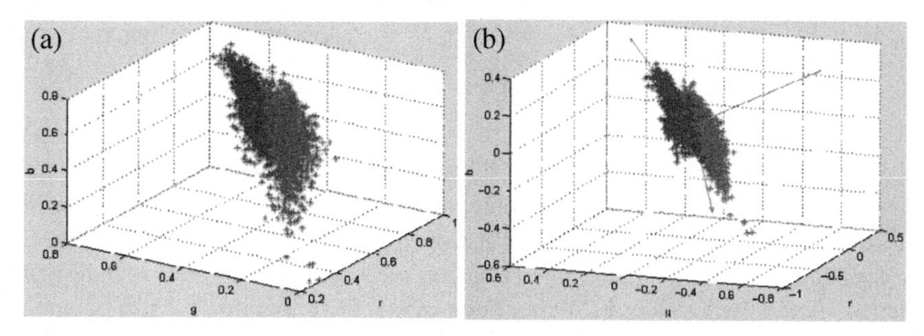

Fig. 1 (**a**) Distribution of lip and non-lip in RGB color space (60 people under different lighting conditions), (**b**) Eigenvectors of distribution in Fig. 1(a)

is obtained form the three dimensional vector and the associated eigenvectors and eigenvalues are determined from the covariance matrix.

$v_3 = (v_1, v_2, v_3)$ is an eigenvector corresponding to the third smallest eigenvalue where lip and non-lip pixels are the least overlapping as shown in Fig. 2(c). Experimentally, $v_1 = 0.2$, $v_2 = -0.6$, $v_3 = 0.3$ are obtained. Thus a new color space, C is defined as

$$C = 0.2 \times R - 0.6 \times G + 0.3 \times B \tag{1}$$

The new color space C is normalized as

$$C_{\text{norm}} = \frac{C - C_{\text{min}}}{C_{\text{max}} - C_{\text{min}}} \tag{2}$$

Note that after normalization, the lip region shows higher value than the non-lip region. By squaring the C_{norm}, we can further increase the dissimilarity between these two clusters as shown in Fig. 3. A similar conversion of RGB values using the Linear Discriminant Analysis (LDA) was employed by Chan [6] to direct the evolution of snake. PCA based method is simpler compared to LDA especially in dealing with three dimensional RGB components.

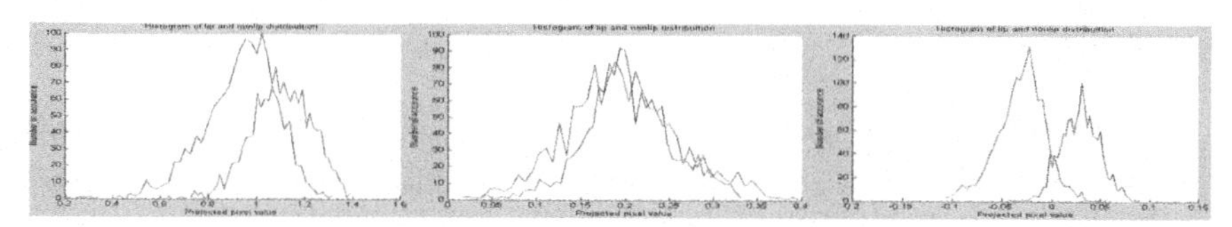

Fig. 2 (**a**) Histogram of projected pixels onto first principle component, (**b**) Histogram of projected pixels onto second principle component, (**c**) Histogram of projected pixels onto third principle component

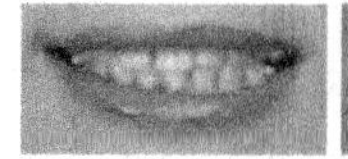
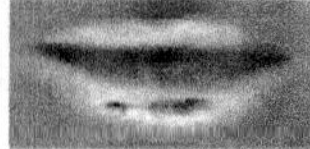
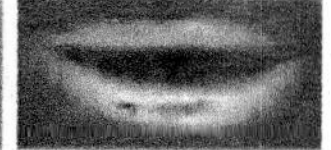

Fig. 3 (**a**) Original image, (**b**) Transformed image, C, (**c**) C squared image

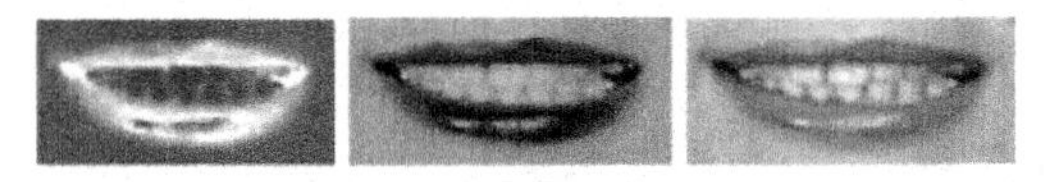

Fig. 4 (**a**) C_{map} image, (**b**) C_{map} negative image, (**c**) Gray-scale image

After the color transformation, the C squared image may still show low contrast in the upper lip region. This problem can be resolved by using the intensity information I. The upper lip region typically consists of lower intensity values. So by combining the C squared image (which is well separable in the lower lip) and intensity image (which has a stronger boundary in the upper lip), we can obtain an enhanced version of the lip color map C_{map} as follows.

$$C_{map} = \alpha C_{squar} + (1 - \alpha)\frac{1}{I} \tag{3}$$

Empirically, $\alpha = 0.75$ are derived. Higher weight is given to the C squared image since it captures most of the lip shape except the upper part and corners of lips.

2.1 Threshold Selection

In this paper, the global threshold is selected based on Otsu [7] method. The optimal threshold T_{opt} is chosen so that between classes variance σ_B^2 is maximized.

$$T_{opt} = \underset{0<T<1}{Arg\max}\ \sigma_B^2(T) \tag{4}$$

Fig. 5 Segmented images

3 Lip Contour Tracking

3.1 Lip Model

Most of the deformable geometric models are established using quadratic fittings (e.g. parabolic) with a prior assumption of lip shape being always symmetric about the center axis. Our lip model is an enhanced version of the proposed method in [8]. In [8] the writers integrated flexibility and constrained deformable template with point distribution model in order to reduce computations. However the geometric model in the paper is described by 15 parameters resulting in significant computation for the parameter updating process. Our proposed lip model is established by six parameters and is composed of three curves defined as follows:

- Lower lip, $\{0 < x < 1\}$

$$y_{low} = \alpha_{low} \cdot x \cdot (\log_2 x) + \beta_{low} \cdot (1 - x) \cdot (\log_2 (1 - x)) + \gamma_{low} \left(\frac{(x - 0.5)^4}{0.5^4} - \frac{(x - 0.5)^2}{0.5^2} \right) \tag{5}$$

- Upper right lip, $\{0.5 \leq x < 1\}$

$$y_{up\_r} = -3.148 \cdot \alpha_{up\_r} \cdot (x - 0.4)^{\frac{1}{2}} \cdot (\log_2 x) + \gamma_{up\_r} \left(\frac{(x - 0.5)^4}{0.5^4} - \frac{(x - 0.5)^2}{0.5^2} \right) \tag{6}$$

- Upper left lip, $\{0 < x \leq 0.5\}$

$$y_{up\_l} = -3.148 \cdot \alpha_{up\_l} \cdot (0.6 - x)^{\frac{1}{2}} \cdot (\log_2 (1 - x)) + \gamma_{up\_l} \left(\frac{(x - 0.5)^4}{0.5^4} - \frac{(x - 0.5)^2}{0.5^2} \right) \tag{7}$$

3.2 Parameters Description

There are six parameters to fully model the lower and upper lips. Parameters α_{low} and β_{low} control vertical height and skewness of lower lip as shown in Fig. 6.

- If $\alpha_{low} = \beta_{low}$, lip (lower) is symmetric with respect to center point, then center height=$\alpha_{low} = \beta_{low}$
- If $\alpha_{low} > \beta_{low}$, lip shape slides to the left
- If $\alpha_{low} < \beta_{low}$, lip shape slides to the right

The parameter γ_{low} controls curvature of lower lip shape with values between 0.15 and 0.15 as shown in Fig. 7.

Compared to the lower lip, upper lip shape remains relatively symmetric. This is due to the fact that the lower lip motion is a result of the mandible movement. The articulation available of the jaw joint allows the mandible movement of left or

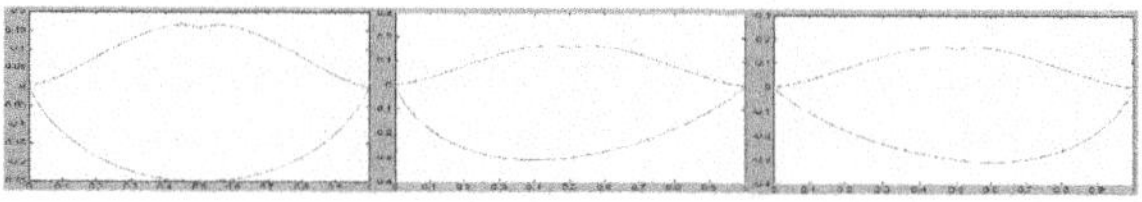

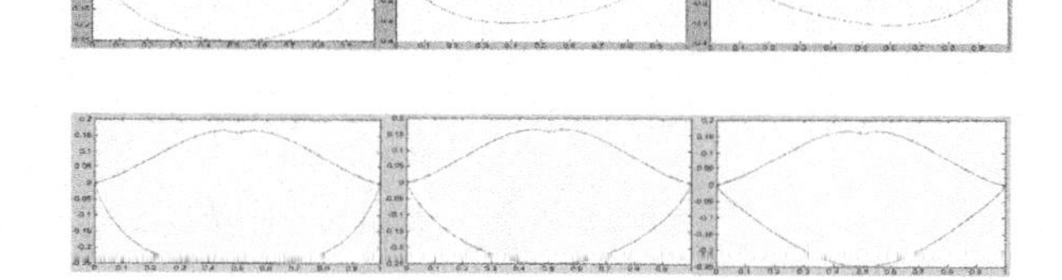

Fig. 6 Lip model for above cases

Fig. 7 **(a)** $\gamma_{low} = -1.5$, (b) $\gamma_{low} - 0$, (c) $\gamma_{low} - 1.5$

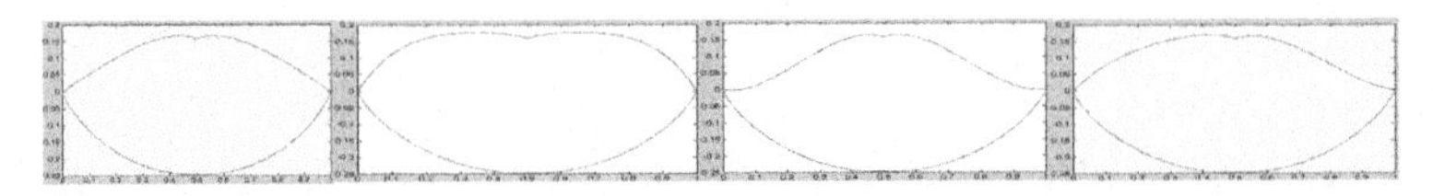

Fig. 8 (**a**) $\gamma_{up\_r} = \gamma_{up\_l} = 0$ (**b**)−0.3,−0.3 (**c**) 0.2, 0.2 (**d**)−0.2, 0.1

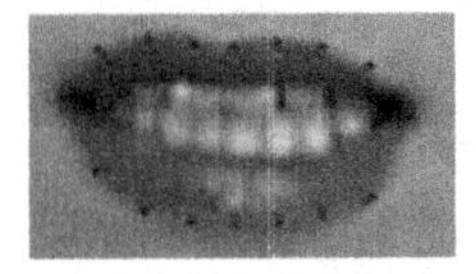

Fig. 9 Lip contour points converge from initial position to optimum position

right of the centerline of the face resulting in asymmetric movement of the lower lip. Thus, we assume the upper lip always remains symmetric

- $\alpha_{up\_l} = \alpha_{up\_r} =$ center height of upper lip
- γ_{up_r} and γ_{up_l} control curvature of upper lip (with value between 0.2 and −0.3)

Lip shapes according to the variations of the upper lip parameters are shown in Fig. 8.

3.3 Model Initialization and Normalization

Our model initialization is based on a segmented lip shape image. In order to reduce the computation time, we limited our model with just 16 points of p $= \{p_1, p_2, ..., p_{16}\}$ where $p_i = (x_i, y_i)$. The lip points are labeled in anti-clockwise direction starting from the left corner. These points are divided into three groups where $p_1, ..., p_9$ describe lower lip, $p_9, ..., p_{13}$ describe upper right points, and $p_{13}, ..., p_{16}$ describe upper left. The contour point normalization process is applied to reduce processing time and to simplify the curve fitting process. The left corner point is fixed as the origin. Rotation and scaling transformations are employed to normalized all points so that p_1 is at (0, 0) and p_9 is at (1, 0). Reverse normalization is applied after curve fitting for obtaining the original coordinates.

3.4 Model Optimization

The optimization procedure is an iterative process and the lip points are adjusted in order to reduce the cost function in each iteration process.

Our cost function F is defined in (8)

$$F = \arg\min_{p} \sum_{i=1}^{16} aE_{\text{int}}(p_i) + bE_{ext}(p_i) + cE_{bal}(p_i) \tag{8}$$

$E_{\text{int}}(p_i)$ is an energy function dependent on the shape of the contour points and it is the continuity energy that enforces the shape of the contour. $E_{ext}(p_i)$ is an energy function based on image properties (we use gradient in this paper). $E_{bal}(p_i)$ is a balloon force that causes the contour to expand (or shrink). In most cases, our binary image provides a good model initialization. Hence, the model usually takes only 5–8 iterations to converge to lip shape in a given image.

3.5 Model Fitting to Contour Points

We used least square approach to fit the model onto optimum contour points. By fitting the model on contour points, we can constraint the deformation of contour points and preserve a legal shape of lip shape. Furthermore, from the fitted model parameters, lip features can be extracted and used in visual speech recognition. The least square fitting is performed for three parts of outer lip separately when optimums lip contour points are obtained from optimization process. For example, in lower lip model that employs three parameters, $\theta = \{\alpha_{low}, \beta_{low}, \gamma_{low}\}$ show a process of least square method to fit model on contour points and parameter values are obtained. Note that p_1 and p_9 are not included since these two points are fixed in the normalization process.

$$H = \begin{bmatrix} x_2 \cdot (\log_2 x_2) & (1-x_2)\cdot(\log_2(1-x_2)) & \left(\frac{(x_2-0.5)^4}{0.5^4} - \frac{(x_2-0.5)^2}{0.5^2}\right) \\ \vdots & \vdots & \vdots \\ x_8 \cdot (\log_2 x_8) & (1-x_8)\cdot(\log_2(1-x_8)) & \left(\frac{(x_8-0.5)^4}{0.5^4} - \frac{(x_8-0.5)^2}{0.5^2}\right) \end{bmatrix}$$

$$\theta = \begin{bmatrix} \alpha_{low} \\ \beta_{low} \\ \gamma_{low} \end{bmatrix}, Y = \begin{bmatrix} y_{low2} \\ \vdots \\ y_{low8} \end{bmatrix}$$

$$\begin{aligned} &\because H\theta = Y \\ &\Rightarrow \theta = (H^T H)^{-1} H^T Y \end{aligned} \tag{9}$$

After least square fitting, new contour points can be found by equally deriving from fitted curves. These contour points are de-normalized by employing reverse scaling and rotation. Before new iteration is processed, sum of distance of new contour points is computed and compare to sum of distance of previous contour points. If the result is less than threshold the iteration process will be terminated.

4 Experiment Result

4.1 Lip Contour Extraction Result

In order to test the performance of our proposed hybrid procedure, we use 2,000 lip images with different sizes over 50 people (not including images that were used in color space training). In the testing images, we also use some images which consist of complex background like mustache and beard. For evaluating the flexibility of model asymmetrical lip shape images are incorporated.

Overall, 97% of the lip contours are accurately extracted. Figures 10 and 11 show examples of such images. We also show that our proposed lip color and intensity mapping have successfully improved the lip contour extraction performance under different lightning conditions. With our algorithm implemented in Matlab, the average computation time for 85×100 size images was approximately 0.9 s.

From experimental results, it had shown that accuracy of lip contour tracking not merely depend upon flexibility of built model but also contingent on preprocessing part. For instance, our proposed lip color and intensity mapping efficiently maximize dissimilarity of lip and non-lip region. This mapping result will be used to localized lip model and also provide clear edge information to derive contour points moving toward lip boundary.

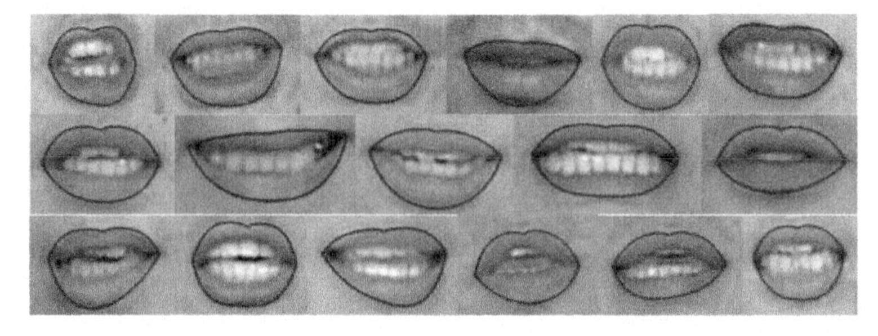

Fig. 10 Lip contour extraction results of female and also male lips

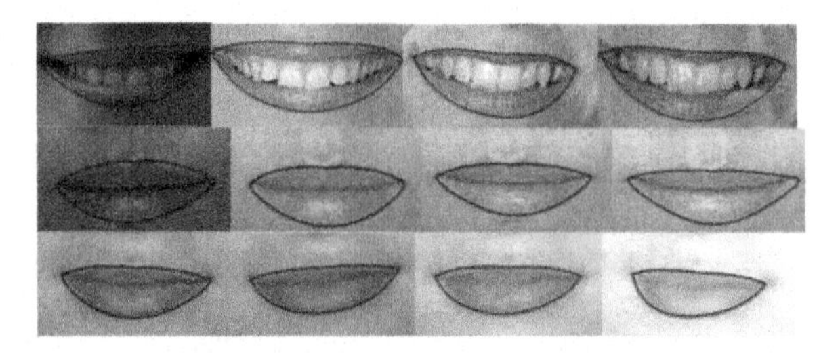

Fig. 11 Lip contour extraction results under different lightning conditions

4.2 Comparative Studies of Lip Mapping

We apply a quantitative technique to evaluate the performance of our color space transformation algorithm. Since no ground truth is available, we manually draw the boundaries of 25 lip images. The first measurement method is the degree of overlap (DOL) between the lip and the non-lip histograms. DOL [9] is used to measure discriminability of the transformed color spaces for differentiating the lip and the non-lip colors. A lower percentage of DOL means a higher contrast between the lip and the non-lip regions.

$$\begin{aligned} DOL &= \sum_{i=0}^{1} \min(P_{lip}(i), P_{nonlip}(i)) \\ \text{where} \quad P_{lip}(i) &= Num_{lip}(i)/\,Total_{lipPixel} \\ P_{nonlip}(i) &= Num_{nonlip}(i)/\,Total_{nonlipPixel}, \{0 \le i \le 1\} \end{aligned} \tag{10}$$

The second method is Classification Error (CE) which is the average of the False Positive (FP) rate and the False Negative (FN) rate FP is error rate of classifying a non-lip as a lip pixel. FN is the error rate in classifying the lip as non-lip pixel.

$$\begin{aligned} CE &= (FN + FP)/2 \\ \text{where } FN &= False_{nonlip}/(False_{nonlip} + True_{lip}) \\ FP &= False_{lip}/(False_{lip} + True_{nonlip}) \end{aligned} \tag{11}$$

From the results, we can see that our proposed lip color transformation method gives the lowest DOL and CE.

Table 1 Comparison of DOL and CE based on five mapping methods for below three images

	Image 1 (Fig. 12)		Image 2 (Fig. 13)		Image 3 (Fig. 14)	
	DOL	CE	DOL	CE	DOL	CE
MM	0.274	0.195	0.202	0.237	0.232	0.229
Our	0.182	0.150	0.201	0.143	0.173	0.144
RE	0.372	0.466	0.423	0.319	0.291	0.5
PH	0.223	0.243	0.294	0.188	0.210	0.162
RG	0.977	0.496	0.318	0.500	0.993	0.499

Table 2 Comparison for average DOL and CE for below three images and additional 22 testing images

	MM	Our	RE	PH	RG
Average DOL (%)	23.8	16.4	36.7	22.9	55.7
Average CE (%)	21.4	12.5	35.8	17.2	36.4

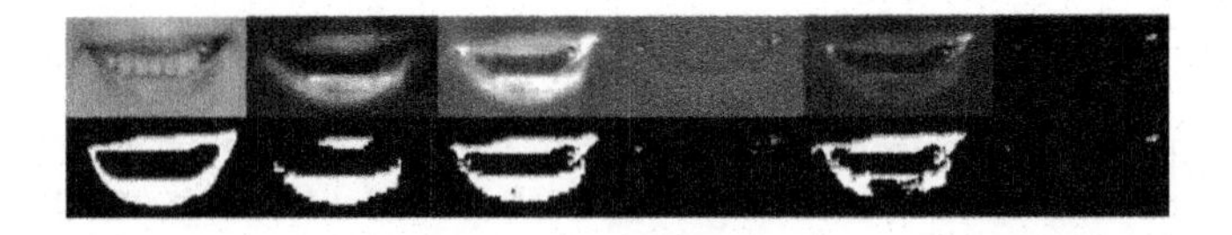

Fig. 12

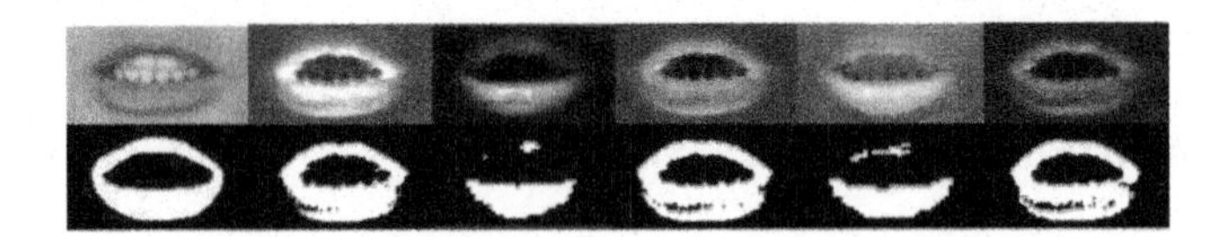

Fig. 13

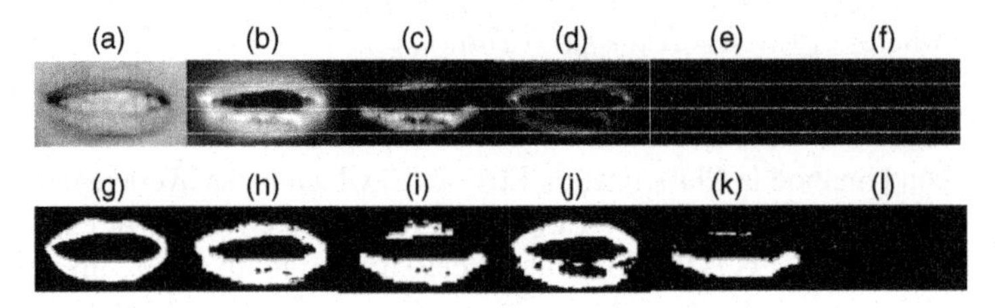

Fig. 14 (**a**) Test image, (**b**) Mouth-map method (MM), (**c**) Our method, (**d**) Red Exclusion method (RE), (**e**) Pseudo Hue method (PH), (**f**) R–G ratio method (RG), (**g**) Ground truth image, (**i**) is segmented image of (c) based on Otsu thresholding, (**h**), (**j**), (**k**), (**l**) are segmented images of (b), (d), (e), (f) with the threshold values proposed by corresponding previous methods

5 Conclusion

In this paper, we describe a new hybrid approach to improve lip localization and tracking. The first part of our proposed algorithm is lip mapping based on color and intensity information. From experimental results, our proposed mapping method successfully enhances the contrast between lip and non-lip regions. Results from the contrast enhancement process allowed more accurate lip region segmentation. In the second part, a new flexible while constrained deformable geometric model is established to accurately locate and track lip shape. Overall, our implemented hybrid approach has shown high reliability and is able to perform robustly under various conditions.

Acknowledgments This research was supported by MKE (Ministry of Knowledge Economy), Korea, under the ITFSIP (IT Foreign Specialist Inviting Program) supervised by the IITA.

References

1. G. Potamianos, C. Neti, G. Gravier, A. Garg, and A. W. Senior, Recent advances in the automatic recognition of audio-visual speech, Invited, IEEE Proc., 91, 1306–1326, 2003.

2. T. W. Lewis.and D. M. Powers, Lip feature extraction using Red Exclusion, Proc. Selected papers from Pan-Sydney Workshop on Visual Information Processing, pp. 61–67, 2000.
3. R. L. Hsu, M. Abdel, A. K. Jain, Face detection in color images, IEEE Trans. Pattern Anal. Mach. Intelli., 2002.
4. S. Igawa, A. Ogihara, A. Shintani, and S. Takamatsu, Speech recognition based on fusion of visual and auditory information using full-frame color image, ZEZCE Trans. Fundam., 1996.
5. A. Hulbert and T. Poggio, Synthesizing a color algorithm from examples, Science, 239, 482–485, 1998.
6. M. T. Chan, Automatic lip model extraction for constrained contour-based tracking, ICIP, 848–851 1999.
7. N. Otsu, A threshold selection method from gray-level histograms, IEEE Trans. Syst. Man Cyber., 62–66, 1979.
8. S. L. Wang, W. H. Lau, and S. H. Leung, A new real-time lip contour extraction algorithm, ICASSP, 217–220, 2003.
9. T. C. Terrillon, M. N. Shirazi, and H. Fukamachi, Comparative performance of different chrominance skin chrominance models and chrominance spaces for the automatic detection of human faces in color images, Proc. IEEE Int. Conf. Autom. Face Gesture Recogn., 54–61, 2000.

Optimal View Selection and Event Retrieval in Multi-Camera Office Environment

Han-Saem Park, Soojung Lim, Jun-Ki Min and Sung-Bae Cho

Abstract Recently, diverse sensor technologies have been advanced dramatically, so that people can use those sensors in many areas. Camera to capture the video data is one of the most useful sensors among them, and the use of camera with other sensors or the use of several cameras has been done to obtain more information. This paper deals with the multi-camera system, which uses the several cameras as sensors. Previous multi-camera systems have been used to track a moving object in a wide area. In this paper, we have set cameras to focus on the same place in an office so that system can provide diverse views on a single event. We have modeled office events, and modeled events can be recognized from annotated features. Finally, we have conducted the event recognition, view selection and event retrieval experiments based on a scenario in an office to show the usefulness of the proposed system.

Keywords Multi-camera system · View selection · Office event

1 Introduction

These days, most people can obtain and use the multimedia data easily as the sensor technologies to obtain those data have been advanced dramatically [1]. In particular, camera comes to a representative and useful sensor device since video data is much more informative than other data types such as text document, sound, and still image in that it contains specific and realistic information even though its analysis is very challenging [2].

In a stage for public performance or in a field for sports broadcasting, a number of cameras are being used to cover a wide area and to display scenes from various angles. Accordingly, the cameras show different image and information, so it is required to analyze those images and select a specific one whose information is the most useful.

H.-S. Park (✉)
Department of Computer Science, Yonsei University, 262 Seongsanno,
Sudaemun-gu, Seoul 120-749, KOREA
e-mail: sammy@sclab.yonsei.ac.kr

S. Lee et al. (eds.), *Multisensor Fusion and Integration for Intelligent Systems*,
Lecture Notes in Electrical Engineering 35, DOI 10.1007/978-3-540-89859-7_4,

This paper deals with a problem that selects the camera of which view is the most informative in multi-camera system set in the office environment. In order to analyze information from input video, we have exploited Bayesian network, which is a model of a joint probability distribution over a set of random variables. Bayesian network is represented as a directed acyclic graph where nodes correspond to variables and arcs correspond to probabilistic dependencies between connected nodes [3]. These models are used to recognize and retrieve office events.

In previous studies, researchers have used the multi-camera system to track a certain object or a person in a wide area [4–6]. Multi-camera systems, however, have another advantage. That is, we can obtain diverse views using multi-camera system. In this paper, we mainly focus on this possibility.

2 Backgrounds

2.1 Multi-Camera Systems

Previously, the multi-camera systems have been used to track a certain object in a wide area. Black and Ellis exploited multi-camera systems to track and detect moving object in outdoor environment [4]. These days, a few research groups have used multi-camera system in indoor environment. Sumi et al. used multiple cameras with other ubiquitous sensors to capture simple interactions between humans in a conference room [5]. Silva et al. presented a system for retrieval and summarization of multimedia data using multi-camera system and other sensors in a home-like ubiquitous environment [6].

All these researches used multi-camera system to cover wide areas, that is, they used only one camera at one place. As mentioned before, we focused on other possibility of the multi-camera system. We have set multi-camera system in office, so most of the cameras focused on the same place. Diverse views obtained from this system have two advantages: recognition of hidden object and higher recognition accuracy with ensemble. We can obtain information we need even if some cameras are hidden accidentally. Besides, higher recognition accuracy is expected due to accurate features extracted from multiple inputs.

2.2 Event Retrieval in Videos

Event retrieval in video data has been researched a lot according as the sensor technologies including cameras have been advanced [7–9].

In particular, one in sport videos is a very popular problem. Li et al. applied the event detection and modeling algorithms to different types of sports videos, and they provided retrieval and summarization of sport events [7]. Ekin et al. proposed a fully automatic and computationally efficient framework for soccer video summarization and analysis using some novel low-level processing algorithms and high level detection algorithms [8]. On the other way, Ersoy et al. presented a framework for the event retrieval in general video. They exploited domain-independent event primitives to provide adaptability of the system [9]. All these works dealt with the video data captured by single camera.

3 Optimal View Selection and Event Retrieval Using Bayesian Network Based Event Modeling

Figure 1 illustrates an overall process of event recognition, view selection, and event retrieval in the proposed multi-camera system. The whole process is divided into three parts: event recognition, view selection, and event retrieval. Low-level features have been annotated by human expert manually based on predefined domain knowledge. In event recognition part, the designed Bayesian network (BN) model recognizes office events with all features from all cameras. This model also recognizes event with features from each camera. View selection part selects the camera that provides the optimal view considering the recognition probability of given event and event priority.

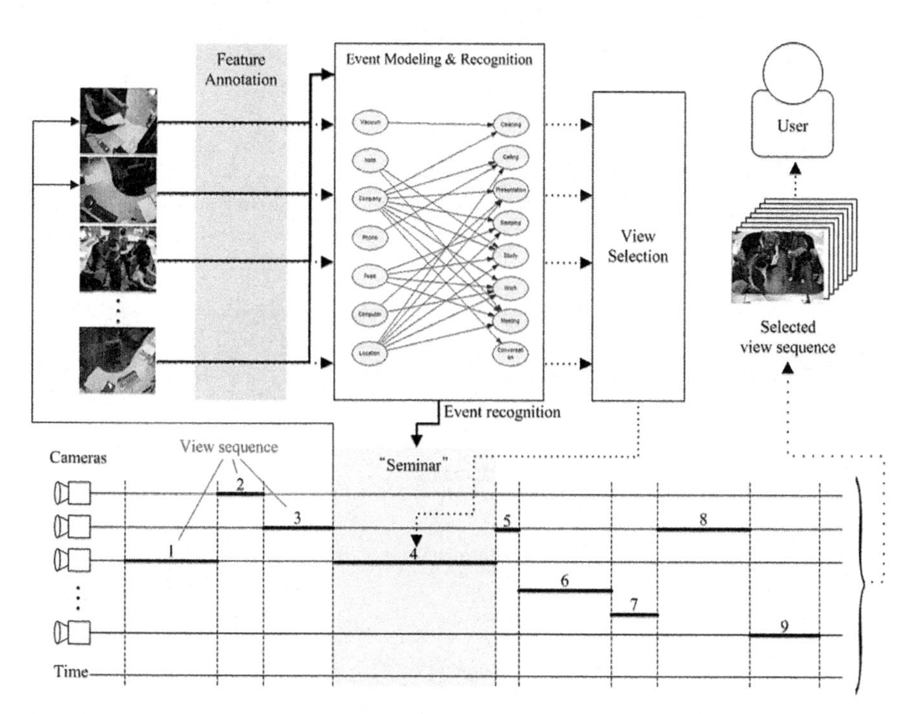

Fig. 1 Overall process for view selection and event retrieval

3.1 *Office Event Modeling*

To model office events, we have defined eight events, and each event is related to proper objects or poses based on event definition. Events, objects and poses used in this paper are as follows.

- Event: Calling, cleaning, conversation, meeting, presentation, sleeping and work
- Object: Computer, phone, note, vacuum, users
- Pose: Stand, sit, rest.

Basic features such as object, person's pose, position in office and person's direction have been annotated by expert based on predefined event, and these annotated features and events have been used to make Bayesian network event model learn.

We have modified this learned model. First, we removed the dependencies among evidences because they are not significant if evidences were set. If we did not find the evidence, it was checked as 'no', meaning there is no evidence. Subsequently, parameters were also modified because they were just based on the learning data so that they can recognize office events generally. Figure 2 shows the designed Bayesian network model.

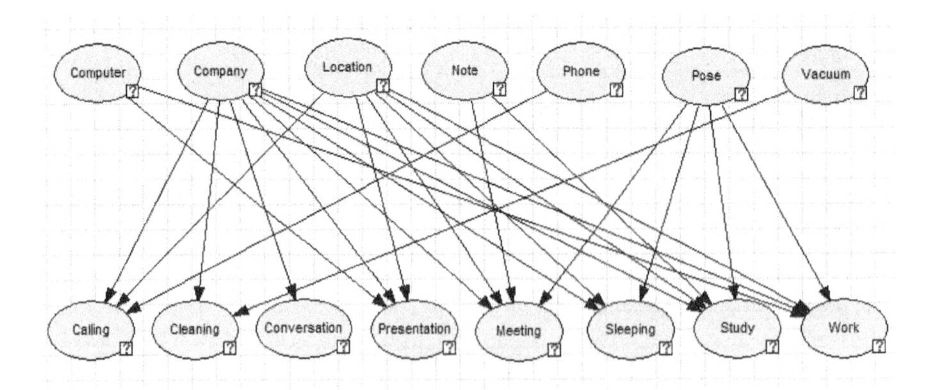

Fig. 2 Bayesian network structure for event modeling

Table 1 Units for magnetic properties

Person A
Person B
Person C
Accuracy (%)
80.0
90.9
86.4

We have used three different models by person. They share the same structure, but the parameters are different. As modeling like this, we can recognize several events happening at the same time. Performance of models by each person is shown in Table 1.

3.2 Optimal View Selection

Once an event at a certain time point is recognized by Bayesian network model, the system selects an optimal camera view at that time considering the probability and priority of a recognized event. Given an event e_k at the segment S_i, an optimal view, V_j is decided as following equation:

$$V_{\mathrm{j}} = \arg\max_{\mathrm{j}} f_{\mathrm{j}}^{S_{\mathrm{i}}}(e_{\mathrm{k}}) = V_{\mathrm{j}} | \mathrm{Max}\,(\mathrm{prob}(e_{\mathrm{k}})) \wedge \mathrm{Max}\,(\mathrm{priority}(e_{\mathrm{k}}))$$

where *prob*(e) is the recognition probability of event e and the *priority*(e) is defined by an expert based on the domain knowledge. A view with the highest probability is selected as an optimal one if only one event is captured at one segment. If two or more events are captured at the same segment, views of events with the highest priorities are selected as candidates first, and then a view with the highest probability is selected among the candidates.

3.3 Event Retrieval

Based on the selected view information and the annotated event information in database, event retrieval is conducted as simply the system retrieves the events, which satisfies the user query.

4 Experiments

4.1 Experimental Environment

For the validation of the proposed system, we have made the experimental environment using 8 cameras in an office. Figure 3 shows the location and coverage of cameras, and Fig. 4 shows an example captured images. We have used Sony network camera (SNC-P5), and video has been saved with the resolution of 320×240 and frame rate of 15 fps using MPEG video format.

For learning of Bayesian network model, we have collected the learning data of three persons in the preset experimental environment. For the event recognition and view selection experiments, we have collected video data based on the designed scenario shown in Fig. 5. The scenario is based on the events happening during office hour-from 9 AM to 6 PM-on one day, and all events have the possibility that can be occurred in the office.

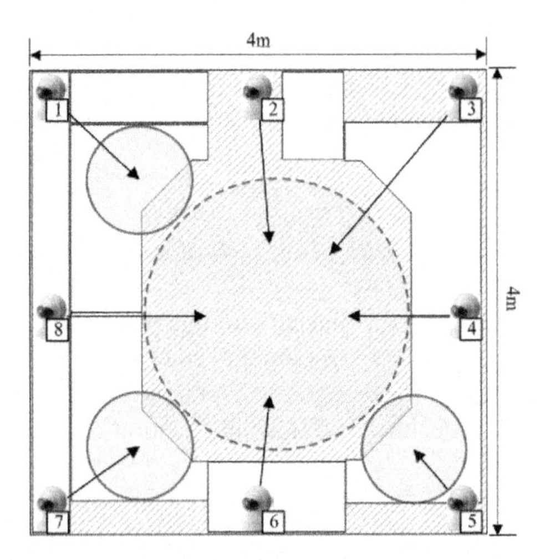

Fig. 3 Multi-camera system (Camera location and coverage)

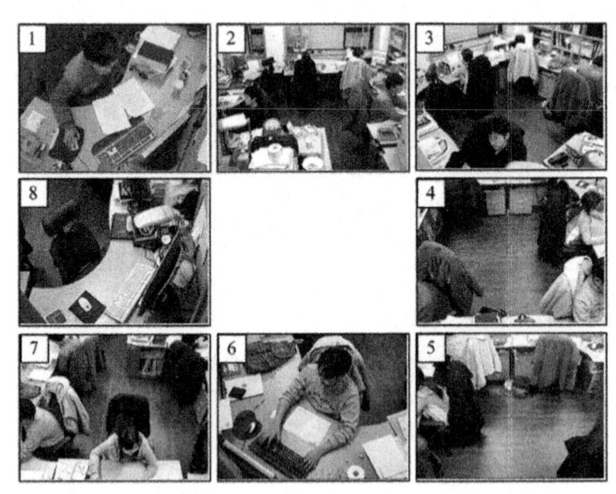

Fig. 4 Multi-camera system (Example of captured images)

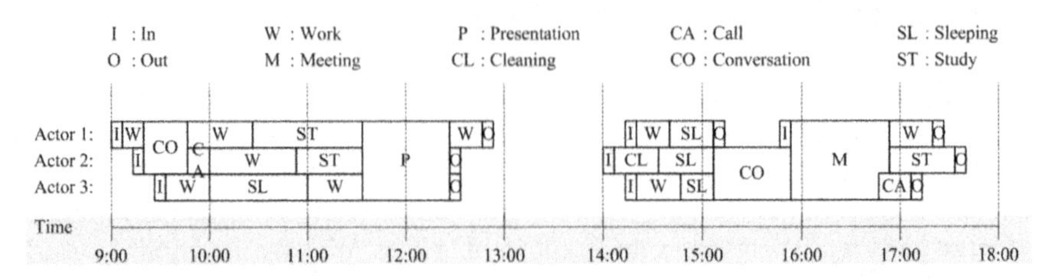

Fig. 5 A scenario in an office

4.2 Event Recognition

We have conducted the event recognition experiment to show the performance of Bayesian network event model. Table 1 summarizes the accuracies by person. It shows high accuracy for each person, and the total accuracy is also high.

There are four incorrectly recognized events: three 'presentation' and one 'meeting'. All 'presentation' events are recognized as 'conversation' because call cameras does not satisfy the definition of 'presentation'. 'Presentation' requires three persons, computer and note, but most cameras do not detect all three persons and note is very difficult to be recognized correctly when three persons are close together.

4.3 View Selection

Figure 6 demonstrates an example of view selection result. In this frame, a person is performing event 'cleaning'. Our system selected a view in camera 7 using view selection method described in Sect. 3.2, and it is reasonable because it shows an event 'cleaning' clearly.

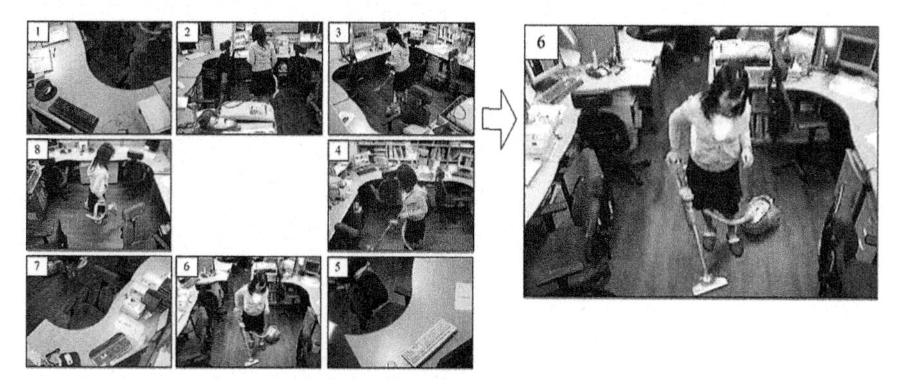

Fig. 6 An example of view selection (Event 'Cleaning')

4.4 Application Implemented and Event Retrieval

We have implemented the application to provide events retrieval in office video. The screen shots are shown in Figs. 7 and 8. In Fig. 7, all views of eight cameras are displayed in normal mode. Figure 8 shows the retrieved scene with the keywords Calling, Conversation, and Meeting. Event 'Meeting' is shown in this shot.

5 Conclusions and Future Works

This paper presented the novel multi-camera system that provided diverse views of each single office event and recognized those events using BN model. Our system also presents the retrieval of events with the application. As shown in experiments section, the proposed system performs acceptable results, and BN model recognizes office events with good accuracies.

There are some limitations that should be solved. Current system considers the recognition probability at a single time point or for a short time, but it may cause

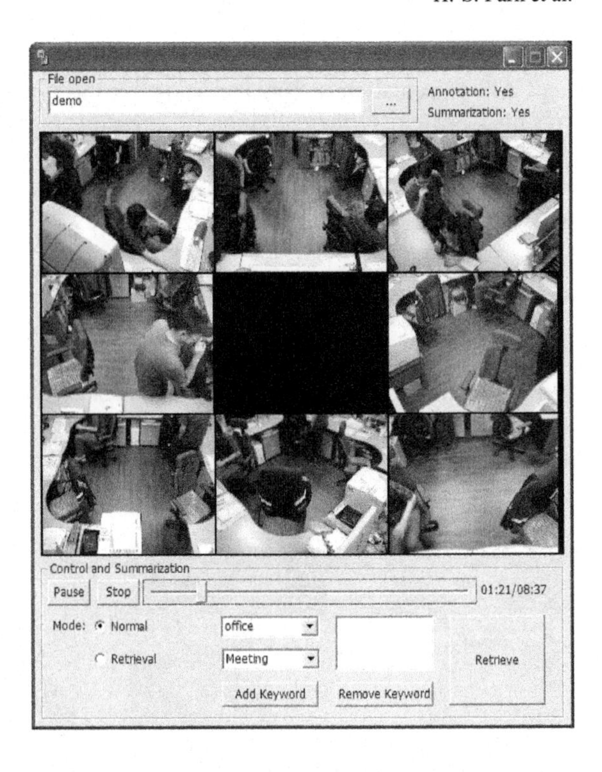

Fig. 7 A screen-shot of the application in normal mode

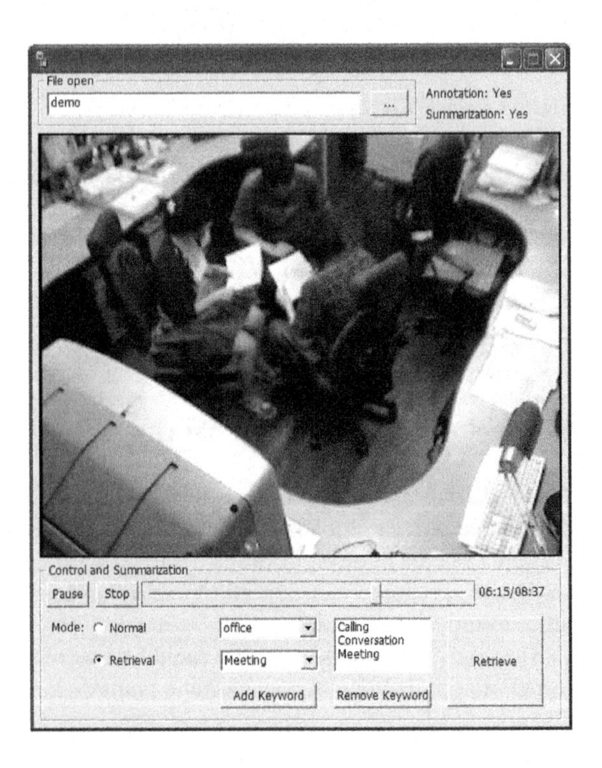

Fig. 8 A screen-shot of the application in retrieval mode (Calling, Conversation, and Meeting are selected as keywords)

a significant problem because the system selects image sequence. Therefore, view selection process should consider the entire scenario. Also, feature annotation part should be replaced by fully automatic annotation.

Future work will focus on applying the proposed system to the video in other domains such as sport videos and comparing the performance of the proposed method with conventional video retrieval methods. The semantic analysis and summarization of videos in diverse domains also can be interesting topics.

Acknowledgments This research was supported by the MKE(Ministry of Knowledge Economy), Korea, under the ITRC(Information Technology Research Center) support program supervised by the IITA(Institute of Information Technology Advancement) (IITA-2008-(C1090-0801-0046)).

References

1. H.-R. TRankler and O. Kanoun, "Recent advances in sensor technology," *Proceedings of the IEEE Instrumentation and Measurement Technology Conference*, vol. 1, pp. 309–316, 2001.
2. X. Zhu, J. Fan, A. K. Elmagarmid, and X. Wu, "Hierarchical video content description and summarization using unified semantic and visual similarity," *Multimedia Systems*, vol. 9, pp. 31–53, 2003.
3. S.-M. Lee and P. A. Abbott, "Bayesian networks for knowledge discovery in large datasets: Basics for nurse researchers," *Journal of Biomedical* Informatics, vol. 36, pp. 389–399, 2003.
4. J. Black and T. Ellis, "Multi camera image tracking," *Image and Vision Computing*, vol. 24, pp. 1256–1267, 2006.
5. Y. Sumi, S. Ito, T. Matsuguchi, S. Fels, and K. Mase, "Collaborative capturing and interpretation of interactions," *Pervasive 2004 Workshop on Memory and Sharing of Experiences*, pp. 1–7, 2004.
6. G. C. de Silva, T. Yamasaki, and K. Aizawa, "Evaluation of video summarization for a large number of cameras in ubiquitous home," *Proceedings of the 13th ACM International Conference on Multimedia*, pp. 820–828, 2005.
7. B. Li, J. H. Errico, H. Pan, and I. Sezan, "Bridging the semantic gap in sports video retrieval and summarization," *Journal of Visual Communication & Image Representation*, vol. 15, pp. 393–424, 2004.
8. A. Ekin, A. M. Tekalp, and R. mehrotra, "Automatic soccer video analysis and summarization," *IEEE Transactions on Image Processing*, vol. 12, no. 7, pp. 796–807, 2003.
9. I. Ersoy, F. Bunyak, and S. R. Subramanya, "A framework for trajectory based visual event retrieval," *International Conference on Information Technology: Coding and Computing*, vol. 2, pp. 23–27, 2004.

Fusion of Multichannel Biosignals Towards Automatic Emotion Recognition

Jonghwa Kim and Elisabeth André

Abstract Endowing the computer with the ability to recognize human emotional states is the most important prerequisites for realizing an affect-sensitive human-computer interaction. In this paper, we deal with all the essential stages of an automatic emotion recognition system using multichannel physiological measures, from data collection to the classification process. Particularly we develop two different classification methods, feature-level fusion and emotion-specific classification scheme. Four-channel biosensors were used to measure electromyogram, electrocardiogram, skin conductivity, and respiration changes while subjects were listening to music. A wide range of physiological features from various analysis domains is proposed to correlate them with emotional states. Classification of four musical emotions is performed by using feature-level fusion combined with an extended linear discriminant analysis (pLDA). Furthermore, by exploiting a dichotomic property of the 2D emotion model, we developed a novel scheme of emotion-specific multilevel dichotomous classification (EMDC) and compare its performance with direct multiclass classification using the pLDA feature-level fusion. Improved recognition accuracy of 95% and 70% for subject-dependent and subject-independent classification, respectively, is achieved by using the EMDC scheme.

Keywords Biosignal · Multisensor data fusion · ECG · EMG · SC · RSP · Emotion recognition · Multichannel biosignals · Pattern recognition · Human-computer interaction · Music and emotion

1 Introduction

Emotional intelligence (understanding and expression of emotions) is indispensable in human communication and facilitates successful interpersonal social interaction. To approach this in human-computer interaction (HCI), the first step is to equip

J. Kim (✉)
Institute of Computer Science, University of Augsburg, Eichleitnerstr. 30, D-86159 Augsburg, Germany
e-mail: kim@informatik.uni-augsburg.de

S. Lee et al. (eds.), *Multisensor Fusion and Integration for Intelligent Systems*, Lecture Notes in Electrical Engineering 35, DOI 10.1007/978-3-540-89859-7_5,

machines with the means to interpret and understand human emotions without the input of a user's translated intention. Hence, one of the most important prerequisites for realizing such affect-sensitive HCI is a reliable emotion recognition system which guarantees acceptable recognition accuracy, robustness against any artifacts, and adaptability to practical applications.

Recently, many works on engineering approaches to automatic emotion recognition have been reported. For an overview we refer to [1]. In particular, many efforts have been deployed to recognize human emotions using audiovisual channels of emotion expression, that is, facial expressions, speech, and gestures. Little attention, however, has been paid so far to using physiological measures, as opposed to audiovisual emotion channels. This is due to some significant limitations that come with the use of multichannel biosignals for emotion recognition. The main difficulty lies in the fact that it is a very hard task to uniquely map physiological patterns onto specific emotional states. As an emotion is a function of time, context, space, culture, and person, physiological patterns may widely differ from user to user and from situation to situation. When using multiple biosensors at the same time, analyzing biosignals is itself a complex multivariate task and requires broad insight into biological processes related to neuropsychological functions. To classify the multichannel variables, we first need to design fusion method for multichannel sensory data and to develop an emotion-specific classification scheme. Most of machine learning algorithms are generalized method based on statistics or linear regression of given data and most suitable for binary classification problems. Therefore they might not be able to capture characteristics of input variables in order to efficiently solve multiclass problems.

In this paper, we deal with all the essential stages of an automatic emotion recognition system, from data collection to the classification, based on four-channel physiological signals: electromyogram (EMG), electrocardiogram (ECG), skin conductivity (SC), and respiration changes (RSP). Generally, fusion of multisensory data can be performed at least at three levels: data, feature, and decision level. When observations are of same type, the data-level fusion might be probably the most appropriate way where we simply combine raw multisensory data. Decision-level fusion is the approach applied most often for *multimodal* sensory data containing time scale differences between modalities. Since, in this paper, we use *multichannel* biosignals that are measured in synchronized time scale and unique dimension, feature-level fusion is the most convincing way to classify by single classifier. Furthermore we develop a novel scheme of emotion-specific multilevel dichotomous classification (EMDC) and compared its performance with direct multiclass classification.

Throughout the paper, we try to provide a focused spectrum for each processing stage with selected methods suitable for handling the nature of physiological changes, instead of conducting a comparison study based on a large number of pattern recognition methods.

2 Related Research

A significant amount of work has been conducted by Picard and colleagues at MIT Lab showing that certain affective states may be recognized by using physiological data including heart rate, skin conductivity, temperature, muscle activity and respiration velocity [2, 3]. Nasoz et al. [4] used movie clips based on the study by Gross and Levenson [5] for eliciting target emotions from 29 subjects and achieved an emotion classification accuracy of 83% using the Marquardt Backpropagation algorithm (MBP). More recently, an interesting user-independent emotion recognition system was reported by Kim et al. [6]. They developed a set of recording protocols using multimodal stimuli (audio, visual, and cognitive) to evoke targeted emotions (sadness, stress, anger, and surprise) from 175 children aged five to eight. A classification ratio of 78.43% was achieved for three emotions (sadness, stress, and anger) and a ratio of 61.76% for four emotions (sadness, stress, anger, and surprise) by adopting support vector machines as pattern classifier.

The physiological datasets used in most of the aforementioned approaches were gathered by using visual elicitation materials in a lab setting. The subjects then "tried and felt" or "acted out" the target emotions while looking at selected photos or watching movie clips that were carefully prearranged to elicit the emotions. In other words, to put it bluntly, the recognition results were achieved for specific users in specific contexts with "forced" emotional states. All the works used simple feature-level fusion to mix features from each sensor and then to classify by using common single classfifier.

3 Setting of Experiment

3.1 Musical Emotion Induction

To collect a database of physiological signals in which the targeted emotions corresponding to the four quadrants in the 2D emotion model (i.e., EQ1, EQ2, EQ3, and EQ4 in Fig. 1) can be *naturally* reflected without any deliberate expression, we decided to use the musical induction method, that is, to record physiological signals while the subjects were listening to different pieces of music.

The subjects were three males aged 25–38 and who all enjoy listening to music in their everyday life. They individually handpicked four songs that were intended to spontaneously evoke emotional memories and certain moods corresponding to the four target emotions. Figure 1[1] shows the musical emotion model referred

[1] Metaphoric cues for song selection: song 1 (positively exciting, energizing, joyful, exuberant), song 2 (noisy, loud, irritating, discord), song 3 (melancholic, sad memory), song 4 (blissful, pleasurable, slumberous, tender).

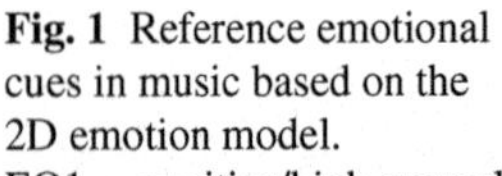

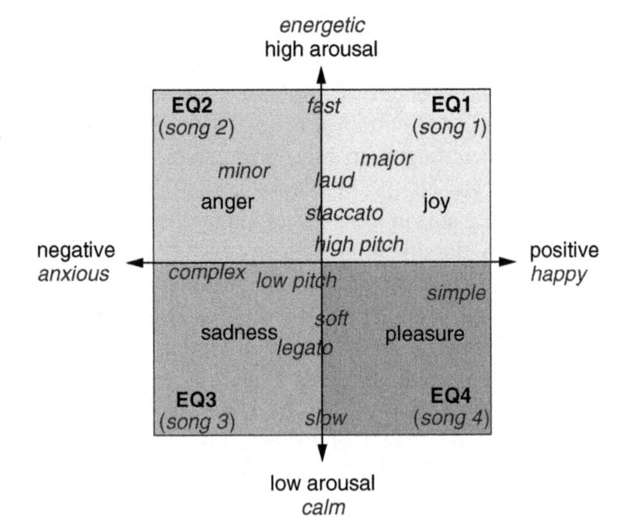

Fig. 1 Reference emotional cues in music based on the 2D emotion model. EQ1 = positive/high arousal, EQ2 = negative/high arousal, EQ3 = negative/low arousal, EQ4 = positive/low arousal

to for the selection of their songs. Generally, emotional responses to music vary greatly from individual to individual depending on their unique past experiences. Moreover, cross-cultural comparisons in literature suggest that emotional responses can be quite differentially emphasized by different musical cultures and training. This is why we advised the subjects to choose themselves the songs they believed would help them recall their individual special memories with respect to the target emotions. Recording schedules were decided by the subjects themselves and the recordings took place whenever they felt like listening to music. They were also free to choose the songs they wanted to listen to. Thus, in contrast to methods used in other studies, the subjects were not forced to participate in a lab setting scenario and to use prespecified stimulation material.

During the three months, a total of 360 samples (90 samples for each emotion) from three subjects were collected. The signal length of each sample was between 3–5 min depending on the duration of the songs.

3.2 Used Biosensors

The physiological signals were acquired using the Procomp[2] Infiniti™ with four biosensors, electromyogram (EMG), skin conductivity (SC), electrocardiogram (ECG), and respiration (RSP). The sampling rates were 32 Hz for EMG, SC, and RSP, and 256 Hz for ECG. The positions and typical waveforms of the biosensors we used are illustrated in Fig. 2.

[2] This is an eight channel multi-modal Biofeedback system with 14 bit resolution and a fiber optic cable connection to the computer. www.MindMedia.nl

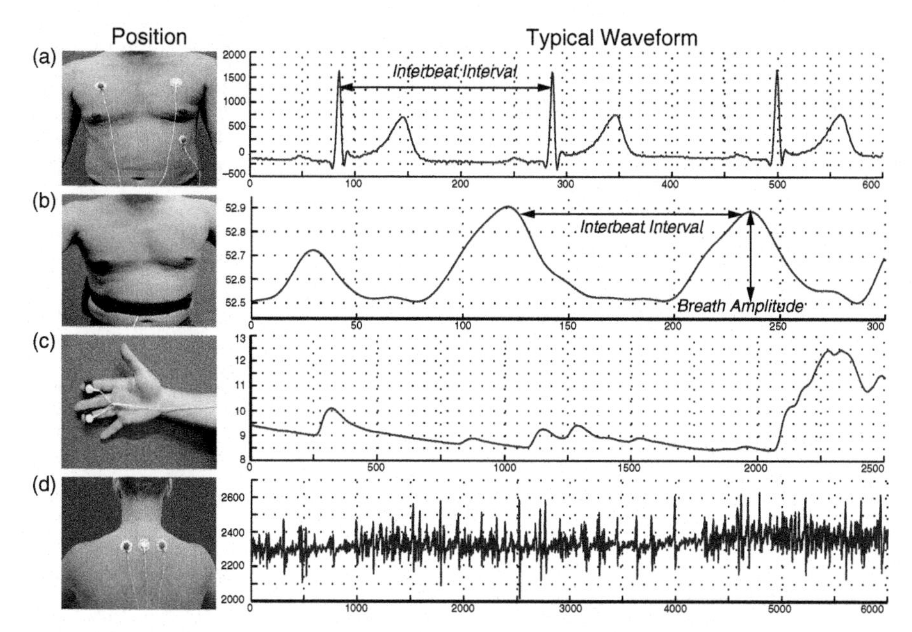

Fig. 2 Position and typical waveforms of the biosensors: **(a)** ECG, **(b)** RSP, **(c)** SC, **(d)** EMG

4 Feature Extraction

4.1 Signal Segmentation

Different types of artifacts were observed in all the four channel signals, such as transient noise due to movement of the subjects during the recording, mostly at the beginning and at the end of each recording. Thus, uniformly for all subjects and channels, we segmented the signals into final samples of a 160 s each, obtained by taking the middle part of each signal.

4.2 Measured Features

From the four channel signals we calculated a total of 110 features from various analysis domains including conventional statistics in time series, frequency domain, geometric analysis, multiscale sample entropy, subband spectra, etc. For the signals with non-periodic characteristics, such as EMG and SC, we focused on capturing the amplitude variance and localizing the occurrences (number of transient changes) in the signals.

4.2.1 Electrocardiogram (ECG)

To obtain the subband spectrum of the ECG signal we used the typical 1,024 points fast Fourier transform (FFT) and partitioned the coefficients within the frequency range 0–10 Hz into eight non-overlapping subbands with equal bandwidth. First, as features, power mean values of each subband and fundamental frequency (F0) are calculated by finding maximum magnitude in the spectrum within the range 0–3 Hz. To capture peaks and their locations in subbands, subband spectral entropy (SSE) is computed for each subband. To compute the SSE, it is necessary to convert each spectrum into a probability mass function (PMF) like form. Equation 1 is used for the normalization of the spectrum.

$$x_i = \frac{X_i}{\sum_{i=1}^{N} X_i}, \qquad \text{for } i = 1 \dots N \tag{1}$$

where X_i is the energy of ith frequency component of the spectrum and $\tilde{\mathbf{x}} = \{x_1 \dots x_N\}$ is to be considered as the PMF of the spectrum. In each subband the SSE is computed from $\tilde{\mathbf{x}}$ by

$$H_{\text{sub}} = -\sum_{i=1}^{N} x_i \cdot \log_2 x_i \tag{2}$$

By packing the eight subbands into two bands, that is, subbands 1–3 as the low frequency (LF) band and subbands 4–8 as the high frequency (HF) band, the ratios of the LF/HF bands are calculated from the power mean values and the SSEs.

To obtain the HRV (heart rate variability) from the continuous ECG signal, each QRS complex is detected and the RR intervals (all intervals between adjacent R waves) or the normal-to-normal (NN) intervals (all intervals between adjacent QRS complexes resulting from sinus node depolarization) are determined. We used the QRS detection algorithm of Pan and Tompkins [7] in order to obtain the HRV time series. Figure 3 shows examples of R wave detection and interpolated HRV time series, referring to the increases and decreases over time in the NN intervals.

In the time-domain of the HRV time series, we calculated statistical features including mean value, standard deviation of all NN intervals (SDNN), standard deviation of the first difference of the HRV, the number of pairs of successive NN intervals differing by more than 50 ms (NN50), the proportion derived by dividing NN50 by the total number of NN intervals. By calculating the standard deviations in different distances of RR interbeats, we also added Poincaré geometry in the feature set to capture the nature of interbeat interval fluctuations. Poincaré plot geometry is a graph of each RR interval plotted against the next interval and provides quantitative information of the heart activity by calculating the standard deviations of the distances of $R-R(i)$ to lines $y = x$ and $y = -x+2*R-R_m$, where $R-R_m$ is the mean of all $R - R(i)$, [8]. Figure 3(e) shows an example plot of the Poincaré geometry. The standard deviations SD_1 and SD_2 refer to the fast beat-to-beat variability and longer-term variability of $R - R(i)$ respectively.

Entropy-based features from the HRV time series were also considered. Based on the so-called *approximate entropy* and *sample entropy* proposed in [9], a multiscale sample entropy (MSE) was introduced [10] and successfully applied to physiological data, especially for analysis of short and noisy biosignal [11]. Given a time series $\{X_i\} = \{x_1, x_2, \dots, x_N\}$ of length N, the number $(n_i^{(m)})$ of similar m-dimensional

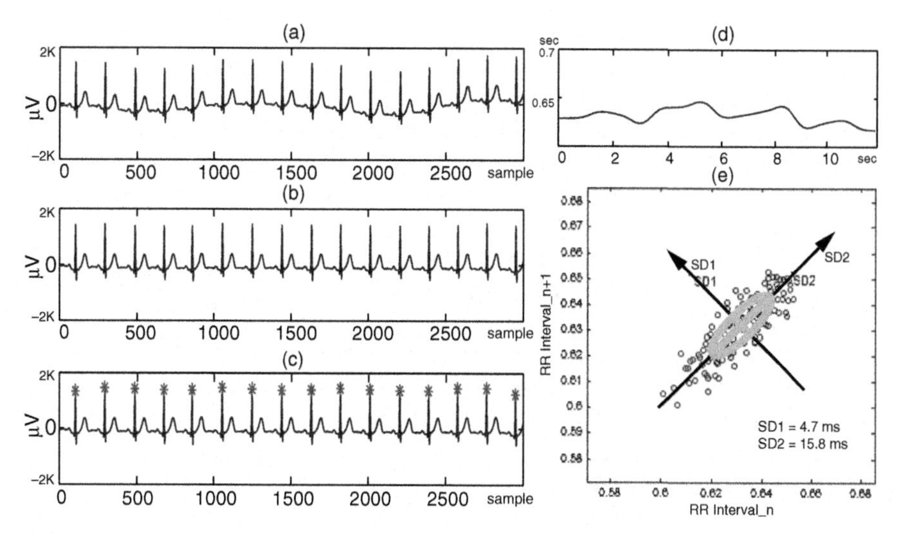

Fig. 3 Example of ECG Analysis: **(a)** raw ECG signal with respiration artifacts, **(b)** detrended signal, **(c)** detected RR interbeats, **(d)** interpolated HRV time series using RR intervals, **(e)** Poincaré plot of the HRV time series

vectors $y^{(m)}(j)$ for each sequence vectors $y^{(m)}(i) = \{x_i, x_{i+1}, \ldots, x_{i+m-1}\}$ is determined by measuring their respective distances. The relative frequency to find the vector $y^{(m)}(j)$ within a tolerance level δ is defined by

$$C_i^{(m)}(\delta) = \frac{n_i^{(m)}}{N - m + 1} \tag{3}$$

The approximate entropy, $h_A(\delta, m)$, and the sample entropy, $h_S(\delta, m)$ are defined as

$$h_A(\delta, m) = \lim_{N\to\infty} [H_N^{(m]}(\delta) - H_N^{(m+1)}(\delta)], \tag{4}$$

$$h_S(\delta, m) = \lim_{N\to\infty} -\ln \frac{C^{(m+1)}(\delta)}{C^{(m)}(\delta)}, \tag{5}$$

where

$$H_N^{(m)}(\delta) = \frac{1}{N - m + 1} \sum_{i=1}^{N-m+1} \ln C_i^{(m)}(\delta), \tag{6}$$

Because it has the advantage of being less dependent on the time series length N, we applied the sample entropy h_S to coarse-grained versions $(y_j^{(\tau)})$ of the original HRV time series $\{X_i\}$,

$$y_j(\tau) = \frac{1}{\tau} \sum_{l=(j-1)\tau+1}^{j\tau} x_i, \quad 1 \le j \le N/\tau, \quad \tau = 1, 2, 3, \ldots \tag{7}$$

The time series $\{X_i\}$ is first divided into N/τ segments by non-overlapped windowing with length of scale factor τ and then the mean value of each segment is calculated. Note that for scale one $y_j(1) = x_j$. From the scaled time series $y_j(\tau)$ we obtain the m-dimensional sequence vectors $y^{(m)}(i, \tau)$. Finally, we calculate the sample entropy h_S for each sequence vector $y_j(\tau)$. In our analysis we used $m = 2$ and fixed $\delta = 0.2\sigma$ for all scales, where σ is the standard deviation of the original time series x_i.

In the frequency-domain of the HRV time series, three frequency bands are of general interest: the very-low frequency (VLF) band (0.003–0.04 Hz), the low frequency (LF) band (0.04–0.15 Hz), and the high frequency (HF) band (0.15–0.4 Hz). From these subband spectra, we computed the dominant frequency and power of each band by integrating the power spectral densities (PSD) obtained by using Welch's algorithm, as well as the ratio of power within the low-frequency band to that within the high-frequency band (LF/HF).

4.2.2 Respiration (RSP)

Including the typical statistics of the raw RSP signal, we calculated similar types of features, such as the ECG features, the power mean values of three subbands (obtained by dividing the Fourier coefficients within the range 0–0.8 Hz into non-overlapped three subbands with equal bandwidth), and the set of subband spectral entropies (SSE).

In order to investigate inherent correlation between respiration rate and heart rate, we considered a novel feature content for the RSP signal. Since an RSP signal exhibits a quasi periodic waveform with sinusoidal properties, it does not seem unreasonable to conduct an HRV-like analysis for the RSP signal, that is, analysis of breathing rate variability (BRV). After detrending using the mean value of the entire signal and lowpass filtering, we calculated the BRV time series, referring to the increases and decreases over time in the peak-to-peak (PP) intervals, by detecting the peaks in the signal using the maxima ranks within each zero-crossing. From the BRV time series, we calculated the mean value, SD, SD of the first difference, MSE, Poincaré analysis, etc. In the spectrum of the BRV, peak frequency, power of the two subbands, the low-frequency band (0–0.03Hz) and the high-frequency band (0.03–0.15 Hz), and the ratio of the power within the two bands (LF/HF) were calculated.

4.2.3 Skin Conductivity (SC)

The mean value, standard deviation, and mean of first and second derivations were extracted as features from the normalized SC signal and the low-passed SC signal using a cutoff frequency of 0.2 Hz. To obtain a detrended SCR waveform without DC-level components, we removed the continuous, piecewise linear trend in the two low-passed signals, that is, the very low-passed (VLP) and the low-passed (LP) signal with a cutoff frequency of 0.08 Hz and 0.2 Hz, respectively (see Fig. 4 (a)–(e)).

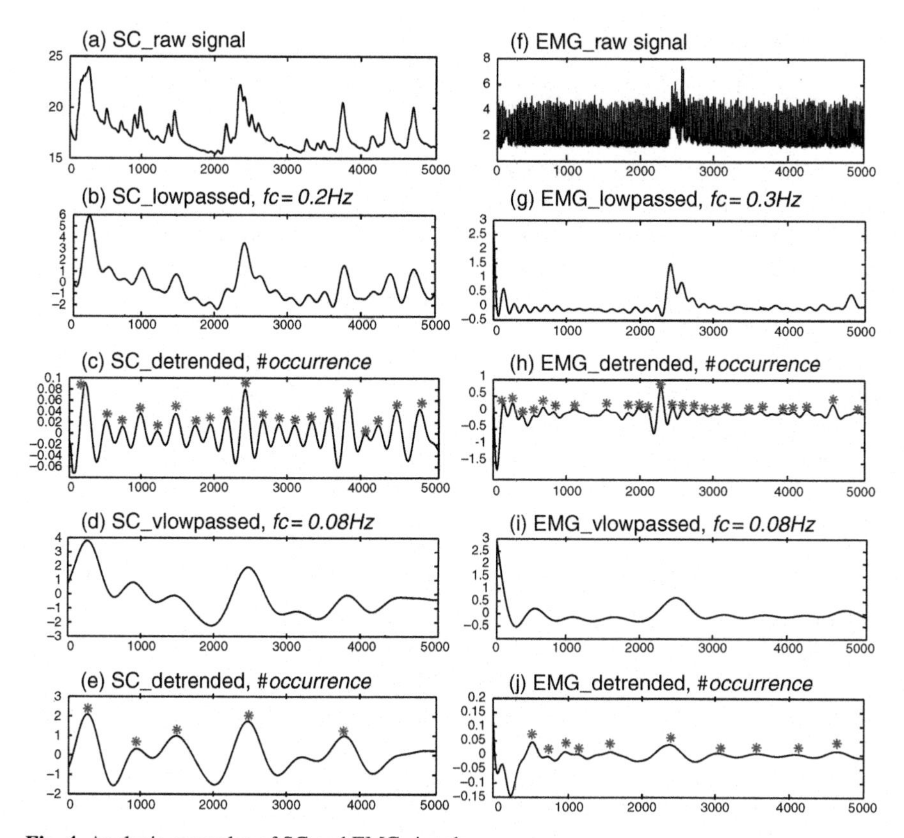

Fig. 4 Analysis examples of SC and EMG signals

The baseline of the SC signal was calculated and subtracted to consider only relative amplitudes. By finding two consecutive zero-crossings and the maximum value between them, we calculated the number of SCR occurrences within 100 s from each LP and VLP signal, the mean of the amplitudes of all occurrences, and the ratio of the SCR occurrences within the low-passed signals (VLP/LP).

4.2.4 Electromyography (EMG)

For the EMG signal, we calculated types of features similar to those of the SC signal. The mean value of the entire signal, the mean of the first and second derivations, and the standard deviation were extracted as features from the normalized and low-passed signals. The occurrence number of myo-responses and the ratio of that within VLP and LP signals were also added to the feature set and were determined in the same way as the SCR occurrence but using cutoff frequencies with 0.08 Hz (VLP) and 0.3 Hz (LP) (see Fig. 4(f)–(j)).

In the end, we obtained a total of 110 features from the 4-channel biosignals; 53 (ECG) + 37 (RSP) + 10 (SC) + 10 (EMG).

5 Classification Result

5.1 pLDA Feature-Level Fusion

Figure 5 shows the feature-level fusion model for multichannel biosensor data. For classification we used the pseudoinverse linear discriminant analysis (pLDA) [12], combined with the sequential backward selection (SBS) [13] to select significant feature subset. The pLDA is a natural extension of classical LDA by applying eigenvalue decomposition to the scatter matrices, in order to deal with the sigularity problem of LDA.

Table 1 with confusion matrix presents the correct classification ratio (CCR) of subject-dependent (Subject A, B, and C) and subject-independent (All) classification where the features of all the subjects are simply merged and normalized. We used leave-one-out cross-validation where a single observation taken from the entire samples is used as the test data and the remaining observations are used for training the classifier. This is repeated such that each observation in the samples is used once as the test data.

The table shows that the CCR varies from subject to subject. For example, the best accuracy was 91% for subject B and the lowest was 81% for subject A. Not only does the overall accuracy differ from one subject to the next, but the CCR of the single emotions varies as well. For example, EQ2 was perfectly recognized for subject C while it caused the highest error rate for subject B. It was three times mixed up with EQ1 which is characterized by opposite valence. As the confusion matrix shows, the difficulty in valence differentiation can be observed for all subjects. Most classification errors for Subject A and B lie in false classification between EQ1 and EQ2 while an extreme uncertainty can be observed in the differentiation between EQ3 and EQ4 for Subject C. On the other hand, it is very meaningful that relatively robust recognition accuracy is achieved for the classification of emotions that are reciprocal in the diagonal quadrants of the 2D emotion model, that is, EQ1 vs. EQ3 and EQ2 vs. EQ4. Moreover, the accuracy is much better than that of arousal classification.

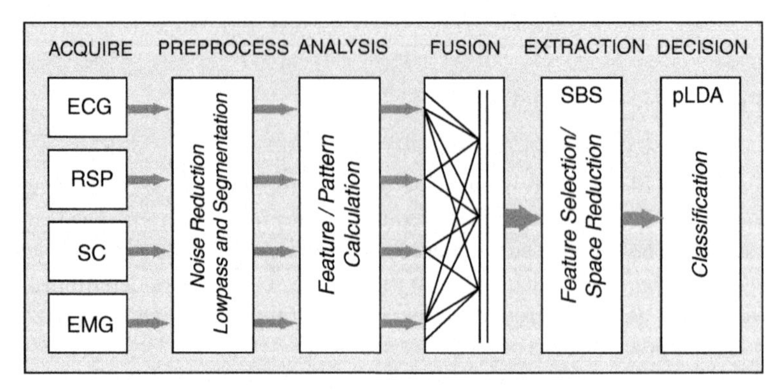

Fig. 5 Feature-level fusion for four-channel biosensor data, combined with SBS and pLDA

Table 1 Recognition results in rates (*error 0.00* = *CCR 100%*) achieved by using pLDA with SBS and leave-one-out cross validation. # of samples: 120 for each subject and 360 for all

	EQ1	EQ2	EQ3	EQ4	Total*	Error
Subject A (*CCR %* = ***81%***)						
EQ1	***22***	4	1	3	*30*	*0.27*
EQ2	3	***26***	1	0	*30*	*0.13*
EQ3	1	2	***23***	4	*30*	*0.23*
EQ4	3	0	1	***26***	*30*	*0.13*
Subject B (*CCR %* = ***91%***)						
EQ1	***27***	3	0	0	*30*	*0.10*
EQ2	3	***25***	1	1	*30*	*0.17*
EQ3	0	2	***28***	0	*30*	*0.07*
EQ4	0	1	0	***29***	*30*	*0.03*
Subject C (*CCR %* = ***89%***)						
EQ1	***28***	0	2	0	*30*	*0.07*
EQ2	0	***30***	0	0	*30*	*0.00*
EQ3	0	0	***24***	6	*30*	*0.20*
EQ4	0	0	5	***25***	*30*	*0.17*
All: subject-independent (*CCR %* = ***65%***)						
EQ1	***62***	9	8	11	*90*	*0.31*
EQ2	15	***57***	13	5	*90*	*0.37*
EQ3	9	6	***58***	17	*90*	*0.36*
EQ4	8	5	21	***56***	*90*	*0.38*

* Actual total # of samples.

We also tried to differentiate the emotions based on the two axes, arousal and valence, in the 2D emotion model. The samples of four emotions were divided into groups of negative valence (EQ2+EQ3) and positive valence (EQ1+EQ4) and into groups of high arousal (EQ1+EQ2) and low arousal (EQ3+EQ4). By using the same methods, we then performed a two-class classification of the divided samples for arousal and valence separately. It turned out that emotion-relevant ANS specificity can be observed more conspicuously in the arousal axis regardless of subject-dependent or independent cases. Classification of arousal achieved an acceptable *CCR* of 97–99% for the subject-dependent recognition and 89% for the subject-independent recognition, while the results for valence were 88–94% and 77%, respectively.

5.2 Emotion-specific Multilevel Dichotomous Classification

By taking advantage of supervised classification (where we know in advance which emotion types have to be recognized) we developed an emotion-specific multilevel dichotomous classification (EMDC) scheme. This scheme exploits the property of the dichotomous categorization in the 2D emotion model and the fact that arousal classification yields higher *CCR* than valence classification or direct multiclass classification. This proves true in almost all previous works and according to our results as well. Figure 6 illustrates the EMDC scheme and provides an example of the dyadic decomposition for an eight-class problem.

First, the entire training patterns are grouped into two opposing *"superclasses"* (on the basis of valence or arousal), $\bar{C}$ consisting of all patterns in some subset of the class categories and C as all remaining patterns, that is, $\bar{C} \cap C = \{\}$. This dyadic decomposition using one of the two axes is serially performed until one subset contains only two classes. The grouping axis can be different from each dichotomous level. Then multiple binary classifiers for each level are trained from the corresponding dyadic patterns. Therefore, the EMDC scheme is obviously emotion-specific and effective for a 2D emotion model. Note that the performance of the EMDC scheme is limited by a maximum *CCR* of first level classification and makes sense only if the *CCR* for one of the two superclasses is higher than that for direct multiclass classification (theoretically this always holds true). Because we used four emotion classes in our experiment, we needed a two-level classification based on arousal and valence grouping for both superclasses in parallel.

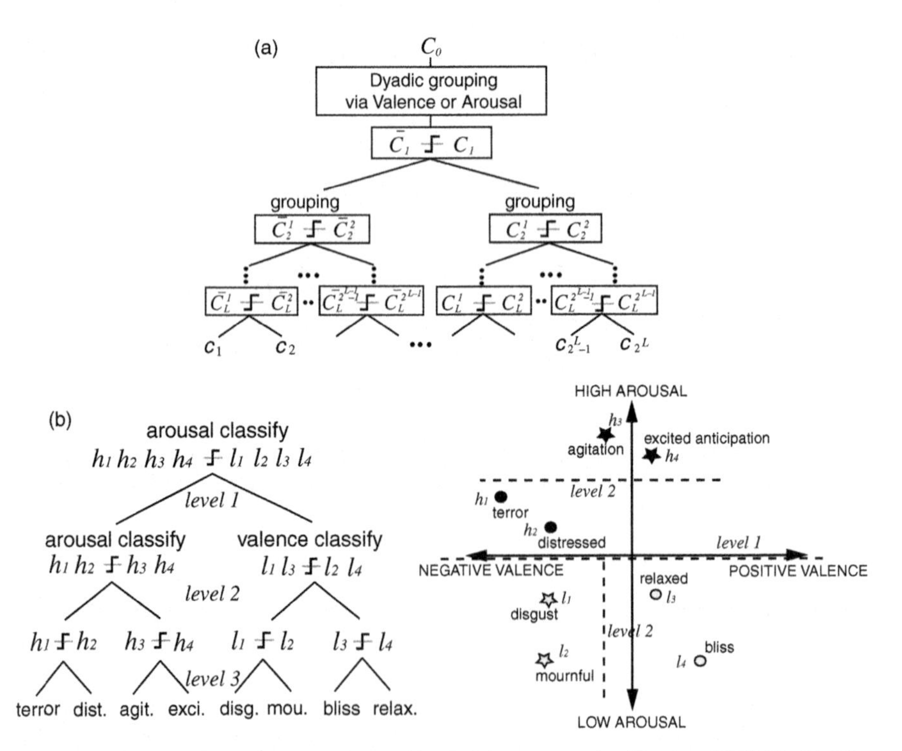

Fig. 6 Framework of emotion-specific multilevel dichotomous classification (EMDC). **(a)** Diagram of decomposition process, **(b)** Decomposition example for an eight-class problem

Table 2 Results using EMDC scheme with the best features

Subject A (*CCR % =* ***94%****, 113/120)*

	EQ1 & EQ2			EQ3 & EQ4		
EQ1 & EQ2	**58**			2		
		EQ1	EQ2			
	EQ1	**27**	1			
	EQ2	1	**29**			
EQ3 & EQ4	2			**58**		
					EQ3	EQ4
				EQ3	**29**	0
				EQ4	1	**28**

Subject B (*CCR % =* ***98%****, 117/120)*

	EQ1 & EQ2			EQ3 & EQ4		
EQ1 & EQ2	**60**			0		
		EQ1	EQ2			
	EQ1	**30**	0			
	EQ2	0	**30**			
EQ3 & EQ4	1			**59**		
					EQ3	EQ4
				EQ3	**29**	1
				EQ4	1	**28**

Subject C (*CCR % =* ***94%****, 113/120)*

	EQ1 & EQ2			EQ3 & EQ4		
EQ1 & EQ2	**60**			0		
		EQ1	EQ2			
	EQ1	**30**	0			
	EQ2	0	**30**			
EQ3 & EQ4	1			**59**		
					EQ3	EQ4
				EQ3	**27**	2
				EQ4	4	**26**

Subject All (*CCR % =* ***70%****, 251/360)*

	EQ1 & EQ2			EQ3 & EQ4		
EQ1 & EQ2	**155**			25		
		EQ1	EQ2			
	EQ1	**62**	13			
	EQ2	15	**65**			
EQ3 & EQ4	16			**164**		
					EQ3	EQ4
				EQ3	**64**	19
				EQ4	21	**60**

Table 2 shows the dichotomous contingency table of recognition results by using the novel EMDC scheme. As expected, the *CCRs* significantly improved for all class problems. For the classification of four emotions, we obtained an average *CCR* of 95% for subject-dependent and 70% for subject-independent classification. Compared to the results obtained for pLDA, the EMDC scheme achieved an overall *CCR* improvement of about 5–13% in each class problem (see Table 3).

Table 3 *CCR* Comparison between pLDA and EMDC

	Subj. A	Subj. B	Subj. C	Subj. All	*Average (ABC)*
pLDA	81%	91%	89%	65%	*87%*
EMDC	**94%**	**98%**	**94%**	**70%**	***95%***

6 Conclusion

In this paper, we dealt with all the essential stages of an automatic emotion recognition system using multichannel physiological measures, from data collection to the classification process. By analyzing a wide range of physiological features from various analysis domains, we found that SC and EMG are linearly correlated with arousal change in emotional ANS activities, and that the features in ECG and RSP are dominant for valence differentiation. Particularly, the HRV/BRV analysis revealed the cross-correlation between heart rate and respiration.

By fusing the multichannel features at the feature-level, we achieved an average recognition accuracy of 98%, 91%, and 87% for arousal, valence, and four emotion classes respectively. In order to further improve the accuracy of the four emotion classes, we developed a new classification scheme (EMDC). Although this new scheme is based on a very simple idea, it significantly improves the recognition accuracy obtained by common feature-level fusion. We actually achieved an average recognition accuracy of 95% improved which also connotes more than a prima facie evidence that there are some ANS differences among emotions. Moreover, the accuracy is higher than that in the previous works reviewed in this paper, even when considering the different experimental settings in the works, such as the number of target classes, number of subjects, naturalness of dataset, etc.

Acknowledgments This research was partially supported by the European Commission (HUMAINE NoE; FP6-IST, METABO; FP7-ICT).

References

1. R. Cowie, E. Douglas-Cowie, N. Tsapatsoulis, G. Votsis, S. Kollias, W. Fellenz, and J. G. Taylor, Emotion recognition in human-computer interaction, *IEEE Signal Process. Mag.*, 18, 32–80, 2001.
2. J. Healey and R. W. Picard, Digital processing of affective signals, in *Proc. IEEE Int. Conf. Acoust., Speech, and Signal Proc.*, Seattle, WA, 1998, pp. 3749–3752.
3. R. Picard, E. Vyzas, and J. Healy, Toward machine emotional intelligence: Analysis of affective physiological state, *IEEE Trans. Pattern Anal. Mach Intell.*, 23(10), 1175–1191, 2001.
4. F. Nasoz, K. Alvarez, C. Lisetti, and N. Finkelstein, Emotion recognition from physiological signals for presence technologies, *Int. J. Cogn., Technol., and Work*, 6(1), 4–14, 2003.
5. J. J. Gross and R. W. Levenson, Emotion elicitation using films, *Cogn. Emot.*, 9, pp. 87–108, 1995.
6. K. H. Kim, S. W. Bang, and S. R. Kim, Emotion recognition system using short-term monitoring of physiological signals, *Med. Biol. Eng. Comput.*, 42, pp. 419–427, 2004.
7. J. Pan and W. Tompkins, A real-time qrs detection algorithm, *IEEE Trans. Biomed. Eng.*, 32(3), 230–323, 1985.
8. P. W. Kamen, H. Krum, and A. M. Tonkin, Poincare plot of heart rate variability allows quantitative display of parasympathetic nervous activity, *Clin. Sci.*, 91, 201–208, 1996.
9. J. Richmann and J. Moorman, Physiological time series analysis using approximate entropy and sample entropy, *Am. J. Physiol. Heart Circ. Physiol.* 278, H2039, 2000.
10. M. Costa, A. L. Goldberger, and C.-K. Peng, Multiscale entropy analysis of biological signals," *Phys. Rev.*, E 71, 021906, 2005.
11. R. Thuraisingham and G. Gottwald, On multiscale entropy analysis for physiological data, *Physica A*, 336, 323–333, 2006.
12. J. Ye and Q. Li, A two-stage linear discriminant analysis via Qr-decomposition, *IEEE Trans. on PAMI*, 27(6), 929–941, 2005.
13. J. Kittler, *Feature Selection and Extraction*. San Diego, CA: Academic Press, Inc., 59–83, 1986.

A Comparison of Track-to-Track Fusion Algorithms for Automotive Sensor Fusion

Stephan Matzka and Richard Altendorfer

Abstract In exteroceptive automotive sensor fusion, sensor data are usually only available as processed, tracked object data and not as raw sensor data. Applying a Kalman filter to such data leads to additional delays and generally underestimates the fused objects' covariance due to temporal correlations of individual sensor data as well as inter-sensor correlations. We compare the performance of a standard asynchronous Kalman filter applied to tracked sensor data to several algorithms for the track-to-track fusion of sensor objects of unknown correlation, namely covariance union, covariance intersection, and use of cross-covariance. For our simulation setup, covariance intersection and use of cross-covariance turn out to yield significantly lower errors than a Kalman filter at a comparable computational load.

1 Introduction

Driver assistance systems (DAS) such as adaptive cruise control (ACC) or lane departure warning (LDW) are being offered by many car manufacturers and are getting more and more popular. While most current systems rely on only one exteroceptive sensor (e.g., radar or laser for ACC and camera for LDW) in addition to proprioceptive sensors such as wheel speed and yaw rate sensors, enhanced versions of the above DAS such as ACC operating at the entire speed range or new systems such as autonomous braking for collision avoidance will require a more sophisticated exteroception that is based upon multiple sensors. Hence exteroceptive sensor fusion will play a crucial role for the performance of future DAS.

Automotive sensors such as radar are produced by automotive suppliers and usually output processed, tracked object lists. While a raw data interface might be provided by suppliers, hardware-specific,proprietary sensor knowledge is necessary

S. Matzka (✉)
Heriot-Watt University, Edinburgh, EH14 4AS, UK
e-mail: sm217@hw.ac.uk

S. Lee et al. (eds.), *Multisensor Fusion and Integration for Intelligent Systems*,
Lecture Notes in Electrical Engineering 35, DOI 10.1007/978-3-540-89859-7_6,

to gain full advantage of raw sensor data. For DAS developed at car manufacturers, fusion of already tracked sensor objects is therefore the more likely type of data fusion.

In general sensor data from different sensor are output at different cycle times and are not synchronized to a common clock. The probably simplest level 1 fusion [1] of such sensor data would be achieved by an asynchronous Kalman filter for "sensor-to-track" fusion, where the fused object list of the Kalman filter is updated asynchronously every time a new sensor object list arrives. However, the application of Kalman filtering to such data is in principle incorrect since sensor objects of each sensor are temporally correlated due to previous filtering and since tracked objects from different sensors are in general also correlated because of for example, common modeling assumptions [2]. In addition, applying a Kalman filter to already (Kalman) filtered data will result in additional phase delays due to the Kalman filter's low-pass characteristics.

In order to overcome the above difficulties several track-to-track fusion algorithms have been proposed. In this paper we focus on three well-established methods, namely use of cross-covariance [3], covariance intersection [4], and covariance union [5]. The goal of this paper is to assess the improvement in tracking accuracy by those algorithms with respect to an asynchronous Kalman filter that is often used despite its known shortcomings, see e.g. [6]. By comparing the root mean squared errors (RMSE) and correlation coefficients for simulated trajectories of the four aforementioned fusion strategies, we want to quantify in detail an assessment of track-to-track fusion algorithms briefly mentioned in [7].

The paper is organized as follows: in Sect. 2 we present an overview of our Simulation Setup. In Sect. 3 we briefly review the Track-to-Track Fusion algorithms to be used. Section 4 contains a detailed discussion of the simulation results. Details of the simulation setup such as the vehicle dynamics or the sensor model are relegated to the appendix.

2 Simulation Setup

In order to assess different fusion strategies by numerical simulation, the following subsystems are needed: the generation of a reference trajectory for the target vehicle to be observed by sensors, the simulation of the sensor measurements and the sensor processing (Kalman filter), implementations of the abovementioned "sensor"-to-track fusion and track-to-track fusion algorithms and an error computation module that compares the fusion results to ground truth – the reference trajectory. In this setup we restrict ourselves to a single target vehicle in order to cleanly assess the performance of the fusion algorithms without additional errors due to false associations. Furthermore, the simulated scenario is static in the sense that the ego vehicle on which the sensors are mounted does not move; hence no "ego-compensation" is necessary and the relative measurements that the sensors provide are also measurements in an inertial system whose origin is taken to be at the ego vehicle. A block diagram of the simulation setup is shown in Fig. 1.

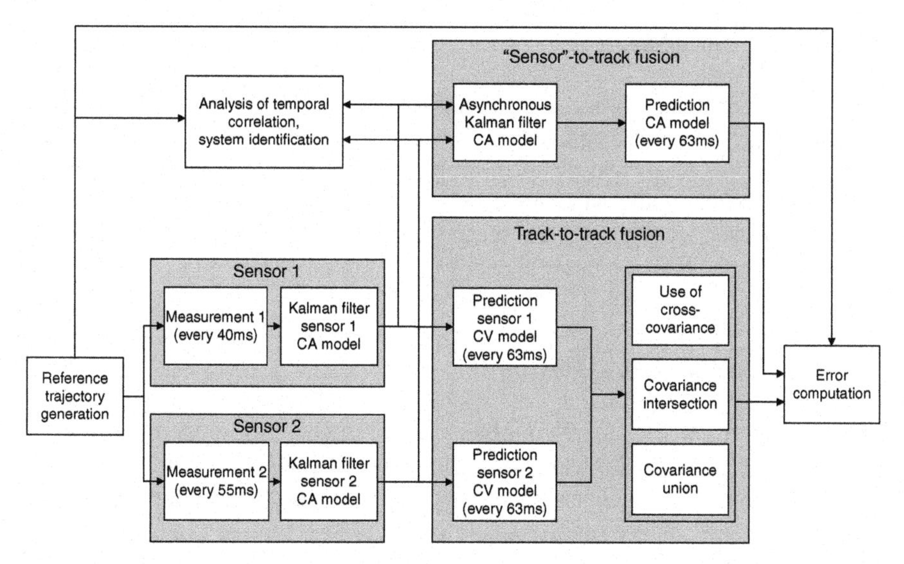

Fig. 1 Block diagram of the simulation setup. CA (constant acceleration) model refers to a white noise jerk model and CV (constant velocity) model refers to a white noise acceleration model both detailed in the appendix

2.1 Reference Trajectory Generation

The trajectory of the target vehicle is generated with a 2D white noise jerk model, for details see Appendix 6.1. It is run at a cycle time of 1 ms. With process noise covariance values of $10\left(m/s^3\right)^2$, trajectories are the result of random walk dynamics and tend to proceed in a relatively straight line. If the target leaves a rectangular region of $\pm$ 200 m in x-direction and $\pm$ 150 m in y-direction around the centre, a constant bias of

$$\dddot{x} = -sgn(x) \cdot 10\frac{m}{s^3}\ , \ \text{if } |x| > 200\,\text{m}$$

$$\dddot{y} = -sgn(y) \cdot 10\frac{m}{s^3}\ , \ \text{if } |y| > 150\,\text{m}$$

is superimposed on the white noise jerk in order to force the object back into the rectangular region. The repulsive force is applied as long as the trajectory proceeds along its path away from the centre and is outside the defined rectangular region.

Parts of the vehicle trajectory where this non-random jerk component is applied are crucial in assessing the different fusion algorithms' performances since the underlying dynamical models of the various Kalman filters in this simulation setup will not be appropriate in these cases.

2.2 Sensor Simulation

Measurements of the two sensors are assumed to be the relative position of the target vehicle with respect to the ego vehicle as would be appropriate for a laser scanner, for example. The measurements are corrupted by Gaussian noise and are then fed into a Kalman filter with a white noise jerk model, for details see Appendix 6.2. For a realistic sensor scenario, the two sensors output data at different cycle times of 40 ms and 55 ms.

2.3 "Sensor"-to-Track Fusion

The "sensor"-to-track fusion of the two sensors is achieved by an asynchronous Kalman filter, that is, a Kalman filter that updates its state every time a new measurement arrives. This Kalman filter – like the sensor Kalman filters – also uses a white noise jerk model (see Appendix 6.1). In our setup we assume that automotive applications that use sensor fusion data require fused objects at a constant cycle time of 63 ms. Hence a prediction module using the same dynamical model as the Kalman filter is activated every time an output is required by the application. Kalman filter and prediction module together constitute the "sensor"-to-track fusion module.

2.4 Track-to-Track Fusion

In track-to-track fusion, the states to be fused must be synchronized, that is, must have the same time stamp. In general, however, sensors output data at different cycle times with local, non-synchronized clocks. While synchronization methods of various degrees of sophistication have been proposed in literature [8], we use the simplest one, namely prediction of all sensor data to the time required by the application. Since the details of sensor preprocessing and tracking are usually kept confidential by automotive suppliers, we make the conservative choice of a white noise acceleration model for the prediction inside the track-to-track fusion module. The synchronized sensor data are then fed into one of the track-to-track fusion algorithms under consideration, namely use of cross-covariance, covariance intersection, and covariance union. A brief review of those three algorithms is provided in Sect. 3.

2.5 Error Computation

Every 63 ms the asynchronous Kalman filter and the three track-to-track fusion algorithms output fused target state vectors and their covariances. Using the target vehicle reference trajectory the RMSE can be computed at every time step. We then average the RMSE of the position coordinates over entire simulation runs of 40 s. This allows us to evaluate the benefits of track-to-track fusion versus Kalman filtering when the dynamical model of the Kalman filter is valid and when it is not.

3 Track to Track Fusion

3.1 Use of Cross-Covariance

The use of cross-covariance method was first described in [3]. Given two state estimates $\widehat{\xi_a}$ and $\widehat{\xi_b}$ and their covariance matrices P_a and P_b, the fused estimate using cross-covariance reads

$$\widehat{\xi_c} = \widehat{\xi_a} + \chi[\widehat{\xi_b} - \widehat{\xi_a}] \tag{1}$$

where

$$\chi = [P_a - P_{ab}]U_{ab}^{-1} \tag{2}$$

$$U_{ab} = P_a + P_b - P_{ab} - P_{ab}^T \tag{3}$$

The covariance matrix for the estimation pair is calculated using

$$P_c = P_a - [P_a - P_{ab}]U_{ab}^{-1}[P_a - P_{ab}]^T \tag{4}$$

The cross-covariance matrix P_{ab} is initially set to $P_{ab}(0|0)$ and subsequentially updates using a recursive relationship

$$P_{ab}(k|k) = \alpha_a(k)\beta(k-1)\alpha_b^T(k) \tag{5}$$

where

$$\alpha_a(k) = I - K_a(k)H_a(k),\ \alpha_b(k) = I - K_b(k)H_b(k)$$
$$\beta(k-1) = F_a P_{ab}(k-1|k-1)F_b^T + Q(k-1)$$

However, for $\alpha_a(k)$ and $\alpha_b(k)$ in (5) the details of both sensors' Kalman filters such as H_i, K_i, and F_i as well as Q have to be known. This is not always the case, especially for industrial sensors with integrated tracking. For this case, an alternative computation of P_{ab} is proposed in [8] where the cross-covariance matrix is approximated by the Hadamard (or entrywise) product of both input covariance matrices.

$$P_{ab} \approx \rho\sqrt{P_a \bullet P_b} \tag{6}$$

where ρ represents an effective correlation coefficient. It can be determined numerically by Monte-Carlo simulations for specific setups; in [8] values of $\rho \approx 0.4$ were found to be optimal for the fusion of 2D Cartesian position vectors (x,y). For the fusion of other state vectors, for example, state vectors including velocity entries, other values of ρ should be used.

3.2 Covariance Intersection

In [4, 9] the use of a fusion rule named *Covariance Intersection* (CI) for combining tracks of unknown correlation is proposed.

Based upon two estimates A and B that can originate both from a sensor or a model of the observed process it is possible to determine a fused estimate C. For consistent estimates A, B, it is possible to represent these as a pairs of estimated state and covariance $A = \{\widehat{\xi_a}, P_a\}$ and $B = \{\widehat{\xi_b}, P_b\}$. Note that these pairs do not contain any information about the cross correlation between A and B.

The fused estimate C can then be determined using

$$P_c^{-1} = \omega P_a^{-1} + (1-\omega) P_b^{-1} \tag{7}$$

$$\widehat{\xi_c} = P_c \left(\omega P_a^{-1} \widehat{\xi_a} + (1-\omega) P_b^{-1} \widehat{\xi_b} \right) \tag{8}$$

where

$$\omega = arg\ min(\det(P_c)) \tag{9}$$

The computation of ω in Eq. (9) is determined through an optimisation process to minimise $\det(P_c)$, which is proposed in [10]. Yet is is also possible to minimise $tr(P_c)$ or other criteria of uncertainty. For all $\omega \in [0, 1]$ the consistency of the estimate is guaranteed. Moreover, in [9] this estimate is shown to be optimal if the cross-covariance is unknown.

3.3 Covariance Union

A common problem in track-to-track data fusion is the resolution of statistically inconsistent states – known as database deconfliction. The covariance union method proposed in [5] allows to unify two tracks, even if the difference of the state estimates exceeds the covariance indicated by at least one track. To obtain a unified estimate, a new state vector estimate $\widehat{\xi_c}$ is determined. The unified covariance matrix P_c is a covariance exceeding both P_a and P_b.

While covariance union is not commonly used as an independent fusion method but only for the fusion of states that are deemed to be statistically inconsistent we use it here as a standalone fusion method in order to cleanly separate its performance from the influence of a criterion for statistical inconsistency.

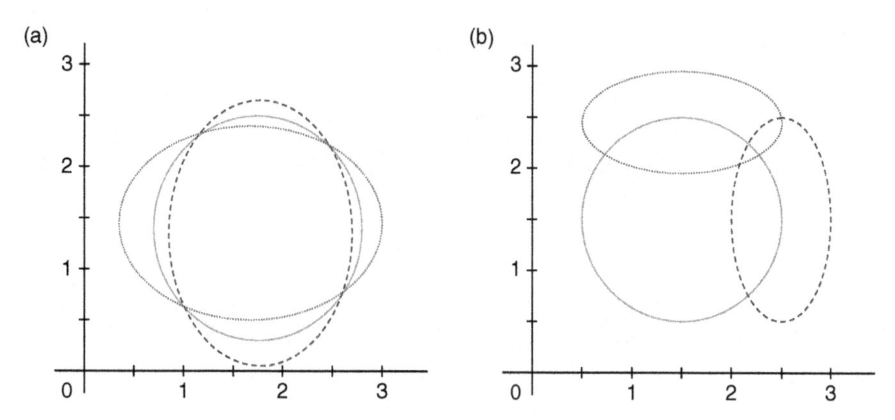

Fig. 2 **(a)** Covariance ellipses (locus of points $x^T P^{-1} x = 1$) of A (dotted red ellipse) and B (dashed blue ellipse) and their covariance intersection fusion result C (solid green ellipse) for $\omega = 0.5$. **(b)** Example of a covariance union for P_a (dotted red ellipse) and P_b (dashed blue ellipse). For $\widehat{\xi_u} = (1\ 1)^T$ det(P_c) is minimal (solid green ellipse)

The unified estimate $C = \{\widehat{\xi_c}, P_c\}$ can be determined by

$$U_a = P_a + (\widehat{\xi_c} - \widehat{\xi_a}) \cdot (\widehat{\xi_c} - \widehat{\xi_a})^T \tag{10}$$

$$U_b = P_b + (\widehat{\xi_c} - \widehat{\xi_b}) \cdot (\widehat{\xi_c} - \widehat{\xi_c})^T \tag{11}$$

$$P_{ab} = max(U_a, U_b) \tag{12}$$

$$\widehat{\xi_c} = arg\ min(\det(P_{ab})) \tag{13}$$

As for the covariance intersection method an optimisation process has to be performed to minimise a criterion of uncertainty such as determinant, trace, etc. For the covariance union method, $\widehat{\xi_c}$ has to be determined, which represents a multi-dimensional optimisation problem. In our simulation setup, we optimise $\widehat{\xi_c} = (x\ y)^T$.

3.4 Computational Costs of Track-to-Track Fusion

Computational costs for the asynchronous Kalman filter and the track-to-track fusion algorithms are measured in our simulation setup running on a 2 GHz Pentium IV processor in eight test runs of 10,000 updates and predictions for each method. The computation times are given in Table 1 as well as the relative time per update and prediction as compared to the asynchronous Kalman filter.

The measured computational costs in Table 1 show that the use of cross covariance approach performs at the smallest computational cost, followed by covariance intersection. The asynchronous Kalman filter requires slightly more computational resources, but far less than the covariance union method. Note that

Table 1 Mean absolute and relative computation times for the asynchronous Kalman filter and the track-to-track fusion algorithms

Method	Absolute [ns]	Relative
Async. Kalman	151.75	1.00
Use of cross cov.	103.95	0.69
Cov. intersection	132.30	0.87
Cov. union	397.50	2.62

the multi-dimensional optimisation for the covariance union has been implemented in a preliminary way using simple gradient descent. It can clearly be sped up using more sophisticated multi-dimensional minimization techniques such as Nelder-Mead, etc.

4 Evaluation

Evaluation of the Track-to-Track Fusion algorithms discussed in Sect. 3 is performed using our simulation setup. In the following two effects are discussed: the effect of sensor cross-correlation and the sensors' auto-correlation.

4.1 Cross Correlation

The adverse effect of correlated input sources is a major problem in track-to-track fusion. We use our simulation setup to evaluate the performance of three different track-to-track fusion algorithms and compare their performances to that of a standard asynchronous Kalman filter. Towards this end, we determine the RMSE at the outputs of both sensor Kalman filters. The inter-sensor (track-to-track) correlation of the errors is computed numerically over entire simulation runs by comparing the sensor outputs to the reference trajectories. Track-to-track fusion results are evaluated in the *error computation* module by comparing the reference trajectory to the fusion results. The resulting overall RMSE values for an individual test run are plotted against the measured track-to-track correlation in Fig. 3

It can be seen from Fig. 3 that the asynchronous Kalman filter performs well, that is with a lower RMSE value than both input tracks, for most test runs with a cross-correlation $\rho < 0.15$. However, for input tracks with a correlation $\rho > 0.15$, the fusion errors become higher than the RMSE of both input vectors. On the other hand the errors of the track-to-track fusion methods are smaller than those of the Kalman filter with the exception of covariance intersection at small correlations; their range is also smaller.

The mean and median RMSE values for 2,000 test runs over 40 s can be seen from Table 2. Considering the mean RMSE values, the use of cross covariance algorithm yields the lowest RMSE, followed by covariance union, covariance intersection, and finally the asynchronous Kalman filter.

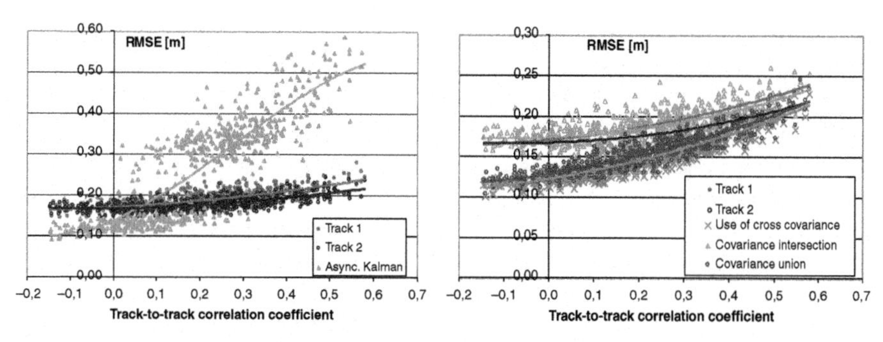

Fig. 3 Root mean squared errors (RMSE) of track fusion results and input tracks *Track 1* and *Track 2* depending on the cross correlation of the latter. Due to considerably differing RMSE values of the asynchronous Kalman filter and the other fusion algorithms two diagrams with differing RMSE ranges are used

Table 2 Mean (median) Root mean squared error of all test runs, test runs with low cross-correlation and high-cross correlation

Method	Overall	$-0.2 < \rho < 0.2$	$0.2 < \rho < 0.6$
Async. Kalman	0.283 (0.317)	0.181 (0.138)	0.369 (0.358)
Use of cross cov.	0.151 (0.150)	0.131 (0.129)	0.167 (0.164)
Cov. intersection	0.189 (0.187)	0.178 (0.176)	0.199 (0.198)
Cov. union	0.156 (0.156)	0.137 (0.134)	0.172 (0.170)

4.2 Auto Correlation

Apart from the cross-correlation between the two Kalman filter tracks, the auto-correlation of each Kalman track is determined numerically assuming ergodicity. The auto-correlation for the position and velocities of both Kalman sensor tracks is drawn against the time difference in Fig. 4.

The auto-correlation plot in Fig. 4 is indicative of a Gauss-Markov process of second order or higher, as is expected from the third order (white noise jerkmodel)

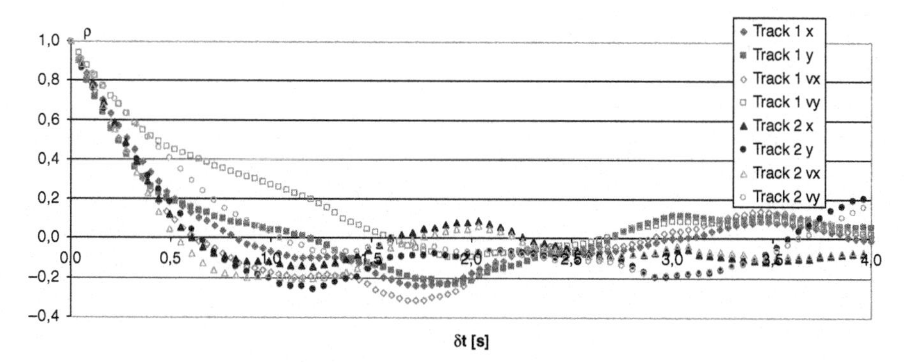

Fig. 4 Auto-correlation coefficient ρ of Kalman filter tracks *Track 1* and *Track 2* respectively drawn against the time difference δt

dynamics of the underlying Kalman filter. The degree of auto-correlation is significant since even after 0.25 s the auto-correlation coefficient exceeds 0.5 for both sensors.

Therefore the assumption of temporal independence does not hold for the Kalman filter tracks which are used in the the track-to-track fusion process. This property does not affect the discussed track-to-track fusion algorithms as these perform the fusion exclusively on the most actual track updates. However, the asynchronous Kalman filter assumes temporal independence which is clearly not the case for the given setup. Those temporally correlated measurements can be "decorrelated" in a Kalman filter by augmenting its dynamical model by a shaping filter (see e.g. [11]). The shaping filter can for example be identified by fitting the autocorrelation function of linear models of increasing order to the numerically determined autocorrelation graphs of Fig. 4.

5 Conclusions and Outlook

We presented an assessment of a standard Kalman filter applied to tracked sensor data of unknown correlation versus several well-known algorithms for track-to-track fusion. Tracked sensor data were evaluated for inter-track correlation as well as for auto-correlation.

Among the three track-to-track algorithms, the use of cross-covariance returned the best quality of fusion results with respect to RMSE. Moreover, the run time of the use of cross-covariance was the lowest of the four fusion methods. Covariance intersection required the second fastest runtime with a RMSE that was lower than the Kalman filter but higher than both use of cross covariance and covariance union. The covariance union algorithm showed RMSE values comparable to the use of cross-correlation albeit at a considerably higher computational cost. The standard Kalman filter performs well for small inter-track cross-correlations however for large cross-correlations the RMSEs are a factor of 2–3 larger than for the track-to-track correlation methods. Its computational cost is between those of use of cross-covariance and covariance intersection, and the cost for covariance union.

The poor performance of the Kalman filter at large correlations was expected. More interesting is the ranking of the track-to-track fusion methods with respect to RMSE: as was observed in [10], covariance intersection is a conservative fusion method that can be improved if partial knowledge of cross-covariance can be incorporated. Hence the better results for the use of cross-covariance. Perhaps more unexpectedly, covariance union yields results better than covariance intersection and comparable to those of use of cross-covariance. Due to the application of non-random jerk during the vehicle motion (cf. Sect. 2.1) statistically inconsistent sensor states[1] might arise that are better taken care of by covariance union than by covariance intersection.

---

[1] Statistical inconsistency of two states representing the same physical entity can for example be characterized by a threshold for the Mahalanobis distance (cf. [5]).

The auto-correlation of both sensor tracks have been evaluated to be significant, exceeding 0.9 for a single fusion cycle of 40 ms and 55 ms, respectively. Auto-correlation does not affect the track-to-track fusion algorithms investigated in this paper, as they do not include past states and covariances. The detrimental effect of autocorrelated measurements on the Kalman filter is due to its assumption of the whiteness of the measurement noise. The Kalman filter can, however, be improved by using an augmented state that includes a model of the autocorrelation dynamics [11]. The exact system identification and its incorporation into the Kalman filter constitutes future work.

While the numerical study presented in this paper gives an indication of performance improvements using track-to-track fusion for automotive applications, many simplifying assumptions about vehicle dynamics, sensor characteristics, sensor filtering, etc have been made. The next step will be the implementation of the above track-to-track algorithms in a test vehicle and their performance assessment in realistic traffic scenarios using real sensor data.

6 Appendix

6.1 Vehicle Dynamics

The reference vehicle trajectory is characterized by a six-dimensional state vector

$$\xi = (x \quad y \quad \dot{x} \quad \dot{y} \quad \ddot{x} \quad \ddot{y})^\top \tag{14}$$

The dynamical model is a discrete-time counterpart white noise jerk model [12]

$$\xi(k+1) = F(k+1,k)\xi(k) + G(k+1,k)\nu(k) \tag{15}$$

where

$$F(k+1,k) = \begin{pmatrix} 1 & 0 & \Delta t_k & 0 & \frac{\Delta t_k^2}{2} & 0 \\ 0 & 1 & 0 & \Delta t_k & 0 & \frac{\Delta t_k^2}{2} \\ 0 & 0 & 1 & 0 & \Delta t_k & 0 \\ 0 & 0 & 0 & 1 & 0 & \Delta t_k \\ 0 & 0 & 0 & 0 & 1 & 0 \\ 0 & 0 & 0 & 0 & 0 & 1 \end{pmatrix}$$

$$G(k+1,k) = \begin{pmatrix} \frac{\Delta t_k^3}{6} & 0 \\ 0 & \frac{\Delta t_k^3}{6} \\ \frac{\Delta t_k^2}{2} & 0 \\ 0 & \frac{\Delta t_k^2}{2} \\ \Delta t_k & 0 \\ 0 & \Delta t_k \end{pmatrix} \tag{16}$$

with $\Delta t_k = t_{k+1} - t_k$. The two-dimensional stochastic process vector $\nu(k) \in I\!R^2$ models the process noise of the vehicle dynamics. It is a white, Gaussian process with $cov(\nu(k)) = V = \text{diag}(10, 10)\left(\frac{m}{s^3}\right)^2 \ \forall k$. In our simulation the state ξ is updated every 1 ms using Eq. (15).

This model is also referred to as CA model in Fig. 4. The CV model used in the track-to-track predictions is a truncated CA model.

6.2 Sensor Measurements

Sensors 1 and 2 are assumed to measure position only, hence the output equation for the Kalman filter reads

$$z_i(k) = H(k)\widehat{\xi}_i(k) + w_i(k) \quad i \in \{1, 2\} \tag{17}$$

where $z_i(k)$ are the measurements, $\widehat{\xi}_i(k)$ are the internal states of the sensor Kalman filters and $w_i(k)$ are two-dimensional stochastic processes simulating the sensor noise. The output function is given by

$$H(k) = \begin{pmatrix} 1\,0\,0\,0\,0\,0 \\ 0\,1\,0\,0\,0\,0 \end{pmatrix}$$

The noise processes are characterized by

$$\begin{aligned} cov\,(w_1(k)) &= W_1 = \text{diag}(0.25, 0.25)\, m^2 \quad \forall k \\ cov\,(w_2(k)) &= W_2 = \text{diag}(0.16, 0.09)\, m^2 \quad \forall k \end{aligned} \tag{18}$$

References

1. A. N. Steinberg, C. L. Bowman, and F. E. White. Revisions to the JDL data fusion model. *Proceedings of SPIE, Sensor Fusion: Architectures, Algorithms, and Applications III*, 3719: 430–441, 1999.
2. S. Blackman and R. Popoli. *Design and analysis of modern tracking systems*. 2nd edition. Artech House radar library. Artech House, Norwood 1999.
3. Y. Bar Shalom. On the track-to-track correlation problem. *IEEE Transactions on Automatic Control*, AC-26(2): 571–572, 1981.
4. S. J. Julier and J. K. Uhlmann. *Handbook of Data Fusion, chapter 12: General decentralized data fusion with covariance intersection (CI)* (pp. 1–25). Boca Raton FL: CRC Press, 2001.
5. J. K. Uhlmann. Covariance consistency methods for fault-tolerant distributed data fusion. *Information Fusion*, 4(3): 201–215, 2003.
6. K. Weiss, D. Stueker, and A. Kirchner. Target modeling and dynamic classification for adaptive sensor data fusion. In *Proceedings of the IEEE Intelligent Vehicles Symposium*, pp. 32–137, 2003.
7. N. Floudas, M. Tsogas, A. Amditis, and A. Polychronopoulos. Track level fusion for object recognition in road environments. *PReVENT Fusion Forum e-Journal*, (2): 16–23, January 2008.

8. Y. Bar-Shalom and X.-R. Li. Multitarget-multisensor tracking: Principles and techniques. Storrs, CT: YBS Publishing, 1995.
9. J. K. Uhlmann. Dynamic map building and localization for autonomous vehicles. PhD thesis, University of Oxford, 1995.
10. S. J. Julier and J. K. Uhlmann. Using covariance intersection for slam. *Robotics and Autonomous Systems*, 55(1): 3–20, 2007.
11. A. Gelb. (ed.) *Applied optimal estimation*, (Chap. 4.5, pp. 133–136). Cambridge, MA: MIT Press, 1974.
12. X. R. Li and V. P. Jilkov. Survey of maneuvering target tracking-part I: dynamic models. *IEEE Transactions on Aerospace and Electronic Systems*, 39(4): 1333–1364, 2003.

Effective and Efficient Communication of Information

Jan Willem Marck, Leon Kester, Miranda van Iersel, Jeroen Bergmans and Eelke van Foeken

Abstract Research fields such as Network Enabled Capabilities, ubiquitous computing and distributed sensor networks, etc. deal with a lot of information and a lot of different processes to fuse data to a desired level of situation awareness. Already quite some research is done in applying information theoretic concepts to sensor management. Effective communication between functional components in the fusion hierarchy can be achieved by combining these approaches. We propose a method that makes the interaction between different fusion processes more efficient and effective. This is particularly useful in situations with costly or overused communication facilities.

1 Introduction

At the moment a great amount of research is done in the fields of Network Enabled Capabilities, ubiquitous computing and distributed sensor networks. A common theme is that a lot of information is available or is extracted from the environment and that there are processes that fuse this information to reach a desired level of situation awareness. There are at least two distinct approaches to these kind of problems. One is data fusion, that develops information fusing architectures and design principles for such systems. The other is sensor management, focusing on information quality and intelligent choices on when to use which sensor. Our work tries to bridge the gap between these two approaches, as they are related and strongly interdependent.

Several architectures [1–3] are developed for describing and designing large, scalable, modular fusion systems, that need to provide fused data according to some (implicitly or explicitly given) information need. The Joint Directors of Laboratories (JDL) data fusion model [3] is perhaps the most well known. The Networked Adaptive Interactive Hybrid Systems (NAIHS) model [1, 2] builds upon this to describe

J.W. Marck (✉)
TNO Defence, Security and Safety, Oude waalsdorperweg 63, The Hague, The Netherlands
e-mail: jan_willem.marck@tno.nl

S. Lee et al. (eds.), *Multisensor Fusion and Integration for Intelligent Systems*,
Lecture Notes in Electrical Engineering 35, DOI 10.1007/978-3-540-89859-7_7,

the interactions between different functional components (human or machine) more completely. We will use the NAIHS model throughout this paper as a guideline.

Extensive research is done in information theoretical approaches to sensor management [4, 5]. Less research is done [6] in managing different functional components e.a. combining the sensor management approach on higher level sensor fusion processes, while this is just as important in systems with distributed processes.

In distributed, networked systems various functional components exist. Various properties determine the interaction between those functional components. This interaction determines how effective the system complies with the required Situation Awareness (SA). Some of these properties are:

- bandwidth,
- update rate,
- dealing with costly or limited resources,
- scalability,
- dealing with irrelevant information/communication,
- when to transmit or request information.

The common denominator in these properties is that transmission and processing of irrelevant information is undesired and causes bad performance.

In Sect. 2 the theory of our method is explained. Section 3 describes two experiments and Sect. 4 discusses results from these experiments. Section 5 concludes this paper.

2 Theory

Consider a component A that has expressed a certain need for data. It requests this service to deliver data from other processes called $B_1 \dots B_N$ (see Fig. 1). The processes A and B_i are for example components in different NAIHS or JDL levels. In this paper we will study how the communication between consumer A and providers B_i can be minimized so that only relevant information is transmitted.

Process A, the consumer has a certain information need. For this A has to acquire relevant information from other processes $B_{1,\dots,N}$, the provider processes. The com-

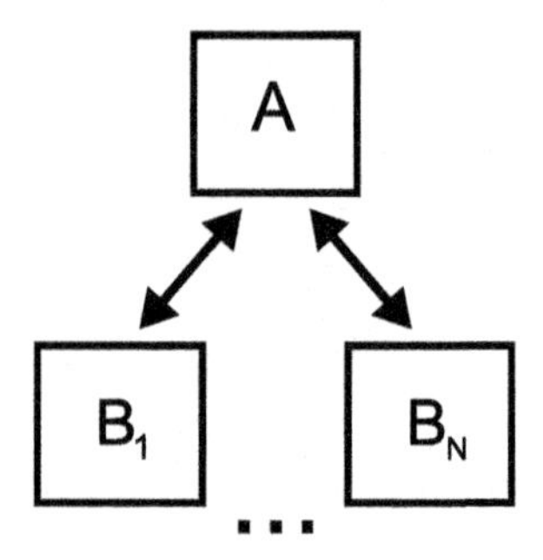

Fig. 1 Process A requests data from multiple providers B_i to get information

munication between A and B_i should be optimized for optimal performance of process A with respect to the communication costs. How should be decided when process B_i sends information to A? Somehow A should express its information need in such a way that B_i can decide when the gains outweigh the costs of transmitting information.

We denote the state of process A by X_A. Z_{B_i} is the information in process B_i in which A is interested. The contributional value of Z_{B_i} to A can be determined by comparing state $P(X_A)$ with and without incorporating Z_{B_i}. A widely used method for this is the Kullback–Leibler divergence (KLD) [7] between the probability functions: $P(X_A)$ and $P(X_A|Z_{B_i})$ Eq. (1):

$$\begin{aligned} D_i &= D_{KL}(X_A|Z_{B_i}||X_A) \\ &= \sum_{z \in Z_{B_i}} \sum_{x \in X_A} P(x|z) \log \frac{P(x|z)}{P(x)}. \end{aligned} \tag{1}$$

D_i is the KLD in process A caused by the knowledge of Z_{B_i} . Z_{B_i} is the probability mass function (PMF) of the information requested by process A. Z_{B_i} is only known in process B_i. B_i should decide when to transmit it to A. X_A is a PMF that describes the state of process A and X_A known in process A only.

If Z_{B_i} does not have impact on state X_A (low KLD), the information is irrelevant to transmit or process. Ignoring irrelevant information makes all concerned processes more cost effective. Z_{B_i} is sent by process B_i to A when $D_{KL}(X_A|Z_{B_i}||X_A)$ exceeds a threshold function $C(\cdot)$ as in Eq. (2):

$$D_{KL}(X_A|Z_{B_i}||X_A) > C(\cdot). \tag{2}$$

This cost function can describe the amount of bytes transmitted, the battery power used for communication or it can be there to allow information with a certain minimum impact on X_A. Or supply and demand could set a value to $C(\cdot)$. For the sake of simplicity we keep $C(\cdot)$ constant throughout this paper and consider it known by all processes.

2.1 The Problem

To calculate the KLD, knowledge of process A, its state X_A and information Z_{B_i} is necessary. The processes B_i and A each contain only a part of the necessary information to calculate the divergence. Process A has up-to-date knowledge of state X_A and about its own process. Process B_i has the up to date knowledge of Z_{B_i} and should know when to send this to A.

It is undesirable for B_i (the provider) to calculate the divergence because this requires intricate knowledge of process A (the consumer) and state X_A. This would mean that the complete functionality of A is replicated in B_i. This will turn the system into a monolithic construct instead of a scalable, adaptive and modular system as suggested by the NAIHS architecture [2]. When A needs multiple processes B_i to fulfill its task, placing the functionality of A in every B_i will not solve the problem, because the information from the other B_i processes is still needed.

Also it clearly is not possible for process A to determine the divergence before Z_{B_i} is transmitted.

The remaining possibility is that A and B_i each calculate the divergence partially. The KLD is a function of X_A and Z_{B_i}. When X_A can be marginalized out of the KLD. Process A can calculate the marginal over X_A and send this over to B_i. Subsequently B_i determines the divergence for a current Z_{B_I} to asses whether it is to be sent.

This way, the divergence in the state space of the consumer is mapped into that of the provider. In other words A paraphrases its request to B_i into a function and sends this to B_i. Using this function B_i calculates the KLD and determines whether or not to send Z_{B_i}.

2.2 Splitting the Divergence

To calculate to divergence in two steps it has to be split in two separate parts. This is done in the following way. Eq. (1) is rewritten by using Bayes' rule and splitting the logarithm. Factors that depend on z only are moved out of the sum over x.

$$\begin{aligned} D_i = & \sum_{z \in Z_{B_i}} \frac{1}{P(z)} \sum_{x \in X_A} P(z|x)P(x) \log P(z|x) \\ & - \sum_{z \in Z_{B_i}} \frac{\log P(z)}{P(z)} \sum_{x \in X_A} P(z|x)P(x) \end{aligned} \tag{3}$$

$$D_i = \sum_{z \in Z_{B_i}} \frac{E_i(z)}{P(z)} - \frac{F_i(z) \log P(z)}{P(z)}. \tag{4}$$

The sums $E_i(z)$ and $F_i(z)$ are the parts of the KLD that are calculated by A and subsequently sent to B_i:

$$E_i(z) = \sum_{x \in X_A} P(z|x)P(x) \log P(z|x) \tag{5}$$

$$F_i(z) = \sum_{x \in X_A} P(z|x)P(x). \tag{6}$$

Using E_i and F_i process B_i can now calculate D_i and determine when to send an update to A. In Eq. (6) state X_A is marginalized out of the joint probability over both X_A and Z_{B_i}. From the result F_i be interpreted as the estimation of Z_{B_i} made by process A. We will use this interpretation later on.

2.3 Updating E_i and F_i

It is the responsibility of process A to keep E_i and F_i up to date in B_i. E_i and F_i change whenever X_A changes. But this change will not always be a large change. It is not always relevant to communicate a new E_i and F_i if they have not changed significantly since their last update. A method is needed to decide when to transmit E_i and F_i.

The following is a working heuristic for the experiments we performed to solve the problem of when to update E_i and F_i. If process A receives an update Z_{B_i} it calculates E_i and F_i for the B_i that sent information Z_{B_i}. Subsequently process A transmits E_i and F_i to B_i. This results in an extra communication overhead. Process A than waits until it gets new information. Then A repeats this process.

3 Experiments

We illustrate the method described above using two experiments. In the first experiment only one process, B_1, is used. In the second experiment two processes, B_1 and B_2, are used to illustrate the method.

Real world problems or more realistic problems usually contain continuous spaces. The experiment is designed to make an as simple as possible proof of concept. For this a discrete environment is used.

3.1 Experimental Setup

A single target is moving in one dimensional (1D) space. The space in which the target is moving is divided into three sectors (Fig. 2). The target cannot move out of the three sectors. The track is generated by combining various random sinusoidal curves to simulate a more or less unpredictable target motion.

Process A has to report in which sector the target is and the state of A, X_A consists of three probabilities, for each sector the probability that the target is in that sector. To decide in which sector the target is A uses process B_i to get the position information Z_{B_i}. To accurately report the sector not every position update from B_i is useful. If the target is in the middle of a sector a small change of position will not

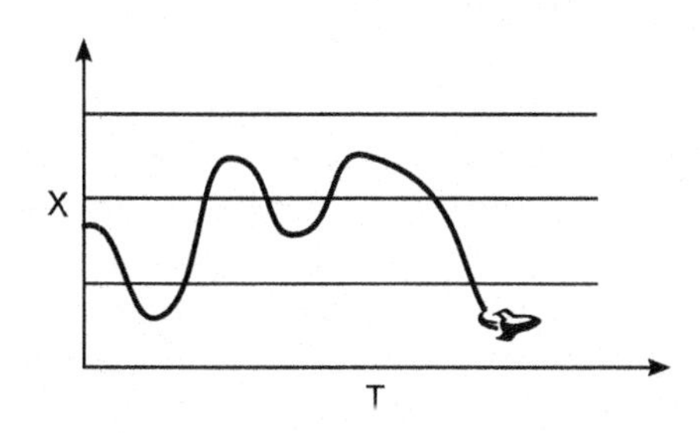

Fig. 2 A target moves through three sectors on a random path

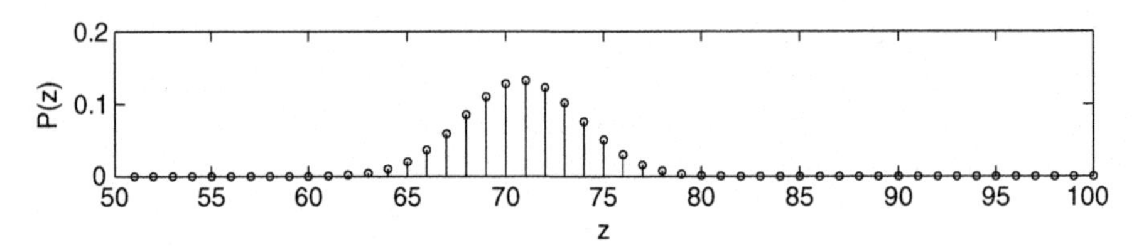

Fig. 3 Part of the sequence of segments that contain the histogram of Z_{B_i}

affect state X_A as much as if the target would be near a sector border. Preferably the B_i processes only communicate information Z_{B_i} to A when it is relevant A.

The 1D space of Z_{B_i} is divided into 150 equal segments. Each Process B_i is capable of monitoring a part the world, a sequence of the segments. The area covered by each process may overlap, like sensors with overlapping field of views. Z_{B_i} is a histogram with each element describing the probability that the target is in this element (Fig. 3)

The state of process A, X_A, consists of the three probabilities that the target is in each respective sector. To calculate X_A at each iteration A maintains also a world size histogram of where the target might be. If A receives Z_{B_i} than this histogram incorporates Z_{B_i} otherwise A increases the uncertainty of the position estimate. In other word process A tries to map Z_{B_i} into its state space, X_A.

In the experiments the partial divergences, E_i and F_i, are both vectors with the length of Z_{B_i}. Both E_i and F_i are immediately calculated and send to B_i by A, when A receives Z_{B_i}. This results in a communication overhead with twice the size of the amount of Z_{B_i} communication.

Both experiments follow the same setup. The following run is done for both. The target moves with a random track over the sectors. Each time step X_A is evaluated with and without communication. This is done to compare the case with "relevant" communication to the one with 100% communication. In both experiments such an example run is used to illustrate the method.

Next a more encompassing variant is done. The run is done multiple times in sets. Each subsequent set has a slightly higher cost function $C(\cdot)$. This is done to illustrate the quality of less but "relevant" communication.

3.2 Experiment 1

In the first experiment one process B_1 observes all three sectors using 150 segments as is shown in Fig. 4. First an example track is given and explained. Next the experiment is done repeatedly with a higher communication cost for each subsequent run. This results in a decrease in communication and an increase in the proportion of sector misclassifications. The amount of sector misclassifications is plotted against the amount of communication cost, both relative to the case where B_1 communicates its data every time step.

Fig. 4 Process A requests information from a single process B_1. Process B_1 covers the entire world using 150 segments

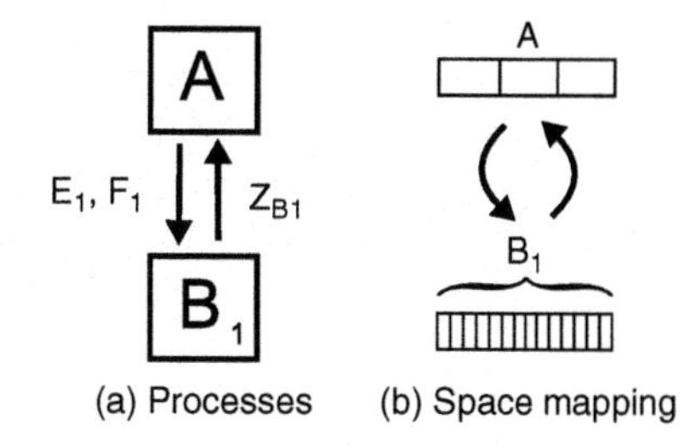

3.2.1 Results Experiment 1

The target motion is shown in Fig. 5. The crosses represent the times that process B_1 communicates information Z_{B_i} to process A. In this run of a 500 iterations B_1 sends a total of 58 updates to A.

For this case, Fig. 6 shows the estimation made by A in which sector the target is located. In a thousand iterations there are two misclassifications. These are small delays of a single iteration in deciding which the most probable sector is.

Figure 7 illustrates the probabilities of X_A of this run with relevant communication (b) set out against the same run but with 100% communication (a).

When the communication cost is varied, the amount of Z_{B_i} transmissions determines the proportion of sector misclassifications. This is shown in Fig. 8. With a

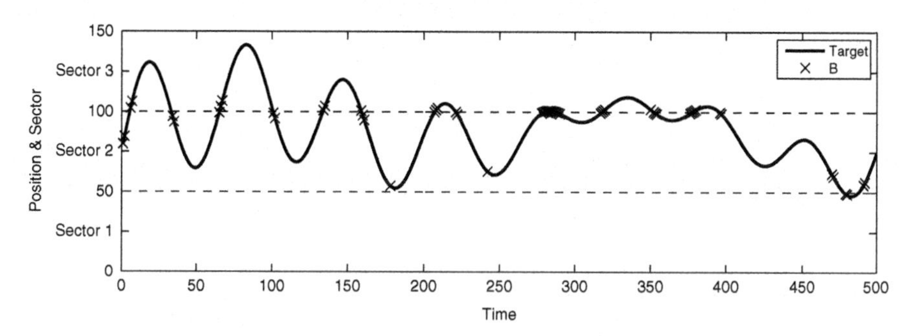

Fig. 5 An example run and the times that process B_1 communicates with process A

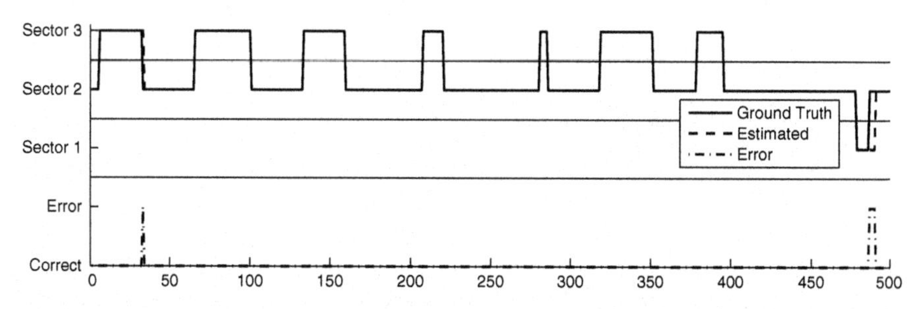

Fig. 6 The correct and faulty sector classifications for the target motion from Fig. 5

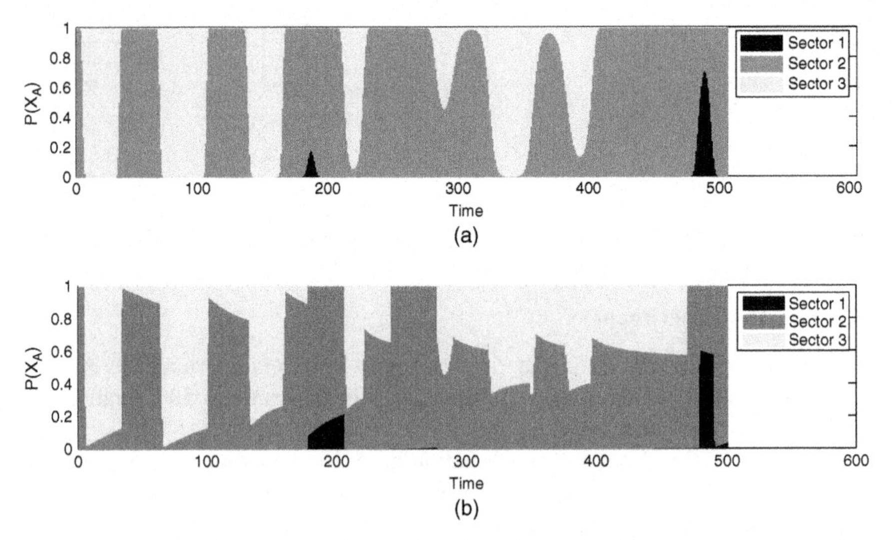

Fig. 7 Sector probabilities with 100% communication (**a**) and with limited communication (**b**) for the target motion from Fig. 5

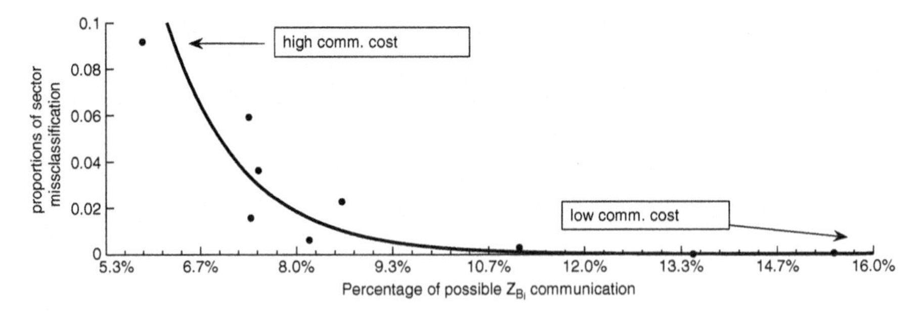

Fig. 8 The communication costs are varied. The amount of communication is plotted against the proportion of erroneous sector classifications

high communication cost there is less communication and more sector misclassifications. With a low communication cost there is more communication and less misclassifications.

3.2.2 Conclusion Experiment 1

The amount of communication of Z_{B_i} can be reduced by at least a factor ten with only small errors in sector classification. Including E_i and F_i overhead this results in a decrease of 70% of communication, compared to the case that Z_{B_i} is always communicated and no E_i and F_i communication is necessary. The method avoids irrelevant communication and decides to communicate when the target is close to sector borders, resulting in a significant reduction of communication.

3.3 Experiment 2

In the second experiment two processes B_i are used (Fig. 9). Both cover two out of three sectors. Firstly we will show the algorithm with two B processes. Next, the same experiment is done repeatedly and the quality of sector classification is plotted against the amount of communication.

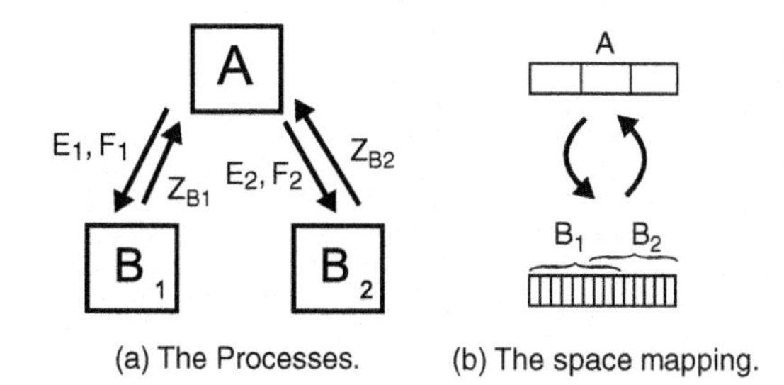

Fig. 9 Process A requests information from two processes B_1 and B_2. Together processes B_1 and B_2 cover all 150 segments

3.4 E_i, F_i Transmission Timing

The heuristic used in the former experiment for deciding when to transmit E_i and F_i does not work always. If the target moves out of the field of view of a given B_i, both vectors E_i and F_i contain only very small values. This results in a small D_i, despite a significant values of Z_{B_i}. This results in a shutdown of the process B_i.

To deal with a stopped process that has become relevant again, process A has to decide at a certain moment that process B_i may supply relevant information once again and that it should be restarted. This is done in following way.

First the target moves out of the field of view of the process B_i. A transmits an E_i and an F_i close to zero, stopping process B_i. A determines from now on in which sector the target is with information from the other and relevant process. The term F_i as calculated in A with Eq. (6) is the estimation from A of Z_{B_i}. If the sum over F_i for a certain stopped B_i grows larger than a given threshold (Process A estimates that the target is moving into the field of view of B_i), E_i and F_i are recalculated and send the process B_i, effectively restarting the process by indicating a more significant information need.

3.4.1 Results Experiment 2

Figure 10 illustrates a track with plotted communications from the processes B_i, and the restarts from A. The processes B_i communicate with process A near the sector borders.

Figure 11 shows the sector classifications for the target motion from Fig. 10. The behavior is very similar to the case of a single process B_1.

Figure 12 shows the sector probabilities, X_A. With 100% communication (a) and limited communication (b). Again this is quite comparable to the case with a single process B_1.

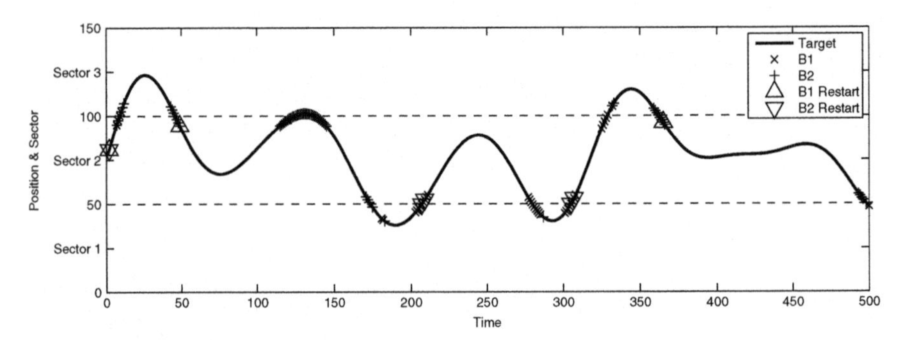

Fig. 10 An example track with communication from processes B_1, B_2 and A. Process B_1 covers the lower two sectors and process B_2 covers the upper two sectors

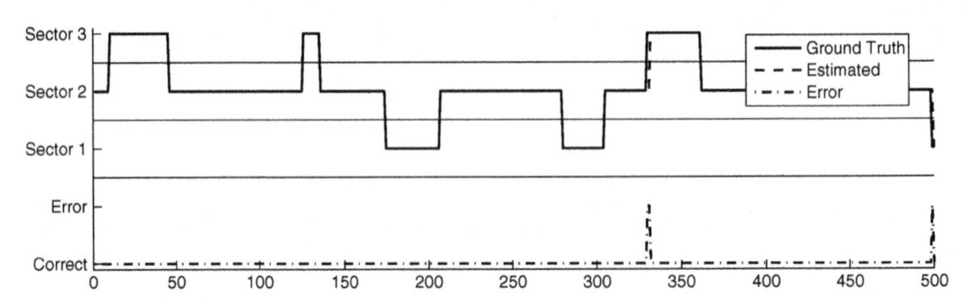

Fig. 11 Good and erroneous sector classifications for the target motion from Fig. 10

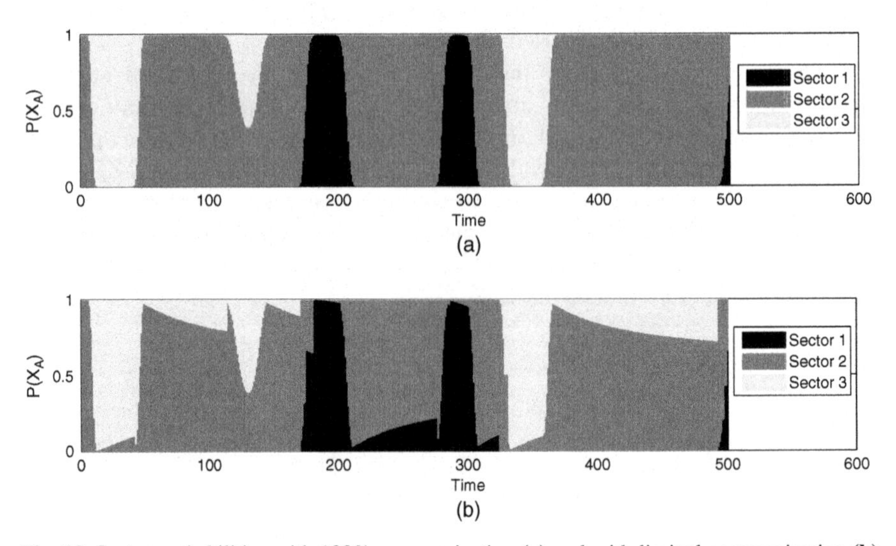

Fig. 12 Sector probabilities with 100% communication **(a)** and with limited communication **(b)** for the target motion from Fig. 10

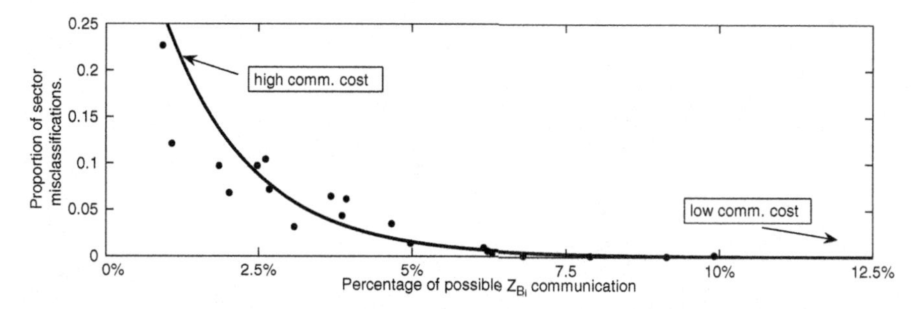

Fig. 13 Like Fig. 8. The communication cost is varied. The proportions in error sector classifications is set out against the percentage of communicated Z_{B_i}. The over head of communicating E_i and F_i adds roughly 1% of communication

The elaborate variant of this experiment is depicted in Fig. 13. The communication cost is varied and the amount of Z_{B_i} transmissions is set out against the proportion of sector misclassifications.

3.4.2 Conclusion Experiment 2

The amount of Z_{B_i} communication can be reduced to 7.5% for each process B_i with only small errors in sector classification. Note however that each process covers only two thirds of the area. The 7.5% is slightly more than two thirds of the 10% of Experiment 1. This is caused by both processes sending the same information. The probabilities over the sectors are negatively influenced by reduced communicating. Process A sends out restarts to either process B_i in 1% of all iterations. As in the first experiment the method prevents unnecessary communication and increases communication when the target is close to sector borders.

Another interesting phenomenon can be observed in Fig. 10 when the target approaches a sector border close to the field of view of B_i. The provider communicates its measurements sooner than it would for a border that lies in its field of view completely. The characteristics of the divergence that causes this will be looked at in future work.

4 Conclusion

In both experiments the method leads to a reduction of the communication of provider to consumer. A downside is that the E_i and F_i overhead is significant as well. We believe that an improved heuristic will be able to limit this overhead.

In both experiments there are some rare degenerate cases where process A does not notice a sector transition for an extended period. This is caused by an imperfect transformation of the KLD in consumer space to provider space. In Fig. 14 a typical degenerate case is depicted. Around iteration 200 the KLD goes up, while D_i goes down. This results in a period where no communication takes place at a point where the provider clearly has to communicate in order to comply with the information need of its consumer.

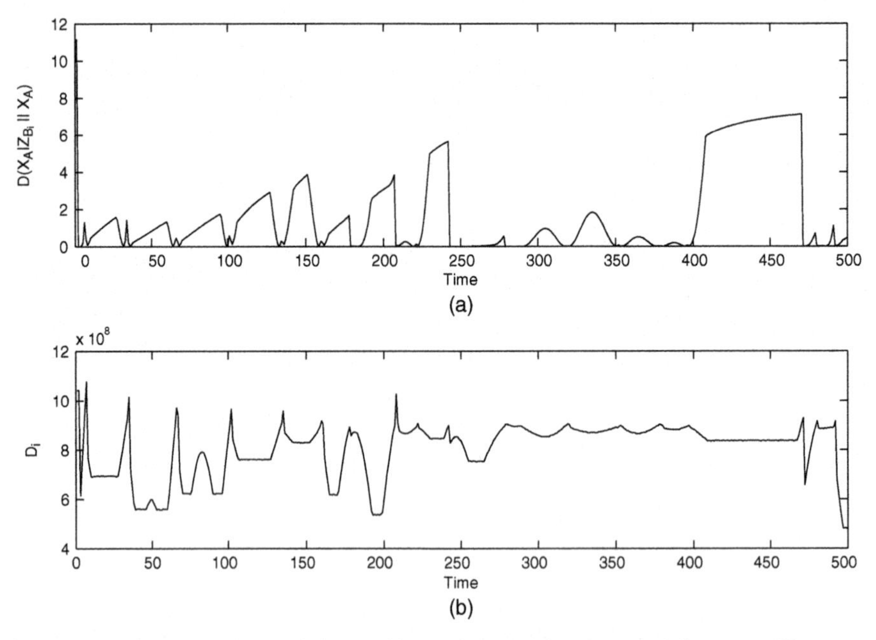

Fig. 14 D_i calculated in space of the provider and compared to the actual divergence. The upper graph **(a)** is the actual divergence. The lower graph **(b)** is the same divergence as calculated by process B_i. The more these are alike, the better the method should work

One important aspect of mapping the divergence from consumer space into provider space is that resolution and/or dimensionality in both spaces will differ. In E_i and F_i Eqs. (5) and (6) the term $P(z|x)$ translates the coarse resolution of X_A into the finer resolution of Z_{B_i}. For a given $x \in X_A$ (one of the three sectors) this results in a uniform piecewise probability distribution in Z_{B_i} space. This uniform distribution is compared with the more precise Z_{B_i} from Eq. (4). The fact that the estimation of Z_{B_i}, F_i, and the actual Z_{B_i} have differently shaped densities causes a big difference in D_i as calculated in B_i and the actual divergence. This is depicted in Fig. 14.

For a particular consumer/provider pair this mapping problem has as result that D_i must be scaled by an unknown factor and it has to be manually set. For the experiment this works, but this makes it hard to design a cost function that is applicable on a general provider/consumer pair.

When the method is better equipped to deal with these transformation errors, the communication needed will decrease.

5 Discussion and Future Work

We have shown that the method proposed provides a way to communicate only relevant information between different functional components. This results in a significant reduction of communication and enables the system to operate more efficiently and effectively.

If both consumer and provider use the method described in this paper, the provider process can determine the relevance of the information it could provide without having to understand the consumer process. This is an important aspect because it allows distributed, modular and scalable systems, without having to deal with an overload of communication. It enables the effective service oriented communication between functional components as the described in the NAIHS model.

Another aspect is that the amount of irrelevant information exchanged in the system is minimized. It is clear that this method is particularly useful in situations with costly or overused communication facilities. In larger systems where information goes through multiple processes, the method will have more impact because if irrelevant information is removed in one functional component, it will not be passed on to the next functional component.

5.1 Future Work

The proposed method can be improved in a number of aspects. A better heuristic for the updates of E_i and F_i, could reduce the amount of communication. More importantly, a better understanding of mapping the divergence into representation space of the provider is necessary to extend the method to more complex applications.

References

1. L.J.H.M. Kester. Model for networked adaptive interactive hybrid systems. In *Cognitive systems with interactive sensors (COGIS)*, 2006.
2. L.J.H.M. Kester. Designing networked adaptive interactive hybrid systems. In *Multi-Sensor Fusion and Integration for Intelligent Systems*, 2008.
3. J. Llinas, C. Bowman, G. Rogova, A. Steinberg, E. Waltz, and F. White. Revisiting the JDL data fusion model II. In *The 7th International Conference on Information Fusion*, June 2004.
4. C. Kreucher, K. Kastella, and A.O. Hero. Multi-target sensor management using alpha-divergence measures. In *3rd Workshop on Information Processing for Sensor Networks*, Palo Alto, CA, 2003.
5. T. Minka. Divergence measures and message passing. In *Microsoft Research Technical Report* (MSR-TR-2005-173), 2005.
6. P. Velagapudi, O. Prokopyev, K. Sycara, and P. Scerri. Maintaining shared belief in a large multiagent team. In *10th Interntional Conference on Information Fusion*, pp. 1–8, July 2007.
7. T.M. Cover and J.A. Thomas. *Elements of information theory*. Wiley-Interscience, New York, 1991.

Most Probable Data Association with Distance and Amplitude Information for Target Tracking in Clutter

Taek Lyul Song

Abstract In this paper, a new target tracking filter combined with data association called most probable and data association (MPDA) is proposed and its performance is evaluated by a series Monte Carlo simulation tests. The proposed MPDA method utilizes both distance to the predicted target measurement and amplitude information as the probabilistic data association with amplitude information (PDA-AI) method however, it is one-to-one association of track and measurement. All the measurements inside the validation gate at the current sampling time are lined up in the order of the closeness to the predicted target measurement. Probability that the measurement is target-originated is evaluated for each measurement by utilizing order statistics. The measurement with the largest probability is selected to be target-originated and the measurement is used to update the state estimate with the probability used as a weighting factor of the filter gain. To accommodate the probabilistic nature of the MPDA algorithm, a filter structure is developed. The proposed MPDA algorithm can be easily extended for multi-target tracking.

Keywords Target tracking · Data association · PDA · MPDAF · Clutter

1 Introduction

Accurate target state estimation in an adverse environment including false alarms and ECM generated by a stand-off jammer is essential for modern missile combat management systems. In this paper, a new target tracking filter combined with data association called most probable data association (MPDA) is proposed and its performance is compared with the probabilistic data association filter (PDAF) with amplitude information.

The nearest neighbor filter (NNF) [1] and the strongest neighbor filter (SNF) [2] are known for simplicity and easiness of implementation. However, performance

T.L. Song (✉)
Department of Control and Instrumentation Engineering, Hanyang University, Sa 1 Dong 1271, Sangnok-Gu, Ansan-Si, Gyeonggi-Do 426-791, Korea
e-mail: tsong@hanyang.ac.kr

S. Lee et al. (eds.), *Multisensor Fusion and Integration for Intelligent Systems*,
Lecture Notes in Electrical Engineering 35, DOI 10.1007/978-3-540-89859-7_8,

of each filter algorithm in a cluttered environment is poor because each filter may conclude unconditionally that the selected measurement is target-originated in spite of clutter. To overcome such defects, the probabilistic nearest neighbor filter (PNNF) [3], probabilistic strongest neighbor filter (PSNF) [4], PNNF with m validated measurements (PNNF-m) [5], and PSNF with m validated measurements (PSNF-m) [6] are proposed. Moreover, The PNNF-m and PSNF-m directly utilize the number of validated measurements so that they have reliable performance even for the case of unknown clutter density. The probabilistic data association filter (PDAF) [7] and the PDAF with amplitude information (PDAF-AI) [8]are known to have superior tracking performance at a cost of high computational complexity. The MPDA algorithm utilizes both distance and amplitude information as the PDA-AI algorithm however, it delves one-to-one track-to-measurement association unlike one-to-many association of the PDA-AI algorithm. The proposed MPDA evaluates the probabilities that the designated measurement is target originated for each measurement in the validation gate. All of the measurements at the present instance are lined up in the order of closeness to the predicted target measurement.

The measurement with the largest probability is selected to be target-originated and the measurement is used to update the target state estimate of a filter structure called the most probable data association filter (MPDAF) with the probability weight. Amplitude information is involved in the probability weight calculation similar to the highest probability data association (HPDA) algorithm [9]. But the HPDA utilizes order statistics based on signal intensity so that resulting algorithms are different. In this paper, first, the MPDA algorithm and the corresponding filter structure, MPDAF, are developed. The MPDAF is applied to single target tracking problems and its performance is compared with that of the PDAF-AI by using a series of Monte Carlo simulation runs.

The proposed MPDAF algorithm can be easily extended to multi-target tracking problems with slightly increased computational complexities and it can also be used to avoid track coalescence phenomenon that prevails when several close tracks move together as is often seen in target tracking with the joint probabilistic data association (JPDA) filter [10]. Finally, the MPDAF is applied to multi-target tracking in clutter and its performance is compared with that of the JPDAF-AI.

2 Most Probable Data Association Filter (MPDAF)

Most probable data association (MPDA) proposed in this paper utilizes a set of validated measurements at the present sampling instance. Similar to the HPDA algorithm [9], it considers both distance and signal amplitude information to evaluate the probabilities that the designated measurement is target-originated however, it utilizes distance to the predicted measurement for ordering the validated measurements whereas amplitude information is used for measurement ordering of the HPDA algorithm. When the distance information alone is used for probability calculation, the NN has the highest probability so that the NN measurement is always regarded to be target-originated. However, this is not the case when the amplitude information is involved in addition to the distance information. Order statistics is applied to evaluate the probabilities for selecting the most plausible target-originated measurement based on probabilistic natures of distance and amplitude distributions for a target and clutter.

The NN is defined as the measurement with the smallest normalized distance squared (NDS) [1, 3, 5] among a set of m validated measurements $Z_k = \{z_k^1, z_k^2, \cdots, z_k^m\}$ at the kth sampling instance. The measurements in the set Z_k are lined up with the NDS so that z_k^1 is the measurement of the NN and z_k^2 is the second nearest measurement and so on. Each z_k^l represents location information.

There are at most two events for each z_k^l, $l = 1, 2, \cdots, m$, except $m = 0$ case where there is no validated measurements. The two events are as follows : (1) z_k^l is target-originated (M_T^l) and (2) z_k^l is from a false target (M_F^l). When $m = 0$, we refer the event of M_0. The location information is n-dimensional and the residual representing the difference between location information and the center of validation gate which is the predicted target measurement, The validation gate used is the ellipsoid

$$G_\gamma(k) = \left\{\nu_k : \nu_k^T S_k^{-1} \nu_k \leq \gamma\right\} \tag{1}$$

where ν_k is a zero-mean Gaussian residual with covariance of S_k for the true measurement. The gate size of $G_\gamma(k)$ is defined as $\sqrt{\gamma}$.

The NDS D of a measurement z_k is defined as $D_k = \nu_k^T S_k^{-1} \nu_k$. The volume of the n-dimensional gate G_γ satisfies

$$V_G = C_n \left|S_k\right|^{1/2} \gamma^{n/2} \tag{2}$$

where $C_1 = 2$, $C_2 = \pi$, $C_3 = (4/3)\pi$, etc.

The following assumptions are used in the paper.

(A1) The probability that the target is detected and inside the validation gate is $P_D P_G$, where P_D is the probability of target detection indicating that the target signal amplitude exceeds a threshold τ; and P_G is the probability that the target falls inside the validation gate.

(A2) The true target signal amplitude is the magnitude-square output of a matched filter, so that the signal is χ^2-distributed with probability density function (pdf)

$$f_1(a) = \frac{1}{1+\rho} e^{-a/(1+\rho)} \tag{3}$$

where ρ is the expected signal-to-noise ratio. The clutter signal amplitude satisfies

$$f_0(a) = e^{-a}. \tag{4}$$

P_D in this case satisfies $P_D = e^{-\tau(1+\rho)}$, and the probability that the false measurement signal exceeds the threshold τ is $P_{fa} = e^{-\tau}$.

(A3) The number of validated false measurements in the validation gate, denoted by m^F, is Poisson distributed with a spatial density λ such that

$$\mu_F(m) = P\left(m^F = m\right) = \frac{(\lambda V_G)^m}{m!} e^{-\lambda V_G}. \tag{5}$$

(A4) The state prediction error $\overline{e}_k = x_k - \overline{x}_k$ for any given time k is a zero-mean Gaussian process with a covariance $\overline{P}_k$. Hence, $\overline{e}_k$ is said to satisfy $N\left(\overline{e}_k; 0, \overline{P}_k\right)$.

(A5) The validated false measurements at any time are independent and identically distributed (i.i.d.) and uniform over the gate.

(A6) The location of a validated false measurement is independent of the true measurement at any time and other validated false measurements at any other time.

(A7) The randomized discrete events M_T^l, M_F^l, and M_0 are considered to be the ones without any correlation with the previous events.

(A8) The target is existing and can be detectable, that is, it is perceivable [11].

2.1 Probability Density Functions (PDFs)

The NDS of the lth measurement with the location information z_k^l satisfies $D^l = \nu^{l^T} S^{-1} \nu^l$ where the subscript k indicating the current time step is omitted for brevity. The lth measurement has the amplitude information denoted here as a^l. The following probability functions derived in Theorems are needed to calculate the probability that each validated measurement is target-originated and to select most plausible measurement. Moreover, they are used to derive the information reduction factor of the proposed filter structure.

Theorem 1. *With assumptions* (A1)–(A8), *$f(D^l, a^l | M_T^l, m)$ is given by*

$$f(D^l, a^l | M_T^l, m) = \frac{1}{P(M_T^l, m)} f(D^l, a^l, M_T^l, m) 1(\gamma - D^l), \tag{6}$$

$$f(D^l, a^l, M_T^l, m) = P_D \frac{n V_D}{2D} N(D) f_1^\tau(a) \frac{(m-1)!}{(l-1)!(m-l)!} \times \left(\left(\frac{D}{\gamma}\right)^{n/2}\right)^{l-1} \cdot \left(1 - \left(\frac{D}{\gamma}\right)^{n/2}\right)^{m-l} \mu_F(m-1)|_{D=D^l, a=a^l} \tag{7}$$

where $f_1^\tau(a)$ *is the pdf of the target signal amplitude which exceeds the threshold* τ *such that* $f_1^\tau(a) = (1/P_D)f_1(a)$*,* $N(D)$ *is the Gaussian pdf of the target residual under* M_T^l*.*

Such that $N(D) = e^{-D/2}/\sqrt{|2\pi S|}$*, and* $1(x)$ *is the unit step function defined 1 as if* $x \geq 0$*, and 0 for elsewhere.*

$P(M_T^l, m)$ *of (6) satisfies*

$$P(M_T^l, m) = \int_0^\gamma \int_\tau^\infty f(D, a, M_T^l, m)dadD. \tag{8}$$

Proof. Omitted.

Note that $(D/\gamma)^{n/2}$ is the probability of the event that a validated false measurement has the NDS smaller than D while $(1 - (D/\gamma)^{n/2})$ is the probability of the complement event. The volume of the ellipse with the gate size $\sqrt{D}$ is denoted as $V_D = C_n |S|^{1/2} D^{n/2}$.

For M_F^l, the lth measurement is from a false target and it has the location information z_k^l and the signal amplitude information a^l. There are four events concerning the target location under M_F^l:

1. The target is located inside G_γ but not detected.
2. The target is located outside G_γ.
3. The target is located and detected inside G_γ but the NDS of the target is bigger than D^l.
4. The target is located and detected inside G_γ but the NDS of the target is smaller than D^l.

Based on these events, one can establish $f(D^l, a^l|M_F^l, m)$ by using order statistics.

Theorem 2. *With assumptions* (A1)–(A8),

$f(D^l, a^l|M_F^l, m)$ *is given by*

$$f(D^l, a^l|M_F^l, m) = \frac{1}{P(M_F^l, m)} f(D^l, a^l, M_F^l, m), \tag{9}$$

$$\begin{aligned} f(D^l, a^l, M_F^l, m) = f_D^\tau(a)[&(1 - P_D P_G) f_{C_l}(D|m)\mu_F(m) \\ &+ P_D(P_G - P_R(D)) f_{C_l}(D|m-1)\mu_F(m-1) \\ &+ P_D P_R(D) f_{C_{l-1}}(D|m-1)\mu_F(m-1)]|_{D=D^l, a=a^l} \end{aligned} \tag{10}$$

where $f_{C_l}(D|m)$ *is the conditional pdf of NDS D of the lth measurement under the assumptions that the lth measurement is from a false target and the number of validated false targets* m^F *is m.* $f_{C_l}(D|m)$ *is expressed as*

$$f_{C_l}(D|m) = \frac{m!}{(l-1)!(m-l)!}\left(\left(\frac{D}{\gamma}\right)^{n/2}\right)^{l-1}\frac{n}{2D}\left(\frac{D}{\gamma}\right)^{n/2}\left(1-\left(\frac{D}{\gamma}\right)^{n/2}\right)^{m-l}. \tag{11}$$

$P_R(D)$ in Eq. (10) *is the probability that the target exists in the ellipsoid with gate size $\sqrt{D}$ such that* [5]

$$P_R(D) = \frac{nC_n}{2^{n/2+1}\pi^{n/2}}\int_0^D q^{n/2-1}e^{-q/2}dq. \tag{12}$$

The joint posterior probability $P(M_F^l, m)$ is obtained by

$$P(M_F^l, m) = \int_0^{\gamma}\int_{\tau}^{\infty} f(D, a, M_F^l, m)dadD. \tag{13}$$

Proof. Omitted.

Note that $f_{C_l}(D|m-1)$ and $f_{C_{l-1}}(D|m-1)$ in Eq. (10) are obtained from $f_{C_l}(D|m)$ in Eq. (11) by replacing l and m with corresponding values.

It can be shown that

$$P(M_T^l, m) + P(M_F^l, m) = (1 - P_D P_G)\mu_F(m) + P_D P_G \mu_F(m-1) \tag{14}$$

The result is equivalent to the probability that the number of validated measurements is m.

The association probability that the lth measurement is from the true target is denoted as β^l and it can be derived from

$$\begin{aligned}\beta^l &= P(M_T^l|D^l, a^l, m)\\ &= \frac{f(D^l, a^l, M_T^l, m)}{f(D^l, a^l, M_T^l, m) + f(D^l, a^l, M_F^l, m)}.\end{aligned} \tag{15}$$

In the proposed algorithm, the measurement with the largest association probability, say β^S, is selected to be target-originated and the corresponding location information z_k^S is used for measurement update with the probability weight β^S that satisfies

$$\beta^S = \max\left\{\beta^l, l \in [1, m]\right\}. \tag{16}$$

In order to obtain the updated error covariance matrices conditioned on M_0 and M_F^s where s is the argument of the selected measurement among the m validated measurements, we need to have distributions of the NDS of the true target under M_0 and M_F^s. Under the event M_0, there is no measurement in G_γ and the pdf of the NDS D of the target-originated measurement satisfies [4, 5]

$$f(D|M_0) = \left(\frac{nV_D}{2D}\right) N(D) \frac{\{(1 - P_D 1(\gamma - D))\}}{1 - P_D P_G}. \tag{17}$$

Under the event M_F^s, the selected measurement with the NDS D is not target-originated. The pdf of D^t, the NDS of the target-originated measurement, conditioned on M_F^s and m is needed for error covariance update.

Theorem 3. *With assumptions* (A1)–(A8) *, the pdf of the D^t conditioned on M_F^s, m, and the NDS D of the selected sth measurement is given by*

$$\begin{aligned} f(D^t|D, M_F^s, m) &= \frac{f(D^t, D, M_F^s, m)}{f(D, M_F^s, m)} \\ &= \frac{\begin{aligned} &\frac{nV_{D^t}}{2D^t} N(D^t)[(1 - P_D 1(\gamma - D^t)) f_{C_l}(D, m) \\ &+ P_D(1(\gamma - D^t) - 1(D - D^t)) f_{C_l}(D, m-1) \\ &+ P_D 1(D - D^t) f_{C_{l-1}}(D, m-1)] \end{aligned}}{\begin{aligned} &(1 - P_D P_G) f_{C_l}(D, m) \\ &+ P_D(P_G - P_R(D)) f_{C_l}(D, m-1) \\ &+ P_D P_R(D) f_{C_{l-1}}(D, m-1) \end{aligned}} \end{aligned} \tag{18}$$

where the joint pdf $f_{C_l}(D, m)$ is expressed from Eqs. (10) and (11) *as $f_{C_l}(D|m)$ $\mu_F(m)$.*

Proof. Omitted

2.2 MPDAF

Under M_0 where there is no validated measurement in G_γ, the updated error covariance for the target estimate $\overline{x}_k$ is obtained from Eq. (17) as [4, 5]

$$\overline{P}_{k,M_0} = \overline{P}_k + \frac{P_D P_G (1 - C_{\tau g})}{1 - P_D P_G} K S K^T \tag{19}$$

where $\overline{P}_k$ is the predicted error covariance and K is the filter gain, and $C_{\tau g}$ satisfies

$$C_{\tau g} = \frac{\int_0^{\gamma} q^{n/2} e^{-q/2} dq}{n \int_0^{\gamma} q^{n/2-1} e^{-q/2} dq}. \tag{20}$$

Furthermore, the update error covariance under M_F^s can be obtained from Eq. (18) and the result is described in Theorem 4.

Theorem 4. *With assumptions* (A1)–(A8), *the updated error covariance for given* M_F^s *with the NDS D of the selected sth measurement and m is given by*

$$\overline{P}_k^{M_F^s}(D) = \overline{P}_k - KSK^T + \alpha KSK^T \tag{21}$$

where α *satisfies*

$$\alpha = \frac{\begin{array}{l}\lambda(1 - P_D P_G C_{\tau g}) V_D (V_G - V_D) \\ +P_D (P_G - P_R(D) C_\tau(D))(m - l) V_D \\ +P_D P_R(D) C_\tau(D)(l - 1)(V_G - V_D)\end{array}}{\begin{array}{l}\lambda(1 - P_D P_G) V_D (V_G - V_D) \\ +P_D (P_G - P_R(D))(m - l) V_D \\ +P_D P_R(D)(l - 1)(V_G - V_D)\end{array}}, \tag{22}$$

and $C_\tau(D)$ *is given by*

$$C_\tau(D) = \frac{\int_0^{D} q^{n/2} e^{-q/2} dq}{n \int_0^{D} q^{n/2-1} e^{-q/2} dq}. \tag{23}$$

Proof. Omitted

Note that the updated error covariance conditioned on M_T^s is equivalent to that of the standard Kalman filter (SKF) such as $\overline{P}_{k,M_T^s} = \overline{P}_k - KSK^T$. With $\overline{P}_{k,M_0}$, $\overline{P}_{k,M_F^s}$, and $\overline{P}_{k,M_T^s}$, we can establish, the updated error covariance for the MPDAF. The following is the summary of the MPDAF algorithm.

The MPDAF algorithm

- Predicted Step
  Identical to the SKF
- Update Step

  A) For the case of M_0
  $\hat{x}_k = \overline{x}_k$
  $\hat{P}_k = \overline{P}_{k,M_0}$

  B) For the case of $\overline{M_0}$
  $\hat{x}_k = \overline{x}_k + K\beta^s \nu^s$
  $\nu^s = z_k^s - \overline{z}_k$
  $\hat{P}_k = (1 - \beta^s)\overline{P}_{k,M_F^s} + \beta^s \overline{P}_{k,M_T^s} + \beta^s(1 - \beta^s) K \nu^s \nu^{sT} K^T$

For the case of $s = 1$ and neglecting the amplitude distributions, the MPDAF algorithm becomes identical to the PNNF-m algorithm of [5].

3 Simulation Results

In this section, performance of the MPDAF is compared with that of the PDAF-AI by a series of Monte Carlo simulation studies. Filter states for planar tracking problems are composed of target position, velocity, and acceleration. The continuous dynamic model is represented by

$$\dot{x} = Ax + Bw \tag{24}$$

where $x = (X, Y, \dot{X}, \dot{Y}, A_{T_X}, A_{T_Y})^T$,

$$A = \begin{bmatrix} 0_2 & I_2 & 0_2 \\ 0_2 & 0_2 & I_2 \\ 0_2 & 0_2 & -\dfrac{1}{\tau_S} I_2 \end{bmatrix}, \quad B = \begin{bmatrix} 0 \\ 0 \\ \dfrac{1}{\tau_S} I_2 \end{bmatrix} \tag{25}$$

and the process noise $w = (w_X, w_Y)^T$ is white Gaussian noise vector with zero-mean and power spectral density of $2\tau_S {\sigma_{A_T}}^2 I_2 (\mathrm{m}^2/\mathrm{s}^4/\mathrm{Hz})$. It is assumed that the Singer model represents target acceleration and the time constant $\tau_S = 15$ s, and $\sigma_{A_T} = 0.2(\mathrm{m}/\mathrm{s}^2)$. The target location information z_k is corrupted by a measurement noise vector such that

$$z_k = (I_2, 0_2, 0_2)x + v_k \tag{26}$$

where v_k is a zero-mean white Gaussian noise vector sequence with covariance of $400\,\mathrm{m}^2 I_2$. The sampling time interval for target tracking is 0.1 s. The initial position of the target is (7 km, 4 km) and the target moves with a speed of 380 m/s and the heading angle of 30° and then executes a 1.5 g maneuver at 5 s. Table 1 shows the track loss percentages obtained from 100 runs of Monte Carlo simulation. Simulation is done with various values of P_D, ρ, and λ. The track loss is declared if the estimation error in location estimation exceeds 10 times the standard deviation of the measurement noise [12]. Judging from Table 1, the MPDAF has similar performance to the PDAF-AI in general though the MPDAF shows even better performance in low P_D, low ρ, and high clutter density cases.

The MPDAF algorithm is easily extended to multi-target tracking environments. The proposed MPDA for multi-target tracking utilizes a track-to-measurement association matrix of which (i, j) element is β_i^j calculated from Eq. (15). The measurements in the matrix are the validated measurements inside the validation gates of

Table 1 Track loss percentages

ρ	P_D	λ	Track loss percentages	
			MPDA	PDA-AI
		0.0001	0	0
		0.00015	0	0
20	0.9	0.00025	0	0
		0.0003	0	0
		0.0001	1	0
		0.00015	0	0
10	0.7	0.00025	0	0
		0.0003	1	0
5	0.7	0.0001	0	0
		0.00015	1	2
		0.00025	1	4
		0.0003	1	2
		0.0001	0	0
		0.00015	1	0
15	0.7	0.00025	0	1
		0.0003	0	0

the tracks which share common measurements. From the matrix, one can establish possible combinations of one-to-one track-to measurement association. The score gain used in this paper is the same as the one in [9] such that

$$J(\theta_n) = \sum_{(i,j)\in\theta_n} \beta_i^j \tag{27}$$

where θ_n is an event in the combinations. The event with the maximum score gain is selected and the track and measurement pairs of the event with corresponding β_i^j's are selected for weight update for the tracks. After finding the pairs, one may reevaluate the association probabilities inside the validation gates, however, estimation performance is almost same as the cases without the reevaluation. For practical purpose, the reevaluation of association probabilities is not mandatory.

Table 2 is the summary of 100 runs of Monte Carlo simulation results of tracking 2 closely located targets. The numbers in the table indicate percentages. The results are obtained for various values of P_D and ρ. The clutter density of $\lambda = 1.5 \times 10^{-4}$ is used for simulation. The initial position of first target is (30.5 km, 28.5 km) and it has a speed of 380 m/s. with the initial heading angle of 40°. The initial position of the second target is (30 km, 29 km) and it has the same speed and initial heading angle as the first target. The first target executes a 0.3 g maneuver at 80 s but the second target does not maneuver. Target trajectories can be seen in Figs. 1 and 2. For most part of the trajectories, the targets are located closely with a separation distance of 350 m, the sampling interval is chosen to be 1 s and the measurement noise covariance is $200^2\,\mathrm{m}^2\, I_2$. Table 2 indicates that the JMPDAF, which is the MPDAF with the data association algorithm for the multi-target tracking environments, outperforms the JPDAF-AI especially in avoiding the track coalescence problem which is a major drawback of the filters employing the PDA-type data association algorithm.

Table 2 Performance comparison of the JMPDAF and the JPDAF-AI

		JMPDAF				JPDAF-AI			
ρ	P_D	Swap	No swap	Loss	Coalescence	Swap	No swap	Loss	Coalescence
20	0.9	6	88	6	0	7	78	10	5
	0.8	8	83	9	0	10	39	12	39
	0.7	9	79	12	0	10	28	15	47
15	0.85	9	82	9	0	11	65	8	16
	0.8	8	82	10	0	10	40	12	38
	0.7	10	78	12	0	13	27	11	49
10	0.8	9	80	11	0	12	35	10	43
	0.75	11	74	15	0	15	26	17	42
	0.7	15	66	16	0	13	22	19	46
5	0.7	20	54	26	0	19	16	28	37
	0.65	23	50	22	0	21	14	30	35

Moreover, the JMPDAF is shown to be a better solution for track-swap problems and it also has better performance in terms of less track loss in general.

Figures 1 and 2 are examples of the true and estimated trajectories generated by the JMPDAF and the JPDAF-AI respectively. For $\rho = 10$, $P_D = 0.7$, and $\lambda = 1.5 \times 10^{-4}$, the JMPDAF tracks the two targets with nice separation while the JPDAF-AI has the coalescence problem. It is shown that the JMPDAF has better performance especially for clutter environments with low P_D and low ρ.

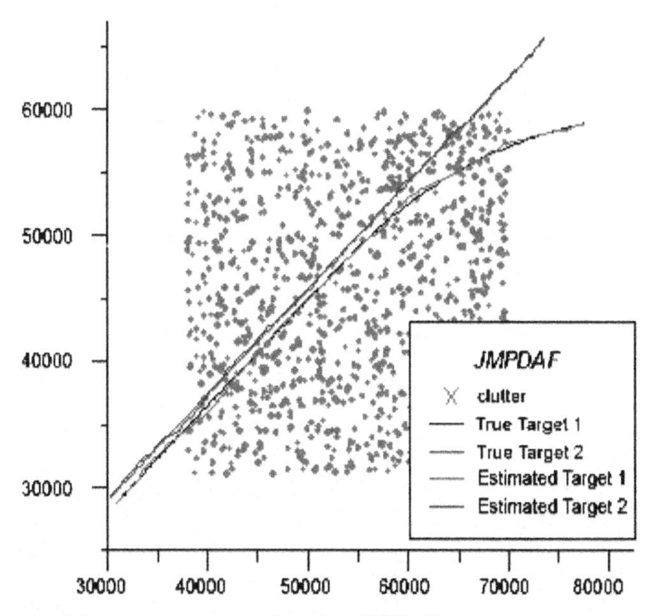

Fig. 1 The true and the estimated trajectories of the JMPDAF

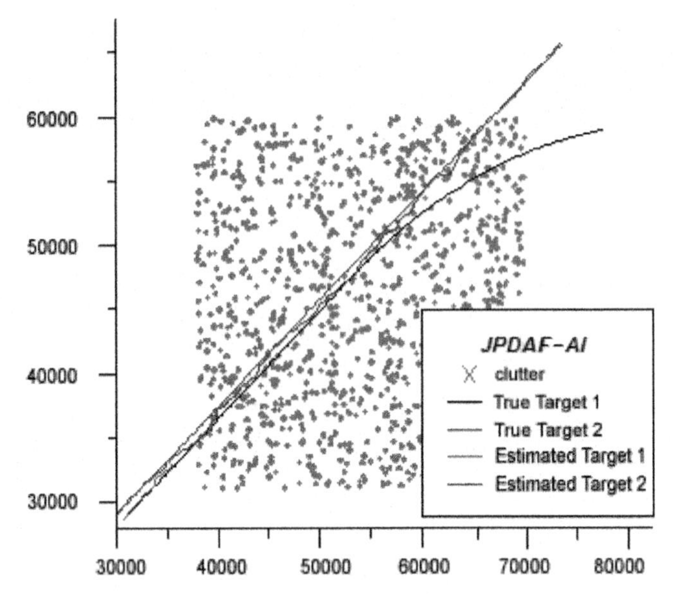

Fig. 2 The true and the estimated trajectories of the JPDAF-AI

4 Conclusion

In this paper a new filter called the most probable data association filter (MPDAF) based on one-to-one track-to-measurement data association with distance and amplitude information is proposed. The MPDAF selects the measurement with the largest association probability whose calculation formula is derived from order statistics established by ordering the validated measurements with distance information. A series of Monte Carlo simulation studies indicates that the MPDAF has similar performance to the PDAF-AI for single target tracking in clutter with less computational complexities.

The proposed data association algorithm is extended to multi-target tracking environments and the performance of the proposed filter is better than the JPDAF-AI especially in avoiding the coalescence and swap problems.

References

1. X. R. Li and Y. Bar-Shalom. (1996). Tracking in clutter with nearest neighbor filters: analysis and performance. *IEEE Transactions on Aerospace and Electronic Systems*, 32(3): 995–1010.
2. X. R. Li. (1998). Tracking in clutter with strongest neighbor measurements – Part I: theoretical analysis. *IEEE Transactions on Automatic Control*, 43: 1560–1578.
3. T. L. Song, D. G. Lee, and J. Ryu. (2005). A probabilistic nearest neighbor filter algorithm for tracking in a clutter environment. *Signal Process*, 95(10): 2044–2053.

4. C. R. Li and X. Zhi. (1996). PSNF: a refined strongest neighbor filter for tracking clutter. In Proceedings of the 35th CDC, Kobe Japan, December 1996, pp. 2557–2562.
5. T. L. song and D. G. Lee. (1996). A probabilistic nearest neighbor filter algorithm for m validated measurements. *IEEE Transactions on Signal Processing*, 54: 2797–2805.
6. T. L. song, Y. T. Lim, and D. G. Lee. (2009). A probabilistic strongest neighbor filter algorithm for m validated measurements. *IEEE Transactions on Aerospace and Electronic Systems*, 45(1).
7. Y. Bar-Shalom and T. E. Fortmann. (1988). *Tracking and data association*. New York: Academic.
8. D. Lerro and Y. Bar-Shalom. (1993). Interacting multiple model tracking with target amplitude feature. *IEEE Transactions on Aerospace and Electronic Systems*, 29: 494–509.
9. T. L. Song and D. S. Kim. (2006). Highest probability data association for active sonar tracking. In Proceedings of the 9th International Conference on Information Fusion, Florence, Italy.
10. S. S. Blackman and R. Popoli. (1999). *Design and analysis of modern tracking systems*. Norwood, MA: Artech House.
11. N. Li and X. R. Li. (2001). Target perceivability and its applications. *IEEE Transactions on Signal Process*, 49(11): 2588–2604.
12. P. Willett, R. Niu, Y. Bar-Shalom. (2001). Integration of Bayes detection with target tracking. *IEEE Transactions on Signal Process*, 49(1): 17–29.

Simultaneous Multi-Information Fusion and Parameter Estimation for Robust 3-D Indoor Positioning Systems

Hui Wang, Andrei Szabo, Joachim Bamberger and Uwe D. Hanebeck

Abstract Typical WLAN based indoor positioning systems take the received signal strength (RSS) as the major information source. Due to the complicated indoor environment, the RSS measurements are hard to model and too noisy to achieve a satisfactory 3-D accuracy in multi-floor scenarios. To enhance the performance of WLAN positioning systems, extra information sources could be integrated. In this paper, a Bayesian framework is applied to fuse multi-information sources and estimate the spatial and time varying parameters simultaneously and adaptively. An application of this framework, which fuses pressure measurements, a topological building map with RSS measurements, and simultaneously estimates the pressure sensor bias, is investigated. Our experiments indicate that the localization performance is more accurate and robust by using our approach.

1 Introduction

Indoor positioning systems recently attracted a lot of research efforts in both academia and industry for their broad applications such as security, asset tracking, robotics, and many others [1–3]. Many promising systems utilize the received signal strength (RSS) of wireless LAN (WLAN) to infer the location information. These systems have a big advantage in the installation and maintenance cost by using the existing communication infrastructure. Numerous research results also indicate that the WLAN positioning system suffers from the noisy characteristics of radio propagation [3]. For example, in large multi-floor buildings, the location error often has a large variance due to the complexity of indoor environment and insufficient number of reachable access points (APs) [4]. Although adding more APs could improve the performance, this solution implies higher installation costs. Another solution is to

H. Wang (✉)
Information and Communications, Corporate Technology, Siemens AG, Otto-Hahn-Ring 6, Munich, Germany
e-mail: hui.wang@siemens.com

S. Lee et al. (eds.), *Multisensor Fusion and Integration for Intelligent Systems*,
Lecture Notes in Electrical Engineering 35, DOI 10.1007/978-3-540-89859-7_9,

integrate other location related information besides the WLAN signal. For instance, in our previous work [4, 5], MEMS sensors were used to enhance the localization performance due to their small sizes and low prices.

To determine the user's location, the model that maps the sensed measurements (e.g., RSS, air pressure and etc.) to the location should be known. Most systems assume that this model has an accurate analytical form and it does not change with time. But in reality, the model parameters could be inaccurate and sometimes vary over time and space. For instance, the air pressure is determined not only by the altitude but also by some unknown environmental change, which could be modeled as a spatial and time varying bias. Besides, the radio distribution should also be accurately known and it could be temporally varying due to the changes of transmission power or the movement of scatters.

This paper is intended as an investigation of using the Bayesian filtering framework to fuse location-related information sources and simultaneously estimate their unknown parameters, that is, to solve a joint state and parameter estimation problem. Since the localization problem is usually nonlinear, the exact Bayesian filter is in general intractable. Different approximate estimators like the Extended Kalman Filter (EKF) [6], Unscented Kalman filter (UKF) [7], Particle Filter [8] or Hybrid Density Filter [9] could solve this problem. Additionally, dual estimation [10] or the expectation-maximization (EM) algorithm [11], which decouples the state and parameter estimation into two different problems in a suboptimal way, can also be used.

There already exists a class of self-estimation and calibration algorithms such as simultaneous floor identification and pressure compensation [4], simultaneous localization and learning (SLL) [3], and simultaneous localization and mapping (SLAM) [12]. These algorithms have a strong requirement with respect to initial conditions to ensure the convergence. Different from them, this paper aims to use different information sources to *teach* each other so that the system is ensured to be robust.

As an example application of the proposed framework, this paper simultaneously fuses RSS measurements, a discrete topological map as well as pressure measurements, and estimate the spatial and time varying pressure bias. The resulting posterior joint probabilities are proven to be in a Gaussian-mixture form. For simplicity, a dual estimation algorithm is applied, which takes a grid-based filter for location estimation and a Kalman filter for parameter estimation respectively. Our experiments in a typical multi-floor office building indicate that the location performance is more accurate and robust by using this approach even given an inaccurate initial condition.

The remainder of this paper is organized as follows: in Sect. 2, the general multi-information fusion and parameter estimation problem is formulated. In Sect. 3, the characteristics of the selected information sources are given. In Sect. 4, the specific form of the Bayesian framework is derived given the information sources in Sect. 3. Its solution by the dual estimation is described. Section 5 presents the experiment setup and discusses the result. Finally, conclusions and an outlook to future work are given in Sect. 6.

2 Problem Formulation

The indoor positioning system can be modeled as a nonlinear and non-Gaussian dynamic system. We use $\boldsymbol{x}_k = [x_k, y_k, z_k]^T \in \mathcal{L}$ to denote the 3-D location at time k. $\mathcal{L} \subset \mathcal{R}^3$ denotes the indoor location domain. $\boldsymbol{o}_k = \left[o_k^1, \ldots, o_k^N\right]^T \in \mathcal{O}$ represents the sensed measurements from sensor 1 to N at time k, including RSS measurements, pressure measurements and so on. $\mathcal{O} \subset \mathcal{R}^N$ denotes the oberservation domain. $\boldsymbol{\gamma}_k = \left[\boldsymbol{\gamma}_k^s, \boldsymbol{\gamma}_k^m\right]^T$ represents parameters in system model and measurement model respectively. $\boldsymbol{\gamma}_k$ belongs to the parameter domain $\mathcal{P} \subset \mathcal{R}^K$, where K is the parameter dimension. The whole system is described by the following system equation

$$\boldsymbol{x}_{k+1} = a_k\left(\boldsymbol{x}_k, \boldsymbol{\gamma}_k^s\right) + \boldsymbol{w}_k, \tag{1}$$

and measurement equation

$$\boldsymbol{o}_{k+1} = h_k\left(\boldsymbol{x}_{k+1}, \boldsymbol{\gamma}_k^m\right) + \boldsymbol{v}_{k+1}, \tag{2}$$

where $a_k(\cdot)$ is the system function, which updates the current state to the next state. $h_k(\cdot)$ is the measurement function, which relates the state to the measurements. $\boldsymbol{w}_k$ and $\boldsymbol{v}_k$ represent system and measurement noise. The system can also be illustrated by a graphical model in Fig. 1.

The Bayesian approach provides a recursive way to estimate the hidden state of dynamic systems with the above form. It has also two steps: prediction step

$$f_{k+1}^p(\boldsymbol{x}_{k+1}) = \int_{\mathcal{L}} f_k^T\left(\boldsymbol{x}_{k+1}|\boldsymbol{x}_k, \boldsymbol{\gamma}_k^s\right) f_k^e(\boldsymbol{x}_k)\, \mathrm{d}\boldsymbol{x}_k \tag{3}$$

and update step

$$f_{k+1}^e(\boldsymbol{x}_{k+1}) = \frac{1}{c_k} f_{k+1}^L\left(\boldsymbol{o}_{k+1}|\boldsymbol{x}_{k+1}, \boldsymbol{\gamma}_k^m\right) f_{k+1}^p(\boldsymbol{x}_{k+1}), \tag{4}$$

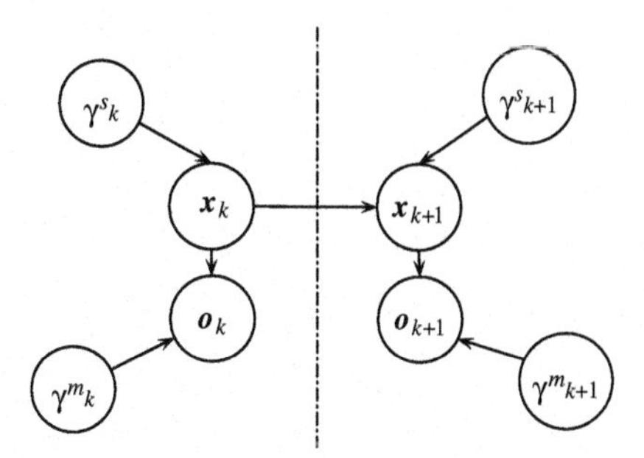

Fig. 1 A sensor fusion graphic model for localization problem

where $f_{k+1}^{P}(\boldsymbol{x}_{k+1})$ is the predicted density at time k. $f_{k+1}^{T}\left(\boldsymbol{x}_{k+1}|\boldsymbol{x}_k, \boldsymbol{\gamma}_k^s\right)$ is the transition density, which is given by

$$f_{k+1}^{T}(\boldsymbol{x}_{k+1}|\boldsymbol{x}_k, \boldsymbol{u}_k) = f_k^{w}\left(\boldsymbol{x}_{k+1} - a_k\left(\boldsymbol{x}_k, \boldsymbol{\gamma}_k^s\right)\right), \tag{5}$$

where $f_k^{w}(\cdot)$ is the density of the system noise at time k. $f_k^{e}(\boldsymbol{x}_k)$ is the posterior density function at time k. c_k is the normalization constant. $f_{k+1}^{L}\left(\boldsymbol{y}_{k+1}|\boldsymbol{x}_{k+1}, \boldsymbol{\gamma}_k^m\right)$ is the conditional likelihood density given by

$$f_{k+1}^{L}(\boldsymbol{y}_{k+1}|\boldsymbol{x}_{k+1}) = f_{k+1}^{v}\left(\boldsymbol{y}_{k+1} - h_{k+1}\left(\boldsymbol{x}_{k+1}, \boldsymbol{\gamma}_k^m\right)\right), \tag{6}$$

where $f_k^{v}(\cdot)$ is the density of the measurement noise at time k.

If the parameters are not accurately known or vary with time, they can be also regarded as states. So Eqs. (1) and (2) become

$$\begin{bmatrix} \boldsymbol{x}_{k+1} \\ \boldsymbol{\gamma}_{k+1} \end{bmatrix} = \begin{bmatrix} a_{x,k}\left(\boldsymbol{x}_k, \boldsymbol{\gamma}_k^s\right) + \boldsymbol{w}_{x,k} \\ a_{\gamma,k}(\boldsymbol{\gamma}_k) + \boldsymbol{w}_{\gamma,k} \end{bmatrix}, \tag{7}$$

and

$$\boldsymbol{o}_{k+1} = h_k\left(\boldsymbol{x}_{k+1}, \boldsymbol{\gamma}_k^m\right) + \boldsymbol{v}_{k+1}. \tag{8}$$

The joint state and parameter probabilities can also be derived by Bayesian framework as the following

$$f_{k+1}^{P}(\boldsymbol{x}_{k+1}, \boldsymbol{\gamma}_{k+1}) = \int_{\mathcal{P}}\int_{\mathcal{L}} f_k^{T}(\boldsymbol{x}_{k+1}, \boldsymbol{\gamma}_{k+1}|\boldsymbol{x}_k, \boldsymbol{\gamma}_k)\, f_k^{e}(\boldsymbol{x}_k, \boldsymbol{\gamma}_k)\,\mathrm{d}\boldsymbol{x}_k \mathrm{d}\boldsymbol{\gamma}_k, \tag{9}$$

and

$$f_{k+1}^{e}(\boldsymbol{x}_{k+1}, \boldsymbol{\gamma}_{k+1}) = \frac{1}{c_k} f_{k+1}^{L}(\boldsymbol{o}_{k+1}|\boldsymbol{x}_{k+1}, \boldsymbol{\gamma}_{k+1})\, f_{k+1}^{P}(\boldsymbol{x}_{k+1}, \boldsymbol{\gamma}_{k+1}) \tag{10}$$

Afterwards, the marginal posterior state and parameter densities are easily obtained by

$$f_k^{e}(\boldsymbol{x}_k) = \int_{\mathcal{P}} f_k^{e}(\boldsymbol{x}_k, \boldsymbol{\gamma}_k)\,\mathrm{d}\boldsymbol{\gamma}_k \tag{11}$$

and

$$f_k^{e}(\boldsymbol{\gamma}_k) = \int_{\mathcal{L}} f_k^{e}(\boldsymbol{x}_k, \boldsymbol{\gamma}_k)\,\mathrm{d}\boldsymbol{x}_k. \tag{12}$$

The above equations provide a general framework for simultaneous state and parameter estimation problem. Theoretically, many problems in the localization area can be solved by this framework, such as dynamical radio map estimation, sensor bias estimation, motion parameter estimation and so on. In the following, we use this framework to simultaneously fuse the RSS measurements, pressure measurements and a topological map, and estimate the spatial and time varying bias for pressure measurement. In this example application, one advantage is that the states are estimated by two different information sources but the unknown parameter is only related to one of them. So different from state and parameter estimation algorithms, the two information sources can teach each other to ensure the convergence and robustness of the final result.

3 Models of Information Sources

3.1 RSS of WLAN Signal

In WLAN based localization systems, RSS is most often used as the input of the positioning algorithm because it is much easier to obtain than the time or the angle information. The relation between the location and the RSS is modeled by the so-called radio map function $R := R(\boldsymbol{x})$, where $\boldsymbol{x} = [x, y, z]^T$ is the 3-D location vector. In theory, the radio map function follows the radio propagation rule. But in reality, due to the complexity of indoor environment, the radio map function has a very complicated form. As illustrated in Fig. 2, a real measured radio map in a office building is hard to be described in an analytic way. So in practice, the radio map function is usually modeled in a non-parametric way by a number of selected grid points, that is, $\boldsymbol{x} \in \left[x^i, y^i, z^i\right]^T$, where $i = 1, 2, \ldots, M$ is the index of grid points. In addition, since the wireless channel is influenced by many factors, for example, measurement noise, changing environment, and moving people, the

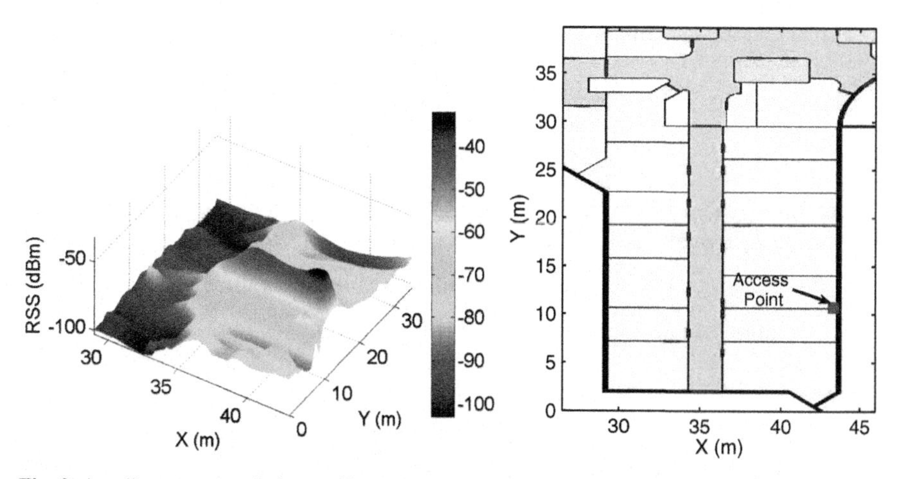

Fig. 2 A radio map example in an office building

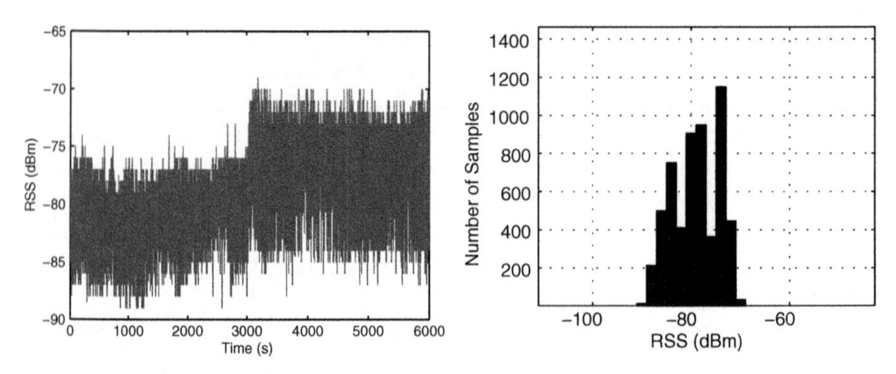

Fig. 3 Temporal variation and histogram of RSS measurements in 6,000 s

measured signal power fluctuates with time as shown in Fig. 3, that is, $r = R(\boldsymbol{x}) + \triangle(\boldsymbol{x}, t) + w$, where $\triangle(\boldsymbol{x}, t)$ could be regarded as an unknown spatial and time varying parameter indicating the inaccuracy of radio map model and its temporal variation; w is the measurement noise, usually regarded as a Gaussian as Fig. 3. In this paper, for simplicity, we assume a non-parametric radio map model that is accurately known and time invariant. Then the measurement equation for the RSS measurement from AP n becomes

$$r^n = R^n(\boldsymbol{x}) + w^n, \boldsymbol{x} \in \left[x^i, y^i, z^i\right]^T, i = 1, 2, \ldots, M. \tag{13}$$

3.2 Pressure Measurement from MEMS Barometric Sensor

As it is well known, the atmospheric pressure is a physical property strongly related to the altitude. Assuming a constant temperature gradient of $\mathrm{d}T/\mathrm{d}z$, the altitude z can be expressed as a function of pressure p using the following standard equation,

$$z = T_0/(-\mathrm{d}T/\mathrm{d}z) \cdot \left[1 - \left(\frac{p}{p_0}\right)^{-\mathrm{d}T/\mathrm{d}z \cdot R/g}\right], \tag{14}$$

where $T_0 = 288.15$ K and $p_0 = 101,325$ Pa are the reference temperature and pressure, respectively. R is in standard conditions equal to 287.052 $\mathrm{m}^2/\mathrm{s}^2/\mathrm{K}$; g is equal to 9.82 m/s^2. If the change in pressure $\frac{\mathrm{d}p}{p_0}$ and temperature $\frac{\mathrm{d}T}{T}$ are small, the above equation can be approximated as

$$\mathrm{d}z = -\frac{\mathrm{d}p \cdot R \cdot T}{g \cdot p_0} \tag{15}$$

which can be further modified to a linear function between p and z as

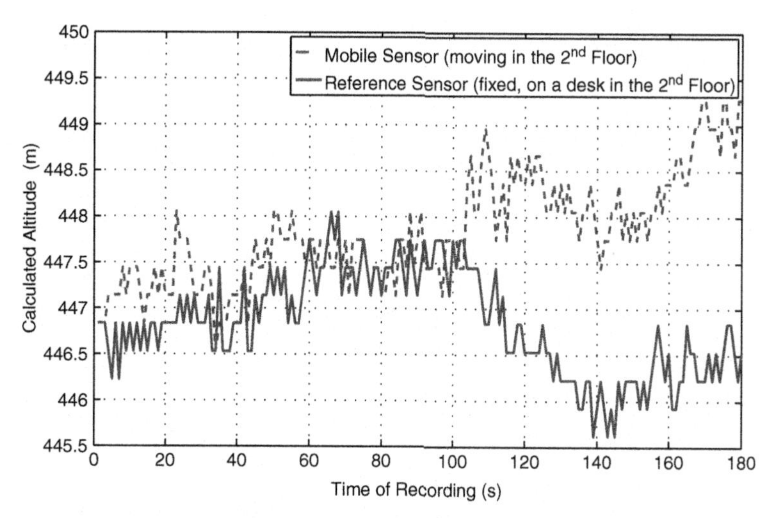

Fig. 4 Estimated altitudes by a static and a mobile barometric sensors

$$p = \underbrace{\left(-\frac{p_o \cdot g}{T \cdot R}\right)}_{=\alpha} \cdot z + \underbrace{p' + \frac{p_o \cdot g \cdot z'}{T \cdot R}}_{=\beta}, \tag{16}$$

where p' and z' are the reference altitude and pressure. Figure 4 shows the estimated altitude using pressure measurements by a static barometric sensor and a mobile barometric sensor, which moves in the same floor. We notice that the pressure is varying both temporally and spatially. Fortunately, when we consider the tracking problem, the pressure variation caused by horizontal moving and temporal environmental change is relatively small and slow. So we can simply assume that the pressure is only related to the height and a bias $\beta(t)$, which is slowly varying, that is,

$$p(t) = \alpha \cdot z + \beta(t). \tag{17}$$

3.3 Building Map

A building map is another very important information source. The positions of obstacles, such as walls or doors, determine the possible routes where people can move. Mathematically, the map influences the prior joint probability of $f(x, y, z)$ and the transition density $f^T(\boldsymbol{x}_{k+1}|\boldsymbol{x}_k)$.

In this paper, we represent the building map by a topological graph consisting of a number of location points as shown in Fig. 5. These points are connected by their spatial relation, that is, points which can *see* each other are connected. In this way, the indoor environmental restriction such as walls, doors or floors can be easily

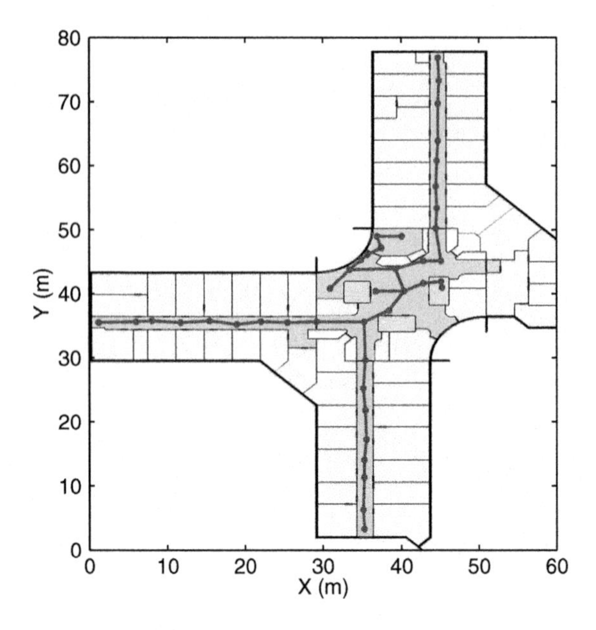

Fig. 5 Example of a topological map

integrated. In addition, the topological graph can reflect the restriction to the location, for example, the height z is limited to discrete floor heights except in some special places such as stairs or elevators. With the topological graph, the location transition density can be represented by a Gaussian distribution

$$f^T\left(\boldsymbol{x}_{k+1}^i|\boldsymbol{x}_k^j\right) = \mathcal{N}\left(d_s\left(i,j\right) - \bar{v}\cdot\Delta t, (\sigma_v\cdot\Delta t)^2\right), \tag{18}$$

where $d_s(i, j)$ is the shortest distance between $\boldsymbol{x}^i$ and $\boldsymbol{x}^j$, which can be calculated offline by Floyd's algorithm [13]. $\bar{v}$ is the mean of moving speed and σ_v is the standard deviation of moving speed.

4 Bayesian Filtering for Simultaneous Localization and Bias Estimation

In this section, the Bayesian framework are used to fuse all the information sources in the last section and simultaneously estimate the parameter $\beta(t)$. This is illustrated by a graph model in Fig. 6. The system is described by the following system function

$$\begin{bmatrix} \boldsymbol{x}_{k+1} \\ \beta_{k+1} \end{bmatrix} = \begin{bmatrix} a\left(\boldsymbol{x}_k, m\right) + \boldsymbol{w}_{\boldsymbol{x},k} \\ \beta_k + w_{\beta,k} \end{bmatrix} \tag{19}$$

and measurement function

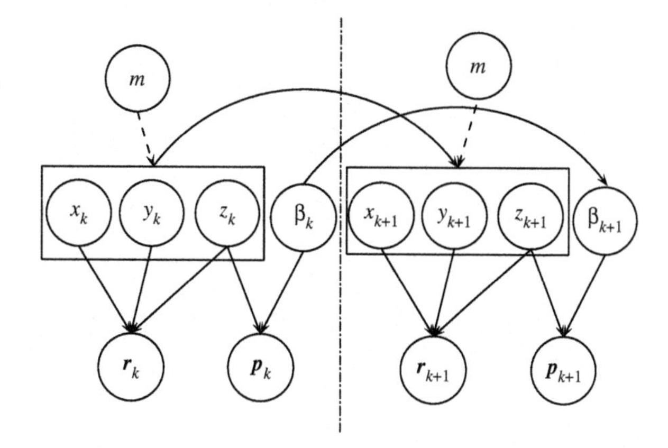

Fig. 6 A graphic model for simultaneous multi-information fusion and pressure bias estimation

$$\begin{bmatrix} r_{k+1}^n \\ p_{k+1} \end{bmatrix} = \begin{bmatrix} R^n(\boldsymbol{x}_{k+1}) + v_{r,k+1}^n \\ \alpha \cdot z_{k+1} + \beta_{k+1} + v_{p,k+1} \end{bmatrix}, \tag{20}$$

where m represents the information from the building map; $n = 1, \ldots, N$ is the index of AP; w_β, v_r and v_p are the noise terms for parameter prediction, RSS and pressure measurements, following Gaussian distribution $\mathcal{N}(0, \sigma_\beta)$, $\mathcal{N}(0, \sigma_r)$ and $\mathcal{N}(0, \sigma_p)$ respectively. The corresponding posterior joint density function is hybrid, which includes both a discrete state $\boldsymbol{x}_k^i$ and a continuous state β_k. This can be derived as

$$f_{k+1}^p\left(\boldsymbol{x}_{k+1}^i, \beta_{k+1}\right) = \sum_j \int_{\mathcal{R}} f_k^T\left(\boldsymbol{x}_{k+1}^i, \beta_{k+1} | \boldsymbol{x}_k^j, \beta_k, m\right) f_k^e\left(\boldsymbol{x}_k^j, \beta_k\right) \mathrm{d}\beta_k \tag{21}$$

and

$$f_{k+1}^e\left(\boldsymbol{x}_{k+1}^i, \beta_{k+1}\right) = \frac{1}{c_k} f_{k+1}^L\left(\boldsymbol{r}_{k+1}, b_{k+1} | \boldsymbol{x}_{k+1}^i, \beta_{k+1}\right) f_{k+1}^p\left(\boldsymbol{x}_{k+1}^i, \beta_{k+1}\right). \tag{22}$$

Combining (18), (19) and (20),

$$\begin{aligned} f_{k+1}^T\left(\boldsymbol{x}_{k+1}^i, \beta_{k+1} | \boldsymbol{x}_k^j, \beta_k, m\right) &= f_{\boldsymbol{x}^i,k}^T\left(\boldsymbol{x}_{k+1}^i | \boldsymbol{x}_k^j, m\right) f_{\beta,k}^T\left(\beta_{k+1} | \beta_k\right) \\ &= \mathcal{N}\left(d_s(i,j) - \bar{v}, \sigma_v^2\right) \cdot \mathcal{N}\left(\beta_{k+1} - \beta_k, \sigma_\beta^2\right), \end{aligned}$$

and

$$\begin{aligned} & f_{k+1}^L\left(\boldsymbol{r}_{k+1}, p_{k+1} | \boldsymbol{x}_{k+1}^i, \beta_{k+1}\right) \\ &= f_{\boldsymbol{r},k+1}^L\left(\boldsymbol{r}_{k+1} | \boldsymbol{x}_{k+1}^i\right) f_{p,k+1}^L\left(p_{k+1} | z_{k+1}^i, \beta_{k+1}\right) \\ &= \prod_{n=1}^N \mathcal{N}\left(r^n - R^n\left(\boldsymbol{x}_{k+1}^i\right), \sigma_r^2\right) \cdot \mathcal{N}\left(p_{k+1} - \alpha \cdot z_{k+1}^i - \beta_{k+1}, \sigma_p^2\right). \end{aligned} \tag{23}$$

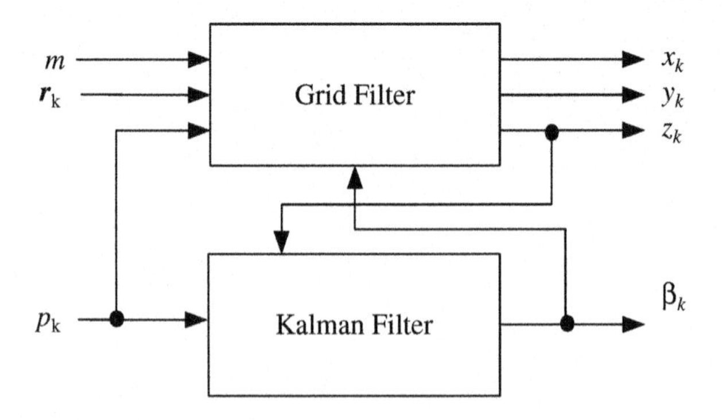

Fig. 7 Illustration for dual estimation

Given the uniformly or Gaussian distributed initial joint density $f_{k+1}\left(\boldsymbol{x}_0^i, \beta_0\right)$, the hybrid posterior density $f_k^e\left(\boldsymbol{x}_k^i, \beta_k\right)$ and $f_k^e\left(\beta_k\right)$ both have Gaussian mixture form. For each recursive step, the number of mixture components increases so that the optimal analytical solution is not tractable. The algorithm proposed in [14] can optimally reduce the number of Gaussian components and get the suboptimal solution. Besides, particle filters are also used by some researchers to solve the similar problem. One problem using particle filters is that when the parameter is part of state, the augmented state space model is not ergodic, and the uniform convergence result does not hold anymore [15].

In this paper, we use another suboptimal algorithm, so-called dual estimation [10, 16]. The idea of dual estimation is to separate joint state and parameter estimation into two independent processes. As illustrated in Fig. 7, the grid based filter is used to estimate the discrete location vector $\boldsymbol{x}_k$ assuming the parameter β_k is known. The expectation of posterior $\boldsymbol{x}_k$ is sent back to the parameter estimator, which is a Kalman filter in our case. Afterwards, the estimated β_k is sent again to the state estimator. The dual estimation at time k stops either after a given number of iterations or if the new state estimation is close enough to the old one. Note that the dual estimation can be regarded as a generalized EM algorithm. Its convergence to the suboptimal solution is guaranteed by its iterative optimization process [16].

Since only the posterior expectations of height and sensor bias instead of the whole probability are exchanged between two filters, the dual estimation might bring a large error if the posterior density function has a complicated form. But in our case in most of the time the height z is limited to the discrete floor height, the posterior of z tends to be unimodal. The pressure bias has also a Gaussian-like shape. That assures the suboptimal dual estimation works well for our application.

5 Experiment Results

5.1 Experiment Setup

We evaluate our algorithm in a typical multi-floor office building depicted in Fig. 5. Each floor has a area of 50 m × 80 m and has a similar structure like the left figure

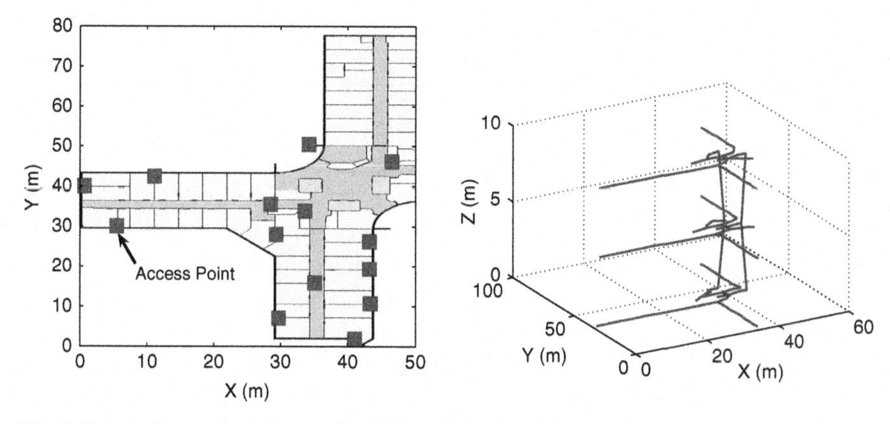

Fig. 8 Test environment and its topological graph for three floors

in Fig. 8. There are 14 access points installed on the first floor. We take three floors for evaluation. Since APs only exist on the first floor, the 3-D localization error could be large when users stay in the higher floors. A topological graph is built automatically based on the chosen reference points. Here we only take the points along the corridor, stairs, and elevator like the graph in Fig. 6. These points represent the basic moving possibilities. The rooms can also be easily added to the graph if some points in rooms are taken. The whole topological graph for three floors is shown in Fig. 8. The RSS values at reference points are measured offline and their noise parameters are also estimated according to the measurements. In online step, we first walked in the corridor of the first floor and then went up to the second floor by stairs, walked around in the second floor and finally went up to the third floor by elevator. While moving, the RSS and pressure measurements by the barometric sensor were recorded simultaneously.

5.2 Results

The parameters taken in our test are list in Table 1. Table 2 shows the comparison of the nearest neighbour (NN) algorithm that only uses RSS measurements, the fusion algorithm (Fusion) that integrates the RSS, air pressure, and topological map using grid filter and an assumed sensor bias, and the algorithm that simultaneously estimates the location and parameter. The criteria for the comparison are the mean

Table 1 The values of parameters

Parameter	Value
σ_r	4 dB
$\bar{v}$	0 m/s
σ_v	2 m/s
σ_p	6 Pa
σ_β	3 Pa

Table 2 3-D localization results

	NN	Fusion	Fusion and bias estimation
Mean of 3-D error (m)	8.2	6.5	6.1
Standard deviation of 3-D error (m)	7.8	6.6	6.3
Mean of altitude error (m)	1.2	0.2	0

and standard deviation of 3-D localization errors as well as the mean altitude error. Figure 9 plots the altitude error by different algorithms. Figure 10 compares the true bias with estimated bias by simultaneous localization and bias estimation algorithm.

From Table 2 it can be seen that both the 2-D and the 3-D localization performances are improved by fusing more information sources. By simultaneously adapting the bias, the altitude error can be reduced to zero, that is, perfect floor identification.

5.3 Sensitivity Analysis

The algorithm in our previous paper [4] can also provide very good performance for floor identification. But since it is actually a self-calibration algorithm, the initial condition is very important. Given the wrong initial condition, the result can be totally wrong. The algorithm in this paper uses the RSS measurement to teach the bias estimation and hence make sure that the bias can be tracked without any limitation for initial value. To validate this, we add an artificial pressure bias −60 Pa to the real pressure measurement so that the wrong floor (10 m higher) is identified if only depending on the pressure. Besides, we start with moving in the third floor so that the intial RSS is also very unaccurate. The altitude errors are given in Fig. 11.

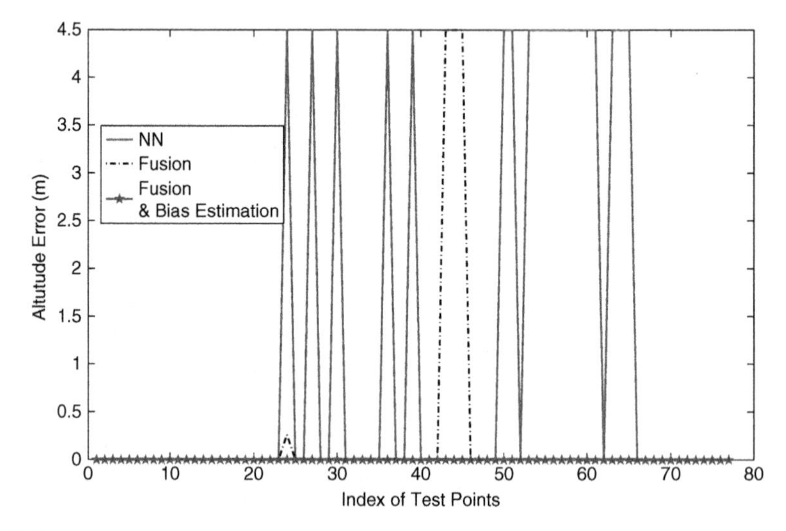

Fig. 9 Altitude error by different algorithms

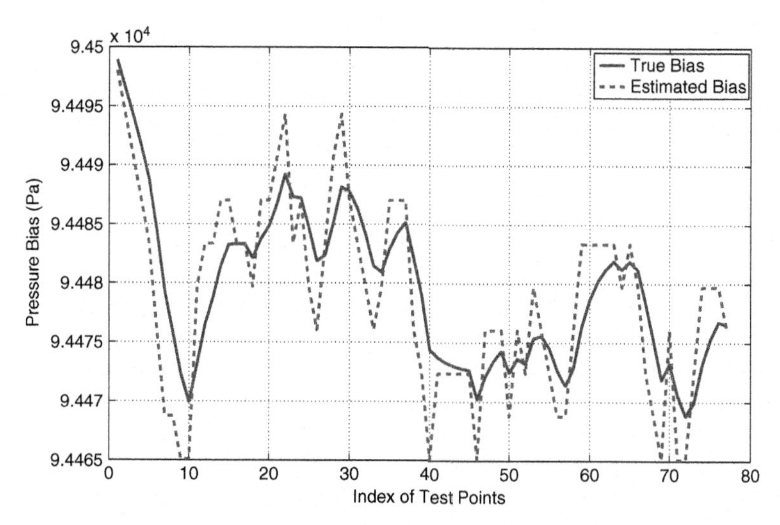

Fig. 10 Comparison between true bias and estimated bias by fusion and bias estimation. The true bias is obtained by filtering the difference between the measured pressures and calculated pressures

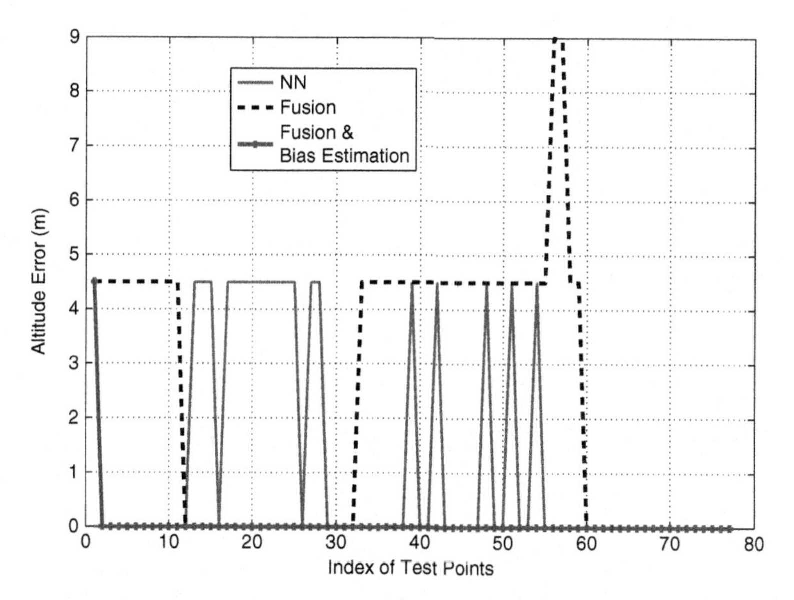

Fig. 11 Altitude error by different algorithms with an artificial bias

We see that compare the result in Fig. 9, the error by fusion is obviously larger. That is because adding the artificial bias, the wrong floor is identified by pressure. If the same wrong floor also happened to be estimated by RSS measurements, the altitude error become bigger. But by simultaneously estimating the bias, the sensor bias can be quickly calibrated and then the altitude error becomes zero.

Additionally, we notice that the height z is the variable that influences both RSS and pressure measurements. Through the inference of RSS and the correlation between (x, y) and z, the estimated z can be used to train the β to ensure the convergence.

6 Conclusions and Future Works

Bayesian framework can be used to solve joint state and parameter estimation problems in indoor positioning systems. In this paper, we apply this framework to a specific problem: simultaneous localization and sensor bias estimation. By fusing pressure measurements, a topological graph with RSS measurements and simultaneously estimating the pressure bias, the WLAN indoor positioning system becomes more robust and more accurate. In the next step, we will investigate the feasibility of using Bayesian framework for more complicated parameters, such as the parameter for the radio map generation and other state and parameter estimation problems in indoor positioning systems. Other more complicated filtering techniques like Gaussian mixture filter or particle filter will also be considered to solve the joint density estimation problems with proper forms.

References

1. P. Bahland and V. N. Padmanabhan. RADAR: An in-building RF-based user location and tracking system. In: *Proceedings of IEEE INFOCOM 2000*, pp. 775–784, 2000.
2. T. Roos, P. Myllymaki, H. Tirri, P. Misikangas, and J. Sievanen. A probabilistic approach to WLAN user location estimation. *International Journal of Wireless Information Networks*, 9(3), 155–164, 2002.
3. H. Lenz, B. B. Parodi, H. Wang, A. Szabo, J. Bamberger, J. Horn, and U. D. Hanebeck. Adaptive localization in adaptive networks. In: *Chapter of Signal Processing Techniques for Knowledge Extraction and Information Fusion*, Springer, 2008.
4. H. Wang, H. Lenz, A. Szabo, U. D. Hanebeck, and J. Bamberger. Fusion of barometric sensors, WLAN signals and building information for 3-D indoor campus localization. In: *Proceedings of International Conference on Multisensor Fusion and Integration for Intelligent Systems* (MFI 2006), pp. 426-432, Heidelberg, Germany, 2006.
5. H. Wang, H. Lenz, A. Szabo, J. Bamberger, and U. D. Hanebeck: WLAN-Based pedestrian tracking using particle filters and low-cost MEMS sensors. In: *Proceedings of 4th Workshop on Positioning, Navigation and Communication 2007* (WPNC'07), Hannover, Germany, 2007.
6. H. W. Sorenson. *Kalman Filtering: Theory and Application*. Piscataway, NJ: IEEE, 1985.
7. S. J. Julier and J. K. Uhlmann. Unscented filtering and nonlinear estimation. *Proceedings of the IEEE*, 92(3), 2004.
8. B. Ristic, S. Arulamplalm, and N. Gordon. *Beyond the Kalman Filter*. Boston: Artech House, 2004.
9. M. F. Huber and U. D. Hanebeck. The hybrid density filter for nonlinear estimation based on hybrid conditional density approximation. In: *Proceeding of the 10th International Conference on Information Fusion* (FUSION), 2007.
10. E. Wan and A. Nelson. Dual extended Kalman filter methods. In: *Kalman Filtering and Neural Networks* (Chap. 5), S. Haykin (ed.). New York: Wiley, 2001.

11. A. Dempster, N. Laird, and D. Rubin. Maximum likelihood from incomplete data via the EM algorithm. *Journal of the Royal Statistical Society, Series B*, 39(1), 1–38, 1977.
12. H. Durrant-Whyte and T. Bailey. Simultaneous localisation and mapping (SLAM): Part I the essential algorithms. *Robotics and Automation Magazine*, 13, 99–110, 2006.
13. R. W. Floyd. Algorithm 97: Shortest Path. *Communications of the ACM*, 5(6), 345, 1962.
14. U. D. Hanebeck and O. Feiermann. Progressive Bayesian estimation for nonlinear discrete-time systems: the filter step for scalar measurements and multidimensional states. In: *Proceedings of the 2003 IEEE Conference on Decision and Control* (CDC 2003), pp. 5366–5371, Maui, Hawaii, December, 2003.
15. D. Crisan, J. Gaines, and T. Lyons. Convergence of a branching particle method to the solution of the Zakai equation. *SIAM Journal on Applied Mathematics*, 58(5), 1568–1598, 1998.
16. Z. Chen. Bayesian filtering: From Kalman filters to particle filters, and beyond. In: *Technical Report*, McMaster University, 2006.

Efficient Multi-Target Tracking with Sub-Event IMM-JPDA and One-Point Prime Initialization

Seokwon Yeom

Abstract This paper addresses an IMM (interacting multiple model)-JPDA (joint probabilistic data association) tracker with sub-event decomposition and one-point prime initialization. In the original JPDA, the number of joint feasible association events increases exponentially along with the target number, which may cause a huge computational burden. The proposed sub-event JPDA can significantly reduce the number of hypotheses while maintaining the same tracking performance. One-point prime initialization method estimates initial velocities using initial range rates and azimuths. A scenario of ten multiple targets is tested and performance is evaluated in terms of the root mean-squared errors of position and velocity and the averaged mode probability. It will be shown that the proposed technique can significantly improved the time efficiency of target tracking.

Keywords Target tracking · State estimation · Data association · Initialization

1 Introduction

There are many challenges to multi-target tracking such as high-maneuvering, false alarms, low detection probability, and closely located target formations. In literature, various researches on state estimation, data association, and target initialization have been performed to overcome those challenges [1–5].

One-point initialization with Doppler information, which we may call "one-point prime initialization," has been proposed in [2]. This initialization technique estimates the initial velocity using a measured range rate and an azimuth by means of the linear minimum mean squared error (LMMSE) estimator. The one-point prime

S. Yeom (✉)
Department of Computer and Communication Engineering, Daegu University, Gyeongsan Gyeongbuk, 712-714 South Korea
e-mail: yeom@daegu.ac.kr

S. Lee et al. (eds.), *Multisensor Fusion and Integration for Intelligent Systems*,
Lecture Notes in Electrical Engineering 35, DOI 10.1007/978-3-540-89859-7_10,

initialization has proven to be superior to the one-point initialization method, especially in clutter environments [2].

The purpose of a state estimator is to properly overcome the uncertainty of the target state caused by sensor noise and unknown maneuvering. The interacting multiple model (IMM) approach is a popular state estimation method to deal with high maneuvering targets [1]. The fine-step IMM estimator has been proposed by Yeom et al. to track a large number of high-maneuvering airborne military targets with a long sampling interval and a low detection probability in heavy clutter environments [2].

Data association is required to maintain multiple tracks given consecutive measurement sets. The probabilistic data association (PDA) method has been proposed to associate a measurement set at the current frame to a single track [3]. It has been extended to the joint probabilistic data association (JPDA) technique to process multiple targets simultaneously [3]. The ND (N-dimension) assignment algorithm has been developed to overcome the N-P (Non-polynomial) hard problem of the Bayesian approach [4].

The JPDA evaluates the association probability between the latest set of measurements and the established tracks to update the current target state. However, the number of the feasible joint association events increases exponentially along with the number of targets, which may cause huge computational load to hinder a real-time processing. Although the gating process may reduce the number of the feasible joint association events by excluding measurements falling outside the validation region, the computational burden may be still high because the feasible joint association event considers all the measurements and the tracks the same time.

In this paper, the IMM–JPDA tracker with the sub-event decomposition is proposed and implemented with the one-point prime initialization. In this scheme, the measurements in each validation region are considered separately and the feasible joint association event is decomposed into multiple sub-events. The scenario of ten multiple targets is tested during 100 Monte-Carlo (MC) runs. The performance of the proposed system is evaluated by the computational time, the root mean squared error (RMSE)'s of position and velocity, and the averaged mode probability. The simulation result shows significantly improved computational efficiency of the proposed method.

The remaining of the paper is organized as follows. In Sect. 2, the procedures of the sub-event IMM-JPDA tracker with the one-point prime initialization are illustrated in details step by step. Also, the performance evaluation metrics are described. A test scenario and the parameter design are described in Sect. 3. The simulation results are presented in Sect. 4. Conclusions follow in Sect. 5.

2 Multi-Target Tracker

The sub-event IMM–JPDA with one-point prime initialization is illustrated in this section. The procedures of the proposed method will be explained with theoretical analysis.

2.1 Dynamic System Modeling

A dynamic state of a target can be modeled with a nearly constant velocity (NCV) model. Targets' maneuvering is modeled by the uncertainty of the process noise, which is assumed to be white Gaussian. The following is the discrete state equation of a target in two-dimensional (2D) Cartesian coordinates:

$$\mathbf{x}(k-1) = F(T)\mathbf{x}(k) - q(T)\boldsymbol{\nu}(k), \tag{1}$$

$$F(T) = \begin{bmatrix} 1 & T & 0 & 0 \\ 0 & 1 & 0 & 0 \\ 0 & 0 & 1 & T \\ 0 & 0 & 0 & 1 \end{bmatrix}, \qquad q(T) = \begin{bmatrix} T^2/2 & 0 \\ T & 0 \\ 0 & T^2/2 \\ 0 & T \end{bmatrix}, \tag{2}$$

where T is the sampling time, $\mathbf{x}(k)$ is a state vector of a target, and $\mathbf{v}(k)$ is a process noise vector which follows a multivariate Gaussian distribution, $N(\mathbf{0}; Q)$, where N denotes the Gaussian distribution, and Q is the covariance matrix of the process noise vector which is $diag\left(\sigma_x^2, \sigma_y^2\right)$, where $diag\,(.)$ indicates a diagonal matrix. A measurement vector is composed of three components: range (r_z), azimuth (θ_z), and range rate $(\dot{r}_z)$. The measurement equation is described as

$$\mathbf{z}(k) = [r_z\ \theta_z\ \dot{r}_z]' = \mathbf{h}[\mathbf{x}(k)] + \mathbf{w}(k), \tag{3}$$

where $\mathbf{h}[\mathbf{x}(k)]$ is a non-linear conversion from 2D Cartesian coordinates to polar coordinates and $\mathbf{w}(k)$ is a measurement noise vector which is assumed to follow a multivariate Gaussian distribution, $N(\mathbf{0}; R)$, where R is the covariance matrix of the measurement noise vector which is $diag(\sigma_r^2, \sigma_\theta^2, \sigma_{\dot{r}}^2)$.

2.2 One-Point Prime Initialization

The conventional one-point initialization assumes zero initial velocity with the standard deviation proportional to a maximum target speed, but one-point prime initialization estimates the initial velocity with Doppler information [2]. The initial velocity vector and covariance matrix are estimated by the LMMSE estimator using the initial range rate and the initial azimuth. The measurement equation of the range rate at the initial frame $(k = 0)$ is, approximately,

$$\dot{r}_z(0) = [\cos(\theta_z)\ \sin(\theta_z)]'(\dot{\mathbf{x}}(0) - \dot{\mathbf{x}}_s(0)) - w_{\dot{r}}(0), \tag{4}$$

where $\dot{\mathbf{x}}(0)$ and $\dot{\mathbf{x}}_s(0)$ are the velocity vectors of the target and the sensor, respectively. The initial velocity vector $\dot{\mathbf{x}}(0)$ is assumed have a covariance matrix $\bar{P}_v$,

$$\bar{P}_v = \begin{bmatrix} 0 & \sigma_s^2 \\ \sigma_s^2 & 0 \end{bmatrix} \tag{5}$$

Therefore, the target's initial velocity can be estimated as follows [2]

$$\hat{\mathbf{x}}(0|0) = [\hat{\dot{x}}(0|0) \quad \hat{\dot{y}}(0|0)]' = \frac{\sigma_s^2}{\sigma_s^2 + \sigma_{\dot{r}}^2} \begin{bmatrix} \cos(\theta_z(0)) \\ \sin(\theta_z(0)) \end{bmatrix} \dot{r}_z(0) + \begin{bmatrix} \dot{x}_s(0) \\ \dot{y}_s(0) \end{bmatrix}, \tag{6}$$

and the covariance matrix of the initial velocity vector becomes

$$P_v = \sigma_s^2 \begin{pmatrix} \left(1 - \frac{\sigma_s^2}{\sigma_s^2 + \sigma_r^2}\right) \cos(\theta_z(0)) & -\frac{\sigma_s^2}{\sigma_s^2 + \sigma_r^2} \cos(\theta_z(0)) \sin(\theta_z(0)) \\ -\frac{\sigma_s^2}{\sigma_s^2 + \sigma_r^2} \cos(\theta_z(0)) \sin(\theta_z(0)) & (1 - \frac{\sigma_s^2}{\sigma_s^2 + \sigma_r^2}) \sin(\theta_z(0)) \end{pmatrix}. \tag{7}$$

Measurements can be converted from polar coordinates to Cartesian coordinates by the unbiased conversion method [5]. Therefore, the initialization is completed as follows

$$\hat{\mathbf{x}}(0|0) = [x_u(0)\, \hat{\dot{x}}(0|0)\, y_u(0)\, \hat{\dot{y}}(0|0)]', \tag{8}$$

$$P(0|0) = \begin{bmatrix} \sigma_x^2(0) & 0 & \sigma_{xy}^2(0) & 0 \\ 0 & P_v(1,1) & 0 & P_v(1,2) \\ \sigma_{xy}^2(0) & 0 & \sigma_y^2(0) & 0 \\ 0 & P_v(2,1) & 0 & P_v(2,2) \end{bmatrix}, \tag{9}$$

where $x_u(0)$ and $y_u(0)$ are the converted measurements in x and y directions, respectively, and $\sigma_x^2(0)$, $\sigma_y^2(0)$, $\sigma_{xy}^2(0)$ are the corresponding measurement noise variances [5].

2.3 Multi-Mode Interaction

The state vectors and the covariance matrices of all of the IMM mode filters at the previous frame $k-1$ are mixed to generate the initial state vectors and the covariance matrices for each of the IMM mode filter at the current frame k [1]. The mixing probability of a target t is described as

$$\mu_{i|j}^t(k-1|k-1) = \frac{p_{ij}\mu_{ti}(k-1)}{\bar{c}_j^t}, \tag{10}$$

$$\bar{c}_j^t = \sum_{i=1}^{r} p_{ij}\mu_{ti}(k-1), \tag{11}$$

where r is the number of filter modes of the IMM estimator, $\mu_j(k)$ is the mode probability of mode j at frame k, and p_{ij} is the mode transition probability. The initial state vector and the covariance matrix of target t for mode j after the mixing are, respectively,

$$\hat{\mathbf{x}}_{0j}^t(k-1|k-1) = \sum_{i=1}^{r} \hat{\mathbf{x}}_{ti}(k-1|k-1)\mu_{i|j}^t(k-1|k-1), \tag{12}$$

$$P_{0j}^t(k-1|k-1) = \sum_{i=1}^{r} \left\{ \begin{array}{l} \mu_{i|j}^t(k-1|k-1)\{P_{ti}(k-1|k-1)+ \\ [\hat{\mathbf{x}}_{ti}(k-1|k-1) - \hat{\mathbf{x}}_{0j}^t(k-1|k-1)]\times \\ [\hat{\mathbf{x}}_{ti}(k-1|k-1) - \hat{\mathbf{x}}_{0j}^t(k-1|k-1)]'\} \end{array} \right\} \tag{13}$$

Equations (12) and (13) are valid except for the first frame where the same initial values for all the IMM mode filters are used from the initialization as

$$\hat{\mathbf{x}}_{0j}^t(0|0) = \hat{\mathbf{x}}_t(0|0), \tag{14}$$

$$P_{0j}^t(0|0) = P_t(0|0), \tag{15}$$

$$\mu_{tj}(0) = \frac{1}{r}, \tag{16}$$

where $\hat{\mathbf{x}}_t(0|0)$ and $P_t(0|0)$ are estimated by the one-point prime initialization as described in the previous section.

2.4 Mode Matched Filtering

The mode matched filtering of the IMM-sub event JPDA is composed of several sub-stages. The main works of these stages are (1) to associate measurements including false alarms at the current frame to the established multi-tracks at the previous frame and (2) to estimate the state of the target corresponding to a certain track at the current frame.

2.4.1 Extended Kalman Filtering (EKF)

The extended Kalman filtering (EKF) is performed for each target and for each mode of the IMM estimator. The first step is to predict the state of each target of which the dynamic state is modeled by mode j [1]:

$$\hat{\mathbf{x}}_{tj}(k|k-1) = F_j(k-1)\hat{\mathbf{x}}_{tj}(k-1|k-1), \tag{17}$$

$$P_{tj}(k|k-1) = F_j(k-1)P_{tj}(k-1|k-1)F_j(k-1) + Q_j(k-1), \tag{18}$$

where $\hat{\mathbf{x}}_{tj}(k|k-1)$ and $P_{tj}(k|k-1)$ are the state prediction and the covariance prediction for target t and mode j, respectively. Next, the residual covariance $S_{tj}(k)$ and the filter gain $W_{tj}(k)$ are obtained as

$$S_{tj}(k) = R(k) + H_{tj}(k)P_{tj}(k|k-1)H_{tj}(k), \tag{19}$$

$$W_{tj}(k) = P_{tj}(k|k-1)H_{tj}(k)'S_{tj}(k)^{-1}, \tag{20}$$

where $H_{tj}(k)$ is the first order derivative function of the measurement model:

$$H_{tj}(k) = \nabla_{\mathbf{x}}\mathbf{h}|_{\hat{\mathbf{x}}_{ij}(k|k-1)}$$
$$= \begin{bmatrix} \frac{(\hat{x}_{ij}(k|k-1)-x_s)}{\sqrt{(\hat{x}_{ij}(k|k-1)-x_s)^2+(\hat{y}_{ij}(k|k-1)-y_s)^2}} & 0 & \frac{(\hat{y}_{ij}(k|k-1)-y_s)}{\sqrt{(\hat{x}_{ij}(k|k-1)-x_s)^2+(\hat{y}_{ij}(k|k-1)-y_s)^2}} & 0 \\ \frac{-(\eta-\eta_0)}{(\hat{x}_{ij}(k|k-1)-x_s)^2+(\hat{y}_{ij}(k|k-1)-y_s)^2} & 0 & \frac{(\hat{x}_{ij}(k|k-1)-x_s)}{(\hat{x}_{ij}(k|k-1)-x_s)^2+(\hat{y}_{ij}(k|k-1)-y_s)^2} & 0 \end{bmatrix} \tag{21}$$

The measurement prediction $\hat{\mathbf{z}}_{tj}(k|k-1)$ and the measurement residual $\boldsymbol{\nu}^{j}_{mt}(k)$ are, respectively,

$$\hat{\mathbf{z}}_{tj}(k|k-1) = \mathbf{h}[\mathbf{x}_{tj}(k|k-1)], \tag{22}$$

$$\boldsymbol{\nu}^{j}_{mt}(k) = \mathbf{z}_m(k) - \hat{\mathbf{z}}_{tj}(k|k-1), \tag{23}$$

where $z_m(k)$ is the m-measurement at the current frame.

2.4.2 Measurement Gating

The measurement gating process reduces the number of the feasible joint association event outcomes by excluding the association between the measurements out of the validation regions and the target of that region [3]. Let Z(k) be a set of m_k measurements at frame k:

$$Z(k) = \{\mathbf{z}_m(k)\},\ m = 1, \ldots, m_k. \tag{24}$$

The measurement gating is Chi-square hypothesis testing assuming the Gaussian measurement residuals:

$$Z^t(k) = \left\{\mathbf{z}_m(k)|\boldsymbol{\nu}^{\hat{j}}_{mt}(k)'[S_{t\hat{j}(t)}(k)]^{-1}\boldsymbol{\nu}^{\hat{j}}_{mt}(k) \le \gamma, m = 1, \ldots, m_k\right\}, \tag{25}$$

$$\hat{j}(t) = \max_{j} |S_{tj}(k)|, \tag{26}$$

$$\boldsymbol{\nu}^{\hat{j}}_{mt}(k) = \mathbf{z}_m(k) - \hat{\mathbf{z}}_{t\hat{j}(t)}(k|k-1), \tag{27}$$

where γ is the gating size. Figure 1 illustrates an example of the gating process. Each ellipsis shows the validation region of each target.

2.4.3 Sub-Event of Feasible Joint Association

Although the gating process may exclude some of candidate associations between tracks and measurements, the feasible joint association event still causes a large computational burden because the measurements in different validated regions can be included in one validation matrix of the JPDA [3]. However, the feasible joint association event can be decomposed into several sub-events in the sub-event JPDA. Figure 2 shows the splitting process of the validation matrix where n_θ is the number

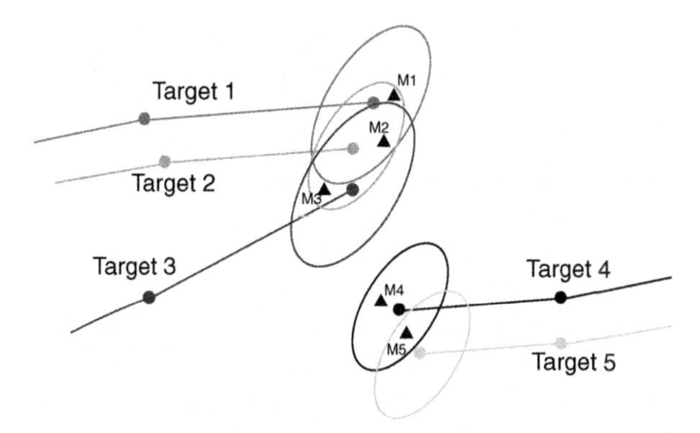

Fig. 1 An example of the measurement gating process

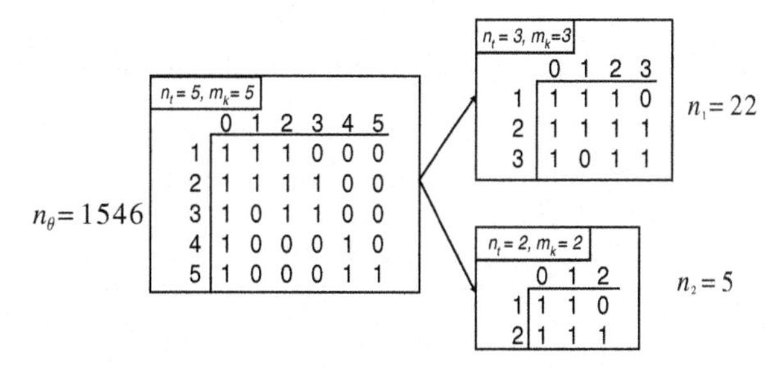

Fig. 2 Validation matrices and $n_\theta' s$ with the sub-event decomposition

of the feasible joint association event outcomes. The validation matrix can be split if it can be arranged to be block-diagonalized. It is noted that the validation matrix on the left side in Fig. 2 corresponds to the track-measurement formation illustrated in Fig. 1. The sub-validation matrices corresponding to the sub-feasible joint association event generate the sub-event outcomes, and n_θ decreases significantly in the sub-validation matrices.

Let Θ be a sample space of the feasible joint association event as

$$\Theta = \{\boldsymbol{\theta}(1), \ldots, \boldsymbol{\theta}(n_\theta)\}, \tag{28}$$

Similarly, Θ_s is defined as the sample space of the s-th sub feasible joint association event:

$$\Theta_s = \{\boldsymbol{\theta}_s(1), \ldots, \boldsymbol{\theta}_s(n_{\theta_s})\}, \tag{29}$$

where n_{θ_s} is the number of elements in the set Θ_s. It can be shown that $\Theta = \Theta_1 \times \Theta_2 \times \ldots \times \Theta_{n_s}$ and $\boldsymbol{\theta} = \boldsymbol{\theta}_1 \cap \boldsymbol{\theta}_2 \cap \ldots \boldsymbol{\theta}_{n_s}$, where $\boldsymbol{\theta}_1, \ldots, \boldsymbol{\theta}_{n_s}$ are the sub feasible joint association events.

$$\boldsymbol{\theta}_s = \bigcap_{m^s=1}^{m_k^s} \theta_{m^s t_{m^s}^s}, \tag{30}$$

where m_k^s is the number of measurements in the s-th sub-event at k frame, and $t_{m^s}^s$ indicates a track in the s-th sub-event which is associated with measurement m^s.

The following is the overall steps performed at this sub-stage.

1. Block diagonalization of the validation matrix: $\Omega \rightarrow \Omega_1, \ldots, \Omega_{n_s}$
2. Partition of the number of measurements and tracks: $m_k \rightarrow m_k^1, \ldots, m_k^{n_s}$, $n_t \rightarrow n_t^1, \ldots, n_t^{n_s}$, where n_t^s is the number of tracks included in the s-th sub-event.
3. Partition of the sets of measurement vectors: $Z(k) \rightarrow Z_1(k), \ldots, Z_{n_s}(k)$ such that $Z_s(k) = \{\mathbf{z}_m^s(k)\}$, $m = 1, \ldots, m_k^s$.

2.4.4 Posterior Probability of the Sub Feasible Joint Association Event

One can refer to [3] for detailed explanations for this sub section. In this sub section, the sub event will replace the event in the original JPDA. The probability of the sub-event is computed as

$$\begin{aligned} P\{\boldsymbol{\theta}_s(k)|Z^k, M_j(k)\} &= P\{\boldsymbol{\theta}_s(k)|Z_s(k), m_k^s, Z^{k-1}, M_j(k)\} \\ &= \frac{1}{c} P[Z_s(k)|\boldsymbol{\theta}_s(k), m_k^s, Z^{k-1}, M_j(k)] P\{\boldsymbol{\theta}_s(k)|m_k^s, M_j(k)\}, \end{aligned} \tag{31}$$

where c is a normalization constant, $M_j(k)$ denotes the j-th mode in the IMM estimator, and Z^{k-1} is the set of all measurement up to the frame $k-1$. The likelihood function of measurements is obtained as

$$\begin{aligned} P[Z_s(k)|\boldsymbol{\theta}_s(k), m_k^s, Z^{k-1}, M_j(k)] &= \prod_{m=1}^{m_k^s} p[\mathbf{z}_{m^s}^s(k)|\theta_{m^s t_{m^s}^s}(k), Z^{k-1}, M_j(k)] \\ &= V^{-\phi(\boldsymbol{\theta}_s(k))} \prod_{m^s=1}^{m_k^s} \{f_{t_{m^s}^s j}[\mathbf{z}_{m^s}(k)]\}^{\tau_{m^s}(\boldsymbol{\theta}_s(k))}, \end{aligned} \tag{32}$$

since

$$p[\mathbf{z}_{m^s}^s(k)|\theta_{m^s t_{m^s}^s}(k), Z^{k-1}, M_j(k)] = \begin{cases} f_{t_{m^s}^s j}[\mathbf{z}_{m^s}^s(k)] \text{ if} \tau_{m^s}[\boldsymbol{\theta}_s(k)] = 1 \\ V^{-1} \quad \text{if } \tau_{m^s}[\boldsymbol{\theta}_s(k)] = 0 \end{cases}, \tag{33}$$

$$f_{t_{m^s}^s j}[\mathbf{z}_{m^s}^s(k)] = N[\mathbf{z}_{m^s}^s(k) - \hat{\mathbf{z}}_{t_{m^s} j}(k|k-1), S_{t_{m^s} j}(k)], \tag{34}$$

where V denotes the volume of surveillance region.

The prior probability of the sub joint association event is

$$P\{\boldsymbol{\theta}_s(k)|m_k^s, M_j(k)\} = P\{\boldsymbol{\theta}_s(k), \boldsymbol{\delta}(\boldsymbol{\theta}_s), \phi(\boldsymbol{\theta}_s)|m_k^s\}$$
$$= P\{\boldsymbol{\theta}_s(k)|\boldsymbol{\delta}(\boldsymbol{\theta}_s), \phi(\boldsymbol{\theta}_s), m_k^s\}P\{\boldsymbol{\delta}(\boldsymbol{\theta}_s), \phi(\boldsymbol{\theta}_s)|m_k^s\}, \quad (35)$$

where $\boldsymbol{\delta}(\boldsymbol{\theta}_s)$ is the target association indicator vector [3]:

$$\boldsymbol{\delta}(\boldsymbol{\theta}_s) = [\delta_1(\boldsymbol{\theta}_s) \cdots \delta_{n_t^s}(\boldsymbol{\theta}_s)], \quad (36)$$

$$P\{\boldsymbol{\theta}_s(k)|\boldsymbol{\delta}(\boldsymbol{\theta}_s), \phi(\boldsymbol{\theta}_s), m_k^s\} = \left(\mathrm{P}^{m_k^s}_{m_k^s-\phi(\boldsymbol{\theta}_s)}\right)^{-1} = \left(\frac{m_k^s!}{\phi(\boldsymbol{\theta}_s)!}\right)^{-1}, \quad (37)$$

where P denotes permutation. The second term in Eq. (35) is

$$P\{\boldsymbol{\delta}(\boldsymbol{\theta}_s), \phi(\boldsymbol{\theta}_s)|m_k^s\} = \mu_F(\phi(\boldsymbol{\theta}_s)) \prod_{t^s=1}^{n_t^s} (P_D)^{\delta_{t^s}(\boldsymbol{\theta}_s)}(1-P_D)^{1-\delta_{t^s}(\boldsymbol{\theta}_s)}, \quad (38)$$

where P_D is the detection probability, and $\mu_F(\phi)$ is the probability mass function of the false alarm. The post probability of the joint association event of the non-parametric model is obtained as

$$P\{\boldsymbol{\theta}_s|Z_s(k), Z^{k-1}, M_j\} = \frac{1}{c^s}\phi(\boldsymbol{\theta}_s)! \prod_{m^s=1}^{m_k^s} \{V f_{t^s_{m^s} j}[z^S_{m^s}(k)]\}^{\tau_{m^s}}$$
$$\prod_{t^s=1}^{n_t^s} (P_D)^{\delta_{t^s}(\boldsymbol{\theta}_s)}(1-P_D)^{1-\delta_{t^s}(\boldsymbol{\theta}_s)}. \quad (39)$$

2.4.5 Marginal Association Event

Marginal probability is essential to update the target's state estimate and covariance matrix. The marginal probability is obtained as

$$\begin{aligned}
\beta^j_{mt}(k) &\equiv P\{\theta_{mt}|Z^k, M_j\} \\
&= \sum_{\boldsymbol{\theta}\in\Theta} P\{\boldsymbol{\theta}|Z^k, M_j\}\delta(1-\theta_{mt}) \\
&= \sum_{\boldsymbol{\theta}_1\in\Theta_1} \cdots \sum_{\boldsymbol{\theta}_{n_s}\in\Theta_{n_s}} P\{\boldsymbol{\theta}_1|Z_1(k), Z^{k-1}, M_j\} \cdots P\{\boldsymbol{\theta}_{n_s}|Z_{n_s}(k), Z^{k-1}, M_j\}\delta(1-\theta_{mt}) \quad (40) \\
&= \sum_{\boldsymbol{\theta}_s\in\Theta_s} P\{\boldsymbol{\theta}_s|Z_s(k), Z^{k-1}, M_j\}\delta(1-\theta_{mt}),
\end{aligned}$$

Since θ_{mt} is the association event which is included in only one sub feasible joint association event $\boldsymbol{\theta}_s$, the following relationship holds: $\sum_{\boldsymbol{\theta}_i\in\Theta_i} P\{\boldsymbol{\theta}_i|Z_i(k), Z^{k-1}, M_j\} = 1$ for $\forall i \neq s$ as in Eq. (40).

2.4.6 PDA Update of State Vectors and Covariance Matrices

The state vectors and the covariance matrices of the targets are updated for each IMM mode. The state estimate is updated as

$$\hat{\mathbf{x}}_{tj}(k|k) = \hat{\mathbf{x}}_{tj}(k|k-1) + W_{tj}(k)\boldsymbol{\nu}_{tj}(k), \tag{41}$$

$$\boldsymbol{\nu}_{tj}(k) = \sum_{m=1}^{m_k} \beta^j_{mt}(k)[\mathbf{z}_m(k) - \hat{\mathbf{z}}_{tj}(k|k-1)]. \tag{42}$$

The covariance matrix is updated as

$$P_{tj}(k|k) = \beta^j_{0t}(k)P_{tj}(k|k-1) + [1 - \beta^j_{0t}(k)]P_{tj}{}^{c}(k|k) + \tilde{P}_{tj}(k), \tag{43}$$

$$P_{tj}{}^{c}(k|k) = P_{tj}(k|k-1) - W_{tj}(k)S_{tj}(k)W_{tj}(k)', \tag{44}$$

$$\tilde{P}_{tj}(k) = W_{tj}(k)\left[\sum_{m=1}^{m_k} \beta^j_{mt}(k)\boldsymbol{\nu}^j_{mt}(k)\boldsymbol{\nu}^j_{mt}(k)' - \boldsymbol{\nu}_{tj}(k)\boldsymbol{\nu}_{tj}(k)'\right]W_{tj}(k)'. \tag{45}$$

The likelihood function of measurements is obtained to update the mode probabilities as follows

$$\begin{aligned}\Lambda_{tj}(k) &\equiv P[Z(k)|m_k, M_j(k), Z^{k-1}] \qquad (46)\\ &= \sum_{m=1}^{m_k} P[Z(k)|\theta_{mt}(k), m_k, M_j(k), Z^{k-1}]P\{\theta_{mt}(k)|m_k, M_j(k), Z^{k-1}\},\end{aligned}$$

where the condition probability of measurement is obtained as

$$P[Z(k)|\theta_{mt}(k), m_k, M_j(k), Z^{k-1}] = \begin{cases} V^{-m_k+1}P_G^{-1}N[v^j_{mt}(k); 0, S_{tj}(k)] & \text{if } m = 1, \ldots, m_k \\ V^{-m_k} & \text{if } m = 0 \end{cases}, \tag{47}$$

$$P\{\theta_{mt}(k)|m_k, M_j(k), Z^{k-1}\} = \begin{cases} \frac{1}{m_k}P_D P_G\left[P_D P_G + (1 - P_D P_G)\frac{\mu_F(m_k)}{\mu_F(m_k-1)}\right]^{-1} & \text{if } m = 1, \ldots, m_k \\ (1 - P_D P_G)\frac{\mu_F(m_k)}{\mu_F(m_k-1)}\left[P_D P_G + (1 - P_D P_G)\frac{\mu_F(m_k)}{\mu_F(m_k-1)}\right]^{-1} & \text{if } m = 0 \end{cases}, \tag{48}$$

where P_G is a compensation factor for the validation region. If the volume of surveillance region for V is used, P_G is set at 1.

2.4.7 IMM Update of Mode Probability, State Vector and Covariance Matrix

At the final stage, the mode probability, state vector, and covariance matrix of each target are updated as follows

$$\mu_{tj}(k) = \frac{1}{c_t}\Lambda_{tj}(k)\sum_{i=1}^{r} p_{ij}\mu_{ti}(k-1), \tag{49}$$

$$\hat{\mathbf{x}}_t(k|k) = \sum_{j=1}^{r} \hat{\mathbf{x}}_{tj}(k|k)\mu_{tj}(k), \tag{50}$$

$$P_t(k|k) = \sum_{j=1}^{r} \mu_{tj}(k)\{P_{tj}(k|k) + [\hat{\mathbf{x}}_{tj}(k|k) - \hat{\mathbf{x}}_t(k|k)][\hat{\mathbf{x}}_{tj}(k|k) - \hat{\mathbf{x}}_t(k|k)]'\}, \tag{51}$$

where c_t is normalization constant. The procedures from Sect. 2.3, 2.4, 2.4.1, 2.4.2, 2.4.3, 2.4.4, 2.4.5, 2.4.6, and 2.4.7 repeat. A target can be terminated when no validated measurement is detected for several frames.

2.5 Performance Evaluation

The averaged root mean-squared error (RMSE) of position of target t at frame k is defined as

$$e_P(k,t) = \sqrt{\frac{1}{M}\sum_{m=1}^{M}\left\{[x_t(k) - \hat{x}_t(k|k)]^2 + [y_t(k) - \hat{y}_t(k|k)]^2\right\}}, \tag{52}$$

where M is the number of Monte Carlo (MC) runs, and $x_t(k)$ and $y_t(k)$ are true positions of target t in x and y directions, respectively. The averaged RMSE of velocity is calculated as

$$e_V(k,t) = \sqrt{\frac{1}{M}\sum_{m=1}^{M}\left\{\left[\dot{x}_t(k) - \hat{\dot{x}}_t(k|k)\right]^2 + \left[\dot{y}_t(k) - \hat{\dot{y}}_t(k|k)\right]^2\right\}}, \tag{53}$$

where $\dot{x}_t(k)$ and $\dot{y}_t(k)$ are true velocities of target t in x and y directions, respectively. The average mode probability over M runs of the j-th IMM mode is defined as

$$m_\mu(k,t;j) = \frac{1}{M}\sum_{m=1}^{M} \mu_{tj}(k|k). \tag{54}$$

3 Scenario Description and Filter Design

3.1 Ground Truth and Measurement Simulation

A scenario is designed to have ten ground targets. The range of the targets covers 300 m–15.9 km, and the speed varies from 2.8 to 40 m/s. Acceleration is set

Table 1 Acceleration (m/s^2) applied to targets

Time (frame)		50 (10)	50 (10)	25 (5)	80 (16)	15 (3)	80 (16)	15 (3)	20 (4)	15 (3)	160 (32)
Target 1,3,5	x	0	0	1	0	−1	0	1	0	0	0
	y	0	0	1	0	0	0	0	0	0	0
Target 2,4,6	x	0	0	1	0	0	0	0	0	−1	0
	y	0	0	1	0	0	0	0	0	0	0
Target 7,9	x	0	0	−1	0.5	0	0	−1	0	0	0
	y	0	0	1	0	0	0	0	0	0	0
Target 8,10	x	0	0	-1	0	0	0	0.5	0	1	0
	y	0	0	1	0	0	0	0	0	0	0

at separately in x and y directions to force the targets into maneuvering motion. Total time duration is 510 s or 102 frames with 5 s sampling interval. The sensor is assumed to be fixed at the origin. The initial velocities of all the targets are set at 5 m/s in y-direction and 0 m/s in x-direction. Table 1 shows the acceleration applied to each target. To simulate corrupted measurements, Gaussian random numbers are generated and added to the ground truths. The standard deviations of the measurement noise are set at 20 m for range, 1.1×10^{-3} radian for azimuth, and 1 m/s for range rate. The detection probability is set at 1.

3.2 Filter and Parameter Design

Two modes are adopted for the IMM estimator. The following is the detailed values of the parameters:

1. One-point prime initialization with unbiased conversion: $\sigma_s = 1\,\text{m/s}$
2. Two NCV models: $\text{NCV}_1 : \sigma_x = \sigma_y = 0.05\,\text{m/s}^2$, $\text{NCV}_2 : \sigma_x = \sigma_y = 2\,\text{m/s}^2$
3. Measurement noise: $\sigma_r = 20\,\text{m}$, $\sigma_\theta = 1.1 \times 10^{-3}\,\text{rad}$, $\sigma_{\dot{r}} = 1\,\text{m/s}$
4. Mode transition probabilities: $p_{11} = p_{22} = 0.8$
5. Volume of surveillance region (V): $10^4 \times 2\pi \times 40 \approx 2.5 \times 10^6\, m^2/s^2$
6. Measurement gating size: $\gamma = 25$

4 Simulation Results

The sub-event IMM-JPDA estimator is able to track all of the targets for 100 MC runs. Figure 3 shows the ground truth, measurements, and estimates for one MC run. Figure 4 shows the averaged position RMSE and Fig. 5 shows the averaged velocity RMSE for 100 MC runs. Figure 6 shows the averaged mode probabilities of several selected targets for 100 MC runs. Figure 7 shows the relative time of the original to the sub-event IMM-JPDA for one MC run which is defined as

$$R(n_t) = \log_{10}\left(\frac{T_{orig}(n_t)}{T_{sub}(n_t)}\right), \tag{55}$$

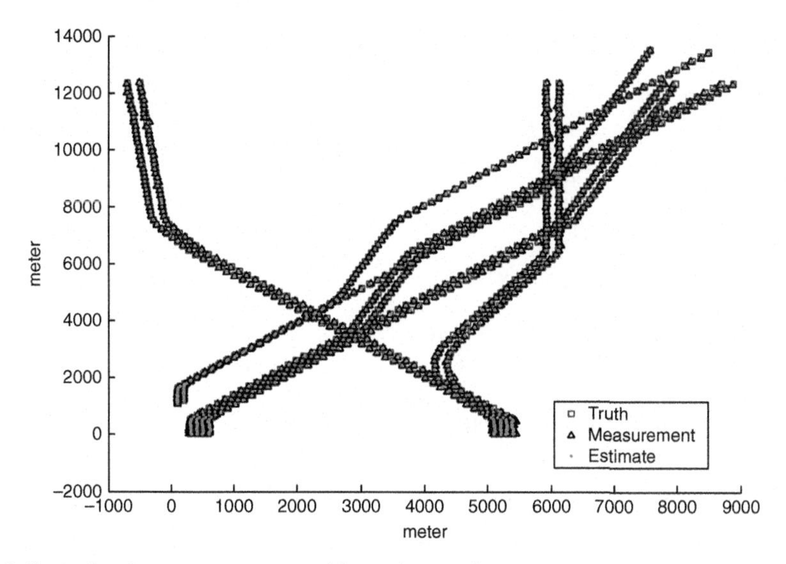

Fig. 3 Ground truths, measurements, position estimates of ten targets

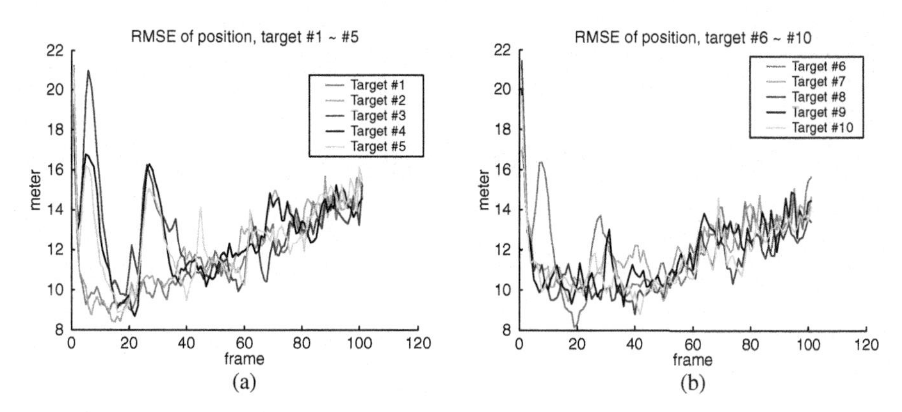

Fig. 4 Averaged position RMSE, (**a**) target 1∼5, (**b**) target 6∼10

where T_{sub} and T_{orig} are the computational times of the sub-event IMM-JPDA and the original method, respectively, and n_t indicates the number of the targets in the scenario.

Table 2 shows the averaged RMSE's of position and velocity for the first MC run to compare the performances between the proposed IMM-JPDA and the original method:

$$\bar{e}_P(n_t) = \frac{1}{n_t K} \sum_{t=1}^{n_t} \sum_{k=1}^{K} e_P(k, t; n_t), \tag{56}$$

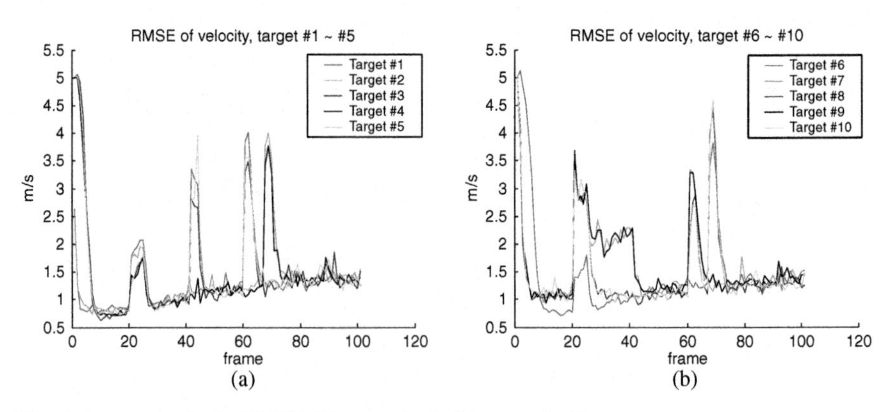

Fig. 5 Averaged velocity RMSE, (**a**) target 1∼5, (**b**) target 6∼10

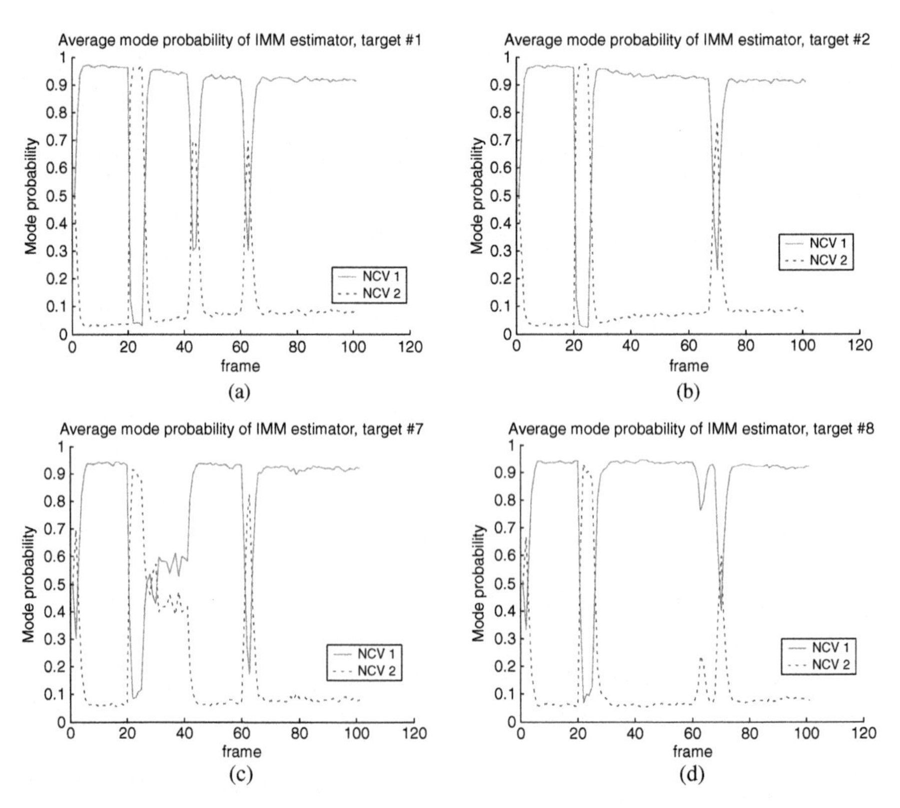

Fig. 6 Averaged mode transition probability of selected targets, (**a**) target 1, (**b**) target 2, (**c**) target 7, (**d**) target 8

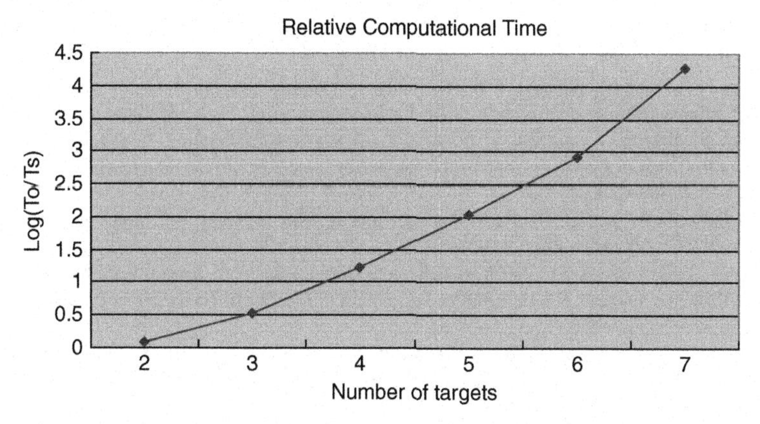

Fig. 7 Relative time of the original IMM–JPDA to the sub-event method

$$\bar{e}_v(n_t) = \frac{1}{n_t K} \sum_{t=1}^{n_t} \sum_{k=1}^{K} e_v(k, t; n_t), \tag{57}$$

where K denotes the total number of frames. As shown in Fig. 7 and Table 2, the proposed algorithm has significant advantage in terms of the computational time.

Table 2 Comparison between sub-event Imm–Jpda and original Imm–Jpda

n_t		2	3	4	5	6	7	8	9	10
$\bar{e}_P(n_t)$ (meter)	Sub	9.48	10.34	10.37	10.39	10.75	10.51	10.42	10.4	10.36
	Original	9.48	10.34	10.36	10.39	10.75	10.51	–	–	–
$\bar{e}_v(n_t)$ (m/s)	Sub	1.11	1.16	1.2	1.17	1.18	1.19	1.23	1.23	1.25
	Original	1.11	1.16	1.2	1.17	1.18	1.19	–	–	–

5 Conclusions

In this paper, the IMM–JPDA approach with the sub-event decomposition and one-point prime initialization has been proposed and implemented. This proposed scheme of the sub-event JPDA decreases the numerical complexity significantly by reducing the number of the feasible joint association event outcomes. It provides significantly reduced time complexity while maintaining the same performance of target tracking. The one-point prime initialization is employed to improve the initial estimates of targets. The research in the clutter environments is planned to examine the robustness of the presented technique.

Acknowledgments This research was supported in part by the Daegu University Research Grant 2008.

References

1. Blom, H.A.P., and Bar-Shalom, Y. The interacting multiple model algorithm for systems with Markovian switching coefficients. *IEEE Trans. Autom. Control*, 33(8), 780–783. (1988).
2. Yeom, S., Kiruba, T., and Bar-Shalom, Y. Track segment association, fine-step IMM and initialization with Doppler for improved track performance. *IEEE Trans. Aerosp. Electron. Syst.*, 40(1), 293–309. (2004).
3. Bar-Shalom, Y., and Li, X. R. Multitarget-multisensor tracking: principles and techniques. Storrs, CT: YBS publishing. (1995).
4. Deb, S., Yeddanapudi, M., Pattipati, K.R., and Bar-shalom, Y. A Generalized S-D assignment algorithm for multisensor-multitarget state estimation. *IEEE Trans. Aerosp. Electron Syst.*, 33(2), 523–538. (1997).
5. Longbin, M., Xiaoquan, S., Yizu, Z., Kang, S.Z., and Bar-Shalom, Y. Unbiased converted measurements for tracking. *IEEE Trans. Aerosp. Electron. Syst.*, 34(3), 1023–1027. (1998).

Enabling Navigation of MAVs through Inertial, Vision, and Air Pressure Sensor Fusion

Clark N. Taylor

Abstract Traditional methods used for navigating miniature unmanned aerial vehicles (MAVs) consist of fusion between Global Positioning System (GPS) and Inertial Measurement Unit (IMU) information. However, many of the flight scenarios envisioned for MAVs (in urban terrain, indoors, in hostile (jammed) environments, etc.) are not conducive to utilizing GPS. Navigation in GPS-denied areas can be performed using an IMU only. However, the size, weight, and power constraints of MAVs severely limits the quality of IMUs that can be placed on-board the MAVs, making IMU-only navigation extremely inaccurate. In this paper, we introduce a Kalman filter based system for fusing information from two additional sensors (an electro-optical camera and differential air pressure sensor) with the IMU to improve the navigation abilities of the MAV. We discuss some important implementation issues that must be addressed when fusing information from these sensors together. Results demonstrate an improvement of at least 10x in final position and attitude accuracy using the system proposed in this paper.

Keywords Vision-aided navigation · GPS-denied narrigation · Sensor fusion

1 Introduction

Recently, Unmanned Aerial Vehicles (UAVs) have seen a dramatic increase in utilization for military applications. In addition, UAVs are being investigated for multiple civilian uses, including rural search and rescue, forest fire monitoring, and agricultural information gathering. Due to their small size, Miniature UAVs (MAVs) are an attractive platform for executing many of the missions traditionally performed by larger UAVs. Some of the primary advantages of MAVs include: (1) they are significantly less expensive to purchase than the large UAVs typically used by the

C.N. Taylor (✉)
459 CB, Department of Electrical and Computer Engineering, Brigham Young University, Provo, UT 84602
e-mail: clark.taylor@byu.edu

S. Lee et al. (eds.), *Multisensor Fusion and Integration for Intelligent Systems*, Lecture Notes in Electrical Engineering 35, DOI 10.1007/978-3-540-89859-7_11,

military, (2) their small size simplifies transport, launch, and retrieval, and (3) they are less expensive to operate than large UAVs.

To enable the utilization of MAVs, the ability to accurately navigate (estimate the location, attitude, and velocity of the MAV) is essential. For example, if an MAV is being utilized as part of a search and rescue operation, knowing the correct location and attitude of the MAV is critical to geo-locate an observed object of interest.

Navigation methods implemented in MAVs today are primarily based on the fusion of measurements from the Global Positioning System (GPS) and an inertial measurement unit (IMU) (e.g. [1–4]). However, there are many scenarios in which an MAV might prove useful but GPS is not available (e.g., indoors, urban terrain, etc.). Therefore, a number of methods have been proposed for fusing visual information with IMU measurements to enable navigation without GPS.

For the most accurate navigation possible, a batch method that analyzes all vision and IMU information from an entire flight was introduced in [5]. While accurate, this method cannot be utilized in real-time due to nature of batch optimization routines. Therefore, the paper also introduces a recursive method which is essentially a SLAM filter. Other implementations of SLAM-based filters for navigation can also be found in [6–8]. While SLAM-based methods are highly effective, there are two bottlenecks to SLAM that make them difficult to implement in the computationally-limited environments that characterize MAVs. First, visual SLAM requires that objects in the video be tracked for an extended period of time. Second, the size of the state grows with the number of landmarks that SLAM is attempting to find the location for, dramatically increasing the computation time required. In [9], the size of the state is limited, but tracking of features over a long period of time is still required.

While it is computationally expensive to track points for an extended period of time in video, it is relatively simple to track points over a small number of video frames. Therefore, we focus in this paper on a method that utilizes only the relationship between objects in two frames of video. Other methods that utilize only two frames of video ([10–12]) have been introduced previously. Ready and Taylor [11] and Andersen and Taylor [12], however, requires that the terrain being observed is planar, while we assume in this paper that the points being tracked from frame to frame are not planar (a better assumption for indoor or dense environments that would obscure GPS signals).

In this paper, we focus on utilizing the *epipolar constraint* for fusing visual measurements with the IMU as described in [10]. Using the epipolar constraint, however, has three significant weaknesses that must be addressed when performing fusion with IMU measurements for MAV navigation. First, the epipolar constraint biases movements of the camera toward the center of points in the image. Second, visual measurements always include a "scale ambiguity" – it is impossible to distinguish between the camera moving quickly and observing an object that is far away and the camera moving slowly and observing an object that is close. Third, when using vision to navigate, the navigation state of the camera can only be determined relative to its previous navigation states.[1]

In this paper, we present methods to overcome each of these three weaknesses in utilizing the epipolar constraint. First, to overcome the bias in the epipolar constraint, we analyze the epipolar constraint equation and propose an alternate algorithm for computing deviations from the epipolar constraint. Second, to overcome the scale ambiguity of vision, we integrate a differential air pressure sensor into the fusion system. On a fixed-wing MAV, the differential air pressure sensor is capable of measuring the airspeed of the MAV. This airspeed can be treated as a direct measurement of the velocity magnitude, allowing the scale ambiguity of vision to be overcome. Third, because vision measurements of motion are relative, we propose using the *minimal* sampling rate at which vision can be effectively fused with IMU data. We demonstrate that sampling at the minimal, rather than maximal, rate increases the accuracy of the overall navigation system. We also discuss limitations on choosing the minimal sampling rate.

Once the weaknesses of epipolar-based fusion are overcome, it is possible to enable on-line estimation of inertial sensor biases. We prove this capability by performing an observability analysis of a simplified system, and demonstrate improved navigation results when estimating biases.

The remainder of this paper is organized as follows. In Sect. 2, we discuss the general framework used for fusing vision and IMU information. In Sect. 3, we describe our three modifications for improving the fusion of visual and IMU information. In Sect. 4, we describe our method for estimating the biases of the inertial sensors. Section 5 demonstrates the improvement in navigation state estimates achieved by implementing our modifications. Section 6 concludes the paper.

2 Epipolar Constraint-Based Fusion

In this section, we first describe the epipolar constraint which is used in our fusion setup. We then describe how the epipolar constraint can be used in a Kalman filter to enable fusion between visual and inertial information.

2.1 The Epipolar Constraint

The epipolar constraint can be utilized whenever a single fixed object is observed by a camera at two locations (or two cameras at different locations). Given that any three points in the world form a plane, a single world point and the two camera projection centers form a plane in the 3-D world – the *epipolar* plane. Similarly, when a world point is observed in two images, the two vectors representing where the point was observed in the image plane ($\mathbf{x}'$ and $\mathbf{x}$), and the translation vector

[1] Note that while it is possible to determine absolute position or attitude using vision, knowledge of pre-existing visual "landmarks" is required. We do not address these techniques in this paper as we are interested in using MAVs to explore new areas, not fly over pre-mapped areas.

between the two camera locations, will all lie on the epipolar plane.[2] By assigning a unique coordinate frame to each camera location, this constraint is represented by the formula

$$\mathbf{x}'[\mathbf{p}_{c1}^{c2}]_\times \mathbf{C}_{c1}^{c2}\mathbf{x} = 0, \tag{1}$$

where $\mathbf{C}_{c1}^{c2}$ is the direction cosine matrix between the camera coordinate frames and $[\mathbf{p}_{c1}^{c2}]_\times$ is the position of camera 1 in the second camera's coordinate frame, put into a skew-symmetric matrix. (In other words, we calculate the cross product between $\mathbf{p}_{c1}^{c2}$ and $\mathbf{C}_{c1}^{c2}\mathbf{x}$.) This constraint enforces that the three vectors $\mathbf{x}'$, $\mathbf{C}_{c1}^{c2}\mathbf{x}$, and $\mathbf{p}_{c1}^{c2}$ all lie within the same plane in the 3-D world.

2.2 *Utilizing the Epipolar Constraint in a Fusion Environment*

To utilize the epipolar constraint in a fusion environment, we created an Unscented Kalman Filter (UKF) framework. The state in the Kalman filter must contain enough information to generate both $\mathbf{C}_{c1}^{c2}$ and $\mathbf{p}_{c1}^{c2}$ during the measurement step. To represent general motion by an MAV between two different locations, the UKF state $\mathbf{X} = \begin{bmatrix} \chi_t \\ \chi_{t-1} \end{bmatrix}$ is used, where χ_t is the navigation state estimate at time t and χ_{t-1} is the navigation state estimate at the previous time. The navigation state at each time contains p_t, the position of the camera at time t in the inertial frame, $\mathbf{C}_t$, the direction cosine matrix relating the inertial frame to the current camera coordinate frame, and v_t, the velocity of the camera at time t. This method for setting up the UKF to enable vision and inertial information fusion was first introduced in [12].

2.2.1 Performing the Time Update

The time update for this UKF implementation takes two different forms. The first form updates χ_t (the first 10 elements of the current state) every time an IMU measurement occurs. This makes the first 10 elements of the state the most recent navigation state estimate. After each measurement update, the state is also updated using the formula

$$\mathbf{X}^+ = \mathbf{A}\mathbf{X}^- \text{ where} \tag{2}$$

$$\mathbf{A} = \begin{bmatrix} I & 0 \\ I & 0 \end{bmatrix}, \tag{3}$$

[2] Note that image locations are typically in pixels, while the discussion of vectors so far assumes all vectors are in an unscaled, Euclidean space. In this paper, we assume that the camera has been calibrated a-priori and that the conversion from image to Euclidean vectors has already occurred using this calibration information.

causing the current state to become $\mathbf{X} = \begin{bmatrix} \chi_t \\ \chi_t \end{bmatrix}$. The first 10 elements of the state vector are then updated by IMU measurements to realize a new current state $\mathbf{X} = \begin{bmatrix} \chi_{t+1} \\ \chi_t \end{bmatrix}$. With this technique, the two most recent χ estimates, corresponding to the time of the two most recently captured images, are always stored as the current state.

2.2.2 Performing the Measurement Update

To utilize the epipolar constraint as a measurement in a UKF framework, the "Dynamic Vision" approach introduced in [10] is used. Assuming a feature has been detected in two images, the locations of the features are represented by $\mathbf{x}'$ and $\mathbf{x}$. Because the epipolar constraint should always be equal to zero, the "measurement" used by the UKF is a vector of zeros in length equal to the number of corresponding features found between the two images. The predicted measurement is $\mathbf{x}'[\mathbf{p}_{c1}^{c2}]_\times \mathbf{C}_{c1}^{c2}\mathbf{x}$ (from Eq. (1)) for each set of features $\mathbf{x}'$ and $\mathbf{x}$, where $\mathbf{p}_{c1}^{c2}$ and $\mathbf{C}_{c1}^{c2}$ are functions of χ_t and χ_{t-1}.

While the Dynamic Vision method yields good results in certain cases, it does exhibit some weaknesses that need to be addressed for use on an MAV. First, the translation direction estimates are biased in the direction the camera is pointing. Second, as with all vision-based approaches, it does not estimate the magnitude of translation. We propose methods for overcoming these weaknesses in the following section.

3 Improving the Fusion of Visual and IMU Sensors

In this section, we propose three modifications to baseline epipolar constraint-based fusion of IMU and visual information to significantly increase the accuracy of navigation state estimation on MAVs. These modifications overcome the centering bias, scale ambiguity, and sampling rate problems discussed in the introduction.

3.1 Overcoming Bias Toward the Center of Image Points

When using the epipolar constraint to fuse inertial and vision sensors together, the fusion system introduces a bias in estimated translation direction toward the center of the points observed from frame to frame. To understand the source of this error, let us analyze the epipolar constraint when the result of Eq. (1) is not zero. The value $x'[\mathbf{p}_{c1}^{c2}]_\times \mathbf{C}_{c1}^{c2}x$ can be rewritten as a cross product of two vectors followed by a dot product of two vectors. The final results of this computation will be

$$||x'||\ ||\mathbf{p}_{c1}^{c2}||\ ||x||\sin(\theta_{\mathbf{p}_{c1}^{c2}\to\mathbf{C}_{c1}^{c2}x})\cos(\theta_{x'\to[\mathbf{p}_{c1}^{c2}]_\times\mathbf{C}_{c1}^{c2}x}), \tag{4}$$

where $\theta_{\mathbf{p}_{c1}^{c2}\to\mathbf{C}_{c1}^{c2}x}$ is the angle between $\mathbf{p}_{c1}^{c2}$ and $\mathbf{C}_{c1}^{c2}x$ and $\theta_{x'\to[\mathbf{p}_{c1}^{c2}]_\times\mathbf{C}_{c1}^{c2}x}$ is the angle between x' and $[\mathbf{p}_{c1}^{c2}]_\times\mathbf{C}_{c1}^{c2}x$.

As mentioned previously, the correct magnitude for these values is 0. Therefore, the UKF will attempt to set χ_t and χ_{t-1} in its state vector such that the resulting $\mathbf{p}_{c1}^{c2}$ and $\mathbf{C}_{c1}^{c2}$ minimizes the set of all measurements. However, there are two ways to push the set of measurements toward zero: (1) the epipolar constraint can be met by setting $[\mathbf{p}_{c1}^{c2}]_\times\mathbf{C}_{c1}^{c2}x$ to be orthogonal to x' (i.e., set $\cos(\theta_{x'\to[\mathbf{p}_{c1}^{c2}]_\times\mathbf{C}_{c1}^{c2}x}) = 0$), or (2) $\mathbf{p}_{c1}^{c2}$ can be set parallel to $\mathbf{C}_{c1}^{c2}x$ (i.e., set $\sin(\theta_{\mathbf{p}_{c1}^{c2}\to\mathbf{C}_{c1}^{c2}x}) = 0$). To meet the first condition, $\mathbf{C}_{c1}^{c2}x$, p_{c1}^{c2}, and x' must all lie in a plane, the original justification behind the epipolar constraint. The second condition, however, can be met by setting the direction of p_{c1}^{c2} equal to x'. Because of this second condition, the results of the epipolar constraint can be pushed to zero by setting $\mathbf{p}_{c1}^{c2}$ to be as close to parallel to the set of $\mathbf{C}_{c1}^{c2}x$ vectors as possible. Therefore, the translation direction after the UKF measurement update is biased toward the center of the feature points that have been tracked in the second image.

To overcome this biasing of the translation direction, we propose modifying the measurement step of the UKF to eliminate the $\sin(\theta_{\mathbf{p}_{c1}^{c2}\to\mathbf{C}_{c1}^{c2}x})$ term from the measurement. To eliminate the effect of $\sin(\theta_{\mathbf{p}_{c1}^{c2}\to\mathbf{C}_{c1}^{c2}x})$, the term $[\mathbf{p}_{c1}^{c2}]_\times\mathbf{C}_{c1}^{c2}x$ is first computed and then normalized to be of length one. The inner product of this term with x' is then taken and returned as the predicted measurement.

To determine the results of this modification, we simulated a 16 s flight of an MAV traveling 200 m in a straight line. (More details on our simulation environment can be found in Sect. 5.) In Table 1, we show the average and standard deviation of the error in the final estimated position of the MAV, both before and after the sin removal modification. Note that before sin removal, the p_z location error mean is a large positive number. This is a result of the bias inherent in the unmodified epipolar fusion environment. After removing the sin as discussed above, the z error is dramatically decreased.

Despite the fact that the p_z error has been decreased by removing the sin term from the epipolar constraint, the p_x error is still quite significant. This error is an artifact of vision-based techniques where there is always a scale ambiguity in the direction of travel (in this case, along the x axis). In the next subsection, we discuss how to reduce the error present in the direction of travel of the MAV due to the scale ambiguity.

Table 1 This table demonstrates the accuracy improvements achieved by removing the sin term from the epipolar constraint computation. Each entry lists the mean and standard deviation of the error, in meters. Note that the average p_z error has decreased from 758 to 12 m, demonstrating the effect of removing the bias from the epipolar computation

	p_x error	p_y error	p_z error
With sin (μ, σ)	$(-607.0, 136.6)$	$(2.83, 76.5)$	$(757.5, 318.1)$
sin removed (μ, σ)	$(23.1, 85.9)$	$(-1.77, 16.0)$	$(11.8, 9.71)$

3.2 Removing the Scale Ambiguity

To remove the large amount of error present in the direction of travel of the MAV, we propose integrating another sensor into the UKF framework discussed above. On a fixed-wing MAV, a pitot tube designed for measuring airspeed can be utilized to measure the current velocity of the MAV. To integrate this measurement in the UKF, we utilize the property discussed in [13] that if measurements are uncorrelated, they can be applied during separate measurement updates of the Kalman Filter. Therefore, whenever the pitot tube is read (approximately 10 Hz), the current magnitude of the velocity in the state is computed as a predicted measurement, with the air speed measured by the pitot tube used as a measurement to the UKF.

By applying this measurement at 10 Hz, significant gains in accuracy were achieved. In Table 2, we present the results of this modification using the same simulation setup as described for Table 1. Note that both the mean and standard deviation of error has decreased in all three location parameters, demonstrating the importance of overcoming the scale ambiguity in visual measurement.

Table 2 This table demonstrates the accuracy improvements achieved by fusing the pitot tube measurements with vision and IMU information. Note that the standard deviation of error on the p_x term has decreased from 86 to 2

	p_x error	p_y error	p_z error
Without pitot tube (μ, σ)	(23.1, 85.9)	(−1.77, 16.0)	(11.8, 9.71)
With pitot tube (μ, σ)	(−0.47, 1.92)	(−1.01, 11.2)	(2.85, 1.08)

3.3 Determining the Optimal Image Sampling Rate

Typically, when fusing information together, the more information that is available, the more accurate the final result will be. However, with epipolar based visual and inertial fusion, this is not the case. In this subsection, we show that it best to sample the imaging sensors at the *minimal* sampling rate allowed. We also discuss what limits the minimal possible sampling rate.

In Figure 1, we show plots of the mean squared error (MSE) in the estimated final location of the MAV for different image sampling rates. The MSE represents the error in estimated location after a 15 s, straight-line flight. Note that the lowest MSE point does *not* lie at the maximal sampling rate. This can be explained by noting that fusing with the epipolar constraint helps to reduce the amount of error present between two navigation state estimates (i.e., relative error). The total error at the end of flight is going to be the summation of all the relative estimation errors during the flight. Therefore, if the same relative error is achieved by each measurement of the epipolar constraint, but fewer measurements occur, the total error will be reduced. This leads to the counter-intuitive fact that the minimal, as opposed to maximal sampling rate, is ideal for epipolar constraint-based information fusion.

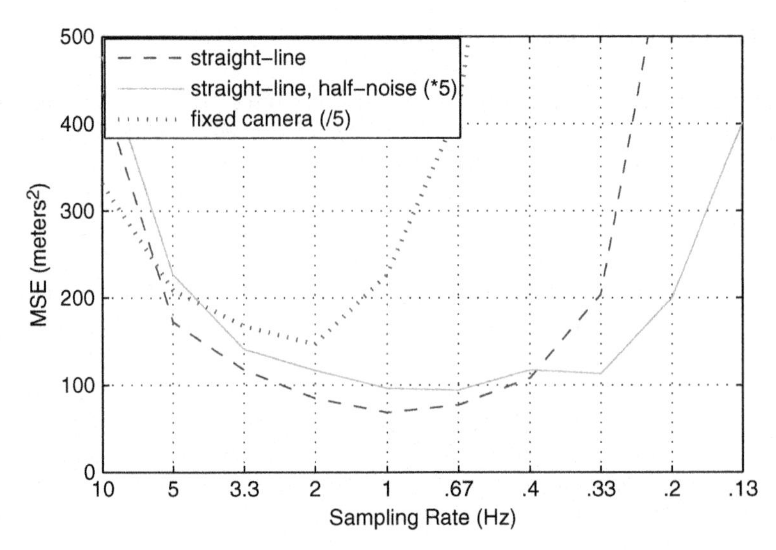

Fig. 1 Results of fusion for different image sampling rates (x-axis) and different fusion setups (different plots). Results are mean squared error (in m^2). To make all plots appear on the same axes, two of the plots have been scaled by 5 (up and down) as denoted in the legend

Despite the general rule that the minimal sampling rate is ideal, there are secondary considerations that must be taken into account when deciding on a sampling rate. Note that in Figure 1, the MSE is *not* monotonically decreasing as the sampling rate decreases. There are two principal causes for the increase in MSE at lower sampling rates.

First, the time update of the IMU introduces error into the estimated navigation states, which the UKF attempts to correct using the epipolar constraint. This correction applied by the UKF is a linear correction. As the distance between the estimated and measured navigation states increase however, the linear assumption becomes invalid. Therefore, if too much noise has been added by the IMU, it will not be possible for the linear update from the epipolar constraint to correct the IMU-introduced noise. This is demonstrated by the "straight-line" and "straight-line, half-noise" plots in Figure 1. The only difference between the simulation setup of the different plots is that the IMU noise was halved for the "straight-line, half noise" plot. Note that as the IMU noise is decreased, the "optimal" sampling rate becomes lower (moving from 1 to .67 Hz), demonstrating that there is a limit placed on the minimal sampling rate by the noise present in the IMU.

Second, the minimal sampling rate is limited by how long the camera can track features in the image. To demonstrate this fact, a simulation was run where, rather than tracking a set of objects throughout the entire flight, as in the "straight-line" simulations of Figure 1, a fixed camera was used so that objects would leave the field of view more quickly. This is the "straight-line, fixed-camera" plot in Figure 1. Note that the sampling rate with the minimal MSE is higher (2 Hz) for this plot than the "straight-line" plot (1 Hz) where the same world points are observed throughout the MAV flight. Therefore, when applying epipolar constraint-based fusion, it is best to apply the minimal sampling rate that is allowed by (1) the noise present in the IMU and (2) the persistence of features across the images.

4 Estimating Inertial Sensor Biases

Using the modifications proposed in the prior section, it is possible to *overcome* a significant amount of error introduced by inertial sensors when navigating an MAV. In this section, we discuss how fusion based on the epipolar constraint can be used to estimate biases of the inertial sensors, thereby *reducing* the noise from those sensors.

Typical low-cost inertial sensors will have several different *types* of noise [14,15]. The Kalman Filter setup described in Sect. 2 essentially assumes that all noise from the inertial sensors is uncorrelated over time. While the existence of correlated noise (*bias* for the remainder of this paper) in the inertial sensors can be overcome by simply increasing the uncorrelated noise covariance estimates, it is preferable to estimate and remove the bias, thereby reducing the amount of noise present in the inertial sensor measurements.

Before describing our method for estimating the bias of the inertial sensors, it is important to demonstrate the feasibility of estimating biases despite the fact that epipolar constraint measurements yield only relative navigation information. In this section, we first demonstrate that it is possible to estimate biases using relative navigation measurements. We then describe our modifications to the Kalman filter framework described in Sect. 2 to enable estimation of the inertial sensor biases.

4.1 Proof of Ability to Estimate Biases

To prove the feasibility of estimating inertial sensor biases, we will perform an observability analysis of a simplified system with a setup that is very similar to our complete MAV navigation system. Our simplified system consists of a state vector with two locations, l_t and l_{t-1}. Similar to the situation where the accelerometers are used to update the current velocity estimates, we use an external rate measurement ($\hat{\dot{l}}_t$) to update the current locations. To estimate the bias on this measurement, we modify the state vector to include the bias, obtaining a state vector of:

$$\mathbf{x}_t = \begin{bmatrix} l_t & l_{t-1} & b \end{bmatrix}^T, \tag{5}$$

where b is the bias of the sensor.

The time update for this state over time Δt is

$$\mathbf{x}_{t+1} = \mathbf{F}\mathbf{x}_t + \mathbf{G}\hat{\dot{l}}_t \tag{6}$$

where

$$\mathbf{G} = [\Delta t \quad 0 \quad 0] \tag{7}$$

and

$$\mathbf{F} = \begin{bmatrix} 0 & 1 & -\Delta t \\ 0 & 1 & 0 \\ 0 & 0 & 1 \end{bmatrix}. \tag{8}$$

If the measurement of the system provides relative measurements (like vision does for MAV motion), then the observation matrix is:

$$\mathbf{H} = [1 \quad -1 \quad 0]. \tag{9}$$

With this simplified system, we can analyze the observability of b for this Kalman Filter setup. To prove observability, the rank and null vectors of the matrix

$$\mathcal{O} = \begin{bmatrix} \mathbf{H} \\ \mathbf{HF} \\ \mathbf{HFF} \\ \cdots \end{bmatrix} \tag{10}$$

must be found. Substituting in for $\mathbf{H}$ and $\mathbf{F}$ in the first three rows of $\mathcal{O}$, we obtain:

$$\mathcal{O} = \begin{bmatrix} 1 & -1 & 0 \\ 0 & 0 & -\Delta t \\ 0 & 0 & -\Delta t \end{bmatrix} \tag{11}$$

By inspection, we find that this system can observe two modes of the system, namely $[1 \quad -1 \quad 0]$ and $[0 \quad 0 \quad 1]$. Because of the second mode, we conclude that biases are observable when relative measurements of the state are used in a Kalman Filter framework.

4.2 Filter Setup for Inertial Bias Estimation

Knowing that biases are observable when relative navigation measurements are used, we can modify the fusion framework developed in Sects. 2 and 3 to estimate biases. First, we modify the state vector to include biases for all sensors, yielding

$$\mathbf{X} = \left[\chi_t, \chi_{t-1}, b_{ax}, b_{ay}, b_{az}, b_{gx}, b_{gy}, b_{gz}\right]^T, \tag{12}$$

where b_{nm} is the bias estimate with the n subscript denoting a for accelerometer and g for gyro, and the m subscript denoting which axis the sensor is measuring (x, y, or z). The $\mathbf{A}$ matrix described in Eq. (3) is modified to

$$\mathbf{A} = \begin{bmatrix} I_{10\times 10} & 0 & 0 \\ I_{10\times 10} & 0 & 0 \\ 0 & 0 & I_{6\times 6} \end{bmatrix}, \tag{13}$$

to maintain bias estimates between measurements from the visual sensor. The time update that utilizes the inertial sensors to update the first 10 elements of the state is modified to subtract out the bias estimate from the inertial measurements. The measurement step for the filter remains unchanged. In the following section, we present results demonstrating the improved navigation performance obtained from estimating the biases of the inertial sensors.

5 Results

To demonstrate the results of fusing visual, air pressure, and inertial sensors together as proposed in this paper, we developed a detailed simulator that generated synthetic MAV flights, together with uncorrupted and corrupted sensor measurements for that flight. In the subsections that follow, we first describe the simulation environment in more detail, followed by results demonstrating the efficacy of the fusion system proposed in this paper.

5.1 Simulation Environment

To enable an evaluation of our epipolar constraint-based fusion environment, we first need to generate the true navigation states of the MAV over time. To generate true location data about the MAV, a Bézier curve representing the true path of the MAV was created. A Bézier curve was chosen due to its inherent flexibility in representing many different types of curves in 3-D space. In addition, Bézier curves are a polynomial function of a single scalar t, yielding two significant advantages. First, the location at any time can be easily determined. Second, by differentiating the polynomial with respect to t, the velocity and acceleration at any point on the curve can be computed in closed form. All quantities are assumed to be in a "navigation frame" which has North as its x axis, East as its y axis, and straight down as the z axis. The origin of this frame was arbitrarily chosen as a location on the ground in Utah, near Brigham Young University (close to our MAV flight test area).

In addition to generating the location, velocity, and acceleration of the MAV, we also need to generate the angular orientation (attitude) of the MAV camera. We have used two basic approaches to generating the attitude of the MAV camera. First, for a "fixed" camera, the angular orientation is always constant within the MAV body frame. Second, we set the attitude of the camera such that a specified world location will always be in the center of the image, representing a gimbaled camera that remains pointed at a specific location. We utilize the second approach for the results presented in this section.

Once the true location and attitude of the camera are known, the inputs to the fusion algorithm are generated. We assume the inputs from the IMU consist of 3-axis accelerometer and gyroscope (gyro) readings. To generate accelerometer readings, the acceleration of the camera is computed from the Bézier curve. The effects of gravity, Coriolis, and the rotation of the earth are then added to the accelerometer readings as described in [16], yielding noise-free accelerometer readings. To generate gyro readings, the attitude at two locations on the Bézier curve is computed. The locations on the curve are separated by the gyro sample time. The difference in attitude is then used to compute the angular rates of the camera, yielding noise-free gyro readings. Noise-free pitot tube readings are computed as the magnitude of the velocity at a point on the Bézier curve.

Once the noise-free readings have been computed, two types of noise are added to the sensor readings. First, Gaussian, zero-mean white noise is added to the computed readings. The variance of the noise values were chosen to approximate measurement errors observed on a Kestrel autopilot [3]. Second, a constant bias is added to the gyro and accelerometer readings. For each run of the simulator, biases were randomly selected from a Gaussian distribution with twice the standard deviation of the white noise for that sensor.

To simulate inputs from the camera, a set of random world points to be imaged are created. Using the locations of the world points and the location and attitude of the MAV over time, a set of feature locations corresponding with time along its flight path are created. Features locations for a specific MAV location and attitude are computed using the formula

$$\lambda \begin{pmatrix} x_i \\ y_i \\ 1 \end{pmatrix} = \mathbf{K}\mathbf{C}_n^c(\mathbf{X}^n - \mathbf{p}^n), \tag{14}$$

where $\mathbf{X}^n$ was the location of the world point (in the navigation frame), $\mathbf{p}^n$ is the position of the camera in navigation frame coordinates (determined from its point on the Bézier curve), $\mathbf{C}_n^c$ is the direction cosine matrix from the navigation frame to the camera frame (also a function of location on the Bézier curve), $\mathbf{K}$ is the calibration matrix of the camera, mapping from Euclidean to pixel locations, λ is a scale factor used for normalizing the third element of the image frame vector to 1, and x_i and y_i are the image coordinates of the point.

After determining the location of the object in the image space, Gaussian white zero-mean noise is added to the image location. We set the standard deviation of the noise equal to a single pixel in the image plane. After adding noise, the pixel values are then "de-calibrated" (multiplied by $\mathbf{K}^{-1}$) to obtain vectors in the same Euclidean space as the MAV navigation state.

5.2 Fusion Results

To test the efficacy of our proposed fusion environment, we use two different "flight scenarios." In the first scenario, the MAV moves in a straight line starting at 100 m above the ground and 100 m south of the navigation frame origin. The camera then moves in a straight line to 100 m north of the navigation frame origin, holding a constant altitude. In the East-West (y) direction, the MAV is always at 0. Along this path, 161 images were captured at a rate of 10 Hz, requiring 16 s to fly this path. These values were chosen to achieve an airspeed (12.5 m/s) typical of MAVs. In addition, 1,601 samples of the gyro and accelerometer readings were collected. Note that while this flight scenario may seem like an overly simplistic maneuver (flying in a straight line), it was chosen because it actually exacerbates one of the fundamental problem of vision, the universal scale ambiguity. Therefore, this scenario is one of the most difficult scenarios for vision-aided navigation. Results for this scenario are shown in Table 3. Note that this scenario was used for the partial results presented earlier in this paper.

The second scenario represents a more generic flight of an MAV. It starts at -100 m north, 100 m in altitude. It then flies an "S" pattern, going northeast before turning to go northwest. While flying northwest, it passes directly over the navigation frame origin, after which it turns back to head northeast, arriving at -30 m east, 100 m north. During the course of the flight, the altitude also drops from 100 m to 60 m. This entire flight takes 19 s. We refer to this scenario as "The S Pattern", with results shown in Table 4.

In both scenarios described above, the world points being observed were distributed using a three-dimensional Gaussian distribution centered about the navigation frame origin. To keep the objects in view, the camera is continuously rotated to "look at" the origin.

To determine the overall accuracy of each fusion technique, we ran each UKF filter setup with each flight path scenario 100 times. In Tables 3 and 4, the mean and standard deviation of the errors across 100 runs of the filter are shown. The mean and standard deviation achieved using only the IMU is also shown for each flight scenario as a reference. The units for the final position error (p_x, p_y, and p_z) are in meters, while the final attitude errors are in degrees. The attitude errors are the amount of yaw (ψ), pitch (θ) and roll (ϕ) that would be required to move from the true location to the estimated locations.

For each of these flight scenarios, five different setups of our UKF environment were used. First, we ran epipolar constraint-based fusion without any of the modifications introduced in Sect. 3 (*Baseline*). Second, we remove the bias in the direction the camera is pointed as discussed in Sect. 3.1 (sin *Removed*). Third, the measurements from the pitot tube are included in the UKF framework (*Pitot Added*). Fourth, a slower sampling rate (2 Hz, as opposed to 10 Hz) is used in addition to all the other modifications (*Min. Sampling*). Finally, the "Min. Sampling" filter is modified to estimate the inertial sensor biases (*Est. Bias*).

As shown in these tables, each modification proposed in this paper significantly reduces the mean and standard deviation of the error. By including all four proposed

Table 3 Mean and standard deviation of error in the final estimated location and attitude of the MAV when flying a straight-line path with a gimbaled camera. Note that the errors in attitude are all under one degree when all four modifications proposed in this paper are implemented

	IMU only	Baseline	sin Removed	Pitot Added	Min. Sampling	Est. Bias
p_x **Error** (μ,σ)	(−4.48, 115.4)	(−607.0, 136.6)	(23.1, 85.9)	(−0.47, 1.92)	(−0.39, 1.19)	(0.03, 1.13)
p_y **Error** (μ,σ)	(8.27, 95.3)	(2.83, 76.5)	(−1.77, 16.0)	(−1.01, 11.2)	(−0.24, 3.05)	(−.08, 1.96)
p_z **Error** (μ,σ)	(15.1, 12.7)	(757.5, 318.1)	(11.8, 9.71)	(2.85, 1.08)	(11.1, 1.17)	(2.19, 0.82)
ψ **Error** (μ,σ)	(−0.69, 14.0)	(2.26, 17.1)	(0.20, 2.52)	(0.21, 2.71)	(0.09, 1.07)	(0.02, 0.58)
θ **Error** (μ,σ)	(0.16, 15.93)	(−143.5, 61.3)	(1.57, 6.59)	(−0.31, 1.70)	(1.23, 1.78)	(0.27, 0.61)
ϕ **Error** (μ,σ)	(−1.05, 12.9)	(2.90, 18.1)	(0.65, 5.67)	(0.26, 4.13)	(0.08, 1.13)	(0.02, 0.72)

Table 4 Mean and standard deviation of error in the final estimated location and attitude of the MAV when flying the "S-curve" path with a gimbaled camera

	IMU only	Baseline	sin Removed	Pitot Added	Min. Sampling	Est. Bias
p_x **Error** (μ,σ)	(0.26, 110.4)	(−128.6, 130.9)	(−22.8, 88.1)	(−1.89, 23.1)	(0.79, 2.39)	(0.08, 1.73)
p_y **Error** (μ,σ)	(−15.4, 132.4)	(78.0, 354.0)	(−11.4, 92.7)	(0.91, 11.8)	(6.20, 2.75)	(2.09, 1.82)
p_z **Error** (μ,σ)	(18.8, 33.6)	(528.3, 278.9)	(41.4, 102.0)	(−2.90, 4.04)	(5.18, 2.79)	(−1.58, 1.36)
ψ **Error** (μ,σ)	(0.91, 14.5)	(39.3, 31.4)	(5.01, 19.4)	(0.38, 3.20)	(−0.23, 2.67)	(0.26, 1.40)
θ **Error** (μ,σ)	(−0.07, 11.5)	(−27.1, 26.9)	(−3.84, 23.0)	(−1.18, 4.91)	(0.78, 1.87)	(−0.41, 0.99)
ϕ **Error** (μ,σ)	(2.32, 13.2)	(−55.7, 47.9)	(−1.40, 25.6)	(1.85, 5.59)	(0.71, 1.28)	(0.65, 0.73)

modifications, more than an order of magnitude decrease in error is achieved from both IMU-only navigation and the baseline fusion approach. This demonstrates the necessity of including the proposed modifications when considering epipolar constraint based fusion for navigation. It also demonstrates the advantages of estimating the inertial sensor biases to help reduce noise.

6 Conclusion

In this paper, we have proposed a system for fusing IMU and visual information together in a UKF framework utilizing the epipolar constraint. However, a naive implementation of this approach yields sub-optimal results. Therefore, we propose three modifications to the baseline fusion setup that significantly improve the overall performance of this system. These include: (1) removing the bias toward the center of tracked points when using the epipolar constraint, (2) using a pitot tube to measure the velocity of th MAV, and (3) minimizing the sampling rate of the image data to achieve minimal error growth. After implementing these improvements it is possible to estimate the biases of the inertial sensors, decreasing the amount of noise present in the system. Gains in accuracy of at least 10X in both the location and attitude estimates were achieved using these improvements.

Acknowledgments This work was funded by an AFOSR Young Investigator Award, number FA9550-07-1-0167. This work began while the author was a summer faculty fellow with the Air Force Research Labs under the supervision of Mikel Miller in the Munitions Directorate. Assistance was also provided by Mike Veth of the Air Force Institute of Technology.

References

1. Beard R, Kingston D, Quigley M, Snyder D, Christiansen R, Johnson W, McLain T, and Goodrich M (2005) Autonomous vehicle technologies for small fixed wing UAVs. *AIAA Journal of Aerospace Computing, Information, and Communication* 2(1): 92–108.
2. Kingston DB and Beard RW (2004) Real-time attitude and position estimation for small uav's using low-cost sensors. In: *AIAA 3rd Unmanned Unlimited Systems Conference and Workshop*, Chicago, IL.
3. Procerus Technologies (2008) Procerus technologies URL http://www.procerusuav.com
4. Micropilot (2008) Micropilot, URL http://www.micropilot.com
5. Strelow D and Singh S (2004) Motion Estimation from Image and Inertial Measurements. *The International Journal of Robotics Research* 23(12): 1157–1195.
6. Veth MJ, Raquet JF, and Pachter M (2006) Stochastic constraints for efficient image correspondence search. *IEEE Transactions on Aerospace and Electronic Systems* 42(3): 973–982.
7. Veth M and Raquet J (2007) Fusing low-cost image and inertial sensors for passive navigation. *Journal of the Institute of Navigation* 54(1): 11–20.
8. Langelaan J (2006) State estimation for autonomous flight in cluttered environments. PhD thesis, Stanford University.
9. Prazenica R, Watkins A, Kurdila A, Ke Q, and Kanade T (2005) Vision-based kalman filtering for aircraft state estimation and structure from motion. In: *2005 AIAA Guidance, Navigation, and Control Conference and Exhibit*, pp. 1–13.

10. Soatto S, Frezza R, and Perona P (1996) Motion estimation via dynamic vision. *IEEE Transactions on Automatic Control* 41(3): 393–413.
11. Ready BB and Taylor CN (2007) Improving accuracy of mav pose estimation using visual odometry. In: *American Control Conference, 2007*. ACC '07, pp. 3721–3726.
12. Andersen ED and Taylor CN (2007) Improving mav pose estimation using visual information. In: *IEEE/RSJ International Conference on Intelligent Robots and Systems*, pp.3745–3750.
13. Sorenson HW (1966) Kalman filtering techniques. In: Leondes CT (ed) *Advances in Control Systems, Theory and Applications*, vol 3, pp. 218–292.
14. Xing Z and Gebre-Egziabher D (2008) Modeling and bounding low cost inertial sensor errors. In: *Proceedings of IEEE/ION Position Location and Navigation Symposium*, pp. 1122–1132.
15. IEEE Std 952 (1997) IEEE Standard specification formate guide and test procedures for single-axis interferometric fiber optic gyros. *IEEE Standard* 952–1997.
16. Titterton D and Weston J. (1997) *Strapdown Inertial Navigation Technology*. Lavenham, United Kingdom: Peter Peregrinus Ltd.
17. Cloudcap (2008) Cloud cap technology. URL http://www.cloudcaptech.com

Part II
Multi-Sensor Fusion and Integration in Robotics and Vision

Sukhan Lee

Multi-sensor fusion and integration play a major role forrobotics and computer vision. The autonomy in navigation and manipulation that robotics pursue as its goal requires ultimately the ability of a robot to recognize and model the environment they are engaged in, often relying on vision. One of the key issues the robotics and vision field is facing today is how to solve the dependability in recognition and modeling against the many real-world variations, such as the variations in illumination, texture, surface reflection, occlusion, form factor, sensor pose, etc., in spite of the presence of fundamental limitations of sensing in, e.g., dynamic range, resolution, measurement error, field of view, etc. Multi-sensor fusion and integration have been regarded as indispensible for solving the issue of dependability addressed above. This chapter presents how multi-sensor fusion and integration can be applied to the dependability in recognition and modeling of environments, in particular, for robotic navigation, manipulation, and interaction with human.

The first two papers address the integration and fusion of two heterogeneous sensors, laser scanners and cameras, to improve the performance of Simultaneous Localization and Map Building(SLAM). The paper entitled "Simultaneous Estimation of Road Region and Ego-Motion with Multiple Road Models," by Yoshiteru Matsushita and Jun Miura addresses the multi-sensor based simultaneous estimation of road region and ego-motion based on a particle filter. A laser range finder and an omni-directional camera system are integrated to detect and fuse the L-shaped curb and the road boundary lines and roadside regions. For the latter, the intensity gradient and the color gradient images are used, respectively. In addition, particles representing the gradual road type change are incorporated in the particle filter. Autonomous driving of a mobile robot is demonstrated as experimentation. The paper entitled "Visual SLAM in Indoor environments using Autonomous Detection and Registration of Objects" by Yong-Ju Lee and Jae-Bok Song presents how a hybrid grid/vision map can be built by integrating vision detected objects with an IR scanner based grip map. Various 2D visual cues are used to distinguish objects from the background for detection. The authors claim that their approach requires a less number of landmarks than the conventional laser scanner based SLAM.

In their paper entitled "The 'Fast Clustering-Tracking' Algorithm in the Bayesian Occupancy Filter Framework," Kamel Mekhnacha, Yong Mao, David Raulo, and Christian Laugier propose clustering the occupancy and velocity grid into an object level report, where the occupancy and velocity grid describes the probability distribution of cell occupancy and cell occupancy velocity. Note that a grid of occupancy probabilities and mean velocity estimates representing environments is from the Bayesian occupancy filter presented as a unified framework for sensor integration and fusion by the authors. A fast clustering algorithm is proposed as a means of avoiding the combinatorial complexity in computation. In their paper entitled "Fusion of Double Layered Multiple Laser range Finders for People Detection from a Mobile Robot," Alexander Carballo, Akihisa Ohya and Shinichi Yuta present a method for simple and accurate detection and tracking of people in an indoor public area based on multi-layered laser range finders. The laser range finders in a double layer configuration provide the 360 degree of surroundings at the human chest and leg levels based on data fusion. The paper has shown that not only the 3D model of people and their positions but also the direction the person is facing at could possibly be obtained.

The paper entitled "Model based Recognition of 3D objects using Intersecting lines" by Hung Q. Truong, Sukhan Lee and Seok-Woo Jang presents the recognition and pose estimation of 3D objects based on perpendicularly intersecting 3D line segments as the cues to match with the model. Probabilities are assigned to all the possible interpretations of the object pose based on the matching scores, such that the probabilities can be updated as more evidences are accumulated in time. The paper entitled "Pedestrian Route Guidance System using Moving Projection based on Personal Feature Extraction" by Takuji Narumi, Yashushi Hada, Hajime Asama, and Kunihiro Tsuji presents a method of the traffic-line guidance for pedestrians by projecting moving images on the site of estimated human gaze by a pan-tile projector. Surveillance cameras are used to estimate the human sight.

In their paper entitled "Behavioral Programming with Hierarchy and Parallelism in the DARPA Urban Challenge and RoboCup," Jesse G. Hurdus and Dennis W. Hong propose a hierarchical state machine for the efficient construction, organization and selection of behaviors in such a way that a robot can exhibit contextual intelligence. Not only the arbitration of competing behaviors but also the assembly of behaviors into an emergent behavior take place in the proposed hierarchical state machine.

Ruben Smits, Tinne de Laet, Kasper Claes, Herman Bruyninckx, Joris de Schutter present "iTASC: a Tool for Multi-Sensor Integration in Robot manipulation" as a unified framework for integrating instantaneous task specification and geometric uncertainty estimation. iTASC helps specify complex tasks for a general sensor-based robot system based on system constraints. A people tracker based on encoders, a force sensor, cameras, a laser distance sensor and a laser scanner is implemented, where kinematic control and uncertainty estimation equations are derived based on 10 primary constraints, 7 uncertainty coordinates, 6 scalar measurements, and 12 secondary constraints. Mario Prats, Philippe Martinet, Sukhan Lee and Pedro J. Sanz show in their paper entitled "Compliant Physical Interaction based on External Vision-Force Control and tactile-Force Combination" that multi-sensor based compliant physical interaction is feasible based on the task frame formalism. They demonstrate the pull-opening of the handle with a parallel jaw gripper, as well as the opening of a sliding door with a three fingered hand, based on the external vision-force coupling and the extracting of a book from a bookshelf by tactile and force integration.

In their paper entitled "Recognizing Human Activities from Accelerometer and Physiological Sensors," Sung-Ihk Yang and Sung-Bae Cho present the recognition of 9 kinds of human activities: walking, running and exercising, eating, reading, studying, playing, sleeping, based on an armband sensor system integrated with accelerometers and physiological sensors. The latter includes heat flux, galvanic skin response, skin temperature sensors and thermometer. The inability of accelerometer to detect near stationery activities is compensated by other sensors. About 74% accuracy is reported with the fuzzy logic used for decision. In their paper entitled "Enhancement of Images Degraded by Fog using Cost Function based on Human Visual Model," Dongjun Kim, Changwon Jeon, Bonghyup Kang, and Hanseok Ko present an estimation of an air-light map generated by fog particles in order to enhance the image quality by subtracting the estimated air-light map from the degraded image. The estimation of an air-light map is based on the estimate of luminance distribution variation under the constraints of the sensitivity derived from the human visual model.

Enhancement of Image Degraded by Fog Using Cost Function Based on Human Visual Model

Dongjun Kim, Changwon Jeon, Bonghyup Kang and Hanseok Ko

Abstract In foggy weather conditions, images become degraded due to the presence of airlight that is generated by scattering light by fog particles. In this paper, we propose an effective method to correct the degraded image by subtracting the estimated airlight map from the degraded image. The airlight map is generated using multiple linear regression, which models the relationship between regional airlight and the coordinates of the image pixels. Airlight can then be estimated using a cost function that is based on the human visual model, wherein a human is more insensitive to variations of the luminance in bright regions than in dark regions. For this objective, the luminance image is employed for airlight estimation. The luminance image is generated by an appropriate fusion of the R, G, and B components. Representative experiments on real foggy images confirm significant enhancement in image quality over the degraded image.

1 Introduction

Fog is a phenomenon caused by tiny droplets of water in the air. Fog reduces visibility down to less than 1 km. In foggy weather, images also become degraded by additive light from scattering of light by fog particles. This additive light is called 'airlight'.

There have been some notable efforts to restore images degraded by fog. The most common method known to enhance degraded images is histogram equalization. However, even though global histogram equalization is simple and fast, it is not suitable because the fog's effect on an image is a function of the distance between the camera and the object. Subsequently, a partially overlapped sub-block histogram equalization was proposed in [1]. However, the physical model of fog was not adequately reflected in this effort.

D. Kim (✉)
School of Electrical Engineering, Korea University, Seoul, KOREA
e-mail: djkim@ispl.korea.ac.kr

S. Lee et al. (eds.), *Multisensor Fusion and Integration for Intelligent Systems*, Lecture Notes in Electrical Engineering 35, DOI 10.1007/978-3-540-89859-7_12,

While Narasimhan and Nayar were able to restore images using a scene-depth map [2], this method required two images taken under different weather conditions.

Grewe and Brooks suggested a method to enhance pictures that were blurred due to fog by using wavelets [3]. Once again, this approach required several images to accomplish the enhancement.

Polarization filtering is used to reduce fog's effect on images [4, 5]. It assumes that natural light is not polarized and that scattered light is polarized. However this method does not guarantee significant improvement in images with dense fog since it falls short of expectations in dense fog.

Oakley and Bu suggested a simple correction of contrast loss in foggy images [6]. In [6], in order to estimate the airlight from a color image, a cost function is used for the RGB channel. However, it assumes that airlight is uniform over the whole image.

In this paper, we improve the Oakley method [6] to make it applicable even when the airlight distribution is not uniform over the image. In order to estimate the airlight, a cost function that is based on the human visual model is used in the luminance image. The luminance image can be estimated by an appropriate fusion of the R, G, and B components. Also, the airlight map is estimated using least squares fitting, which models the relationship between regional airlight and the coordinates of the image pixels.

The structure of this paper is as follows. In Sect. 2, we propose a method to estimate the airlight map and restore the fog image. We present experimental Results and Conclusions in Sects 3 and 4 respectively. The structure of the algorithm is shown in Fig. 1.

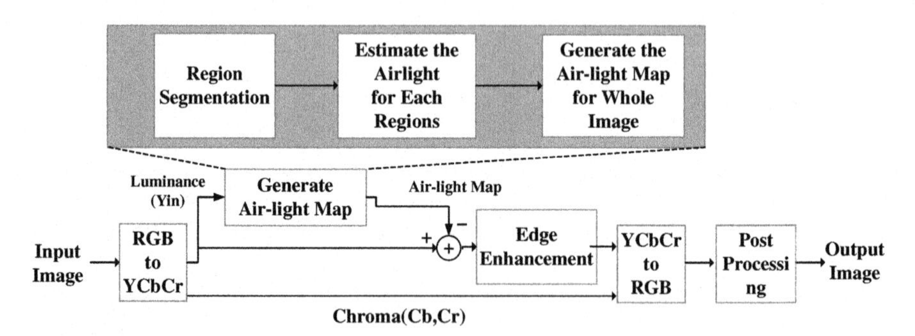

Fig. 1 Structure of the algorithm

2 Proposed Algorithm

2.1 Fog Effect on Image and Fog Model

The foggy image is degraded by airlight that is caused by scattering of light with fog particles in air as depicted in Fig. 2 (right).

Fig. 2 Comparison of the clear image (left) and the fog image (right)

Airlight plays the role of being an additional source of light as modeled in [6] and Eq. (1) below.

$$I'_{R,G,B} = I_{R,G,B} + \lambda_{R,G,B} \tag{1}$$

where $I'_{R,G,B}$ is the degraded image, $I_{R,G,B}$ is the original image, and $\lambda_{R,G,B}$ represents the airlight for the Red, Green, and Blue channels. This relationship can be applied in the case where airlight is uniform throughout the whole image. However, the contribution of airlight is not usually uniform over the image because it is a function of the visual depth, which is the distance between the camera and the object. Therefore, the model can be modified to reflect the depth dependence as follows.

$$I'_{R,G,B}(d) = I_{R,G,B}(d) + \lambda_{R,G,B}(d) \tag{2}$$

Note that "*d*" represents depth. Unfortunately, it is very difficult to estimate the depth using one image taken in foggy weather conditions, so we present an airlight map that models the relationship between the coordinates of the image pixels and the airlight. In this paper, since the amount of scattering of a visible ray by large particles like fog and clouds are almost identical, the luminance component is used alone to estimate the airlight instead of estimating the R, G, and B components. The luminance image can be obtained by a fusion of the R, G, and B components. Subsequently, the color space is transformed from RGB to YCbCr. Therefore Eq. (2) can be re-expressed as follows.

$$Y'(i,j) = Y(i,j) + \lambda_Y(i,j) \tag{3}$$

where Y' and Y reflect the degraded luminance and clear luminance images respectively at position (i,j). Note that λ_Y is the estimated airlight map for the luminance image. The shifting of mean(Y) can be confirmed in Fig. 3.

In order to restore the image blurred by fog, we need to estimate the airlight map and subtract the airlight from the foggy image as follows.

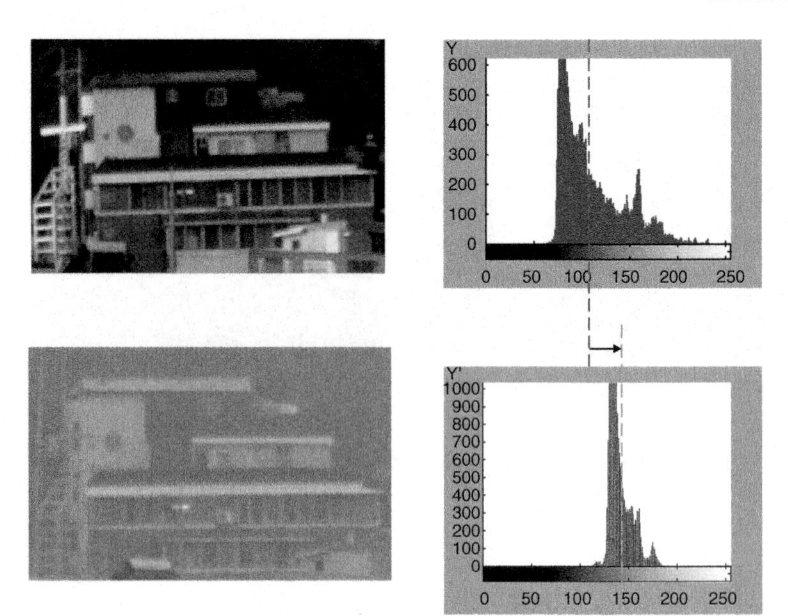

Fig. 3 Comparison of the *Y* histogram

$$\hat{Y}(i, j) = Y'(i, j) - \hat{\lambda}_Y(i, j) \tag{4}$$

In this model, $\hat{Y}$ represents the restored image and $\hat{\lambda}_Y$ is the estimated airlight map.

2.2 Region Segmentation

In this paper, we suggest estimating the airlight for each region and modeling the airlight for each region and the coordinates within the image to generate the airlight map. In the case of an image with various depth, the contribution of airlight can be varied according to the region. Estimating the airlight for each region can reflect the

Fig. 4 Region segmentation

variation of depth within the image. Regions are segmented uniformly to estimate the regional contribution of airlght.

2.3 Estimate Airlight

In order to estimate the airlight, we improved the cost function method in [6] using a compensation that is based on the human visual model.

In Eq. (3), the airlight is to be estimated to restore the image degraded by fog. To estimate the airlight, the human visual model is employed. As described by Weber's law, a human is more insensitive to variations of luminance in bright regions than in dark regions.

$$\Delta S = k\frac{\Delta R}{R} \tag{5}$$

where R is an initial stimulus, ΔR is the variation of the stimulus, and ΔS is a variation of sensation.

In the foggy weather conditions, when the luminance is already high, a human is insensitive to variations in the luminance.

We can estimate the existing stimulus in the image signal by the mean of the luminance within a region. The variation between this and foggy stimulus can be estimated by the standard deviation within the region. Thus the human visual model would estimate the variation of sensation as

$$\frac{STD(Y)}{mean(Y)} = \frac{\sqrt{\frac{1}{n}\sum_{i=1}^{n}(y_i - \overline{Y})^2}}{\overline{Y}} \tag{6}$$

Where $\overline{Y}$ means that mean value of Y. Note that the value of Eq. (6) for a foggy image, $STD(Y')/mean(Y')$, is relatively small since the value of numerator is small and the value of denominator is large.

$$A(\lambda) = \frac{STD(Y' - \lambda)}{mean(Y' - \lambda)} \tag{7}$$

In Eq. (7), increasing λ causes an increase in $A(\lambda)$, which means that a human can perceive the variation in the luminance. However, if the absolute value of the luminance is too small, it is not only too dark, but the human visual sense also becomes insensitive to the variations in the luminance that still exist. To compensate for this, a second function is generated as follows.

$$B'(\lambda) - (mean(Y' - \lambda)) \tag{8}$$

Eq. (8) indicates information about mean of luminance. In a foggy image, the result of Eq. (8) is relatively large. And, increasing λ causes a decrease in $B(\lambda)$ which means that overall brightness of the image decreases.

Functions (7) and (8) reflect different scales from each other. Function (8) is re-scaled to produce Eq. (9) to set 0 when input image is Ideal. Note that "Ideal" represents the ideal image having a uniform distribution from the minimum to the maximum of the luminance range. In general, the maximum value is 235 while the minimum value is 16.

$$B(\lambda) = (mean(Y') - \lambda) \times \frac{STD(Ideal)}{mean(Ideal)^2} \tag{9}$$

For dense foggy image, the result of $A(\lambda)$–$B(\lambda)$ is relatively large when λ is small. Increasing λ causes a decrease in $A(\lambda)$–$B(\lambda)$. If λ is too large, it cause an increase in $|A(\lambda)$–$B(\lambda)|$ which means the image becomes dark. The λ satisfying Eq. (10) is the estimated airlight.

$$\hat{\lambda} = \arg\min_{\lambda}\{|A(\lambda) - B(\lambda)|\} \tag{10}$$

2.4 Estimate Airlight Map Using Multiple Linear Regression

Objects in the image are usually located at different distances from the camera. Therefore, the contribution of the airlight in the image also differs with depth. In most cases, the depth varies with the row or column coordinates of the image scene.

This paper suggests modeling between the coordinates and the airlight values that are obtained from each region. The airlight map is generated by multiple linear regression using least squares (Fig. 5).

2.5 Restoration of Luminance Image

In order to restore the luminance image, the estimated airlight map is subtracted from the degraded image as Eq. (4).

To correct the blurring due to fog, edge enhancement is performed.

$$\hat{Y}_{de-blurr}(i, j) = \hat{Y}(i, j) + s \cdot g(i, j) \tag{11}$$

where $g(i,j)$ is the reverse Fourier transformed signal that is filtered by a high pass filter, s is a constant that determines the strength of enhancement, and $\hat{Y}_{de-blurr}(i,j)$ is the de-blurred luminance image.

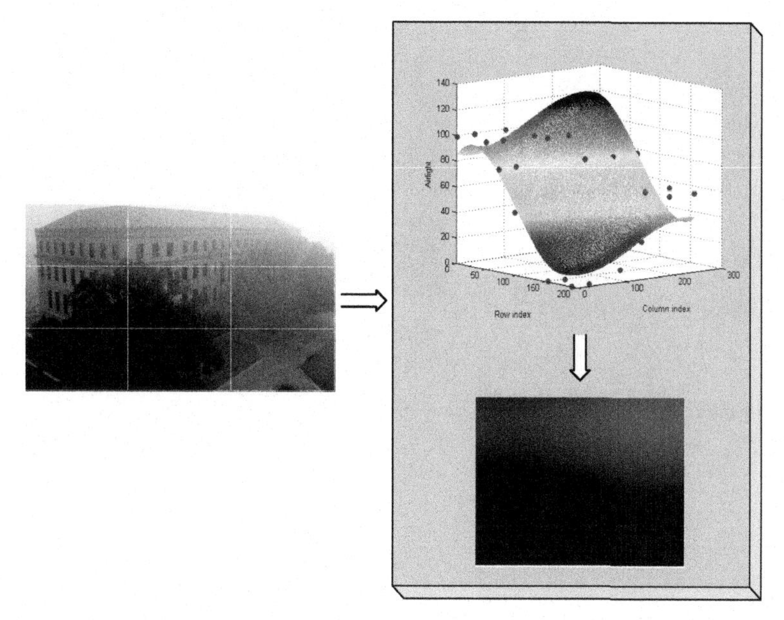

Fig. 5 Generation of airlight map

2.6 Post-Processing

The fog particles absorb a portion of the light in addition to scattering it. By changing the color space from YCbCr to RGB, $I_{R,G,B}$ can be obtained. Therefore, after the color space conversion, histogram stretching is performed as a post-processing step.

$$\tilde{I}_{R,G,B} = 255 \times \frac{I_{R,G,B} - \min(I_{R,G,B})}{\max(I_{R,G,B}) - \min(I_{R,G,B})} \tag{12}$$

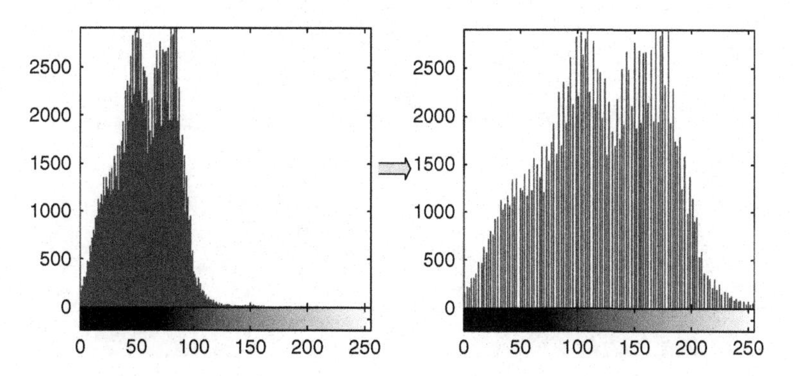

Fig. 6 Histogram stretching

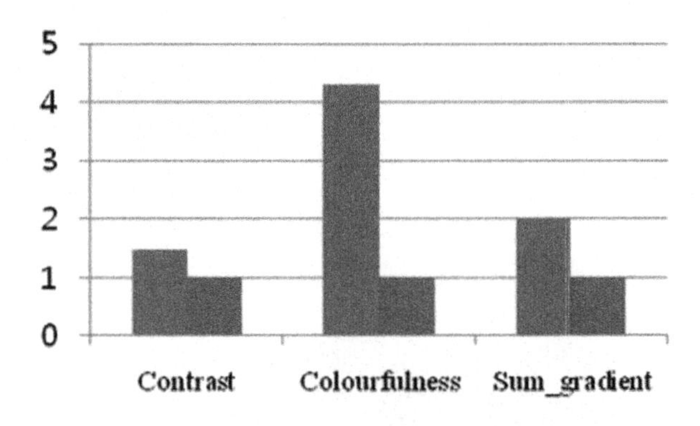

Fig. 7 Result of the evaluation

Fig. 8 Results of image enhancement by defogging

where $\tilde{I}_{R,G,B}$ is the result of histogram stretching, $\max(I_{R,G,B})$ is the maximum value of $I_{R,G,B}$ that is an input for post-processing, and $\min(I_{R,G,B})$ is the minimum value of $I_{R,G,B}$.

3 Result

The experiment is performed on a 3.0 GHz Pentium 4 using MATLAB. The experiment results for images taken in foggy weather are shown in Fig. 8.

In order to evaluate the performance, we calculate contrast, colorfulness, and the sum of the gradient that is based on the important of edges measurement. Contrast and colorfulness are improved by 147% and 430% respectively over the foggy image. In addition, the sum of the gradient is also improved 201% compared to the foggy image.

4 Conclusion

In this paper, we propose to estimate the airlight using cost function, which is based on human visual model, and generate airlight map by modeling the relationship between coordinates of image and airlight. Blurred image due to fog is restored by subtracting airlight map from degraded image.

In order to evaluate the performance, we calculated contrast, colorfulness and sum of gradient. The results confirm a significant improvement in image enhancement over the degraded image. In the future, we plan to investigate a methodology to estimate the depth map from single image. In addition, enhancement of degraded image in bad weather due to non-fog weather will be investigated.

Acknowledgments This work is supported by Samsung Techwin CO., LTD, Thanks to Professor Colin Fyfe for his valuable advices and comments.

References

1. Y. S. Zhai and X. M. Liu, An improved fog-degraded image enhancement algorithm, International Conference on Wavelet Analysis and Pattern Recognition, 2007, ICWAPR'07, vol. 2, 2007.
2. S. G. Narasimhan and S. K. Nayar, Contrast restoration of weather degraded images, *IEEE Transactions on Pattern Analysis and Machine Intelligence*, 25, 713–724, 2003.
3. E. Namer and Y. Y. Schechner, Advanced visibility improvement based on polarization filtered images," *Proceedings of SPIE*, 5888, 36–45, 2005.
4. Y. Y. Schechner, S. G. Narasimhan, and S. K. Nayar, Polarization-based vision through haze, *Applied Optics*, 42, 511–525, 2003.
5. J. P. Oakley and H. Bu, Correction of simple contrast loss in color images, *IEEE Transactions on Image Processing*, 16, 511–522, 2007.
6. Y. Yitzhaky, I. Dror, and N. S. Kopeika, Restoration of atmospherically blurred images according to weather-predicted atmospheric modulation transfer functions, *Optical Engineering*, 36, 3064, 1997.

Pedestrian Route Guidance by Projecting Moving Images

Takuji Narumi, Yasushi Hada, Hajime Asama and Kunihiro Tsuji

Abstract We propose a novel route guidance system for pedestrian in public space by using moving projection and personal feature extraction from computer vision. In our system, we extract visitor's features from the images captured by a surveillance camera. And, the system generates a motion of guidance information based on the features and presents the guidance information with the image moving by using moving projector. By this moving image, the system guides him to the correct way which we aimed. Moreover, for the effective design, we made models are based on how a person perceives information through sight. Thorough two experiments, we demonstrate the effectiveness of our method to easily design traffic lines.

1 Introduction

In the present world flooded with information, the ability to select pertinent information from all the available information is imperative. According to changes in the information environment surrounding us, we require a paradigm shift from the one-way unchanging presentation of a large amount of information to an individual presentation that is interactive and adaptive. In this paper, we consider the problem of route guidance for pedestrian in public spaces as an example of the adaptive and individual presentation. As an application of the individual presentation of information in public spaces, we propose a new method for route guidance system by using moving projection and personal feature extraction from computer vision.

Route guidance for pedestrian is important for an effective utilization of space. For example, directions to some places, advertising for shops, and directions to a safe place for evacuation. Although route guidance is important like these, there are some problems in conventional guidance system. To date, guidance information for pedestrian has generally been provided in the form of arrows and signboards. These conventional methods present information uniformly to all people. Since each individual is different with regard to their age, height, hobbies, etc., these methods

T. Narumi (✉)
Graduate School of Engineering, The University of Tokyo/JSPS, 7-3-1 Bunkyo-Ku Hongo, Japan
e-mail: narumi@cyber.t.u-tokyo.ac.jp

S. Lee et al. (eds.), *Multisensor Fusion and Integration for Intelligent Systems*, Lecture Notes in Electrical Engineering 35, DOI 10.1007/978-3-540-89859-7_13,

do not always work effectively. Our motivation is to use space more effective by developing a more effective route guidance system.

As problems in conventional guidance system, there are three problems. At first, there is oversight problem. The person who should be guided often cannot notice the guidance information. Secondly, there is misdirection problem. An indistinct guidance results in misdirection. Third, presented information is static in most case. It is not easy to change contents of the guidance even if it is necessary. Moreover, there is some limitation in the setting. There are two limitations: physical limitation and designing limitation. We cannot put signboard on the middle of the way. And, we don't want to put some object which doesn't match the spatial design.

According to Hillstrom and Yantis [1], people are more attentive to moving rather than stationary objects. This information has valuable implications in the context of our research. In other words, if the movement of guidance information shows a route to walk, it will attract more attention than the presentation of stationary information.

The contents of guidance information are decided from the relationship between a person's position and the target position. Therefore, many researchers have studied various methods to sense a person's position. GPS has been used by almost all the existing methods to locate a person's position outdoors [2, 3]. Further, tags or markers are generally placed on a person to sense his/her position indoors [4].

In addition, the conventional methods of information presentation require using a portable device or multiple fixed devices [5]. However, since the methods that rely on devices or tags limit the number of people who can utilize the service, they are unsuitable for traffic line guidance in public spaces, which are visited by a large number of individuals.

As the existing method that devices or markers were not putted on a person with, there were studies to show the suitable information for the situation in robotic room with Pan-Tilt projector [6]. However, the studies only showed the stationary information at a place that was chosen from several short listed places. They did not consider a way to attract the attention of people to the presented information.

Therefore, we propose a new method for pedestrian route guidance; this method uses an unspecified personal feature extraction by employing intelligent space technology and moving information projection.

Moreover, we propose a design methodology to realize the effective guidance intended by service providers. We propose models for the design methodology. In order to make a person easily notice the guidance information and for him/her to be easily guided, the models are based on the knowledge of how a person perceives images through sight.

2 Pedestrian Route Guidance by Moving Image Projection

2.1 An Analysis of Traffic Line Guidance by Moving Information

First, we analyze the general methods of presenting moving information for guidance so that we may design a method by sensing a person's feature; we then use the feature for adapting the guidance information to the person and present adaptive information with moving information. From the viewpoint of service engineering [7], we resolved the service of pedestrian route guidance through the presentation of moving information into five phases from the standpoint of the service recipient.

(i) Pre-recognition
This phase occurs before a visitor notices the guidance information.
(ii) Paying Attention
In this phase, the visitor notices the guidance information and pays attention.
(iii) Inducing Eye Movement
When moving information is presented, the visitor visually tracks the information with his/her eyes before following it. This is defined as the inducing eye movement phase.
(iv) Inducing Body Movement
In the next phase, the visitor follows the guidance information. This is defined as the inducing body movement phase.
(v) Service End
In this phase, the guidance information completes the task of leading the visitor to his/her destination.

We can resolve the conventional guidance service by presenting static information in phases from the standpoint of the service recipient.

$$\text{Pre-recognition (i)} \rightarrow \text{Paying Attention (ii)} \rightarrow \text{Service End (v)}$$

By presenting moving information, the inducing eye movement phase (iii) and inducing body movement phase (iv) are newly introduced.

If the appropriate service is performed in all the phases, the service of route guidance will be successful. In this paper, we provide guidelines for effectively designing the service and phase transitions. Five phases and service designs are listed in Table 1.

In the pre-recognition phase (i), the visitor's usual behavior is observed. If the individual features of each service recipient need to be used to adapt the guidance information, you can extract them most appropriately in this phase ((a) in Table 1).

Table 1 Analysis of route guidance by presenting moving information

Service from the standpoint of the visitor	Service design in each phase
(i) Pre-recognition	(a) Sensing
↓	(b) Information Presentation
(ii) Paying Attention	
↓	
(iii) Inducing Eye Movement	Presentation of a Movement Route
↓	(c) Presentation of a Movement Route
(iv) Inducing Body Movement	(d) Presentation of a Movement Route
↓	
(v) Service End	(e) Clarification of the Destination

To change from the pre-recognition phase (i) to the paying attention phase (ii), it is essential to present information that will attract the visitor's attention. To attract his/her attention while he/she is walking, it is preferable to present information at the position where the visitor is looking ((b) in Table 1).

In the paying attention phase (ii), the visitor attempts to understand the semantic content of the presented information. For route guidance, if information is presented in a manner that is easily understood by the visitor, such as an arrow, the guidance will be more effective.

It can be predicted that the change from the paying attention phase (ii) to the inducing eye movement phase (iii) will occur smoothly. However, if the speed of moving information is not appropriate, the visitor will lose track of the information. Therefore, it is necessary to move the information at a suitable speed so that the visitor can track it. Moreover, it is favorable that we present information larger. In the inducing eye movement phase (iii), this guideline is the same.

To change from the pre-recognition phase (i) to the paying attention phase (ii), it is necessary to have the visitor move along the route of moving information. It is necessary that a service provider designs an appropriate route and speed for the moving information so that a visitor may not only track the moving information with his/her eyes but also follow it. Service designers should understand the characteristics of a person's sight in order to design the route and speed ((c) in Table 1).

In the inducing body movement phase (iv), it is necessary that the service provider designs an appropriate route and speed for the moving information so that the service recipient may understand the information and continue walking as usual by watching the moving information ((d) in Table 1).

In the ending service phase (v), if the guidance service is abruptly terminated, that is, it suddenly disappears, the visitor will be puzzled and confused. This confusion can be prevented by providing a presentation that will inform his/her final destination ((e) in Table 1).

2.2 Proposed Method for Pedestrian Route Guidance by Projecting Moving Images Basis on Personal Feature Extraction

On the basis of the design guideline obtained from the analysis in Sect. 2, we propose a method for pedestrian route guidance; this method comprises two methods – one for the presentation of information adapted to each individual and the other for the presentation of moving information.

First, for adapting information to each individual, we extract a person's feature such as his/her height, movement route, and speed in real time using surveillance cameras by employing the background difference method.

Second, for presenting the moving information and guiding a visitor along the target route, we use a pan-tilt projector, which can project an image at an arbitrary position. To adapt the guidance information to each individual, a movement route of the projected information, information content, and timing of the presentation are varied on the basis of the features of the service recipient (Fig. 1).

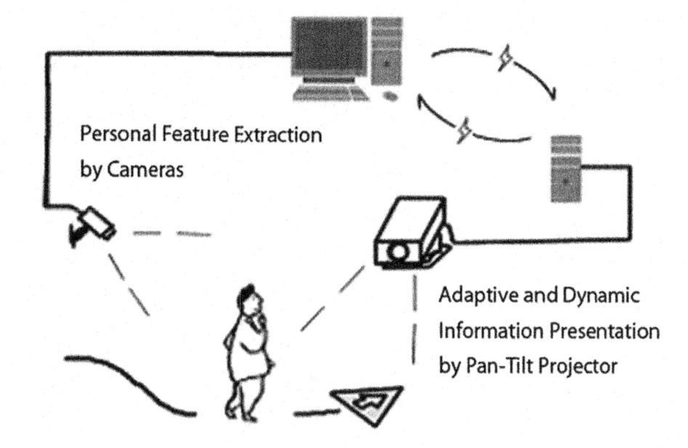

Fig. 1 Method for pedestrian route guidance using moving projection based on personal features

2.3 Pan-Tilt Projector

To apply the proposed method, it is necessary to project an arbitrary image onto an arbitrary place such as the floor or wall. We developed a pan-tilt projector can change the projected area by swinging a platform. It can move from −90° to 90° in pan direction and from −60° to 60° in tilt direction (Fig. 2). By using this projector, we present images with moving. Figure 3 shows the moving image presented by the pan-tilt projector.

Fig. 2 Image of pan-tilt projector

2.4 Personal Feature Extraction from Camera Image

For the proposed method, we determine a person's features including his/her height, movement route, and passing speed by using a camera.

The image center is (u_c, v_c), κ is the distortion coefficient, and the relation between point B and point C is given in Eq. (1). The parameter κ depends on the camera.

Fig. 3 Moving image presented by pan-tilt projector

$$
\begin{aligned}
u - u_c &= \left(u' - u_c\right)\left(1 + \kappa r^2\right) \\
v - v_c &= (v' - v_v)(1 + \kappa r^2) \\
r^2 &= (u - u_c)^2 + (v - v_v)^2
\end{aligned}
\tag{1}
$$

(u', v') is converted into the line $p = (x_{\mathrm{wp}}, y_{\mathrm{wp}}, z_{\mathrm{wp}})^T$ in the world coordinate by Eq. (2).

$$
\underline{p} = \begin{pmatrix} x_{wp} \\ y_{wp} \\ z_{wp} \end{pmatrix} = \alpha R A^{-1} \begin{pmatrix} u' \\ v' \\ 1 \end{pmatrix} + \begin{pmatrix} T_x \\ T_y \\ T_z \end{pmatrix}
\tag{2}
$$

$(T_x, T_y, T_z)^T$ is the positional vector of the camera in the world coordinate. A is the coordinate transformation matrix, it depends on an internal parameter of the camera. R is the rotation matrix of the camera. α is a real number parameter. If one point of the world coordinate corresponding to (u,v) on the image is obtained, α and the other two points can be determined. We define

$$
R A^{-1} \begin{pmatrix} u' \\ v' \\ 1 \end{pmatrix} = \begin{pmatrix} q_{xu} \\ q_{yu} \\ q_{zu} \end{pmatrix}
\tag{3}
$$

In addition, α in the $z_w = 0$ plane of Eq. (2) is set as α_1. Since $z_u = 0$ is already known, we obtain Eq. (4) from Eqs. (2) and (3).

$$
\alpha_1 = \frac{-T_z}{q_{zu}}
\tag{4}
$$

As a result, we obtain the position of a person standing in the world coordinate.

$$\begin{pmatrix} x_u \\ y_u \\ z_u \end{pmatrix} = \begin{pmatrix} -\frac{q_{xu}T_z}{q_{zu}} + T_x \\ -\frac{q_{yu}T_z}{q_{zu}} + T_y \\ 0 \end{pmatrix} \tag{5}$$

We assume that the person stands vertically on the floor and his/her head is above the position of his/her feet. The value of x_t in (x_t, y_t, z_t) corresponding to the person's uppermost point (u_t, v_t), which is extracted from a camera image, is set to x_u.

We define

$$RA^{-1} \begin{pmatrix} u' \\ v' \\ 1 \end{pmatrix} = \begin{pmatrix} q_{xt} \\ q_{yt} \\ q_{zt} \end{pmatrix} \tag{6}$$

from Eq. (2). In addition, α in the $z_w = x_u$ plane of Eq. (2) is set as α_2.

Since $x_t = x_u$ is already known, we obtain Eq. (7) from Eqs. (2) and (6):

$$\alpha_2 = \frac{x_u - T_x}{q_{xt}} \tag{7}$$

The coordinates of the parietal region are obtained from Eqs. (2), (6), and (7). We obtain

$$h = \frac{x_u - T_x}{q_{xt}} q_{zt} + T_z \tag{8}$$

as the person's height.

From the time series data of the person's positions that were obtained using the abovementioned methods, we obtain his/her movement route and average speed. Then, a service recipient's height, movement route, and the walking speed are used for adapting the information of each individual.

3 Information Presentation Considered the Characteristic of a Person's Sight

Since the method proposed in Sect. 2, uses projected moving information, the guidance by the method is based on a person's sight. We propose models based on the knowledge of how a person perceives images through his/her sight to design the guidance by the method effectively and naturally for a visitor.

3.1 Perceiving Images Through the Recipient's Sight

A person perceives detailed images by gazing at them. In this paper, on the basis of [8], we define the field of view of a person to be within 30° in the right and left direction, 20° in the upper direction, and 40° in the lower direction. We devised the method for the presentation of information on the basis of this range.

3.2 Models Based on the Knowledge of How a Person Perceives Images Through His/Her Eyesight

As described in Sect. 2.1, in order to provide an effective guidance service by presenting information that attracts the visitor's attention, in the initial stages, it is preferable to present information at the position where the recipient can gaze without problems. In this section, we decide the position where we present information in the initial stages on the basis of the knowledge of how a person perceives images through his/her eyesight.

Figure 4 shows the field of view of a person on the ground, as defined in Sect. 3 A. It is desirable to present information by the proposed method in the area in the initial stages. Moreover, the larger the information in the eyesight, the more it attracts the visitor's attention. It is preferable to present information at the position where a person looks. We defined L as a criterion for deciding the presentation position. We obtain

$$L = h \cdot \tan^{-1} \theta_l \tag{9}$$

θ_l, 40°; h, Height of Person's Eyes.

Next, we appropriately design the movement route and speed of the moving information so that the visitor may not only track the moving information with his/her eyes but also follow it. A person can track the moving information with his/her eyes only when it is in the defined area. To guide a person effectively, we propose moving the information leaving the defined area at the initial stages. Route B in Fig. 4 shows the movement of the information.

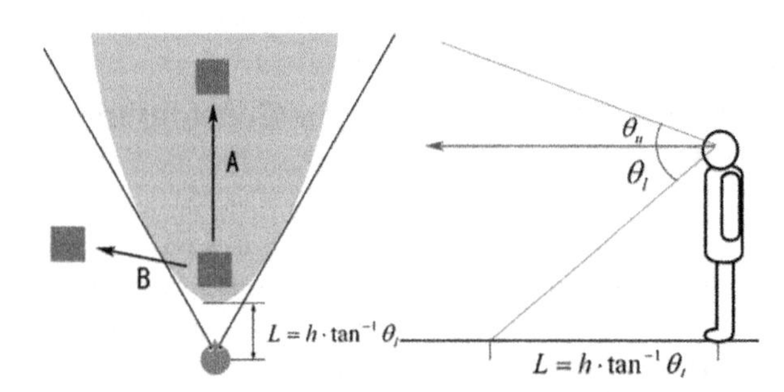

Fig. 4 Field of view of a person on the ground

3.3 Movement Models of a Person Who Follows the Moving Information on the Ground

In the inducing body movement phase, controlling the movement route of the presented information can control a service recipient's movement route. In this section, we propose movement models for a person who follows the moving information when the information moves only on the ground.

If we assume that the visitor considers the position of the information as a temporary destination point each time and walks naturally from the present position in the direction of the destination point, we obtain model Eq. (10).

$$x_{t+1} = x_t + \frac{y_t - x_t}{|y_t - x_t|} v \Delta t \tag{10}$$

x_t, A person's position at time t; y_t, Presented information's position at time t; v, Walking speed of a person; Δt, Sampling duration

Space designers usually design 1–1.5 times distance of a picture to a person who looking at it as long as the diagonal length of the picture [9]. We defined the final destination area from target thing to 1–1.5 times distance as long as the diagonal length of it. If x_t is in the final destination area, we can forecast that a service recipient stops.

When we input v and the time series data of the position to which we want to guide the service recipient in Eq. (10), we obtain the time series data of the movement route of information. Moreover, when we input v, the time series data of the movement route of information, and a person's position at the beginning of the moving information projection in Eq. (10), we obtain the time series data of the position to which a visitor will move. We can design an appropriate route for the moving information for each service recipient by using his/her walking speed that is measured in v.

4 Experiments

4.1 Traffic Line Guidance Experiment in Art Museum

We performed traffic line guidance experiments at the Suntory Museum in Osaka in June 2005 to evaluate the guidance by the provided proposed method. We show the entire composition in Fig. 5. After visitors go downstairs from 5F to 4F, they reach a narrow area. The traffic line is often confused here. Though a service designer designs the way to look at painting A in Fig. 5, some service recipients first look at painting B or painting C. We implemented four methods for the guidance in this place, and we compared their effectiveness. First, we calculated the percentage of visitors who followed the correct way without guidance. Next, as the conventional method, we measured the percentage of visitors who walked along the correct way

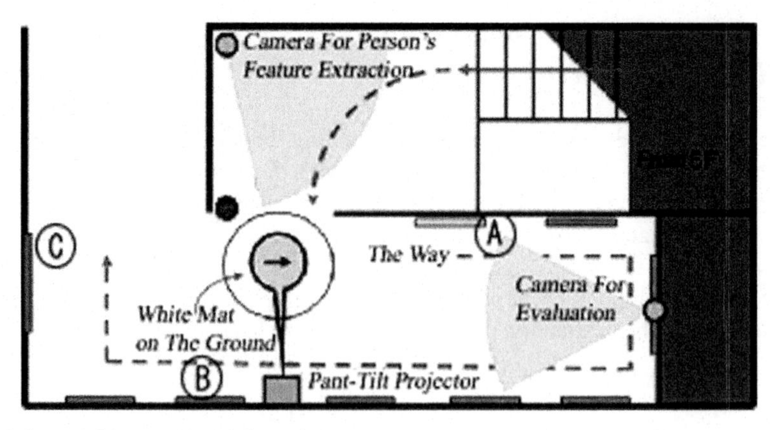

Fig. 5 Suntory Museum, fourth floor chart

with the help of a signboard. Furthermore, we guide visitors by projecting moving information whose speed is always constant. Further, we guide the visitors by projecting moving information whose speed is adapted to each visitor according to the proposed method. We compared the effectiveness by evaluating the success rates of guidance and the questionnaire for these four cases.

4.2 Results

The experimental results are shown in Table 2. The percentage of visitors who followed the correct way when there was no guidance was 84.7%. Using the presentation of moving information, that is, the proposed method, we measured the success percentage of guidance to be 96%. The result shows that the proposed method is more effective than the conventional one. The main cause of the failure of guidance by the proposed method was that the visitor did not notice the presented information; the visitor paid attention to a sound coming from the opposite direction. In this experiment, we excluded the case when the projected information was hidden behind other guests from a parameter. In such a case, the guest who went to the front was guided properly, and the guest who came from behind tended to go the same way. Even if we cannot show the guidance information to all people, we can

Table 2 Success rate of guidance and ease of view

			Guidance success rate	Ease of view	Number of data
		No guidance	84.7%	–	170
Guidance		Signboard	90.6%	–	203
	Moving image	Constant speed	96.2%	2.67	215
		Adapted speed	96.4%	3.48	292

*Ease of view = Four-stage evaluation using a questionnaire.

create a flow of traffic line. Moreover, we compared the guidance by projecting moving information whose speed was constant with the guidance provided by projecting moving information whose speed is adapted to each recipient. There was no significant difference between the success rates of the guidance provided in the two cases. In addition, in the questionnaire, the service recipients stated that the guidance provided by projecting moving information with a speed adapted to each visitor was easy to follow. (Significance level is 1%.)

4.3 *Experiment for Presentation of Information Considered Characteristic of a Person's Sight*

We performed traffic line guidance experiments at the Kashiwa Campus of the University of Tokyo in October 2005 to evaluate the proposed models. We arranged A1 size posters in the room and observed the action of the guided visitors.

For the evaluation of proposed model (a), we compared the success rate of the guidance provided by moving information within the area where a person can gaze (left-hand side of Fig. 6) with the success rate of the guidance provided by moving information outside the area (right-hand side of Fig. 6). Moreover, we compared the guided person's real movement route that was obtained by image data processing with the movement route obtained from Eq. (10).

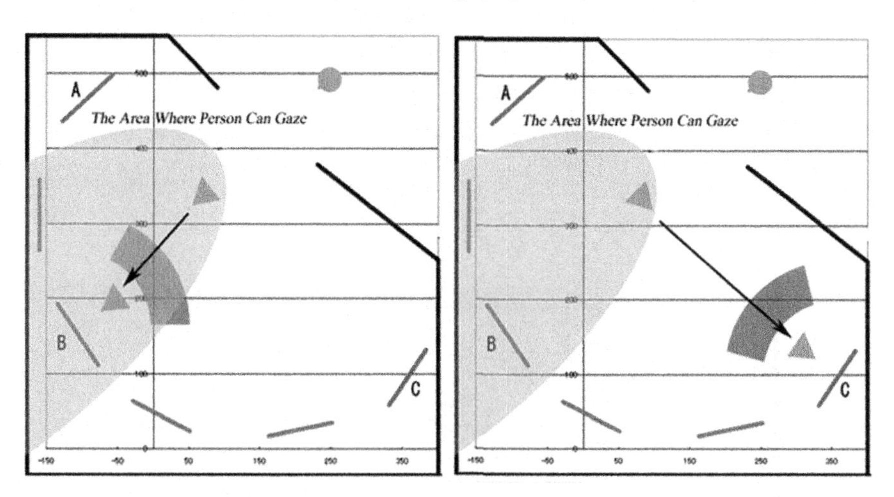

Fig. 6 Two movement routes

4.4 *Results*

The experimental results are shown in Table 3. The success rate of the guidance provided by moving information within the defined area after the information was presented at the position where information is viewed at a maximum size in the

Table 3 Success rate of guidance for each movement route

	Route to poster A	Route to poster B	Route to poster C	Stopped
No guidance	75.0%	0.0%	0.0%	25.0%
Guidance by moving information within the defined area	25.0%	50.0%	0.0%	25.0%
Guidance by moving information outside the defined area	22.7%	0.0%	74.2%	3.0%

*Stopped = the case in which the person halted during guidance.

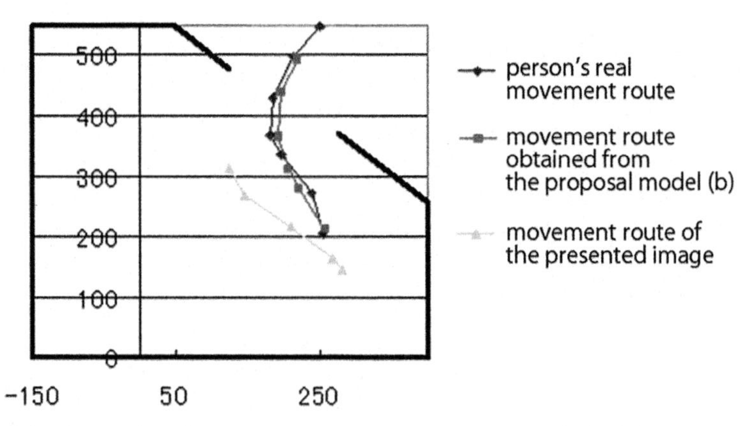

Fig. 7 Guided person's real movement route and movement route obtained from the proposed model (b) (positioning errors are small)

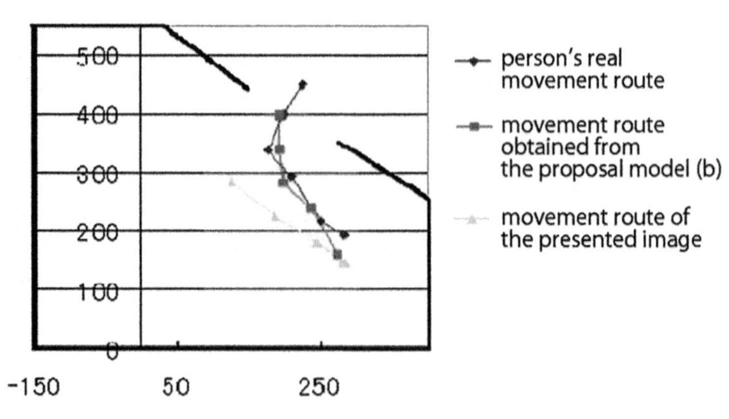

Fig. 8 Guided person's real movement route and movement route obtained from the proposed model (b) (positioning errors are large)

defined area was 50.0%. The success rate of the guidance provided by moving information outside the defined area was 74.2%. While being guided by the proposed model, the visitors hardly halted. The results show that the proposed model is effective.

The percentage of cases in which the differences between the guided person's real movement routes and the movement routes obtained from the proposed model (b) (Ex. Figs. 7 and 8) were less than 10 cm was 20.8%; for differences less than 20 cm and 30 cm the percentages were 50.6% and 87.0%, respectively.

Additionally, the visitors of 80.0[%] stopped at the final destination area defined by the proposal model.

5 Discussions

In the above mentioned experiment, the extraction of the features and natural behavior of each visitor in the pre-recognition phase and their use in deciding the movement speed of the presented information increased the success rate of guidance. This is because we could guide the visitors without disturbing their natural walking pattern. Furthermore, we compared the guidance provided by presenting moving information with a movement speed that is always constant with the guidance provided by presenting moving information with a movement speed adapted to each recipient. The visitors stated that the latter guidance was easy to follow. One reason was because the satisfaction level of the visitors who were contented with the conventional presentation of information designed for an average person was enhanced. Another reason was attributed to the presentation of information by the proposed method was easy to watch for persons whose features are well below or well above average.

In the experiment pertaining to the presentation of information by considering the characteristics of a person's sight, we calculated the position of the person's feet from the most inferior point of the person's domain that was captured by image processing. Therefore, a positioning error related to the size of the service recipient human size occurs depending on the state of his/her legs: open or close. We think that most of the positioning errors are caused due to the errors in the image processing step.

Since the error that is less than 30 cm is not greater than the size of the human body, the proposed model is sufficiently effective forecasting the position of a walking person. It was mainly in the last stage of the guidance service that the positioning error became greater than 30 cm. The visitor stopped at the last stage of the guidance service. When a person is about to stop walking, it is expected that he/she will reduce his/her walking speed. In the experiment, v for the proposed model is set to the average walking speed of a visitor. Large positioning errors were attributed to the fact that the speed value we used for predictions of the position in the last stage of the guidance was greater than the actual walking speed. When we observed the actual walking action of a visitor, we found that the nearer a visitor was to the final destination area at the time before stopping, the lesser is the distance he walks in the last stage. This was due to positioning errors. We show the guided person's real movement route and the movement route obtained from the proposed model (b) in Figs. 7 and 8. Figure 7 shows a case in which the positioning errors are small, and Fig. 8 shows one in which the positioning errors are large.

6 Conclusion

In this paper, we proposed the novel guidance system using a moving projection. The method uses adaptive and dynamic information projection for the purpose of information presentation in public spaces. We developed a route guidance system that uses personal feature extraction of visitor's height, walking speed, and direction of walking by estimating from camera images and that projects moving information by using a pan-tilt projector. Moreover, we propose models to design the route guidance effectively. The models are based on the knowledge of how an individual information by sight. By performing two experiments, we showed the effectiveness of the proposed method in effectively designing route guidance.

In future, we can develop a service that offers extra information for guidance by varying the information content for making it more attractive for commercial advertisement. We can improve the service by using not only personal features but also environmental conditions; personal features include a person's height, walking speed, age, and hobbies, and environmental situations include the place, time, and a natural disaster.

References

1. A. P. Hillstorm and S. Yantis, Visual motion and attentional capture, *Perception & Psychophysics*, 55(4), 399–411, 1994.
2. U. Kiencke et al., The impact of automatic control on recent developments in transportation and vehicle systems, Preprints of the 16th IFAC World Congress, 2005.
3. Z. Hu and K. Uchimura, VICNAS – A new generation of on-road navigation system based on machine vision and augmented reality technology, 2nd ITS Symposium, pp. 119–124, 2003.
4. T. Koiso, K. Hattori, T. Yoshida, and N. Imasaki, Behavior analysis of customers and sales lerks in a home appliance retail store using a trail analyzing system, IPSJ 2nd UBI, pp. 61–66, 2003.
5. K. Kurumatani, CONSORTS: architecture for ubiquitous agents – toward mass-user support, Japanese Society for Artificial Intelligence, 102(603), pp. 13–17, KBSE2002-36, 2003.
6. T. Mori, H. Noguchi, and T. Sato, Daily life experience reservoir and epitomization with sensing room, in Proceedings of the Workshop on Network Robot Systems: Toward Intelligent Robotic Systems Integrated with Environments, 2005.
7. T. Tomiyama, Service engineering to intensify service contents in product life cycles, in Proceedings of the Second International Symposium on Environmentally Conscious Design and Inverse Manufacturing (EcoDesign 2001), pp. 613–618, 2001.
8. K. Noro, *Illustration ergonomics*, Japanese Standards Association, 1990.
9. Japan Lighting Information Service. Available: http://www.city.yokosuka.kanagawa.jp/speed/mypage/m-imajo/jlisseminar/museum/museum2.html

Recognizing Human Activities from Accelerometer and Physiological Sensors

Sung-Ihk Yang and Sung-Bae Cho

Abstract Recently the interest about the services in the ubiquitous environment has increased. These kinds of services are focusing on the context of the user's activities, location or environment. There were many studies about recognizing these contexts using various sensory resources. To recognize human activity, many of them used an accelerometer, which shows good accuracy to recognize the user's activities of movements, but they did not recognize stable activities which can be classified by the user's emotion and inferred by physiological sensors. In this paper, we exploit multiple sensor signals to recognize user's activity. As Armband includes an accelerometer and physiological sensors, we used them with a fuzzy Bayesian network for the continuous sensor data. The fuzzy membership function uses three stages differed by the distribution of each sensor data. Experiments in the activity recognition accuracy have conducted by the combination of the usages of accelerometers and physiological signals. For the result, the total accuracy appears to be 74.4% for the activities including dynamic activities and stable activities, using the physiological signals and one 2-axis accelerometer. When we use only the physiological signals the accuracy is 60.9%, and when we use the 2 axis accelerometer the accuracy is 44.2%. We show that using physiological signals with accelerometer is more efficient in recognizing activities.

Keywords Activity recognition · Physiological sensor · Fuzzy Bayesian network

1 Introduction

The recognition of human activity is a concerning problem to provide interactive service with the user in various environments. Research about activity recognition is emerging recently, using various sensory resources. There are studies using cameras or GPS based on the user's movement by using pattern recognition techniques.

S.-I. Yang (✉)
Department of Computer Science, Yonsei University 262 Seongsan-ro, Sudaemoon-gu, Seoul 120-749, Korea
e-mail: unikys@sclab.yonsei.ac.kr

S. Lee et al. (eds.), *Multisensor Fusion and Integration for Intelligent Systems*, Lecture Notes in Electrical Engineering 35, DOI 10.1007/978-3-540-89859-7_14,

Many of the other research use accelerometers as the main sensor placing the accelerometer on the user's arm, leg or waist. They recognize activities like walking or running, which has movements in the activity and can be optimized for the recognition based on the accelerometer.

Accelerometers show a high accuracy in recognizing activities with lots of movements. But they are weak to recognize the activities with little movements. Accelerometers also provide continuous sensory data which is hard to separate to the exact boundary of several states. Other research use physiological sensors to recognize the user's context, as the user's body status can represent the user's activity and also the user's emotion which depends on the activity. Therefore using the physiological sensor with accelerometers will help to recognize the user's activity.

In this paper, we use not only an accelerometer but also physiological sensors to help the recognition of activities. Accelerometers have advantages to measure the user's movement and the physiological signals have advantages to measure the user's status of the body. Most of these sensors calibrate continuous sensor data, and if the data is near the boundary of a specific state, the evidence variable will show a radical difference in small changes. To lessen these changes in this situation we propose to preprocess the data with fuzzy logic. Fuzzy logic can represent ambiguous states in linguistic symbols, which is good in continuous sensory data. As sensory data include uncertainty of calibrating data and also human activity itself has it too, we use Bayesian network to do the inference, and modify the learning and inference methods that fit to the fuzzy preprocessing.

2 Background

2.1 Related Works

There are many research groups studying about human activity recognition using various sensors like cameras, GPS or accelerometers. Tapia used a simple state-change sensor to detect the objects which the user is using at home [1], and Han used an infrared camera to contrast the silhouette of the user and recognized the activity by using the sequence of the images [2]. When using these kinds of sensors such as state-change sensor, cameras or microphones, the research uses pattern recognition and tracks the user's position to recognize the user's activity.

Using motion detection sensors or cameras can only collect log data about the user's activities in an abstract way, and using a camera needs a large consumption of calculation. There are other research using sensors which can represent the user's activity like accelerometers or physiological sensors. These are sensors that mostly use continuous data values. As these sensors' measurements are continuous, the research using these sensors are using various methods to quantize the continuous measurements. Meijer used a motion sensor and accelerometers for measurements, calculated the difference with each activity and compared it with a rule [3]. Ravi uses three axes accelerometers with several classifiers, naïve Bayes, decision tree and SVM [4]. Parkka also used three axes accelerometers, and in addition, they added physiological sensors and used a decision tree for the classifier [5].

Subramanya used a GPS and a light sensor to detect the location and an accelerometer to recognize the related activity with the location [6]. They used binning for preprocessing and dynamic Bayesian network for classification. These research use sensors which collect continuous measurements and classify activities by using binning or decision tree. Binning and decision tree are useful to recognize activities but there can be some problems of segmenting the continuous measurements, especially if the measurements are ambiguous between activities. In this paper, we use fuzzy logic which has advantages to represent continuous data in symbolic states for preprocessing [7], and a fuzzy Bayesian network, which is compromised with the preprocessed fuzzy data, to solve the problem between ambiguous measurements. We use a sensor, Armband, which has an accelerometer and physiological sensors as well, and collect the user's activity log information by using a PDA.

2.2 Pysiological Sensor

Various sensors are used to recognize the user's activity such as GPS, cameras, microphones, accelerometers and physiological sensors. Among those various sensors, Bodymedia's Armband is a sensor which can measure the user's physiological signals. It has five kinds of sensory resources inside, a two axes accelerometer, a heat flux sensor, a galvanic skin response sensor, a skin temperature sensor, and a thermometer [8]. With these five sensors it calibrates the sensory data and combines to 24 kinds of data.

Thus, the Armband can recognize not only the dynamic activities by the accelerometer but also the stable and static activities by the physiological signals, too. The Armband has a maximum of 32 Hz sampling rate, and uses the Innerview professional 5.0 [8] to collect the data. Innerview professional 5.0 is able to show the collected data in a graph or convert the data to several formats of files. Figure 1

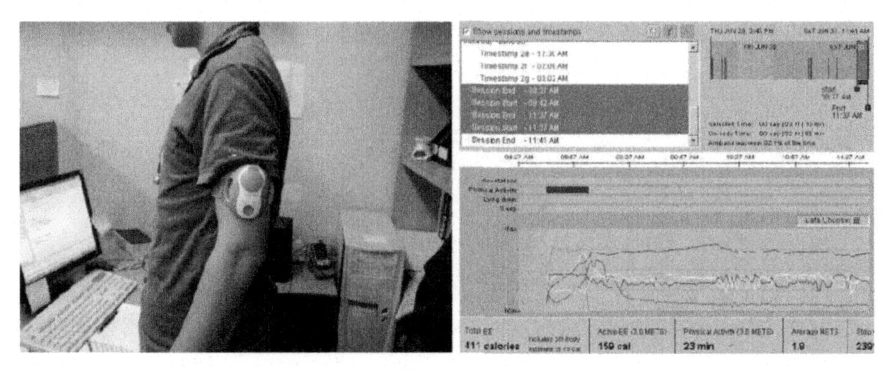

Fig. 1 The Armband worn on the left upper arm (*left*) and the screen of Innerview professional 5.0 (*right*)

shows the appearance of the Armband when it is worn on the upper left arm and the screen of the Innerview professional 5.0 showing the data in a graph.

In this work, from the 24 kinds of measurements by the Armband, we use the two axes accelerometer values, galvanic skin response, heat flux, skin temperature, energy expenditure, metabolic equivalants, and step counts.

2.3 Fuzzy Logic

Fuzzy logic has been used in a wide range of problem domains in process control, management and decision making, operations research, and classification. The conventional crisp set model makes decision in black and white, yes or no, and it has a typical boundary in the classification of several stages. The representation of the low stage will have a value of 1 until the upper bound, and the moment the value goes over the upper bound, the value of the low stage will suddenly change into 0. This crisp set model of classification is simple to implement but when it is used in continuous values in real number, like sensory data, it is a hard problem to decide the boundary. It is also a hard problem for the inference models when the data is nearby the boundary, keeping the robustness of a little change of the data. By using fuzzy logic, the decision becomes flexible and can keep each stage's representation near by the boundary of the data. It helps to represent the vagueness of human intuition in a linguistic modeling which is hard in the crisp model [7].

A fuzzy membership function calculates the fuzzy membership for each stage with a specific value. The most popular function type is a trapezoidal membership function and a triangular membership function which the graph is shown in Fig. 1. These functions are easy to implement, have low consuming calculations like formula (1) and (2) [9]. The triangular membership function(MF_{tri}) requires three parameters and the trapezoidal membership function(MF_{trap}) requires 4. The value is simple divided with a rule of the range of data x. The parameter can be chosen and modified by a direct view of a graph. There are also a membership using the Gaussian distribution and a sigmoidal membership function.

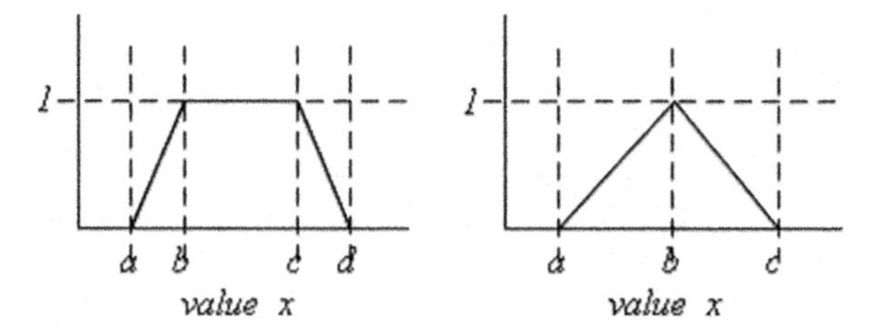

Fig. 2 The graph of the trapezoidal (*left*) and triangular (*right*) fuzzy membership function

$$MF_{tri}(x) = \begin{cases} 0, & x < a \\ \frac{x-a}{b-a}, & a \leq x < b \\ \frac{x-c}{b-c}, & b \leq x < c \\ 0, & c \leq c \end{cases} \tag{1}$$

$$MF_{trap}(x) = \begin{cases} 0, & x < a \\ \frac{x-a}{b-a}, & a \leq x < d \\ 1, & b \leq x < c \\ \frac{x-d}{c-d}, & c \leq x < d \\ 0, & d \leq x \end{cases} \tag{2}$$

2.4 Bayesian Network

A Bayesian network is a graph with probabilities for representing random variables and their dependencies. It efficiently encodes the joint probability distribution of a large set of variables. This representation consists of two components. The first component is a directed acyclic graph (DAG) and the second component describes a conditional distribution for each variable. The nodes of the graph represent random variables and its arcs represent dependencies between random variables with conditional probabilities at nodes. As the graph is a directed acyclic graph all edges are directed and there is no cycle when edge direction are followed.

Each node has a initial probability table which can be measured by a statistical method or by a expert's knowledge, so that the network can make a robust result even if there is a missing value in a uncertain environment. This characteristic is especially important in the usage of sensory resources. Calculating the probability of a node is based on the Bayes rule. If a node does not have a parent the probability is as the initial probability table, and if it has, it is calculated by adding the multiplication of each probability of the state in the variable node and the conditional probability of the variable node's state in the parent's probability table. The calculation of the probability is like formula (3) [10].

$$P(A) = \sum_i P(A|B_i)P(B_i) \tag{3}$$

The joint probability of random variables $\{x_1, \ldots, x_n\}$ in a Bayesian network is calculated by the multiplication of local conditional probabilities of all the nodes. Let a node x_i denote the random variable x_i in the Bayesian network, and $parent_i$ denote the parent nodes of x_i, from which dependency arcs come to node x_i, Then, the joint probability of $\{x_1, \ldots, x_n\}$ is given as the following formula (4).

$$P(x_1, \ldots, x_n) = \prod_i P(x_i|parent_i) \tag{4}$$

It is not a simple task to get the exact conditional probability distribution when the variables have continuous values and high order dependencies. As conditional probability table is not suited for continuous values because the values should be quantized and the table size will grow larger with the dependency order.

So we used fuzzy Bayesian network which can make more flexible inferences by preprocessing the continuous variable data in fuzzy logic, and train the conditional probability table by a fuzzy training method which can be differed with the conventional discrete training model.

3 The Fuzzy Bayesian Network for Activity Recognition

As directly using the sensor data is an issue to consider, we need a framework to quantize the data. The flow of the system is like Fig. 3. The Armband collects the log data based on the user's activity, and a PDA is used for the user to annotate the current activity he or she is doing. These two log data are collected simultaneously and save the same format of the current time. Then, the log integrator will merge these data into entire integrated log data. The preprocessor will use this data and generate a fuzzy integrated data for each sensor log. With this data the fuzzy Bayesian network will train the conditional probability tables, and the inference also uses this fuzzy integrated data.

3.1 Preprocessing with Fuzzy Logic

As measurements from Armband are continuous a step of preprocess is necessary. Figure 4 shows the distribution of the measurements and the continuous measurements with the discrete function results.

Segmenting continuous data is a considerable issue when the data lay on one side, like Fig. 4 (*top*), because a little difference of segmentation makes a huge difference of the result. We made a fuzzy membership function depending on the distribution using the mean and standard deviation of the each sensory measurement. As measurements from Armband are continuous a step of preprocess is necessary. Figure 2 shows the distribution of the measurements and the continuous measurements with the discrete function results.

We separated the data in half by the mean value, and calculated the distribution of each side with the standard deviation in three stages. If the distribution of a side is wide it uses a Gaussian membership function, or if it's distribution is moderate it uses a trapezoidal, and if the distribution is very narrow it uses a triangular fuzzy membership function. Each side can have these three kinds of different fuzzy membership function, so by combining the fuzzy membership of each side, there

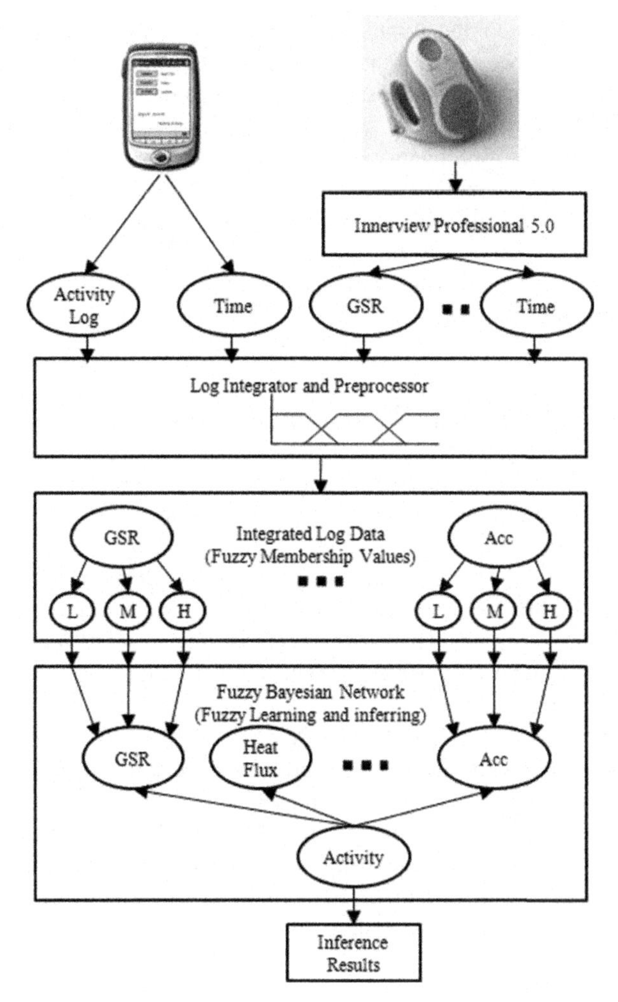

Fig. 3 The overview of the system flow

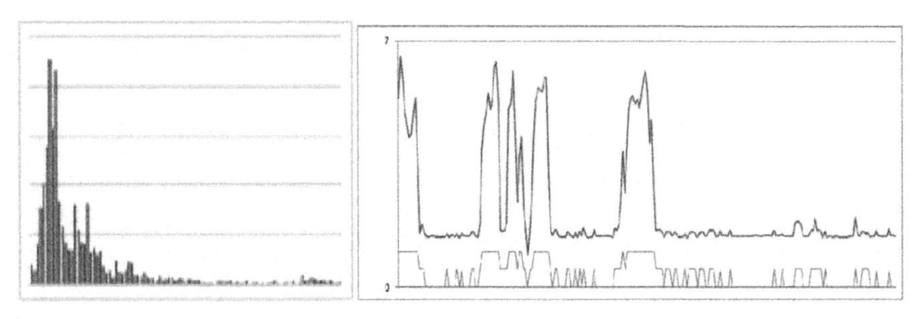

Fig. 4 The distribution of the measurements (*left*) and continuous data with the discrete function results (*right*)

can be nine kinds of combinations of a full complete membership function. For the discrete function we used a function based on the fuzzy membership function which segments the data into three states. The state of the discrete function chooses the state which has the maximum fuzzy membership value between other states. But this function can have a problem when the data is like Fig. 4 (*bottom*). The curve above represents the continuous data from the sensor and the curve below shows the results of a discrete function. The higher values have no problem, but the lower values' discrete function result shows a shaking result with a small transition. This can be a problem in the inferring phase, because the evidence will change with just a small amount of transition.

3.2 Training of the Fuzzy Bayesian Network

Using the same training method with the discrete model will not work perfectly as the input of each variable are not the same as the discrete model which only chooses a single state of the variable, but it contains the fuzzy membership value of each state. As we used fuzzy logic to preprocess the measurements, we used a learning method which fits to the fuzzy data, differed with the discrete model of training based on the Bayes theorem. The training method of the discrete model only counts a state of the data, and the fuzzy method will count all of the state which has a fuzzy membership value.

As sensory resources are independent to each other without any dependencies, we used the structure of the naïve Bayes classifier, and make the variables to give no influence or have any interactions between each other [11]. Naïve Bayes classifier is based on the Bayes theorem calculating $P(B|A)$. This can be calculated by formula (5) in a discrete model. In formula (5), μ stands for the discrete function(D) result of an input of the evidence variable(A) about the state of the evidence(*state*) with the input x. The discrete function only allows one state to be counted for the conditional probability.

$$\mu_{state} = \begin{cases} 1, & D(A) \in state \\ 0, & D(A) \notin state \end{cases}$$
$$P(state \mid A) = \sum_{x} \frac{\mu_{state}}{\sum_{A} \mu^{A}_{state}} \quad (5)$$

Differed with the discrete model the fuzzy learning method allows more than one states to be counted for the conditional probability. The fuzzy membership function(MF) will produce the fuzzy membership value of each state of the input evidence, and all of the value is considered to be used.

$$P(state \mid A) = \sum_{x} \frac{(MF(x) = \mu_{state})}{\sum_{A} \sum_{x} (MF(x)^{A} = \mu^{A}_{state})} \quad (6)$$

This makes the values, on the edge of the discrete function which is ambiguous of determining the state, to be more flexible and give probability to both of the ambiguous states.

3.3 Inference of the Fuzzy Bayesian Network

When we use the fuzzy Bayesian network to infer the status of a variable, we use a discriminant which chooses the state which has the largest probability. The discriminant function *f* is like formula (7) in a naïve Bayes classifier. It will multiply each evidences of each variable's conditional probability. In this formula, *E* is the evidence values from the environment, and *k* is the inferring status of the Bayesian network. v_{ik} means the value of the evidence variable A_j [11].

$$f_i(E) = P(C_i)\prod_j P(A_j = v_{jk} + C_i) \tag{7}$$

In the discrete method of the naïve Bayes classifier, formula (7), only one of the evidence conditional probability of each variable only gives influence to the result of the discriminant. As a very little difference of data will change the results of the inference of the Bayesian network, in the discrete method there are problems when the data is like Fig. 4 (*bottom*). So we modified the discriminant function like formula (8).

$$f_i(E) = P(C_i)\prod_j \sum_k (MF(E_j)_k \times P(A_j = v_{jk} + C_i) \tag{8}$$

In formula (8), when the result of the membership function(MF) is not zero, which means the data is near the boundary of the status, the conditional probability each of the status' above and below the boundary gives influence to the result. This dissolves the radical difference when the data moves over the boundary and helps the classifier to keep the changes calm about the data which is ambiguous to divide into bins.

4 Experiment

4.1 Experiment Method

We collected the data using the Armband sensor for the physiological signals which was worn on the right upper arm and a PDA for the labeling of the activity by the user. The user selects an activity saved in the PDA. The program for the labeling was programmed by Embedded Visual C++ based on the PocketPC standard development kit. As physiological signals change moderately the data collection term of the Armband and the PDA was set for a data once a minute. After the log collected the data from the Armband was converted by the Innerview professional 5.0 program to an excel format. The converted data is integrated with PDA labeling data by a integration program, and it is used for the training and testing of the fuzzy Bayesian network. There are nine kinds of activity which can occur in a real life and office environment. The activities which can be differed by the movements were walking, running, and exercising. The other activities were eating, reading, studying, playing, resting and sleeping. Totally nine activities were collected, in the real life, freely with no restricts to the user.

The raw sensor data was collected like Fig. 5. The graph shows the sequence of the sensor signals. By only using the accelerometer signals, the activities which are dependent of the user's movements show they are significantly different with the others, but other activities, which are stable, are hard to recognize the difference between each other. As the physiological signals in Fig. 5 shows, using these additional signals can lessen the difficulty only using the accelerometer signals.

The first experiment was in four kinds of combinations of the discrete model and fuzzy model for the training and testing method. We compared the training method between the Bayes theorem based method and the fuzzy logic based method. The testing method was compared between the naïve Bayes classifier and the naïve Bayes classifier using the fuzzy inference. The total data set has a size of 5,500 samples and placed randomly for the cross validation which is divided to ten folds.

The second experiment was to show the accuracy of the activities when using different sensor data. The first set uses all of the sensor data, both the physiological signals and accelerometer signals. The second set only used the physiological signals, and the third set used the 2 axis accelerometer signals. Each set was trained and tested in the fuzzy Bayesian network and was compared the accuracy between activities.

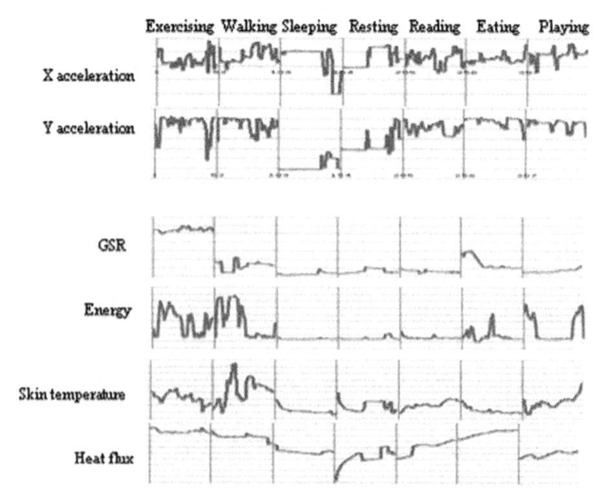

Fig. 5 The Raw Sensor Data from the Accelerometer and Significant Physiological Signals of Each Activity

4.2 Experimental Results

Table 1 is the result of the first experiment's average of the ten folds cross validation. The discrete methods had 70.0% of accuracy and the combination of the discrete training and the fuzzy inferring had 71.3% of accuracy. Even though the training method is a discrete model the fuzzy inference had a higher and more stable accuracy than the discrete inference. The fuzzy methods show 74.4% of accuracy and when the discrete inference is used with the fuzzy training the accuracy is low by 62.3%. This shows that the discrete inference is not capable for the fuzzy training, but the combination of fuzzy methods has higher accuracy than the discrete methods.

Table 1 The average results of the experiment

Learning	Discrete		Fuzzy	
Inferring	Discrete	Fuzzy	Discrete	Fuzzy
Accuracy	70.0(±2.3)	71.3(±2.0)	62.3(±2.7)	74.4(±1.4)

The following Fig. 6 is a chart of the result of the cross validation with ten folds. The bar represents the combinations of training and inferring methods, discrete training and discrete inferring, discrete and fuzzy, fuzzy and discrete, and fuzzy and fuzzy from the left. As the chart shows, the fuzzy method has higher accuracy in all ten folds.

Tables 2–4 shows the result of the second experiment, in three different data sets. Table 2 shows the confusion matrix when using all of the sensor data, and Tables 3 and 4 each shows when using only the physiological signals and accelerometer signals. The rows show the activity and the columns show the recognition results.

The accuracy of Table 2 is all better than the other tables, Table 3, which only used the physiological signals, show that it is more efficient when recognizing the stable activities than Table 4, and in the opposite Table 4 shows that the accelerometer is more efficient when recognizing the dynamic activities. Because there is only

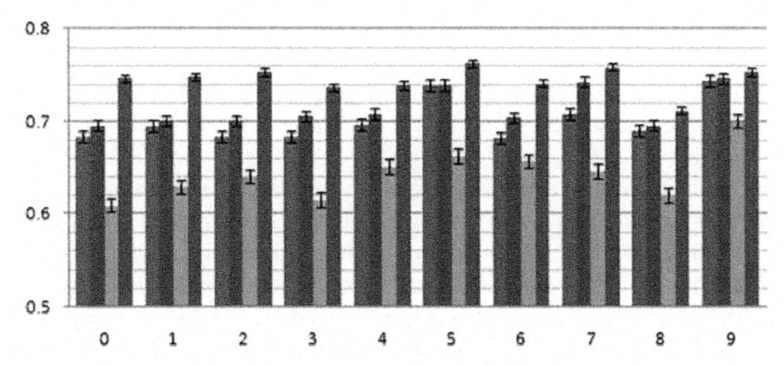

Fig. 6 The results of the cross validation with ten folds. The bars from the left are discrete learning–discrete inferring, discrete–fuzzy, fuzzy–discrete, fuzzy–fuzzy

Table 2 Accuracy of the method using both physiological and accelerometer signals (total accuracy = 74.2%)

	Ea	P	Rea	Res	Ru	Sl	St	Wa	Ex
Eating	43.7	12.1	10.4	0	0	0	14.2	18.9	0
Playing	0	92.7	0	0	0	0	0	0.07	6.6
Reading	16.3	1.0	73.5	5.1	0	0	3.1	0	1.0
Resting	4.0	0	0.5	86.3	0	0	7.6	0	1.6
Running	0	0	1.6	0	84.1	0	0	12.7	1.6
Sleeping	2.4	0.8	9.9	0	0	85.8	0	0.3	0.8
Studying	11.7	0	1.3	5.2	0	0	74.7	2.6	34.5
Walking	6.0	2.7	0.7	0	2.4	0	3.1	75.8	9.3
Exercising	4.7	9.9	8.3	0	1.1	0	1.6	31.2	43.2

Table 3 Accuracy of the method only using physiological signals (total accuracy = 60.9%)

	Ea	P	Rea	Res	Ru	Sl	St	Wa	Ex
Eating	30.9	13.3	0	17.1	0	0	18.3	20.4	0
Playing	0	90.6	0	0	0	0	0	0	9.4
Reading	78.6	1.0	1.0	13.3	0	0	6.1	0	0
Resting	3.0	0	0	55.9	0	0	41.1	0	0
Running	1.6	0	0	0	58.7	0	0	38.1	1.6
Sleeping	13.1	1.1	0	0	0	85.5	0	0.3	0
Studying	13.6	0	0	7.1	0	0	73.5	5.8	0
Walking	1.3	2.2	0	3.5	15.3	0	3.8	57.7	16.2
Exercising	14.6	10.9	0	1.0	3.1	0	0	29.2	41.2

Table 4 Accuracy of the method only using accelerometer signals (total accuracy = 44.7%)

	Ea	P	Rea	Res	Ru	Sl	St	Wa	Ex
Eating	24.2	32.5	0.8	0	0	0	19.2	20.0	3.3
Playing	17.1	67.9	0	2.1	0	0	3.5	3.5	5.9
Reading	5.1	70.4	12.3	2.0	0	0	6.1	3.1	1.0
Resting	1.0	44.7	0.5	48.8	0	0	2.0	0.5	2.5
Running	0	1.6	0	0	82.5	0	0	14.3	1.6
Sleeping	0.8	47.4	1.9	42.2	0	0	6	0.3	1.4
Studying	5.8	7.8	0.6	12.3	0	0	39.7	2.6	31.2
Walking	5.5	3.8	0	0	4.2	0	2	76.3	8.2
Exercising	7.3	13.5	3.1	0	0.5	0	2.1	28.6	44.9

one 2-axis accelerometer worn on the upper right arm the results are not so good, but the dynamic activity recognition accuracy is as good as or better than Table 2's results. As Table 3's accuracy of each activity is a little lower than Table 2's accuracy, because that even an activity is stable, the accelerometer helps the recognition. The dynamic activities in Table 3 shows that they are confused with each other because the there is no accelerometer information. Table 4's accuracy of walking is much higher than Table 2 and running and exercising is similar with Table 2. This means that dynamic activities can be recognized only with the accelerometer. The reading activity was confused with the eating activity and resting activity, but when the physiological signals and accelerometer signals are combined it showed 73.5% of accuracy.

5 Conclusion

In this paper, we showed that using a fuzzy Bayesian network is more efficient than the discrete model when using continuous data for recognizing the user's activity. We used the Armband sensor which calibrates physiological signals and includes a 2-axis accelerometer. The first experiment results are, when we used the discrete model of the naïve Bayes classifier has shown 70.0% of accuracy and the fuzzy Bayesian network has shown 74.4% of accuracy. The second experiment has shown that each activity has an efficient sensor to recognize using the physiological signals or the 2 axis accelerometer depend on the vitality, and combining these two kinds of sensor helps to recognize all of those activities. In the future work, we will need to integrate more sensors for the context-aware service about the user activity, change the frequency of the data collection time, and improve the classifier for a temporal inference model to analyze the sensory data's alterations.

Acknowledgments This research has been supported in part by KIST and Samsung Electronics, Co.

References

1. E. Tapia, S. Intille, and K. Larson, Activity recognition in the home using simple and ubiquitous sensors, *Lecture Notes in Computer Science, International Conference on Pervasive Computing*, vol. 3001, pp. 158–175 (2004).
2. J. Han and B. Bhanu, Human activity recognition in thermal infrared imagery, *IEEE Computer Society Conference on Computer Vision and Pattern Recognition*, vol. 3, pp. 17–25 (2005).
3. G. Meijer, K. Westererp, F. Verhoeven, H. Koper, and F. Hoor, Methods to asses physical activity with special reference to motion sensors and accelerometers, *IEEE Transactions on Biomedical Engineering,* 38(3), 221–229 (1991).
4. N. Ravi, N. Dandekar, P. Mysore, and M. Littman, Activity recognition from accelerometer, In *Proceedings of the Seventeenth Innovative Applications of Artificial Intelligence Conference*, pp. 11–18 (2005).
5. J. Parkka, M. Ermes, P. Korpipaa, J. Mantyjarvi, J. Peltola, and I. Korhonen, Activity classification using realistic data from wearable sensors, *IEEE Transactions on Information Technology in Biomedicine*, 10(1), 119–128 (2006).
6. A. Subramanya, A. Raj, J. Bilmes, and D. Fox, Recognizing activities and spatial context using wearable sensors, In *Proceedings of Conference on Uncertainty in Artificial Intelligence* (2006).
7. E. Cox, Fuzzy fundamentals, *IEEE Spectrum*, 29(10), 58–61 (1992).
8. Bodymedia©, Sensewear behavior prescribed, http://www.sensewear.com/images/bms_brochure.pdf
9. J. Jang, C. Sun, and E. Mizutani, *Neuro-fuzzy and soft computing*, Prentice Hall, New Jersey (1997).
10. K. Korb and A. Nicholson, *Bayesian artificial intelligence*, Chapman and Hall/Crc Computer Science and Data Analysis, Boca Raton, FL (2004).
11. P. Domingos and M. Pazzani, On the optimality of the simple Bayesian classifier under zero-one loss, *Machine Learning*, 29(2-3), 103–130(1997).

The "Fast Clustering-Tracking" Algorithm in the Bayesian Occupancy Filter Framework

Kamel Mekhnacha, Yong Mao, David Raulo and Christian Laugier

Abstract It has been shown that the dynamic environment around the mobile robot can be efficiently and robustly represented by the Bayesian occupancy filter (BOF) [(Tay et al. 2008)]. In the BOF framework, the environment is decomposed into a grid-based representation in which both the occupancy and the velocity distributions are estimated . In such a representation, concepts such as objects or tracks do not exist. However, the object-level representation is necessary for applications needing high-level representations of obstacles and their motion. To achieve this, we present in this paper a novel algorithm which performs clustering on the BOF output grid. The main idea is to use the prediction result of the tracking module as a form of feedback to the clustering module, which reduces drastically the complexity of the data association. Compared with the traditional joint probabilistic data association filter (JPDAF) approach, the proposed algorithm demands less computational costs, so as to be suitable for environments with large amount of dynamic objects. The experiment result on the real data shows the effectiveness of the algorithm.

Keywords Bayesian filtering · BOF · Tracking · Autonomous vehicle

1 Introduction

Perceiving of the surrounding physical world reliably is a major demanding of the driving assistant systems and the autonomous mobile robots. The dynamic environment needs to be perceived and modeled according to the sensor measurements which are noisy. Normally, this problem is treated within the estimation framework. The major requirement for such a system is a robust target tracking system. Most of the existing target tracking algorithms [1] use an object-based representation of the environment. However, these existing techniques have to take into account

K. Mekhnacha (✉)
Probayes SAS, 38330 Montbonnot, France
e-mail: kamel.mekhnacha@probayes.com

S. Lee et al. (eds.), *Multisensor Fusion and Integration for Intelligent Systems*, Lecture Notes in Electrical Engineering 35, DOI 10.1007/978-3-540-89859-7_15,

explicitly data association and occlusion problems which are major challenges to the performances. In view of these problems, a grid based framework, the Bayesian occupancy filter (BOF) [2, 3] has been presented in our previous works.

In the BOF framework, concepts such as objects or tracks do not exist. It decompose the environment into a grid based representation. A Bayesian filter [4] based recursive prediction and estimation paradigm is employed to estimate an occupancy probability and a velocity distribution for each grid cell. Thanks to the grid decomposition, the complicated data association and occlusion problems do not exist. The BOF is extremely convenient for applications where no object-level representation is needed. The second advantage of the BOF is that the multiple sensor fusion task could be easily achieved. In some situations, using information from multiple sensors could provide more reliable and robust information of the environment. However, in traditional multiple sensor fusion techniques, data association problem could be further complicated. The associations between the two consecutive time instances from the same sensor as well as the associations among the tracks of different sensors will have to be take into account at the same time. Fortunately, these difficulties do not exist in the grid based approaches [5] which deal with the data association problems in a more feasible way. Uncertainties of multiple sensors are specified in the sensor models and are fused into the BOF grid naturally with solid mathematical ground.

Despite of the aforementioned advantages, a lot of applications demand the explicit object-level representation. In our former work, a joint probabilistic data association filter (JPDAF) [1] based object detecting and tracking method was implemented above the BOF layer. However, when there are enormous amount of dynamic objects in the environment, the number of hypothesises generated by the JPDAF increases rapidly, which makes the method suffers from the computational cost. Regarding to this problem, a novel fast object detecting and tracking algorithm is proposed. The algorithm is hierarchical in which two filtering levels are used (Fig. 1) namely the robust grid-level sensing and the robust object-level tracking. The output grid of the BOF grid filter (i.e, the probability distribution on the occupancy of the cell, and the probability distribution on the velocity of the cell occupancy) is used as the input from which object hypothesises are extracted. By taking the prediction result of the tracking module as a form of feedback to the clustering module, the clustering algorithm avoids searching in the entire grid which guarantees the performance. A re-clustering and merging strategy is employed whenever the ambiguous data association occurs. The computational cost of this approach is linear to the number of dynamic objects detected, so as to be suitable for cluttered environment. Our approach has been tested on the real data collected on real cars in highway and cluttered urban environments (Fig. 2), and also on our Cycab experimental platform (Fig. 3).

The paper is organized as follows. In the next section, the Bayesian Occupancy Filter(BOF) framework is described. The object detecting and tracking approach is presented in Sect. 3. In Sect. 4, the experimental result of our approach on the real data collected by the Cycab platform is provided. Finally, conclusions are drawn in Sect. 5.

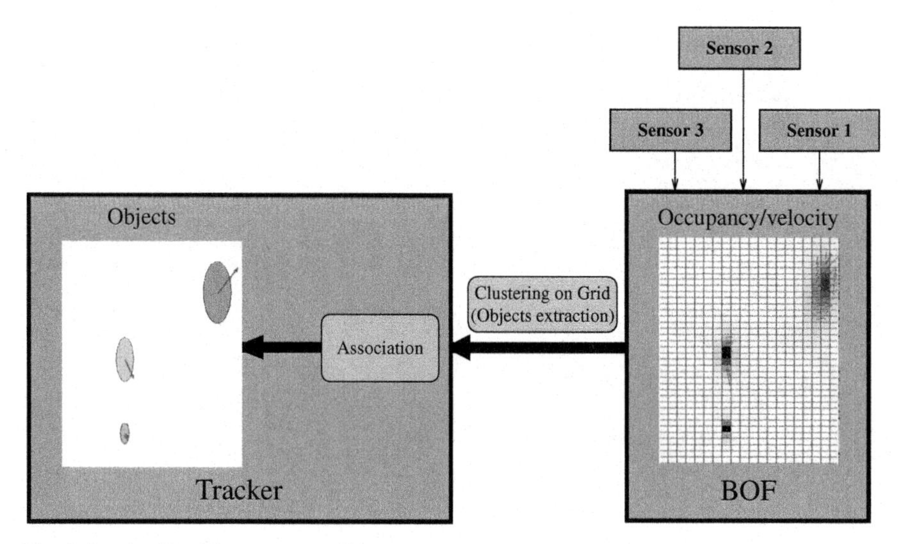

Fig. 1 Sensing/Tracking system architecture

Fig. 2 Example of BOF output using a computer vision car detector as input (red boxes): The images are provided by a camera mounted on the moving ego-vehicle. The BOF output is projected back on the image. It represents a grid of occupancy probability (blue-to-red mapped color) and the mean velocity (red arrows) estimates

Fig. 3 The Cycab platform

2 Bayesian Occupancy Filter (BOF)

The Bayesian Occupancy Filter (BOF) is represented as a two-dimensional grid-based decomposition of the environment. Each cell of the grid contains two probability distributions: (i) the probability distribution over the occupancy of the cell (ii) and the probability distribution over its velocity. Given a set of input sensor readings, the BOF algorithm allows to update the occupancy/velocity estimates of each grid cell.

BOF is a special implementation of the Bayesian filter approach [4, 6]. This approach addresses the general problem of recursively estimating the posterior probability distribution $P(X^k \mid Z^k)$ of the state X of a system conditioned on its observation Z. This posterior distribution is obtained in two stages: prediction and estimation. The prediction stage computes an a priori prediction of the target's current state known as the prior distribution. The estimation stage then computes the posterior distribution by using the prediction with the current measurement of the sensor.

In the case of the BOF, using this prediction/estimation scheme allows filtering out false alarms, miss-detections, and localization errors in sensors data readings. Figure 3 shows an example of BOF output using a computer vision car detector.

The Bayesian model presented in the following text is a reformulation of the one we presented in [2]. The aim of this reformulation is to make clearer the strong link between the discretization of the space and the discretization of the velocity, which reduces the number of the used random variables and makes the model easier to explain. The key idea of the model is to represent the 2D space using a regular grid. Given this space discretization and assuming that objects do not overlap, the velocity of a given c cell at a time t is directly linked to the identity of its antecedent cell A_c

from which the content of cell c moved between $t-1$ and t. In other words, we can define the velocity of a given cell by providing the index of its antecedent. Therefore, estimating the velocity of a given cell is equivalent to estimating a probability table over all its possible antecedents. Possible antecedents of a cell are defined by providing a neighbourhood from which the cell is reachable in a time step. This model applies also to velocities needing more than one time step for a neighbour cell to reach c. However, for simplicity we will assume only one-step velocities (neighbours reaching c in one time step).

The BOF model is shown graphically in Fig. 4 and is described as follows:

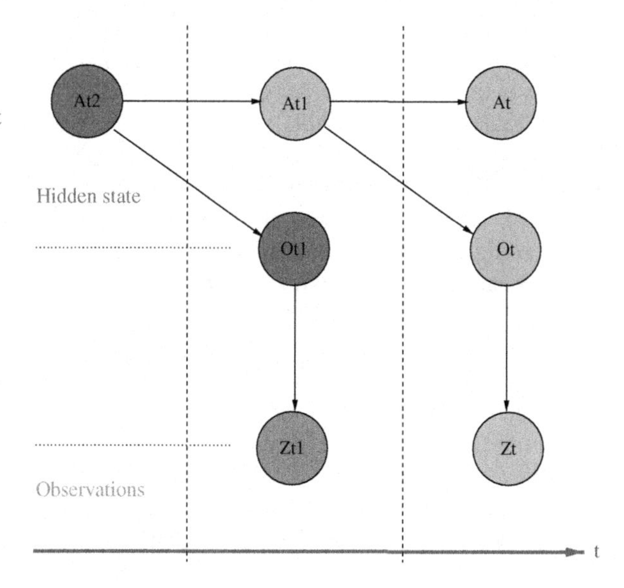

Fig. 4 The Dynamic Bayesian Network corresponding to the BOF model. Here, we suppose that only occupancy sensors are available

2.1 Variables

For a given cell having $c \in \mathcal{Y}$ as index in the grid, let:

- $A_c^t \in \mathcal{A}_c \subset \mathcal{Y}$ represents each possible antecedent of cell c over all the cells in the grid domain $\mathcal{Y}$. The set of antecedent cells of cell c is denoted by $\mathcal{A}_c$ and is defined as a neighbourhood of the cell c.
- $A_c^{t-1} \in \mathcal{A}_c \subset \mathcal{Y}$ the same as A_c^t but for the previous time step.
- $O_c^t \in \mathcal{O} \equiv \{0, 1\}$ is a boolean variable representing the state of the cell in terms of occupancy at time t, either $[O_c = 1]$ if occupied, $[O_c = 0]$ if empty. Given the independency hypothesis, the occupancy of each cell at time t is considered apart from the occupancy of its neighbouring cells at time t.
- $Z_i^t \in \mathcal{Z}, 1 \leq i \leq S \in \mathbb{N}$, is a generic notation for measurements yielded by each sensor i, considering a total of S sensors yielding a measurement at the considered time instant.

2.2 Joint Distribution Factors

The following expression gives the decomposition of the joint distribution of the relevant variables according to Bayes' rule and dependency assumptions:

$$P(A_c^{t-1}\, A_c^t\, O_c^t\, Z_1^t \cdots Z_S^t) = P(A_c^{t-1})P(A_c^t \mid A_c^{t-1})P(O_c^t \mid A_c^{t-1})\prod_{i=1}^{S} P(Z_i^t \mid A_c^t\, O_c^t). \tag{1}$$

The parametric form and semantics of each component of the joint decomposition are as follows:

- $P(A_c^{t-1})$ is the probability for a given neighbouring cell A_c to be the antecedent of c at time $t-1$. In order to represent the fact that cell c is a priori equally reachable from all possible antecedent cells in the considered neighbourhood, this probability table is initialized as uniform and is update in each time step.
- $P(A_c^t \mid A_c^{t-1})$ is the distribution over antecedents at time t given the antecedent of cell c at $t-1$. It represents the prediction (dynamic) model over velocity. If we assume a perfect *constant velocity hypothesis* between the two time frames $t-1$ and t, this distribution is simply:

  $$P(A_c^t \mid A_c^{t-1}) = P(A^{t-1}_{A_c^{t-1}}).$$

  In other words, the predicted probability is simply the probability at the preceding time instant for the antecedent at $t-1$.
  Considering imperfect *constant velocity hypothesis* is possible by introducing the predicate $E \in \{0, 1\} \equiv$ "There was an erroneous prediction", and assuming a probability $P(E) = \varepsilon$. This value is a parameter of the system and corresponds of the probability of not respecting the *constant velocity hypothesis*. We have:

  $$P(A_c^t \mid A_c^{t-1}\, \neg E) = P(A^{t-1}_{A_c^{t-1}}),$$
  $$P(A_c^t \mid A_c^{t-1}\, E) = \mathcal{U}(A_c^t).$$

  where $\mathcal{U}(A_c^t)$ denotes a uniform distribution on A_c^t to say that all possible antecedents have the same probability when *constant velocity hypothesis* is not respected. Thus, $P(A_c^t \mid A_c^{t-1})$ may be written as a mixture:

  $$P(A_c^t \mid A_c^{t-1}) = P(\neg E)P(A_c^t \mid A_c^{t-1}\, \neg E) + P(E)P(A_c^t \mid A_c^{t-1}\, E).$$

  Which leads to:

  $$\begin{aligned} P(A_c^t \mid A_c^{t-1}) &= (1-\varepsilon)P(A^{t-1}_{A_c^{t-1}}) + \varepsilon\, \mathcal{U}(A_c^t) \\ &= (1-\varepsilon)P(A^{t-1}_{A_c^{t-1}}) + \varepsilon / \|\mathcal{A}_c\|, \end{aligned}$$

  where $\|\mathcal{A}_c\|$ is the cardinality of the considered antecedents set $\mathcal{A}_c$.

- $P(O_c^t \mid A_c^{t-1})$ is the distribution over occupancy given the antecedent of cell c at $t-1$. It represents the prediction (dynamic) model over occupancy. Similarly to $P(A_c^t \mid A_c^{t-1})$, the term $P(O_c^t \mid A_c^{t-1})$ may be written as a mixture:

$$\begin{aligned} P(O_c^t \mid A_c^{t-1}) &= (1-\varepsilon)P(O_{A_c^{t-1}}^{t-1}) + \varepsilon\, \mathcal{U}(O_c^t) \\ &= (1-\varepsilon)P(O_{A_c^{t-1}}^{t-1}) + \varepsilon/2. \end{aligned}$$

- $P(Z_i^t \mid A_c^t\, O_c^t)$ is the *direct model* for sensor i. It yields the probability of a measurement given the occupancy O_c^t and the antecedent (velocity) A_c^t of cell c. Measurements for all sensors are assumed to have been taken *independently from each other*. For sensors providing measurements depending exclusively of occupancy, this distribution can be written as $P(Z_i^t \mid O_c^t)$. In the same manner, for sensors providing measurements depending exclusively of velocity, this distribution can be written as $P(Z_i^t \mid A_c^t)$.

2.3 Occupancy and Velocity Estimation Using the BOF Model

At each time step, the estimation of the occupancy and velocity of a cell is answered through Bayesian inference on the model given in (1). This inference leads to a Bayesian filtering process (Fig. 5). In this context, the prediction step propagates cell occupancy and antecedent (velocity) distributions of each cell in the grid to get the prediction $P(O_c^t\, A_c^t)$. In the estimation step, $P(O_c^t\, A_c^t)$ is updated by taking into account the observations yielded by the sensors $\prod_{i=1}^{S} P(Z_i^t \mid A_c^t\, O_c^t)$ to obtain the a posteriori state estimate $P(O_c^t\, A_c^t \mid [Z_1^t \cdots Z_S^t])$. This allows, by marginalization, to compute $P(O_c^t \mid [Z_1^t \cdots Z_S^t])$ and $P(A_c^t \mid [Z_1^t \cdots Z_S^t])$ that will be used for prediction in the next iteration.

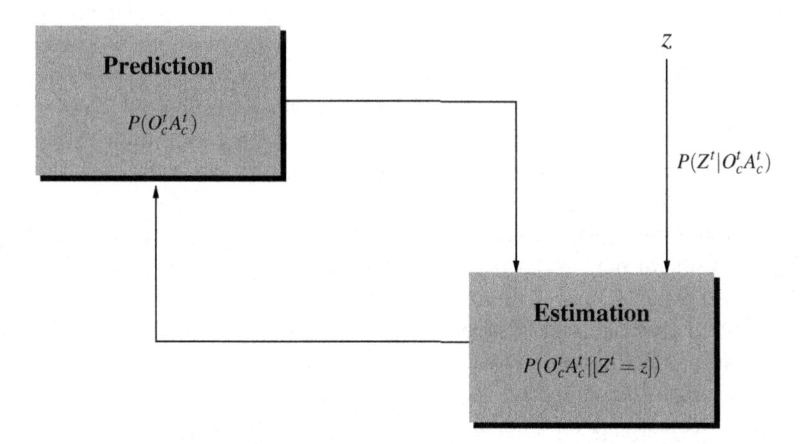

Fig. 5 Bayesian filtering in the estimation of occupancy and velocity distribution in the BOF grids

It's important to notice that the distribution $P(A_c^t)$ over antecedents (velocity) is updated even when no velocity sensors are available. Indeed, suppose we have only one occupancy sensor described by the model $P(Z_{\mathrm{O}}^t \mid O_c^t)$. The a posteriori distribution $P(A_c^t \mid [Z_{\mathrm{O}}^t])$ leads to the formula:

$$P(A_c^t \mid [Z_{\mathrm{O}}^t]) \propto \sum_{A_c^{t-1} \in \mathcal{A}_c} P(A_c^{t-1}) P(A_c^t \mid A_c^{t-1}) \sum_{O_c^t \in \{0,1\}} P(O_c^t \mid A_c^{t-1}) P([Z_{\mathrm{O}}^t] \mid O_c^t) \tag{2}$$

In this case, the update is based exclusively on the occupancy observations.

When an additional velocity sensor $P(Z_{\mathrm{V}}^t \mid A_c^t)$ is available, it should be used to update the estimate (2) as follows:

$$P(A_c^t \mid [Z_{\mathrm{O}}^t \; Z_{\mathrm{V}}^t]) \propto P(A_c^t \mid [Z_{\mathrm{O}}^t]) P([Z_{\mathrm{V}}^t] \mid A_c^t).$$

Finally, as the relationship between the velocities and antecedent IDs (indexes) is deterministic (an antecedent id corresponds to a (vx, vy) vector), the probability $P(V_c^t)$ over the velocity is summarised as a 2D Gaussian distribution using the distribution $P(A_c^t)$. Therefore, the output of the BOF algorithm at each time step t is a grid in which each cell c contains (i) an occupancy probability $P(O_c^t)$ and (ii) a 2D Gaussian distribution $P(V_c^t)$ over velocity.

3 The "Fast Clustering-Tracking" Algorithm

The object-level representation is mandatory for applications needing high-level representations of obstacles and their motion. This needs a robust multi-target tracking system allowing to estimate of the position and the velocity of each moving object.

The main difficulty of multi-target tracking is known as the "data association" problem. It includes observation-to-track association and track management problems. The main goal of observation-to-track association is to decide whether a new sensor observation corresponds to an existing track. Track management includes deciding whether existing tracks should be maintained, deleted, or if new tracks should be created. Numerous methods exist to perform data association. The reader is referred to [7] for a complete review of the existing tracking methods with one or more sensors.

In the BOF framework, we proposed in [7] to use a layered architecture as shown in Fig. 2 to obtain the object-level representation. In [7], the data association is implemented using a classical JPDA algorithm within the proposed architecture.

In cluttered environments with large numbers of moving objects, the JPDA suffers from the combinational explosion of hypotheses. To overcome this problem, we propose the "Fast Clustering-Tracking" algorithm. This algorithm could be roughly divided into a clustering module, an ambiguous association handling module, and a track management module.

3.1 Clustering

The clustering module takes the occupancy/velocity grid of the BOF as the input and extracts object-level reports from it. Its main ideas are:

1. using the prediction result of the tracking module to define a region of interest (ROI) allowing to avoid searching in the complete grid,
2. using both occupancy and velocity estimates in order to better separate/associated the extracted clusters.

A natural algorithm to extract a cluster in a grid is to start from a given cell (pixel) and expand the cluster by deciding, for each eight-neighbor cell, whether the neighbour is to be included or not according to a connectivity criterion.

In order to avoid searching for clusters in the whole grid, we use the predicted targets' states as a form of feedback. For a given target T_i, the predicted state is used to define a region of interest ROI_{T_i} in the BOF grid. ROI_{T_i} is used as the search region in which the clustering module will try to extract a cluster (report) to be associated implicitly to T_i.

To take advantage of occupancy and velocity estimates provided by the BOF grid, the used connectivity criterion is as follows. Starting for a given cell c for which $P(O_c^t) \geq occ\_threshold$, a neighbouring cell n is added to the cluster if and only if:

$$\begin{cases} -\ \text{Cell } n \text{ is not associated yet (included in no cluster)}, \\ -\ P(O_n^t) > occ\_threshold, \\ -\ MahDist\left(P(V_c^t), P(V_n^t)\right) < vel\_threshold, \end{cases}$$

where $MahDist\left(P(V_c^t), P(V_n^t)\right)$ is the Mahalanobis distance between the velocity distributions of a two neighbour cells.

If the "Cell n is not associated yet" predicate is not respected, the corresponding cell is tagged as an ambiguous-associated one. This corresponds to an ambiguous association case which need to be dealt with in a special manner (3.2). If such a situation is not encountered, the extracted cluster is implicitly associated to the target T_i defining the used ROI_{T_i}.

Simple statistics are then performed on the extracted cluster in order to obtain a 4-dimentional observation corresponding to cluster's position and velocity. Both the position and velocity components are represented as 2D Gaussian distributions (mean vector and covariance matrix).

3.2 Re-Clustering and Tracks Merging

During the clustering process, three possible situations need to be considered (Fig. 6).

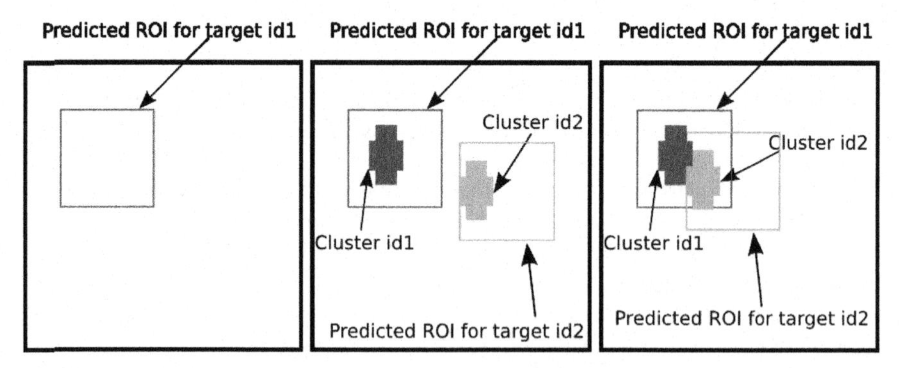

Fig. 6 Situations occurring during the clustering/association process

- **Case 1:** no cell with $P(O_c^t) > occ\_threshold$ is found in ROI_{T_i} defined by the considered target T_i. The target T_i has not been observed and no association is needed.
- **Case 2:** a cluster C is extracted. It's implicitly associated to the target T_i defining the used ROI_{T_i}. This situation occurs when there is no ambiguity in the association. This is an advantageous situation allowing a fast clustering-association procedure. Fortunately, this case is the most frequent one when using the algorithm to the real data sets.
- **Case 3:** a set of cells c_i having $P(O_{c_i}^t) > occ\_threshold$ and $MahDist(\cdot) < vel\_threshold$ are extracted in ROI_{T_i}. However, they have already been assigned to other targets. In this conflicted case, an observation (cluster) could be possibly generated by two (or more) different targets.

The first two cases are the normal cases, however, the third case is referred as an ambiguous association case which need to be dealt with in a special manner. The ambiguous association could occur in the following two situations:

- Different targets are being too close to each other and the observed cluster is in fact the union of the more than one observations generated by different targets.
- The different tracked targets are corresponding to a single object and should be merged into one.

We take a re-clustering strategy to deal with the first situation and a cluster merging strategy to deal with the second one.

When an ambiguous association occurs, a set of tracks $T_1, T_2, \cdots, T_m$ are identified as the potential candidates to be associated to the extracted cluster. We have to cut up this cluster and generate a sub-cluster (possibly empty) for each candidate. A Cartesian distance based K-means [6] algorithm is applied to re-cluster the ambiguous region.

To deal with the second cause of the ambiguous association, we introduce a concept of "alias" which is in the form of a two-tuples to represent the duplicated tracks. When an ambiguous association between two tracks T_i and T_j is detected, an alias $ALIAS(T_i, T_j)$ is initialized and added to the potential aliases list. At each time frame, the tracker updates this list by confirming or disproving the existence of each alias hypothesis $ALIAS(T_i, T_j)$ according to the observation of the ambiguous association. If an ambiguous association occurs between T_i and T_j, and the alias $ALIAS(T_i, T_j)$ is found in the potential alias list, the probability $P^t\left(S(T_i, T_j)\right)$ is updated by a confirming step using a Bayesian filtering approach as follows:

$$P^t(S \mid F) = \frac{P^{t-1}(S) \times P(F \mid S)}{P^{t-1}(S) \times P(F \mid S) + \left[1 - P^{t-1}(S)\right] \times P(F \mid \neg S)}$$

where:

- $S \equiv$ "the T_i and T_j tracks are alias for the same object".
- $F \equiv$ "an ambiguous association between the tracks T_i and T_j is observed".

The probability values $P(F \mid S)$ and $P(F \mid \neg S)$ are constant parameters of the tracker. The former denotes the probability of observing an ambiguous association when the two concerned tracks are alias of the same object and is set to a constant value 0.8. The second denotes the probability of falsely observing an ambiguous association and is set to 0.1.

When $ALIAS(T_i, T_j)$ is found in the potential alias list but is not observed as an ambiguous association, its probability is disproved in a similar manner:

$$P^t(S \mid \neg F) = \frac{P^{t-1}(S) \times P(\neg F \mid S)}{P^{t-1}(S) \times P(\neg F \mid S) + \left[1 - P^{t-1}(S)\right] \times P(\neg F \mid \neg S)}.$$

Then, according to the probability $P^t\left(S(T_i, T_j)\right)$, the decision of merging of tracks T_i and T_j could be considered. The merging decision is done by comparing the actual Mahalanobis distance between T_i and T_j to a given threshold.

3.3 New Tracks Creation

For new targets creation, we introduce a concept "cluster seed" to define a cell in the BOF grid where we will try to find, for each step, a new (non-associated) cluster. Indeed, the searching for potential new targets is performed after all the existing tracks are processed. Thus, only non-associated cells will be processed to extract clusters as the observations for the potential new targets. The "cluster seed" concept is general and can be implemented via various strategies. The simplest strategy is to insert a possible seed in each cell of the grid. However, more sophisticated strategies could be more efficient. For example, cluster seeds could be inserted only in entrance regions of the monitored area.

3.4 Tracks Updating and Deleting

The prediction and estimation of the targets are accomplished by attaching a Kalman filter [8] with each track. Once associated to a given track, a report (Gaussian distributions for both position and velocity) corresponding to an extracted cluster is used as an observation to re-estimate the position and velocity of the track in a prediction-update step. For non-observed tracks, only a prediction step is taken by applying the dynamic model to the estimation result of the precedent time step.

The deleting of tracks is also achieved in a Bayesian manner. If an existing track T is associated with a given report (cluster), its existence probability is increased using the following formula:

$$P^t(E \mid O) = \frac{P^{t-1}(E) \times P(O \mid E)}{P^{t-1}(E) \times P(O \mid E) + \left[1 - P^{t-1}(E)\right] \times P(O \mid \neg E)}$$

where:

- $E \equiv$ "the target T exists".
- $O \equiv$ "the target T has been observed (associated)".

The parameters $P(\neg O \mid E)$ and $P(O \mid \neg E)$ are the tracker miss-detections and false alarms probabilities respectively.

If an existing target is not associated with any report (cluster), its existence probability is decreased in the similar way:

$$P^t(E \mid \neg O) = \frac{P^{t-1}(E) \times P(\neg O \mid E)}{P^{t-1}(E) \times P(\neg O \mid E) + \left[1 - P^{t-1}(E)\right] \times P(\neg O \mid \neg E)}$$

According to the existence probability, the track deleting operation is achieved by applying a deleting threshold on it.

4 Experimental Result

The proposed approach has been applied in several driving assistance projects and achieved satisfied results in conditions including both highway and cluttered urban environments. The used sensor modalities include:

- multi-layer lidars,
- Computer Vision detection algorithms (Fig. 2),
- Stereovision-based 3D sensors.

However, according to the confidentiality agreements of the on-going projects, the results could not be published. Here, we provide some recent experiment results on our Cycab platform.

The Cycab platform is equipped with SICK lidar, GPS, mono-camera and odometer. To demonstrate the accuracy of the tracking algorithm, we used several GPS which are carried by pedestrians or vehicles. Because the experiments were carried out in the parking area of Inria Rhone-Alpes, within this limited range, the precision of the GPS is highly reliable, which provides us the ground truth of the locations of the moving objects. During the experiment, the SICK lidar served as the main sensor. The camera data was not used by the algorithm right now.

The object of the proposed algorithm is to track the moving objects in scene robustly and efficiently. However, in a normal environment (except the extremely cluttered environment) most of the sensor readings come from static objects. Thus, if we update the BOF with all the lidar data and apply the tracking algorithm directly, large amount of static objects will be detected and tracked. Basically we could apply two different straightforward methods to overcome this problem. The first idea is to remove objects with a speed below a given threshold. This could be achieved by making use of the velocity estimates of the objects given by the tracker and the ego-motion estimate provided by the odometer model. Unfortunately, the combination of the estimated ego-velocity and the targets velocities is not accurate enough, which leads to removing the low speed moving objects, i.e. pedestrians, by mistake.

The second idea is to divide the lidar data into a static set and a dynamic set, and only use the dynamic set to update the BOF. We applied the second method in the experiment. The division of the lidar data is achieved by maintaining a well discretized occupancy grid map [4] centred at the Cycab and moved along with it. Each cell in this map represents an occupancy probability. If the occupancy probability exceeds an given threshold, this cell is regarded as a static cell, and the lidar data fall into this cell are removed from updating of the BOF. Different from the BOF, the occupancy grid map is implemented in the global coordinates. Thus, the ego-motion of the Cycab is also needed to be estimated. We applied the odometer motion model to predict the location and used an iterative closest point (ICP) [9] algorithm to update it. In our experiment, this scheme has shown high accuracy.

We first apply our algorithm to detect and track a car which moves in front of the Cycab in the same direction as shown in Fig. 7. The result of dividing of the lidar data into dynamic data set and static data set is shown in Figs. 9 and 10. The first row of the figures corresponds to the data set of the SICK lidar used to update the BOF. The second row of the figures corresponds to the visualization of the Bayesian occupancy filter. The colour of the cells represents the occupancy probability. The third row gives out the tracker output from the fast clustering-tracking algorithm. The scale of the ellipse represents the uncertainty of the tracked target position. The small arrow start from the centre of the eclipse gives the estimated velocity of the target relative to the Cycab. The first column in Figs. 8, 9 and 10 shows the results using the full dataset as the input to the BOF and tracker. There exist about 10 targets being detected and tracked. However, only one of them is the real moving object we are interested in. The second column shows the results using only the dynamic dataset obtained from the aforementioned data division algorithm. It is clearly shown that all the static objects are removed, and the moving object is correctly tracked.

Fig. 7 Experiment scene

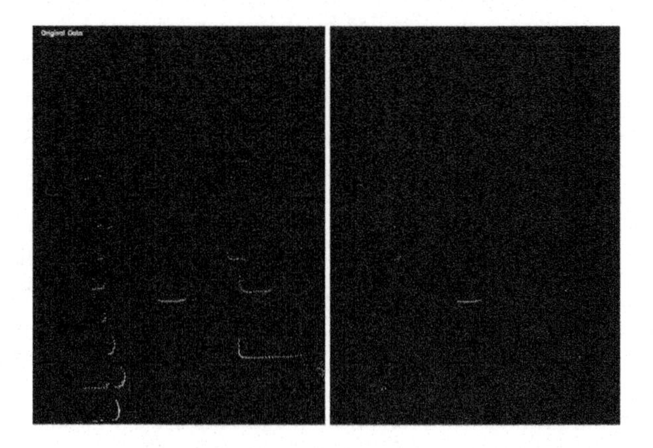

Fig. 8 Comparison of the laser sensor measurement data with and without being divided into stational and dynamic data sets

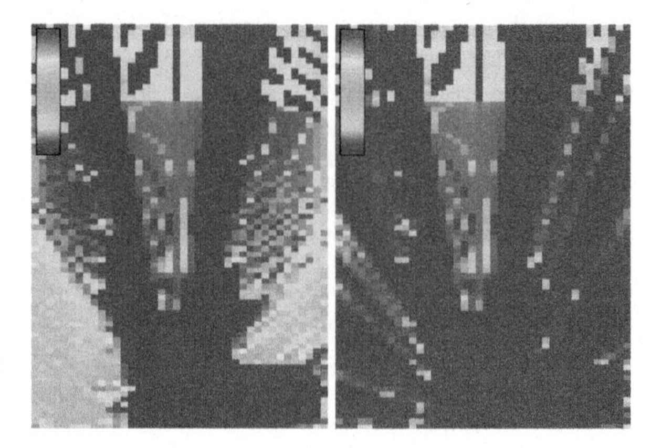

Fig. 9 Comparison of the BOF output with and without dividing the lidar data into static and dynamic data sets

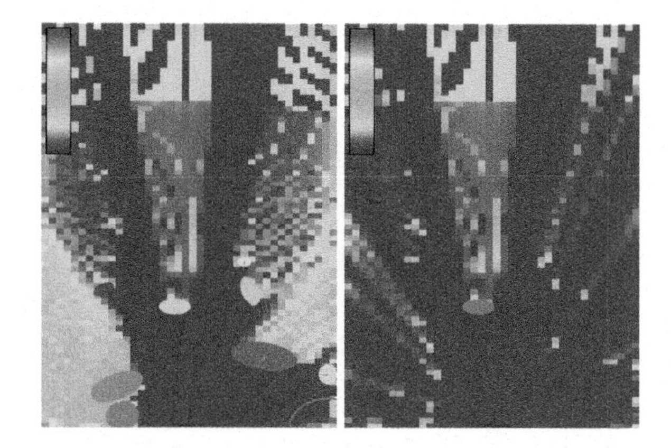

Fig. 10 Comparison of the tracking results with and without dividing the lidar data into static and dynamic data sets

The result shown in Figs. 11, 12 and 13 demonstrates the accuracy of the tracker compared with a NNJPDA tracker. The moving car before the Cycab is detected and tracked consistently for 35 s. The relative position between the Cycab and the target comes from the GPS data is taken as the ground truth and is compared with that estimated by the trackers. The comparison of x relative position is shown in Fig. 11, while the y relative positions are compared in Fig. 12. The distance error of the estimated target position to the GPS data is shown in Fig. 13. As could be seen in the figures, our algorithm succeeded in tracking of the target consistently. However,

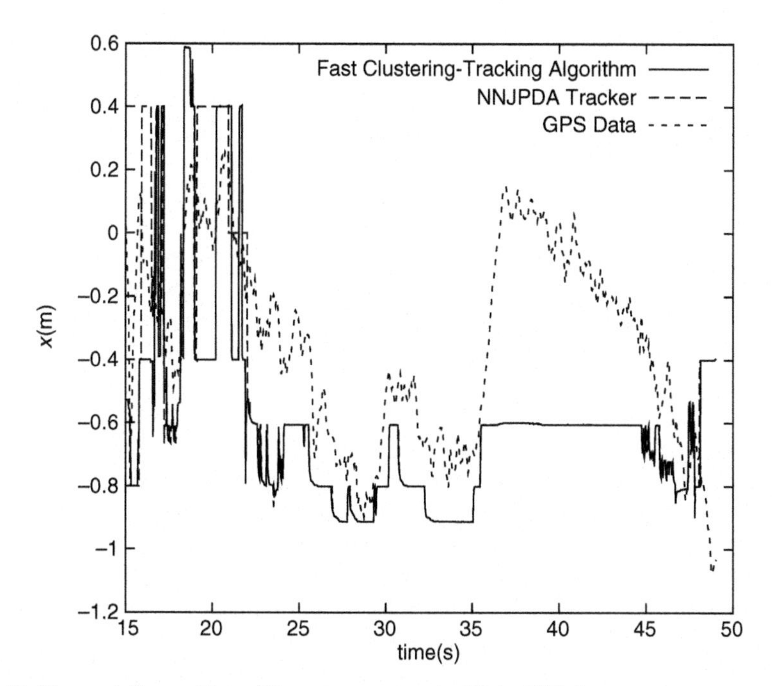

Fig. 11 The x relative position of the target compared with the GPS data

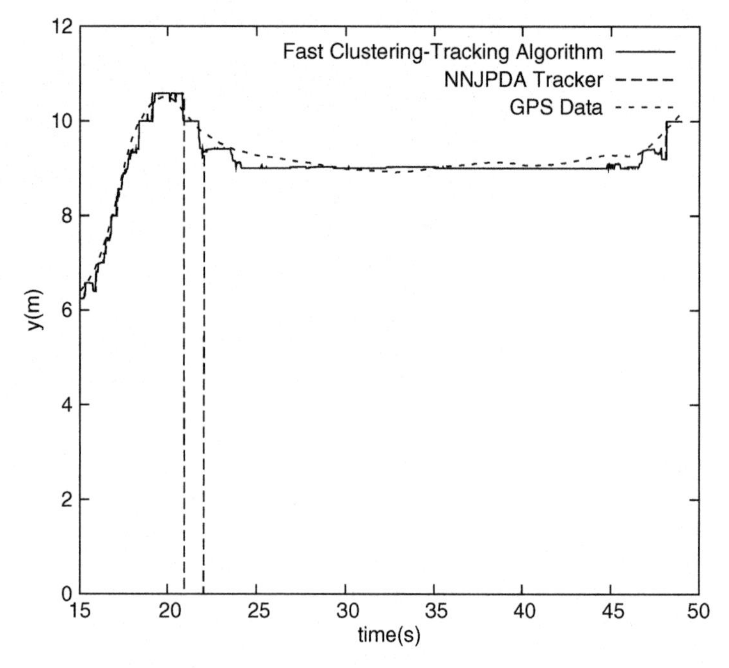

Fig. 12 The y relative position of the target compared with the GPS data

from 21 s to 22 s, the NNJPDA tracker lost the target. The average distance error is 0.39 meters for our algorithm compares with 0.37 m of the NNJPDA tracker. Consider the scale of the target car (roughly 1.5 m by 2.5 m), these results show that the precision of both the algorithms are satisfied in this application.

The implementation of the BOF and the tracker is in the C++ programming language without optimization. Experiments were performed on a laptop with an Intel Centrino processor with a clock speed of 1.6 GHz. The time consumption of the grid map methods depends on the discretization and the discretization of the velocities. In our experiment, we represent the ground plane with a dimension of 30 m by 16 m, with a discretization resolution 0.4 m by 0.4 m. The occupancy grid map represents the ground plane around the vehicle with a dimension of 30 m by 30 m, with a discretization resolution 0.15 m by 0.15 m. The algorithm processes with an average frame rate of 6.2 frames/s. The BOF consumes with an average of 0.11 s frame, while the ICP algorithm uses 0.017 s and the updating of the occupancy grid map uses up to 0.078 s frame. The average time consumption of the fast clustering-tracking algorithm is roughly 0.0003 s frame and increases linearly with the number of targets in scene which could be discarded compared with that of the NNJPDA tracker. The time efficiency of the NNJPDA tracker is highly depended on the number of the targets being tracked and the clusters extracted from the output of the BOF. When there exist an average of 11 targets and 18 clusters, the NNJPDA consumes 0.075 s frame. However, this number increases drastically to 5 s frame, when the number of the targets and the clusters increase up to 22 and 28 accordingly.

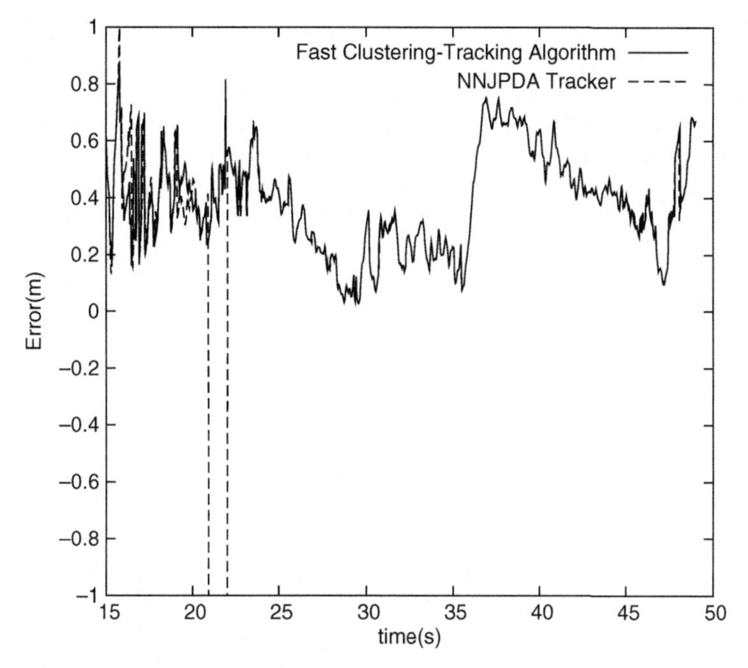

Fig. 13 The distance error of the target to the GPS data

This phenomenon is caused by the combination explosion in the algorithm which produces sparingly hypotheses that need to be processed. The experimental results show that both our algorithm and the classical NNJPDA tracker managed to track the targets accurately and consistently. However, compared with the classical NNJPDA tracker, the fast clustering-tracking algorithm is far more efficient so as to be suitable for cluttered environment.

Another experiment is shown from Figs. 14 to till 16 in time sequence. Two pedestrians walked in the direction perpendicular to the moving direction of the

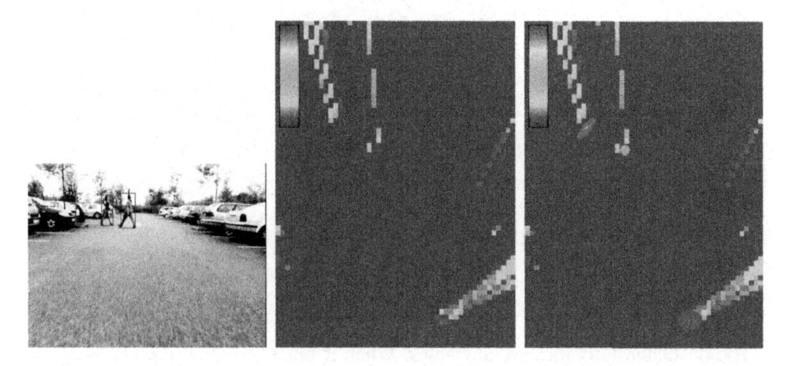

Fig. 14 The two pedestrians are walking towards each other in the perpendicular direction to the Cycab

Cycab. The first columns of the figures are the corresponding camera images in which the target is shown by the bounding box schematically (because of the limitation of the camera's field of view, there exists a third pedestrian which can not be seen in the image). The second columns show the outputs of the Bayesian occupancy filter. The third columns are the tracking results. The uncertainty of the tracked target is also fitted into a Gaussian distribution and is represented by a ellipse. In Fig. 14 the pedestrians are properly tracked. In Fig. 15, an occlusion occurs, one of the targets begin to disappear because of not being associated with any extracted clusters. In Fig. 16, after the occlusion occurs for several frames and finishes, both of the targets are detected and tracked again. Note that, the ID of the occluded object remains the same before and after the occlusion, which is shown by the same colour of the drawn target. This means the BOF framework and the proposed tracker are able to manage the targets properly during the short time occlusion, which is an important characteristic for a wide range of applications.

Fig. 15 An occlusion takes place. The further pedestrian is occluded and not observed by the Cycab

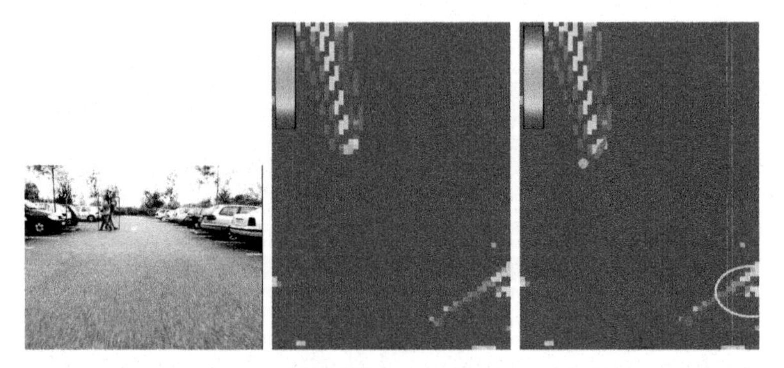

Fig. 16 The occlusion lasts for several frames. When it finishs, both of the two pedestrians are observed and tracked again. Note that the color of the targets are not changed which indicates the shortly occluded target is tracked consistently by the tracker

5 Conclusions

In this paper, we presented a novel object detecting and tracking algorithm for the BOF framework. This algorithm takes the occupancy/velocity grid of the BOF as input and extracts the objects from the grid with a clustering module which takes the prediction of the tracking module as a feedback to reduce the computational cost. A re-clustering and merging module is proposed to deal with the ambiguous data associations. The extracted objects are then tracked and managed in a probabilistic way. The experiment results show that the presented algorithm is robust as well as computationally efficient. Future work involves adding richer sensory data including the IBEO multi-layer laser range finder to the proposed framework.

References

1. Y.B. Shalom and T.E. Fortman. *Tracking and Data Association*. Boston, MA: Academic Press, (1988)
2. M.K. Tay, K. Mekhnacha, C. Chen, M. Yguel, and C. Laugier. An efficient formulation of the Bayesian occupation filter for target tracking in dynamic environments. *International Journal of Autonomous Vehicles*, 6(1–2), 155–171, (2008).
3. C. Coue, C. Pradalier, C. Laugier, T.H. Fraichard, and P. Bessiere. Bayesian occupancy filtering for multitarget tracking: an automotive application. *International Journal of Robotics Research*, 25(1), 19–30, January (2006).
4. S. Thrun, W. Burgard and D. Fox. *Probabilistic Robotic*. Cambridge, MA: The MIT Press, September (2005).
5. H.P. Moravec. Sensor fusion in certainty grids for mobile robots. *AI Magazine*, 9(2), (1988).
6. C.M. Bishop. *Pattern Recognition and Machine Learning*. New York: Springer, pp. 424–429, (2006).
7. S.S. Blackman and R. Popoli. *Design and Analysis of Modern Tracking Systems*. Norwood, MA: Artech House, (1999).
8. G. Welch and G. Bishop. *An Introduction to the Kalman Filter*. Available via: http://www.cs.unc.edu/welch/kalman/index.html
9. P.J. Besl and N.D. McKay. A method for registration of 3-D shapes. *IEEE Transactions on Pattern Analysis and Machine Intelligence*, 14(2), 239–256, (1992).

Compliant Physical Interaction Based on External Vision-Force Control and Tactile-Force Combination

Mario Prats, Philippe Martinet, Sukhan Lee and Pedro J. Sanz

Abstract This paper presents external vision-force control and force-tactile integration in three different examples of multisensor integration for robotic manipulation and execution of everyday tasks, based on a general framework that enables sensor-based compliant physical interaction of the robot with the environment. The first experiment is a door opening task where a mobile manipulator has to pull the handle with a parallel jaw gripper by using vision and force sensors in a novel external vision-force coupling approach, where the combination is done at the control level; the second one is another vision-force door opening task, but including a sliding mechanism and a different robot, endowed with a three-fingered hand; finally, the third task is to grasp a book from a bookshelf by means of tactile and force integration. The purpose of this paper is twofold: first, to show how vision and force modalities can be combined at the control level by means of an external force loop. And, second, to show how the sensor-based manipulation framework that has been adopted can be easily applied to very different physical interaction tasks in the real world, allowing for dependable and versatile manipulation.

1 Introduction

Management of uncertainty is one of the big challenges in the design of robot applications able to operate autonomously in unstructured envirnoments like human or outdoor scenarios. Robot grasping and manipulation of objects is not an exception, but one of the fields in robotics more affected by the uncertainties of the real world. The use of data coming from multiple sensors is a valuable tool to overcome these difficulties.

In particular, vision and force are the most important sensors for task execution. Whereas vision can guide the hand towards the object and supervise the task,

M. Prats (✉)
Robotic Intelligence Lab, Jaume-I University, Castellón, Spain
e-mail: mprats@icc.uji.es

S. Lee et al. (eds.), *Multisensor Fusion and Integration for Intelligent Systems*, Lecture Notes in Electrical Engineering 35, DOI 10.1007/978-3-540-89859-7_16,

force feedback can locally adapt the hand trajectory according to task forces. When dealing with disparate sensors, a fundamental question stands: how to effectively combine the measurements provided by these sensors? One approach is to combine the measurements using multi-sensor fusion techniques [1]. However, such methods are not well adapted to vision and force sensors since the data they provide measure fundamentally different physical phenomena, while multi-sensory fusion is aimed at extracting a single information from disparate sensor data. Another approach is to combine visual and force data at the control level, as we propose in this paper, where a novel vision-force control law [2], based on the concept of external control [3], does the coupling in sensor-space, which allows to control vision and force on all the degrees of freedom, whereas only the vision control law is directly connected to the robot.

In the literature we can found several applications of robots performing physical interaction tasks in real life environments, such as for example [4–6]. However, very few approaches consider multiple sensors in a general framework. Instead, ad-hoc applications are usually implemented, leading in specialized robots unable to perform many different manipulation tasks.

In [7], we presented a general framework for enabling compliant physical interaction based on multiple sensor information. The purpose of this paper is twofold: first, to show how vision and force modalities can be combined at the control level by means of external vision-force control. And, second, to show how this framework can be used for the fast implementation of sensor-based physical interaction tasks in very different robotic systems, as well as its versatility, allowing to perform very different tasks in household environments, without having specific models of them, and without being specifically programmed for a particular task. For this, three different applications are described in different scenarios, and involving several robots and sensors: the first one is a door opening through external vision-force control; the second one is another vision-force door opening task, but including a sliding mechanism and a different robot, endowed with a three-fingered hand; finally, the third task is to grasp a book from a bookshelf by means of tactile and force integration.

The paper is organized as follows: the sensor-based physical interaction framework detailed in [7] is outlined in Sect. 2. Sections 3, 4 and 5 describe three different examples of the framework application, involving different robots and sensor combinations. The novel vision-force coupling approach is introduced in the first experiment. Conclusions and future lines are given in Sect. 5.

2 A Framework for Sensor-Based Compliant Physical Interaction

A framework for describing physical interaction tasks, based on multisensor integration is presented in detail in [7]. Our approach is based on the *Task Frame Formalism* [7, 8], where a *task frame* is defined as a cartesian coordinate system, given in object coordinates, where the task is defined in terms of velocity and force references, according to the natural constraints imposed by the environment.

We describe the task by the following elements (see [7] for a complete description):

- The *task frame*, T, where the task motion can be naturally described in terms of a velocity/force reference.
- The *hand frame*, H, defined in hand coordinates, and the *grasp frame*, G, defined in object coordinates, which indicate, respectively, the part of the hand used for performing the task, and the part of the object where to perform the task.
- The task velocity, $\mathbf{v}^*$, and the task force, $\mathbf{f}^*$, given in the task frame. The velocity reference is suitable for tasks where a desired motion is expected, whereas the force reference is preferred for dynamic interaction with the environment, where no object motion is expected, but a force must be applied (for polishing a surface, for example). A 6×6 diagonal selection matrix, $\mathbf{S}_f$, is used to choose whether a particular task direction needs a velocity or a force reference. A suitable force controller must convert the force references on force-controlled degrees of freedom (DOFs) to velocities, so that the task is finally described as a desired velocity given in the task frame: τ_T^*.

In general, the task frame is not rigidly attached to the robot end-effector frame. The task frame, according to its definition, must be always aligned with the natural decomposition of the task. Therefore, sensors must be integrated in order to provide an estimation of the task frame position and orientation during task execution (sensor-based tracking of the task frame [8]). This estimation is represented by the homogeneous transformation matrix $\widehat{{}^E\mathbf{M}_T}$, so that the desired task velocity, τ_T^*, can be transformed from the task frame to the robot end-effector frame, according to:

$$\tau_E = \widehat{{}^E\mathbf{W}_T} \cdot \tau_T^* \tag{1}$$

where $\widehat{{}^E\mathbf{W}_T}$ is the 6×6 screw transformation matrix [6] associated to $\widehat{{}^E\mathbf{M}_T}$, which is computed from the kinematic chain linking the robot end-effector with the object mechanism, i.e. $\widehat{{}^E\mathbf{M}_T} = {}^E\mathbf{M}_H \cdot \widehat{{}^H\mathbf{M}_G} \cdot {}^G\mathbf{M}_T$ [17].

The relative pose between the robot end-effector and the task frame depends on the particular execution and must be estimated on-line by the robot sensors, because it can vary during execution due to the particular object mechanism, or due to task redundancy, where a particular DOF is controlled by a secondary task. The robot must always estimate the hand-to-object relationship during task execution by means of the model, world knowledge, vision sensors, tactile sensors, force feedback, etc. so that the task frame is always known with respect to the end-effector frame, thus allowing the robot to perform the desired task motion.

3 Example I: Pulling Open a Door Through External Vision-Force Control

In this section, the framework for sensor-based compliant physical interaction is applied to the task of pulling open the door of a wardrobe, using a mobile manipulator composed of an Amtec 7DOF ultra light-weight robot arm mounted on an ActivMedia PowerBot mobile robot. The hand of the robot is a PowerCube parallel jaw gripper. This robot belongs to the Intelligent Systems Research Center (ISRC, Sungkyunkwan University, South Korea), and is already endowed with recognition and navigation capabilities [11], so that it is able to recognise the object to manipulate and to retrieve its structural model from a database.

3.1 Planning the Task, Hand and Grasp Frame

The structural model of the door is shown in Fig. 1. The task of pulling open the door can be specified naturally as a rotation around Y axis of frame O, but also as a negative translation velocity along Z axis of the frame G. The second alternative has the advantage that we can set ${}^{G}\mathbf{M}_{T} = \mathbf{I}_{4\times4}$, without the need to know the door model. We adopt this approach in order to make the solution valid for other doors. Thus, $T = G$, and we set $\mathbf{v}^*$ to be a negative translation velocity along Z axis (the desired opening velocity). As there is no need for force references for this task, $\mathbf{f}^* = \mathbf{0}$ and $\mathbf{S}_f = \mathbf{0}_{6\times6}$.

For the parallel jaw gripper, there are very few manipulation possibilities. We consider only one possible task-oriented hand preshape, which is the precision

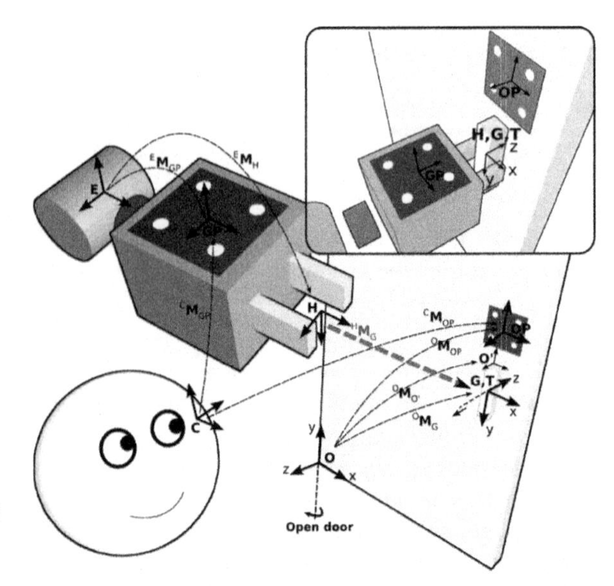

Fig. 1 The different frames used for manipulation. The vision task is to align hand frame H, set at the middlepoint between the fingertips, and the grasp frame G, set to the door handle. The task of pulling open the door is specified as a negative velocity along Z axis of the task frame, T

preshape. The hand frame is set to the middle point between both fingertips, as shown in Fig. 1.

As the door contains a handle, the grasp frame is set to the handle, so that the grasp is performed on it. More concretely, the grasp frame is set centered at the handle main axis, as shown in Fig. 1. Then, according to the specification of the hand and grasp frames, the desired relationship between both is ${}^{H}\mathbf{M}_{G}^{*} = \mathbf{I}_{4\times4}$, i.e. the identity: when grasping, the hand frame must be completely aligned with the grasp frame (the handle must lie in the middle point between both fingertips).

3.2 Task Execution

For this task, a position-based vision-force servoing closed-loop approach has been adopted. A robot head observes both the gripper and the object and tries to achieve the desired relative pose between both.

3.2.1 Estimating Hand-Handle Relative Pose

Virtual visual servoing [12] is used to estimate the pose of the hand and the handle, using a set of point features drawn on a pattern whose model and position is known. One pattern is attached to the gripper, in a known position ${}^{E}\mathbf{M}_{GP}$. Another pattern is attached to the object, also in a known position with respect to the object reference frame: ${}^{O}\mathbf{M}_{OP}$. As future research we would like to implement a new feature extraction algorithm in order to use the natural features of the object instead of the markers, as in [13] or [14]. Figure 1 shows the different frames involved in the relative pose estimation process and the task.

The matrix $\widehat{{}^{H}\mathbf{M}_{G}}$, which relates hand and handle, is computed directly from the pose estimation of the gripper and the object, according to the following expression:

$$\widehat{{}^{H}\mathbf{M}_{G}} = \left({}^{C}\mathbf{M}_{GP} \cdot {}^{E}\mathbf{M}_{GP}^{-1} \cdot {}^{E}\mathbf{M}_{H}\right)^{-1} \cdot {}^{C}\mathbf{M}_{OP} \cdot {}^{O}\mathbf{M}_{OP}^{-1} \cdot {}^{O}\mathbf{M}_{G} \qquad (2)$$

where ${}^{C}\mathbf{M}_{GP}$ is an estimation of the pose of gripper pattern, expressed in the camera frame, and ${}^{C}\mathbf{M}_{OP}$ is an estimation of the object pattern pose, also in the camera frame. ${}^{E}\mathbf{M}_{H}$ and ${}^{O}\mathbf{M}_{G}$ are the hand and grasp frame positions with respect to the end-effector and the object reference frame respectively, as set in the previous points.

3.2.2 Improving the Grasp

After pose estimation, a measure of the error between the desired (${}^{H}\mathbf{M}_{G}^{*}$) and current ($\widehat{{}^{H}\mathbf{M}_{G}}$) hand-handle relative pose is obtained. It is desirable to design a control strategy so that the grasp is continuously improving during task execution. With a vision-based approach, any misalignment between the gripper and the handle (due to sliding, model errors, etc.) can be detected and corrected through a position-based visual servoing control law [15]. We set the vector $\mathbf{s}$ of visual features to be $\mathbf{s} = (\mathbf{t} \quad \mathbf{u}\theta)^T$, where $\mathbf{t}$ is the translational part of the homogeneous matrix $\widehat{{}^H\mathbf{M}_G}$, and $\mathbf{u}\theta$ is the axis/angle representation of the rotational part of $\widehat{{}^H\mathbf{M}_G}$. The velocity in the hand frame τ_H is computed using a classical visual servoing control law:

$$\tau_H = -\lambda \mathbf{e} + \frac{\widehat{\partial \mathbf{e}}}{\partial t} \tag{3}$$

where $\mathbf{e}(\mathbf{s}, \mathbf{s}^d) = \widehat{\mathbf{L}_\mathbf{s}^+}(\mathbf{s} - \mathbf{s}^d)$ (in our case, $\mathbf{s}^d = 0$, as ${}^H\mathbf{M}_G^* = \mathbf{I}_{4\times4}$). The interaction matrix $\widehat{\mathbf{L}_\mathbf{s}}$ is set for the particular case of position-based visual servoing:

$$\widehat{\mathbf{L}_\mathbf{s}} = \begin{pmatrix} -\mathbf{I}_{3\times3} & \mathbf{0}_{3\times3} \\ \mathbf{0}_{3\times3} & -\mathbf{L}_w \end{pmatrix}$$

$$\mathbf{L}_w = \mathbf{I}_{3\times3} - \frac{\theta}{2}[\mathbf{u}]_\times + \left(1 - \frac{\text{sinc}(\theta)}{\text{sinc}^2(\frac{\theta}{2})}\right)[\mathbf{u}]_\times^2$$

where $[\mathbf{u}]_\times$ is the skew anti-symmetric matrix [10] for the rotation axis $\mathbf{u}$. Finally, the end-effector motion is computed as $\tau_E = {}^E\mathbf{W}_H \cdot \tau_H$. Figure 2 shows the kinematic screw, computed by Eq. (3) from the error between the desired (${}^H\mathbf{M}_G$) and the current ($\widehat{{}^H\mathbf{M}_G}$) relative pose of the hand and the grasp frame during reaching the handle.

3.2.3 Task Motion and Coping with Uncertainties

The end-effector velocity that the robot has to achieve in order to perform the task motion, is computed by transforming the task velocity, from the task frame to the end-effector frame, according to Eq. (1).

Even if the relative pose between the hand and the handle, $\widehat{{}^H\mathbf{W}_G}$, is estimated and corrected continuously, this estimation can be subject to important errors, considering that it is based on vision algorithms, that can be strongly affected by illumination, camera calibration errors, etc. Due to this fact, the robot motion is also subject to errors, and cannot match exactly the desired motion for the task. As the hand is in contact with the environment, any deviation of the hand motion regarding the task trajectory will generate important forces on the robot hand that must be taken into account.

We adopt a novel external vision-force control law (see [2] for details) for integrating vision and force and coping with uncertainties. With this approach, the force vector, with current external forces, is used to create a new vision reference according to:

$$\mathbf{s}^* = \mathbf{s}^d + \widehat{\mathbf{L}_\mathbf{s}} \cdot \widehat{\mathbf{L}_\times^{-1}} \cdot \mathbf{K}^{-1}(\mathbf{f}^* - \mathbf{f}) \tag{4}$$

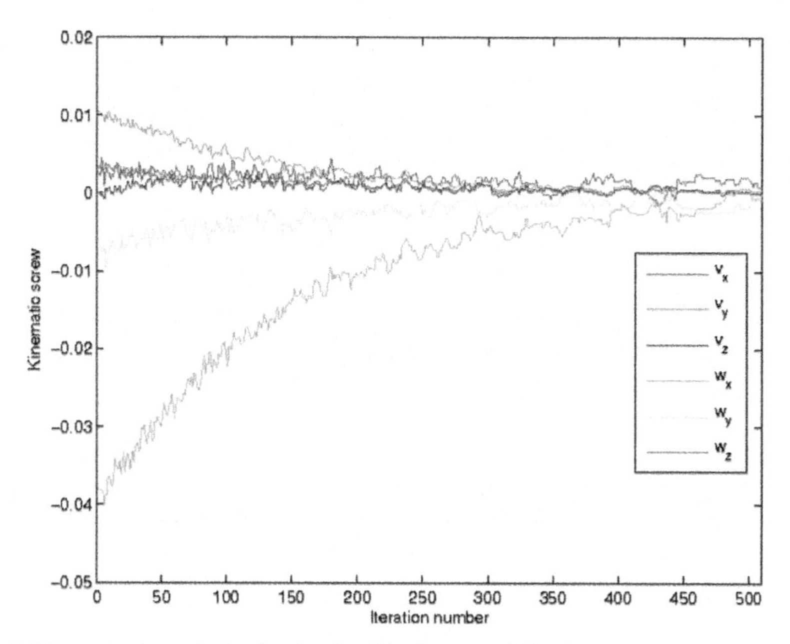

Fig. 2 Kinematic screw during hand-to-handle alignment, following an exponential decrease, which is the classical behaviour on visual servoing tasks. The kinematic screw converges to zero, when $\widehat{{}^H\mathbf{M}_G} = {}^H\mathbf{M}_G^*$

where $\mathbf{f}^*$ is the desired wrench, added as input to the control loop (null in this particular case), $\mathbf{K}$ is the environment stiffness matrix, and $\mathbf{s}^*$ is the modified reference for visual features. $\widehat{\mathbf{L}_\times}$ relates τ_E and $\dot{\mathbf{X}}_E$ according to $\dot{\mathbf{X}}_E = \widehat{\mathbf{L}_\times} \cdot \tau_E$ [15]. Then, the visual servoing control law, described in the previous point, takes as visual reference the new computed reference, $\mathbf{s}^*$. Unlike most of existing approaches, our approach for vision-force coupling does the coupling in sensor-space, which allows to control vision and force on all the degrees of freedom, whereas only the vision control law is directly connected to the robot, thus avoiding local minima [2].

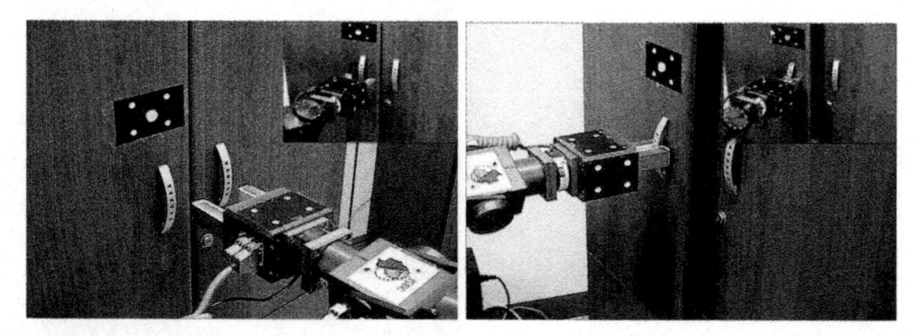

Fig. 3 The mobile manipulator at ISRC opening a door by means of external vision-force control. The left image shows a frame of the reaching process. The right image shows the interaction phase. Small snapshots at the top-right corner of each image show the robot camera view in each case

In conclusion, there are two simultaneous end-effector motions: one, computed by Eq. (1), which is in charge of performing the task motion, and another one, computed by Eq. (3), in charge of continuously aligning the hand with the handle by external vision-force control. Figure 3 shows two snapshots of the real robot performing the task, taken at reaching and during interaction. For more experimental results of the vision-force-based door opening task, along with a detailed analysis and a demonstration video, please refer to [16].

4 Example II: Opening a Sliding Door Through External Vision-Force Control

This experiment is very similar to the previous one in the sense that two complementary sensors (vision and force) are used for a door opening task. However, in this case, we deal with a sliding door and a more complex robot: a mobile manipulator composed of a PA-10 arm, endowed with a three-fingered Barrett Hand, and mounted on an ActivMedia PowerBot mobile robot (the UJI Service Robot).

4.1 Planning the Task, Hand and Grasp Frame

The structural model of the new door is shown in Fig. 4. The only difference with the previous case is that the task is now specified as a negative velocity along X axis of frame O. However, as before, we choose to specify it as a positive translation velocity along Z axis of the frame G, so that we can set ${}^{G}\mathbf{M}_{T} = \mathbf{I}_{4\times4}$, without the need to know the door geometric model.

The Barrett Hand offers more advanced capabilities than the parallel jaw gripper. A task-oriented grasp planner [17] selects a hook preshape as the more suitable hand

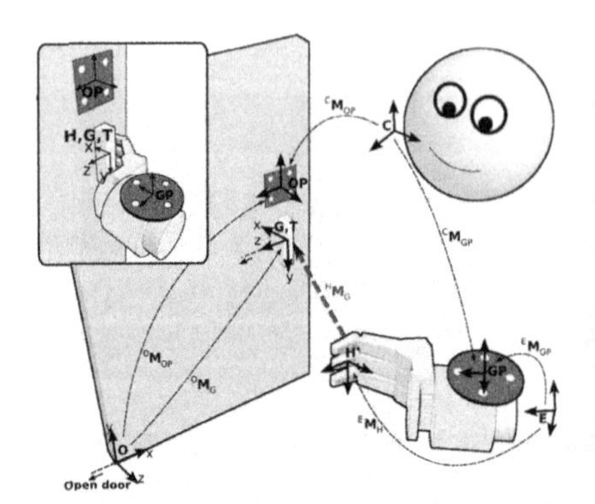

Fig. 4 Specification of the sliding door opening task with the adopted framework (left), and the mobile manipulator at Jaume-I University opening a sliding door with force-vision control (right)

configuration for the intended task, and the hand frame is set to the inner part of the fingertips, as shown in Fig. 4.

The grasp frame is also set by the task-oriented grasp planner to the right part of the handle. Then, according to the specification of the hand and grasp frames, the desired relationship between both is ${}^{H}\mathbf{M}_G = \mathbf{I}_{4\times4}$, which means that the robot has to use the fingertips to make contact with the right face of the handle.

4.2 Task Execution

Once the task has been specified, it is performed by the same methods explained in the previous experiment, supporting the claim that the robot is not specifically programmed for one particular task. Instead, the same algorithms are applied, but its execution depends on the task specification and the multisensor information that the robot receives during execution. As in the previous example, an external camera tracks the robot hand and the object simultaneously and a position-based visual servoing control is performed in order to reach and keep the desired relative hand-object configuration, ${}^{H}\mathbf{M}_G^*$. At the same time, force control is used for ensuring a successful execution of the task, even in the presence of uncertainties and errors.

5 Example III: Grasping a Book Through Force-Tactile Combination

Now, the sensor-based compliant physical interaction framework is applied to the task of taking out a book from a bookshelf, using the UJI Service Robot and force-tactile combination. The goal of the task is to extract a book from a shelf, while standing among other books. The approach is to do it as humans do: only one of the fingers is used, which is placed on the top of the target book and is used to make contact and pull back the book, making it turn with respect to the base, as shown in Fig. 6. In this task, the force/torque sensor is used to apply a force towards the book and avoid sliding, whereas a tactile array sensor provides detailed information about the contact, and helps estimating the hand and grasp frame relationship. This sensor consists of an array of 8×5 cells, each one measuring the local pressure at that point.

5.1 Planning the Task, Hand and Grasp Frame

In Fig. 5, a representation of the book grasping task, including the necessary frames, is shown. There are two possibilities for the task frame in this case. The first is to set it to the book base (frame T' in Fig. 5), so that the task is described as a rotation velocity around this frame. The second possibility is to set the task frame to the top edge of the book (frame T in Fig. 5), so that the task is described as a negative

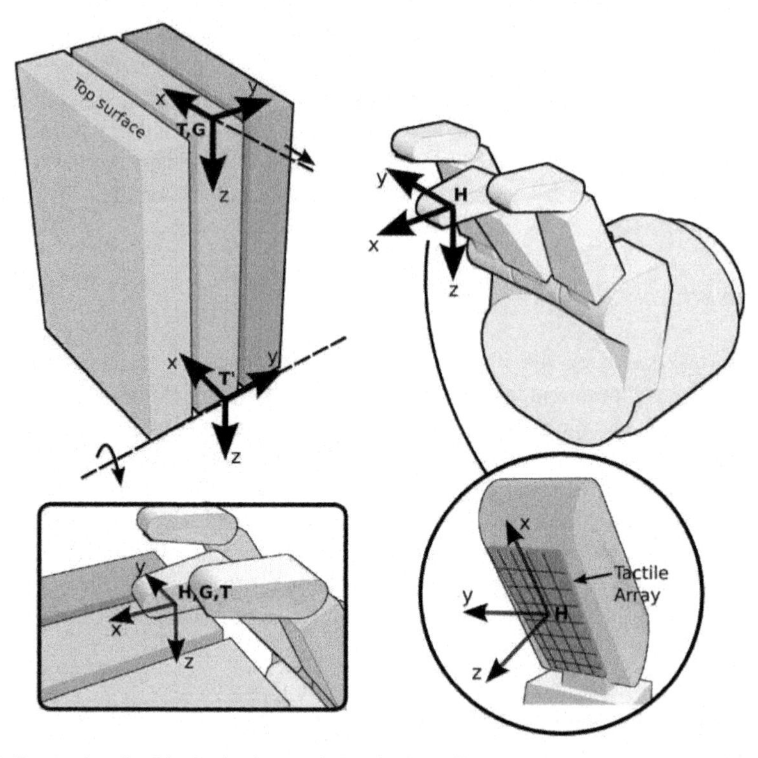

Fig. 5 Frames involved in the book grasping task. A tactile array sensor, placed on the inner part of the fingertip, is used to estimate the relationship between the hand and the grasp frame, $\widehat{^{H}\mathbf{M}_{G}}$. The task is specified as a hand velocity along negative X axis of the task frame, T, while applying a force along Z axis

translational velocity along X direction. We have opted for the second solution, because, in this case, the task frame coincides with the grasp frame, and, then, there is no need to know the book model. In the first case, the height of the book should be known in order to transform the task velocity from the task frame to the hand frame. By adopting the second solution, we make the approach general for any book size. Two references are set in the task frame, $\mathbf{v}^{*}$ and $\mathbf{f}^{*}$. The first one is set to a negative velocity in X axis, in order to perform the task motion, whereas $\mathbf{f}^{*}$ is set to a force along Z axis. This force is needed in order to make enough pressure on the book surface and avoid slip. We have set it to 10 N for our particular system, but it depends on the friction coefficient between the fingertip and the book. For small friction, a bigger force would be needed. Therefore, $\mathbf{S}_f$ is set to $\mathbf{S}_f = \mathbf{diag}(0, 0, 1, 0, 0, 0)$.

For this task, we define a special hand posture where one of the fingers is slightly more closed than the other ones, so that we can easily make contact on the top of the book with one finger, as shown in Fig. 5. The hand frame is set to the inner part of the middle finger fingertip, just in the centre of the tactile sensor. The hand frame pose with respect to the robot end-effector, $^{E}\mathbf{M}_{H}$, is computed from hand kinematics.

The fingertip has to make contact on the top of the book. Therefore, we set the grasp frame to the book top surface, which could be located by vision or range sensors. The desired relationship between the hand and the grasp frame, ${}^{H}\mathbf{M}_{G}^{*}$, is set to the identity.

5.2 *Task Execution*

In this case, the task is performed by combining force and tactile feedback. Tactile information is used to estimate and improve the contact between the hand and the book, whereas force feedback is used in order to cope with uncertainties and ensure that a suitable force is performed on the book surface so that there is no slip.

5.2.1 Estimating Hand-Book Relative Pose

Contact on the book is performed with the tactile array. Depending on the sensor cells that are activated, the relative pose between the sensor surface and the book can be estimated. It is not possible to compute the complete relative pose only with tactile sensors, because they only provide local information when there is contact. However, we can obtain a qualitative description of the relative pose. For example, if there is contact with the upper part of the sensor, but not with the lower part, we can deduce that the sensor plane is rotated around Y axis with respect to the book top plane.

All the tactile cells lie in the XY plane of the hand frame. We consider that the finger is completely aligned with the book surface when there are cells activated on each of the four XY quadrants of the hand frame, i.e., all the tactile sensor surface is in contact. If there is contact on the upper half of the sensor, but not on the lower half, or vice versa, we consider that there is a rotation around Y axis, between the sensor (hand frame) and the book surface (grasp frame). Similarly, a rotation around X axis can be detected.

5.2.2 Improving the Grasp

The goal of this process is to align the finger (tactile sensor) surface with the book surface, taking as input the qualitative description of the relative pose, described in the previous point. We follow a reactive approach, where the fingertip rotation around X and Y axis of the hand frame is continuously controlled, in order to obtain contact on each of the XY quadrants of the hand frame. With this approach, the behaviour of the robot is completely reactive to the tactile sensor readings. The goal is to keep the sensor plane always parallel to the book top plane, thus ensuring that $\widehat{{}^{H}\mathbf{M}_{G}} = {}^{H}\mathbf{M}_{G}^{*} = \mathbf{I}_{4\times4}$.

5.2.3 Task Motion and Coping with Uncertainties

According to the task description, the task motion is performed by moving the hand along negative X axis of the task frame, while applying a force along Z axis. This motion makes the book turn with respect to the base, as shown in Fig. 6. Note that, as the fingertip moves backwards and the book turns, the tactile sensor may lose contact with the lower part. This situation is detected by the qualitative pose estimator, and corrected with the control strategy described in the previous point, so that the hand frame is always aligned with the grasp frame, ensuring that task motion can successfully be transformed to end-effector coordinates by equation 1. Figure 6 shows a sequence of the robot performing the task.

Fig. 6 The robot grasping the book by means of force and tactile-based continuous estimation of hand-to-object relative pose

6 Conclusion

We have shown three different examples of robotic execution of everyday chores, built on top of a new vision-force controller [2] and a general framework for specifying multisensor compliant physical interaction tasks [7]. Two door-opening tasks with different robotic systems and a book grasping task have been implemented making use of external vision-force control and force-tactile integration. The three examples exhibit a reasonable degree of robustness, in the sense that the use of force feedback allows to deal with uncertainties and errors. External vision-force control allows to avoid local minima which is one of the main drawbacks of the existing approaches. The implementation of these examples in very different robotic systems during a short period of time shows the suitability of the framework for versatile specification of disparate multisensor physical interaction tasks.

As future research, we would like to use the proposed framework for the specification and compliant execution of several tasks, based on the integration of visual, tactile and force feedback. We think that the combination of multiple and disparate sensor information for hand-to-object pose estimation is a key point for successful and dependable robotic physical interaction.

Acknowledgments The authors would like to thank Universitat Jaume-I/Bancaja, under project PI1B2005-28, and Generalitat Valenciana, under projects CTBPRB/2005/052 and GV-2007-109 for their invaluable support in this research.

References

1. Bajcsy, R.: Integrating vision and touch for robotic applications. *Trends and Applications of AI in Business* (1984).
2. Mezouar, Y., Prats, M., Martinet, P.: External hybrid vision/force control. In: *Intl. Conf. on Advanced Robotics (ICAR'07)*, pp. 170–175. Jeju, Korea (2007).
3. Perdereau, V., Drouin, M.: A new scheme for hybrid force-position control. *Robotica* **11**, 453–464 (1993).
4. Hillenbrand, U., Brunner, B., Borst, C., Hirzinger, G.: The robutler: a vision-controlled hand-arm system for manipulating bottles and glasses. In: *35th International Symposium on Robotics*. Paris, France (2004).
5. Ott, C., Borst, C., Hillenbrand, U., Brunner, B., Buml, B., Hirzinger, G.: The robutler: Towards service robots for the human environment. In: *Video Proc. Int. Conf. on Robotics and Automation*. Barcelona, Spain (2005).
6. Petersson, L., Austin, D., Kragic, D.: High-level control of a mobile manipulator for door opening. In: *IEEE/RSJ Int. Conf. on Intelligent Robots and Systems*, pp. 2333–2338, vol 3. Takamatsu, Japan (2000).
7. Prats, M., Sanz, P., del Pobil, A.: A framework for compliant physical interaction based on multisensor information. In: *IEEE Int. Conf. on Multisensor Fusion and Integration for Intelligent Systems*, pp. 439–444. Seoul, Korea (2008).
8. Bruyninckx, H., Schutter, J.D.: Specification of force-controlled actions in the "task frame formalism": A synthesis. *IEEE Transactions on Robotics and Automation* **12**(5), 581–589 (1996).
9. Mason, M.: Compliance and force control for computer-controlled manipulators. *IEEE Transactions on Systems, Man, and Cybernetics* **11**(6), 418–432 (1981).
10. Khalil, W., Dombre, E.: Modeling identification and control of robots. *Hermes Penton Science* (2002).
11. Lee, S., Lee, S., Lee, J., Moon, D., Kim, E., Seo, J.: Robust recognition and pose estimation of 3d objects based on evidence fusion in a sequence of images. In: *IEEE Int. Conf. on Robotics and Automation*, pp. 3773–3779. Rome, Italy (2007).
12. Marchand, E., Chaumette, F.: Virtual visual servoing: a framework for real-time augmented reality. In: *EUROGRAPHICS 2002*, vol. 21(3), pp. 289–298. Saarebrücken, Germany (2002).
13. Drummond, T., Cipolla, R.: Real-time visual tracking of complex structures. *IEEE Transactions on Pattern Analysis and Machine Intelligence* **24**(7), 932–946 (2002).
14. Comport, A.I., March, E., Chaumette, F.: Robust model-based tracking for robot vision. In: *IEEE/RSJ Int. Conf. on Intelligent Robots and Systems*, IROS04, pp. 692–697 (2004).
15. Martinet, P., Gallice, J.: Position based visual servoing using a nonlinear approach. In: *IEEE/RSJ Int. Conf. on Intelligent Robots and Systems*, vol. 1, pp. 531–536. Kyongju, Korea (1999).
16. Prats, M., Martinet, P., del Pobil, A., Lee, S.: Vision/force control in task-oriented grasping and manipulation. In: *IEEE/RSJ Int. Conf. on Intelligent Robots and Systems*, pp. 1320–1325. San Diego, USA (2007).
17. Prats, M., del Pobil, A., Sanz, P.: Task-oriented grasping using hand preshapes and task frames. In: *IEEE Int. Conf. on Robotics and Automation*, pp. 1794–1799. Rome, Italy (2007).

iTASC: A Tool for Multi-Sensor Integration in Robot Manipulation

Ruben Smits, Tinne De Laet, Kasper Claes, Herman Bruyninckx and Joris De Schutter

Abstract iTASC (acronym for 'instantaneous task specification using constraints) [1] is a systematic constraint-based approach to specify complex tasks of general sensor-based robot systems. iTASC integrates both instantaneous task specification and estimation of geometric uncertainty in a unified framework. Automatic derivation of controller and estimator equations follows from a geometric task model that is obtained using a *systematic task modeling* procedure. The approach applies to a large variety of robot systems (mobile robots, multiple robot systems, dynamic human-robot interaction, etc.), various sensor systems, and different robot tasks. Using an example task, this paper shows that iTASC is a powerful tool for multi-sensor integration in robot manipulation. The example task includes multiple sensors: encoders, a force sensor, cameras, a laser distance sensor and a laser scanner. The paper details the systematic modeling procedure for the example task and elaborates on the task specific choice of two types of task coordinates: feature coordinates, defined with respect to object and feature frames, which facilitate the task specification, and uncertainty coordinates to model geometric uncertainty. Experimental results for the example task are presented.

Keywords Constraint-based programming · Multi-sensor fusion · Sensor-based roboties

1 Introduction

The goal of our research is to develop programming support for the implementation of complex, sensor-based robotic tasks in the presence of geometric uncertainty. Examples of complex tasks include sensor-based navigation and 3D manipulation in partially or completely unknown environments, using redundant robotic systems

R. Smits (✉)
Department of Mechanical Engineering, Katholieke Universiteit Leuven, Celestijnenlaan 300B, B-3001 Leuven, Belgium
e-mail: ruben.smits@mech.kuleuven.be

S. Lee et al. (eds.), *Multisensor Fusion and Integration for Intelligent Systems*, Lecture Notes in Electrical Engineering 35, DOI 10.1007/978-3-540-89859-7_17,

such as mobile manipulator arms, cooperating robots, robotic hands or humanoid robots, and using multiple sensors such as vision, force, torque, tactile and distance sensors.

The foundation for this programming support is iTASC, a generic and systematic approach [1] to specify and control a task while dealing properly with geometric uncertainty.

Previous work on specification of sensor-based robot tasks, such as force controlled manipulation [2–5] or force controlled compliant motion combined with visual servoing [6], was based on the concept of the *compliance frame* [7] or *task frame* [8]. In this frame, different control modes, such as trajectory following, force control, visual servoing or distance control, are assigned to each of the translational directions along the frame axes and to each of the rotational directions about the frame axes. The drawback of the task frame approach is that it only applies to task geometries for which separate control modes can be assigned independently to three pure translational and three pure rotational directions along the axes of a *single* frame.

A more systematic approach is to assign control modes and corresponding constraints to *arbitrary* directions in the six dimensional manipulation space. This approach, known as *constraint-based programming*, opens up new applications involving a much more complex geometry and/or involving multiple sensors that control different directions in space simultaneously.

Seminal theoretical work on constraint-based programming of robot tasks was done by Ambler and Popplestone [9] and by Samson and coworkers [10]. Our own preliminary work on iTASC was presented in [11], while iTASC, the mature framework of which this paper shows an application, is thoroughly discussed in [1]. Other applications of iTASC were presented in [12] and [13].

This paper is organized as follows. Section 2 introduces the example task and states the contribution of the paper. Section 3 provides a brief overview of the generic control and estimation approach. Section 4 applies the task modeling procedure of [1] to the example task. Sections 5 and 6 provide details on the used control and estimation equations. Finally, Sections 7 and 8 present experimental results and state the conclusions.

2 Example Task

This paper shows the application of iTASC to a multi-robot manipulation task involving multiple sensors, underconstrained specification as well as estimation of uncertain geometric parameters.

2.1 Robot System

The robot system consists of two six degrees of freedom (dof) robots.

2.2 Robot Task

The task consists of three subtasks which are executed simultaneously:

1. the contour of an unknown 2.5D workpiece, held by robot 1, is tracked by a probe mounted on robot 2, as shown in Fig. 1(a). Both the contact force between probe and contour and the tangential velocity along the contour are controlled. In addition, the orientation of the probe with respect to the contour tangent and the distance of the probe with respect to the front plane of the workpiece are kept constant, while the probe axis remains perpendicular to the front plane of the workpiece. This subtask specifies six (equality) constraints.
2. the end effector of robot 1 keeps a specified minimum distance to the closest person as shown in Fig. 1(b). This subtask specifies one (inequality) constraint.
3. a camera (camera 1) attached to robot 1, shown in Fig. 1(c), keeps the closest person in vertical position in the middle of the image. This subtask specifies three (equality) constraints.

All together the contour following task in the human populated environment specifies 10 (or 9)[1] constraints for the 12 dof robot system. The two (or three) remaining dofs of the robot system are used to satisfy 12 secondary constraints: to keep the robots as close as possible to their nominal working position, specified in their respective joint spaces.

2.3 Geometric Uncertainty

In this application following sources of geometric uncertainty are explicitly modeled: *(1)* the shape of the describing contour of the 2.5D workpiece; *(2)* the height and orientation of the front plane of the workpiece relative to the end effector of robot 1, and *(3)* the position of the closest person, expressed in x, y-coordinates of the world frame. All together these sources of geometric uncertainty are modeled by seven coordinates. All other geometric transformations are supposed to be known, even though there may be other sources of geometric uncertainty. For example, the relative position between both robots is only poorly calibrated, which introduces important geometric errors.

2.4 Sensors

Robot 1 is equipped with a camera (camera 1, Fig. 1(c)) mounted on its end effector. This camera provides an image of the closest person, but this image is not used for visual servoing. Robot 2 is equipped with three sensors mounted on its end effector (Fig. 1(a)): *(1)* a six axis force/torque sensor is used to control the contact force

[1] If the inequality constraint is not active.

Fig. 1 Loops

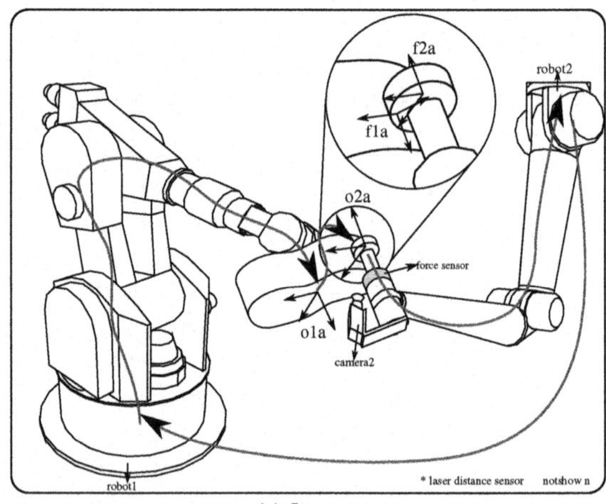

(a) Loop a.

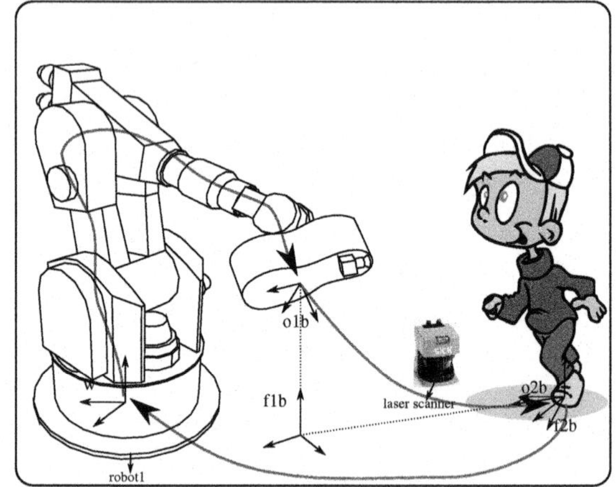

(b) Loop b.

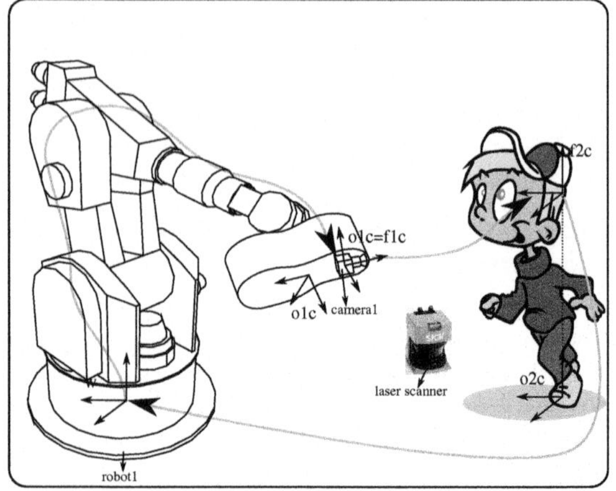

(c) Loop c.

between the probe and the contour, and to estimate the orientation of the contour tangent and normal. Only two force components of the sensor are used in this application; *(2)* a camera (camera 2) directed towards the contact point provides a second estimate of the contour tangent and normal; *(3)* (not shown in Fig. 1(a:) a laser distance sensor measures the distance of the end effector to the front plane of the workpiece. This information is used to control the distance of the probe with respect to the front plane of the workpiece and to keep the probe axis perpendicular to this plane. Additionally, the robot system contains a horizontal laser scanner (Fig. 1(b) and (c)), fixed to the environment, which is used to estimate the motion of people in the robot environment. All together six scalar signals are measured and fed back to the robot system.

2.5 Contribution of the Paper

The main contributions of the paper are: *(1)* to show how to derive the control and estimation equations for this task involving ten primary constraints, seven uncertainty coordinates, six scalar measurements and twelve secondary constraints, *(2)* to present experimental results for this integrated task, *(3)* the development of a laser scanner based people tracker, Sect. 6.1, *(4)* extension of the $2D$ contour tracking approach [1] to a $3D$ context, and *(5)* presentation of experimental results for this contour tracking approach ([1] only contains a $2D$ simulation).

3 Control and Estimation Scheme

Figure 2 shows the general control and estimation scheme presented in [1] and used throughout this paper. This scheme includes the Plant P, the Controller C, and the Model Update and Estimation block $M+E$. The Plant P represents both the robot system (where q represents the robot joint positions) and the environment.

The *control input* to the plant is u, in the case of a velocity-based control scheme, this input corresponds to the set of desired joint velocities. The *system output* is y, which represents the controlled variables. Task specification consists of imposing constraints to the system output y. These constraints take the form of desired values $y_d(t)^2$. The plant is observed through *measurements* z. Not all system outputs are directly measured, and an estimator is needed to generate estimates $\hat{y}$. These estimates are needed in the control law C.

In general the plant is disturbed by various disturbance inputs. Here we focus on *geometric disturbances*, represented by coordinates χ_u. These coordinates represent modeling errors, uncontrolled degrees of freedom in the robot system or geometric disturbances in the robot environment. As with the system outputs, not all these

[2] Further on we omit time dependency in the notation.

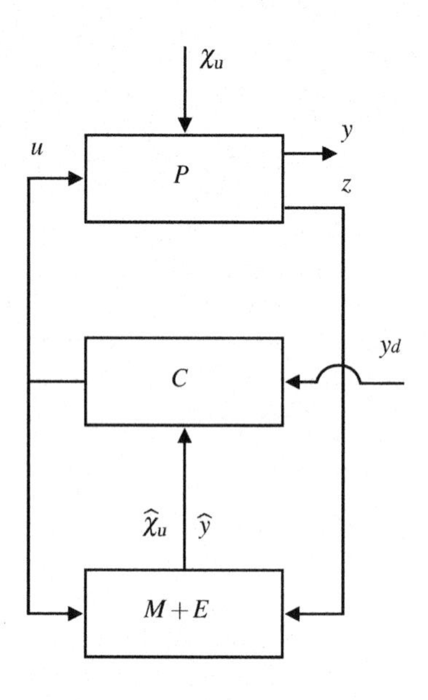

Fig. 2 General control and estimation scheme including plant P, controller C, and model update and estimation block $M + E$

disturbances can be measured directly, but they can be estimated by including a disturbance observer in the estimator block $M + E$.

4 Task Modeling

A typical robot task accomplishes relative motion between objects. The relative motion between two objects is *specified by imposing constraints* on the relative motion between one feature on the first object and a corresponding feature on the second object. Each such constraint needs four frames: *two object frames* (called $o1$ and $o2$, each attached to one of the objects), and *two feature frames* (called $f1$ and $f2$, each attached to one of the corresponding features of the objects). For an application in $3D$ space, there are in general six degrees of freedom between $o1$ and $o2$. The connection $o1 \to f1 \to f2 \to o2$ forms a *kinematic chain*, that is, the degrees of freedom between $o1$ and $o2$ are distributed over three submotions: the relative motion of $f1$ with respect to $o1$ (I), the relative motion of $f2$ with respect to $f1$ (II), and the relative motion of $o2$ with respect to $f2$ (III). The three submotions are modeled using *feature coordinates* χ_{fI}, χ_{fII} and χ_{fIII}, respectively, with

$$\chi_f = \left(\chi_{fI}\ \chi_{fII}\ \chi_{fIII}\right)^T . \tag{1}$$

We consider two kinds of geometric uncertainty: *(1)* uncertainty in the pose of an object with respect to the world, and *(2)* uncertainty in the pose of a feature with respect to its corresponding object. *Uncertainty coordinates* χ_u are introduced to represent the pose uncertainty of a real frame with respect to a modeled frame:

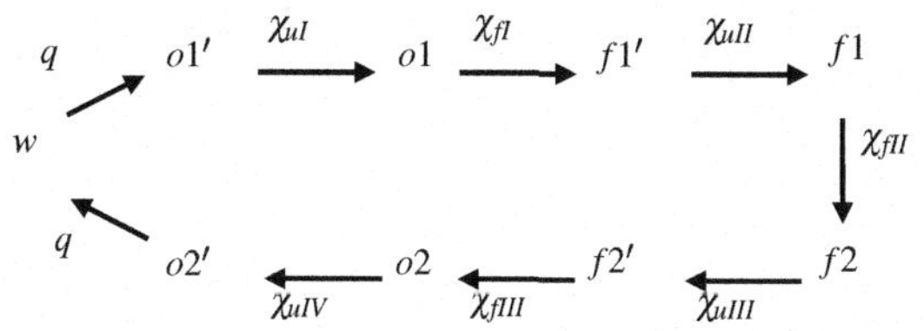

Fig. 3 Feature and uncertainty coordinates. The primed frames represent the modelled frame poses while the others are the actual ones

$$\chi_u = \left(\chi_{uI}{}^T\ \chi_{uII}{}^T\ \chi_{uIII}{}^T\ \chi_{uIV}{}^T\right)^T, \tag{2}$$

in which:

- χ_{uI} represents the pose uncertainty of $o1$,
- χ_{uII} represents the pose uncertainty of $f1$ with respect to $o1$,
- χ_{uIII} represents the pose uncertainty of $f2$ with respect to $o2$, and
- χ_{uIV} represents the pose uncertainty of $o2$.

Figure 3 summarizes the definitions of the object and feature frames, and of the feature and uncertainty coordinates. In this figure w represents the world frame.

For the example task four kinematic loops as defined in Fig. 3 are recognized, one for following the contour of the unknown object (a), one for keeping a minimum distance to the closest person (b), one for pointing the camera to the head of the closest person (c), and one for measuring the distance of the end effector of robot 2 to the front plane of the workpiece (d). Each of these loops is further detailed below.

4.1 Contour Following

The contour following task was described in [1] for the two-dimensional case. This paper extends the contour following to the three-dimensional context of the example task. The object and feature frames are defined as follows (Fig. 1(a)):

- frame $o1^a$ is fixed to the workpiece held by robot 1, with the z-axis perpendicular to the front plane of the workpiece;
- frame $o2^a$ is fixed to the probe held by the robot 2, with its z-axis along the probe's symmetry axis;
- frame $f1^a$ is located at the current contact point between the contour and the probe. The frame's z-axis is parallel to the z-axis of $o1^a$, its x-axis is parallel to the tangent of the contour.
- frame $f2^a$ has the same position and orientation as $o2^a$.

In the case of a *known* contour a minimal set of feature position coordinates exists representing the six degrees of freedom between $o1^a$ and $o2^a$:

$$\chi_{fI}{}^{a} = \left(z^{a}\ s^{a}\right), \tag{3}$$

$$\chi_{fII}{}^{a} = \left(y^{a}\ \theta^{a}\ \phi^{a}\ \psi^{a}\right)^{T}, \tag{4}$$

$$\chi_{fIII}{}^{a} = (-), \tag{5}$$

where z^a is expressed in $f1^a$ and represents the distance from the contact point to the front plane of the workpiece, s^a is the arc length along the contour, y^a is expressed in $f1^a$ and represents the distance of the robot end effector to the contour perpendicular to the contour, and θ^a, ϕ^a and ψ^a are ZXY-Euler angles expressed in $f1^a$ and represent the orientation of the probe with respect to the contour. Using s^a and the planar contour model, the pose of frame $f1^a$ with respect to $o1^a$ in a plane parallel to the front plane of the workpiece can be determined.

In the case of an *unknown* contour no set of minimal position coordinates exists to model the relative position of $f1^a$ with respect to $o1^a$. A minimal set of coordinates however exists at velocity level: $\dot{\chi}^a_{fI} = (\dot{z}^a\ \dot{s}^a)$. Instead of integrating $\dot{s}^a$, the homogeneous transformation matrix between $f1^a$ and $o1^a$ has to be updated at each time step using $\dot{\chi}^a_{fI}$.

Since the real contour is not known, the modeled contour frame $f1^a$ may deviate from the real contour frame. Therefore, uncertainty coordinates are introduced:

$$\chi_{uI}{}^{a} = \left(y^{a}_{u}\ \theta^{a}_{u}\right)^{T}, \tag{6}$$

with y^a_u the distance between the modeled and the real contour, and θ_u the orientation error between the tangents of the modeled and the real contour.

For this subtask constraints are specified on the following outputs, and are easily expressed using the feature coordinates:

$$\begin{aligned} y_1 &= z^a, \quad \dot{y}_2 = \dot{s}^a, \quad y_3 = y^a - R, \\ y_4 &= \theta^a, \ y_5 = \phi^a, \text{ and } \quad y_6 = \psi^a, \end{aligned} \tag{7}$$

where R is the radius of the probe following the contour.

The measurement equations for the magnitude of the contact force and the orientation of the tangent to the contour are easily expressed using the feature coordinates:

$$z_1 = K(y^a - R), \ z_2 = \theta^a, \text{ and } z_3 = \theta^a, \tag{8}$$

where z, represents the normal contact force with K the contact stiffness, z_2 represents the orientation of the tangent obtained from the force measurement, while z_3 represents the orientation of the tangent obtained using the image of camera 2.

4.2 Minimum Distance to Closest Person

For the minimum distance loop (Fig. 1(b)):

- frame $o1^b = o1^a$;
- frame $o2^b$ is at the position of the closest person and with the same orientation as w;
- frame $f1^b$ is located on the ground, just below $o1^b$ and with its z-axis perpendicular to the floor;
- frame $f2^b$ has the same position as $o2^b$, its z-axis is perpendicular to the floor and its x-axis pointing towards the origin of $f2^b$.

$$\chi_{fI}{}^{b} = \left(\alpha^b \; \beta^b \; \gamma^b \; z^b\right), \tag{9}$$

$$\chi_{fII}{}^{b} = \left(x^b\right)^T, \tag{10}$$

$$\chi_{fIII}{}^{b} = \left(\theta^b\right), \tag{11}$$

where α^b, β^b and γ^b are the ZXZ-Euler angles expressed in $f1^b$ that describe the orientation of $o1^b$ with respect to $f1^b$. z^b, expressed in $f1^b$, represents the height of the end effector with respect to the ground. x^b, expressed in $f2^b$, represents the distance of the closest person to the robot end effector. θ^b, expressed in $o2^b$, represents the direction in which the closest person is located.

The position of the closest person is unknown. Therefore, uncertainty coordinates are introduced:

$$\chi_{uIV}{}^{b,c} = \left(x_u^{b,c} \; y_u^{b,c}\right)^T, \tag{12}$$

with $x_u^{b,c}$ and $y_u^{b,c}$ the x- and y-position of the closest person in the world reference frame.

An inequality constraint is specified on:

$$y_7 = x^b. \tag{13}$$

The measurement equations for the position of the closest person as determined by the laser scanner based people tracker are easily specified using the uncertainty coordinates:

$$z_4 = x_u^{b,c} \text{ and } z_5 = y_u^{b,c}. \tag{14}$$

4.3 Camera Pointing to Closest Person

For the camera pointing loop (Fig. 1(c)):

- frame $o1^c = o1^a$;
- frame $o2^c$ is at the position of the closest person and with the same orientation as w,

- frame $f1^c$ is fixed to camera 1 held by the robot 1 and with the z-axis along the camera's principal ray (the line through the principal point and the origin of the pinhole model);
- frame $f2^c$ is located at the closest person's head and is rotated $\frac{\pi}{2}$ around x with respect to $o2^c$ (to avoid singularities).

The feature coordinates expressing the submotions are:

$$\chi_{fI}{}^{c} = (-)\,, \tag{15}$$

$$\chi_{fII}{}^{c} = \left(x^c\ y^c\ z^c\ \phi^c\ \theta^c\ \psi^c\right)^T, \tag{16}$$

$$\chi_{fIII}{}^{c} = (-)\,, \tag{17}$$

where x^c, y^c and z^c are expressed in $f1^c$ and x^c and y^c represent the position of the closest person's head in the camera image and z^c the distance of the person's head to the camera. ϕ^c, θ^c and ψ^c are XYZ-Euler angles expressed in $f1^c$ and represent the pan and tilt of the camera and the orientation of the closest person's head in the camera image respectively.

The position of the closest person is modeled as in Sect. 4.2, while constraints are specified on:

$$y_8 = x^c,\ \ y_9 = y^c,\ \text{and}\ y_{10} = \psi^c. \tag{18}$$

4.4 Laser Distance Sensor

This kinematic loop is modeled in detail in [1], and is only briefly described here. Three uncertainty coordinates are introduced corresponding to the errors on the height, z_u^d and the orientation, α_u^d and β_u^d of the front plane of the workpiece with respect to their modeled values:

$$\chi_{uI}{}^{d} = \left(z_u^d\ \alpha_u^d\ \beta_u^d\right)^T, \tag{19}$$

The laser distance measurement is denoted by z_6. If the laser distance measurement is not available, the uncertainty coordinates are kept constant.

4.5 Nominal Working Position

As secondary constraints the robot has to keep the robot as close as possible to its nominal working position, see Sect. 2.2. To this end, constraints can be set to the robot joint positions:

$$y_{si} = q_i, \tag{20}$$

where s indicates the secondary nature of the constraints and with $i = 1, \ldots, 12$ and q_i the i'th joint of the robot system.

4.6 Overview

Figure 4 provides an overview of the four kinematic loops showing the object and feature frames, the feature coordinates and the uncertainty coordinates.

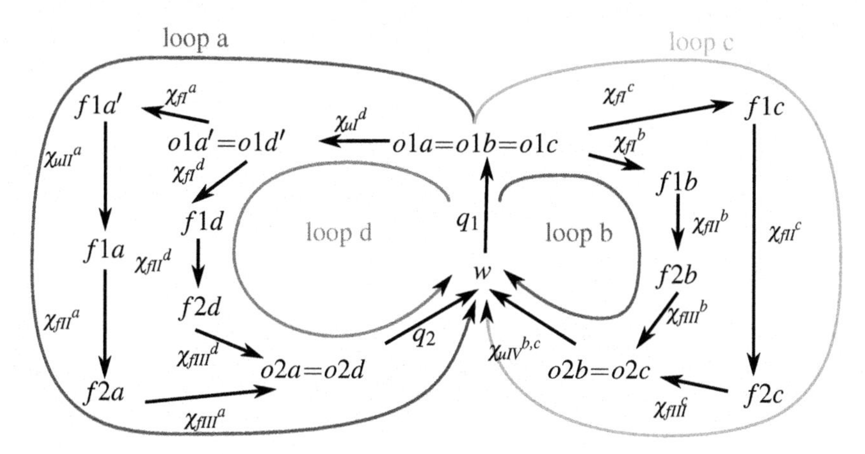

Fig. 4 Object and feature frames and feature coordinates

All feature coordinates are collected into a single vector χ_f, all uncertainty coordinates are collected into a vector χ_u, while all measurements are collected into a vector z.

5 Control Block

The equations for the control block follow automatically from the task model of Sect. 4. While [1, 12, 13] show the derivation of a velocity based control scheme, other control schemes are discussed in detail in [14].

Since we use a velocity based control scheme, the control input for the robot corresponds to:

$$\dot{q}_d = A_W^{\#}\left(\dot{y}_d^{\circ} + B\widehat{\chi}_u\right), \tag{21}$$

where $_W^{\#}$ denotes the weighted pseudoinverse [15, 16] with weighting matrix W and

$$\dot{y}_d^{\circ} = \dot{y}_d + K_p\left(y_d - y\right). \tag{22}$$

The first term of the desired joint velocities (21) controls the system outputs to their desired values, while the second term corresponds to a feedforward term accounting for the rate of change of the (estimated) uncertainty coordinates. The first term of control Eq. (22) corresponds to a feedforward of the time derivative of the desired constraint values (7,13,18), while the second term is a feedback term to compensate for drift, modeling errors and disturbances[3].

The expressions for A and B are given in [1]. We use the Orocos Kinematics and Dynamics Library (KDL, http://www.orocos.org/kdl) to derive the jacobians necessary to build A and B based on the kinematic loops defined in Sect. 4.

Secondary constraints are only realized to the extent that they do not conflict with the primary constraints [17]. The inclusion of secondary constraints (20) modifies the control input (21) to:

$$\begin{aligned} \dot{q}_d = & A_{p\,W_p}^{\#}\left(\dot{y}_{d,p}^{\circ} + B_p\widehat{\dot{\chi}}_u\right) + \\ & \left(I - A_{p\,W_p}^{\#}A_p\right)\left(A_s\left(I - A_{p\,W_p}^{\#}A_p\right)\right)_{W_s}^{\#} \\ & \left(\dot{y}_{d,s}^{\circ} + B_s\widehat{\dot{\chi}}_u - A_s A_{p\,W_p}^{\#}\left(\dot{y}_{d,p}^{\circ} + B_p\widehat{\dot{\chi}}_u\right)\right), \end{aligned} \tag{23}$$

where the subscripts p and s denote primary and secondary, respectively. A_p, B_p and $\dot{y}_{d,p}^{\circ}$ are constructed using the primary constraints (7), (13) and (18), while A_s, B_s and $\dot{y}_{d,s}^{\circ}$ are constructed using the secondary constraints (20).

6 Model Update and Estimation Block

The goal of model update and estimation is threefold: *(1)* to provide an estimate for the system outputs y to be used in the feedback terms of constraint equations (22), *(2)* to provide an estimate for the uncertainty coordinates χ_u and their derivatives, to be used in the control input (21), and *(3)* to maintain the consistency between the joint and feature coordinates q and χ_f based on the loop constraints.

Model update and estimation makes use of a prediction/correction procedure [1] and is based on an extended system model and on the measurement equations. The extended system model for this example is constructed according to [1], while the measurement equations are given by (8), (14) and the measurement equation for z_6.

The extended system model contains a motion model for the uncertainty coordinates. For example, if a constant velocity model is used for the uncertainty coordinates, the motion model is expressed as: $\dot{\chi}_u = C^{te}$ or $\frac{d}{dt}\dot{\chi}_u = 0$. The estimated $\widehat{\dot{\chi}}_u$ can be used as feedforward in (21).

The uncertainty coordinates χ_u, $\dot{\chi}_u$, ...can also be estimated by a dedicated, external estimator. In the example task, both the position of the closest person and the vision-based estimation of the contour tangent and curvature are provided by

[3] There is no feedback term for $y_2 = \dot{s}^a$, only a feedforward term.

external estimators. The next two subsections outline the operation of the people tracker and the vision-based estimator.

6.1 People Tracker

The people tracker is estimating the position and velocity of multiple persons. Only the position and velocity of the closest person, $\chi_{uIV}{}^{b,c}$ and $\dot{\chi}_{uIV}{}^{b,c}$ are used by iTaSC. $\dot{\chi}_{uIV}{}^{b,c}$ is used in (21) as feedforward.

Estimating the position of multiple moving persons is significantly harder than estimating the position of a single person. First, one has to determine the number of persons that are currently in the field of view. Furthermore, the update process is harder, since observations may result in ambiguities, features may not be distinguishable, objects may be occluded, or there may be more features than persons. Hence, a system for tracking multiple moving persons must be able to estimate the number of persons and must be able to assign the observed features to the persons being tracked.

Figure 5 presents the process flow of the people tracker. A laser scanner measures the range of the objects over 180°. From the measured distances, the measurement resulting from the environment is selected if the probability that the measurement does result from the environment is higher than a threshold: $P\,(z \notin \text{environment} \mid \geq)\, p_{ud}$, with p_{ud} a user-defined threshold. From the selected measurements, low level features are extracted using a Variational Bayesian cluster finding algorithm (VBC) [18]. The VBC provides automatic relevance detection, that is, it automatically selects the most probable number of clusters.

To keep track of multiple moving persons requires estimation of the joint probability distribution of the state of all persons. In practice, however, this is already intractable for a small number of persons, since the size of the state space grows exponentially with the number of persons. To overcome this problem, a common approach is to track the different persons independently, using factorial representations for the individual states. A general problem in this context is to determine which measurement is caused by which person. In this paper we apply a Joint Probabilistic Data Association Filter (JPDAF) [19] and/or a Sequential Joint Probabilistic Data Association Filter (SJPDAF) for this purpose [20]. The data association algorithm computes a Bayesian estimate of the correspondence between the low level features and the different persons to be tracked. Using these correspondences the individual filter for each person is updated. The JPDAF uses a Kalman filter to track the individual persons while SJPDAF uses a particle filter. To improve the tracking of individual persons, even in case of occlusions, a motion model is used. In this paper a constant velocity model was incorporated in the estimation.

The data association filters assume that the number of persons to be tracked is known. In our application, however, the number of objects often varies over time. As suggested by [20] this is handled by an extra discrete estimator estimating the number of persons from the low level features.

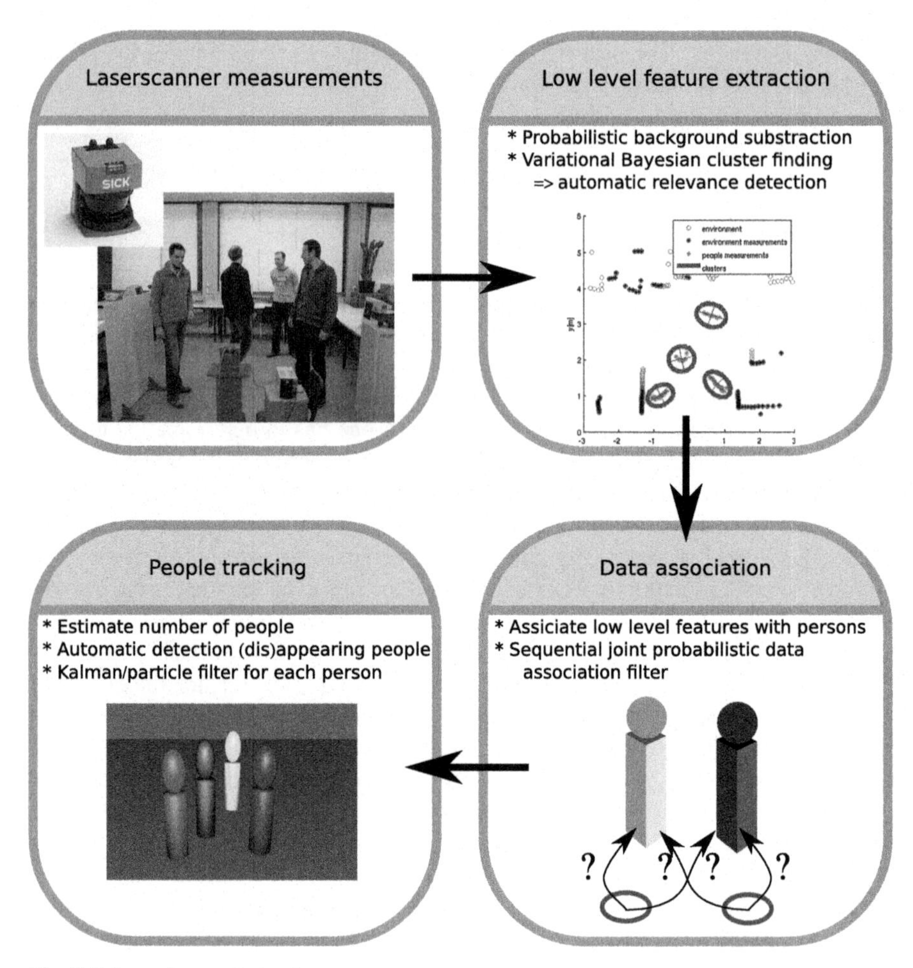

Fig. 5 Scheme for people tracker

6.2 Vision-Based Estimator

The visual contour estimator estimates the tangent and curvature of the unknown contour at the current contact point using camera 2 mounted on robot 2 (Fig. 1(a)). The tangent and curvature can be used to give an estimate of θ_u^a (6) and the estimate $\hat{\theta}_u^a$ can be used as feedforward in Eq. (21). At the beginning of the contour following task, once contact is established between contour and probe, an initialization is carried out in which the user indicates the foreground, that is, a point on the object. Next, information on properties of foreground and background are gathered using cue integration through voting for defocus, hue and saturation cues. Then, based on the integrated cues a floodfill segmentation is carried out using an automatic optimal threshold through a quality number. This quality number is obtained through a voting procedure in which the cues are: the total number of segmented pixels (the more the better), the increase in number of segmented pixels between two thresholds (if the floodfill segmentation overflows in the next valley, the previous threshold has reached an intensity ridge), the average pixel value of the image obtained through cue integration, with the segmented pixels as a mask and an indicator to check whether the segmentation reaches the edges of the camera image. The image segmented with this optimal threshold is smoothed to find a two dimensional edge and subsequentially, a natural cubic spline is fitted to the obtained edge.

For efficiency reasons, the initialization is not repeated on line: during the motion, the spline knots are corrected using line segments perpendicular to the spline [21]. Within each of these 1D search spaces, another voting procedure (cues: a minimal average correct hue/saturation, a maximal average difference in correct hue/saturation, an ISEF edge detector [22], and a weight on the distance to the previous knot position.) determines the optimal new knot position. A base for this procedure is the cue integration of hue and saturation along these line segments.

7 Experimental Results

In our experimental setup robot 1, holding the workpiece, is a Kuka K160 robot while a Kuka K361 robot is holding the probe. The force sensor is a JR3 100M40A3-I63-DH. A Sick laserscanner (LMS200, range $8m$) is used to track the persons in the neighborhood of the robot.

We carried out two different experiments. During the *first experiment* different persons, one standing still and two walking, are present in the robot environment while all constraints are switched on. In the *second experiment*, only the contour following constraints (7) are activated, by putting the weights of the constraints for the minimum distance to the closest person (13) and the camera tracking of the closest person (18) to zero.

A constant velocity along the contour of $0.01\frac{m}{s}$ ($\dot{y}_2$) was applied in both experiments, while a contact force of $30N$ was desired, resulting in a desired y_3 of $0.006m$. The desired angles between contour and follower were set to $0°$ ($y_4 = y_5 = y_6 = 0°$). A minimum distance of $3.7m$ to the closest person was commanded (y_7), while the closest person was kept straight in the middle of the image ($y_8 = y_9 = 0m$, $y_{10} = 0°$).

The contour of the $2.5D$ workpiece estimated during the two experiments is shown in Fig. 6(a). The difference between the estimated contour in the two experiments is due to the imperfect kinematics and relative position of the two robots. Figure 7 shows the estimated uncertainty on the height of the frontplane of the workpiece. The variation of $\sim 0.01\,m$ is partially due to the imperfect kinematics and relative position of the two robots. Figure 8 shows the output on the distance from the contact point to the front plane of the workpiece. This uncertainty is very well compensated by the feedback controller. Figure 9 shows that the inequality constraint on the distance of the closest person is only active when $y_7 < 3.7m$. As expected, the orientation error of the probe with respect to the contour in the

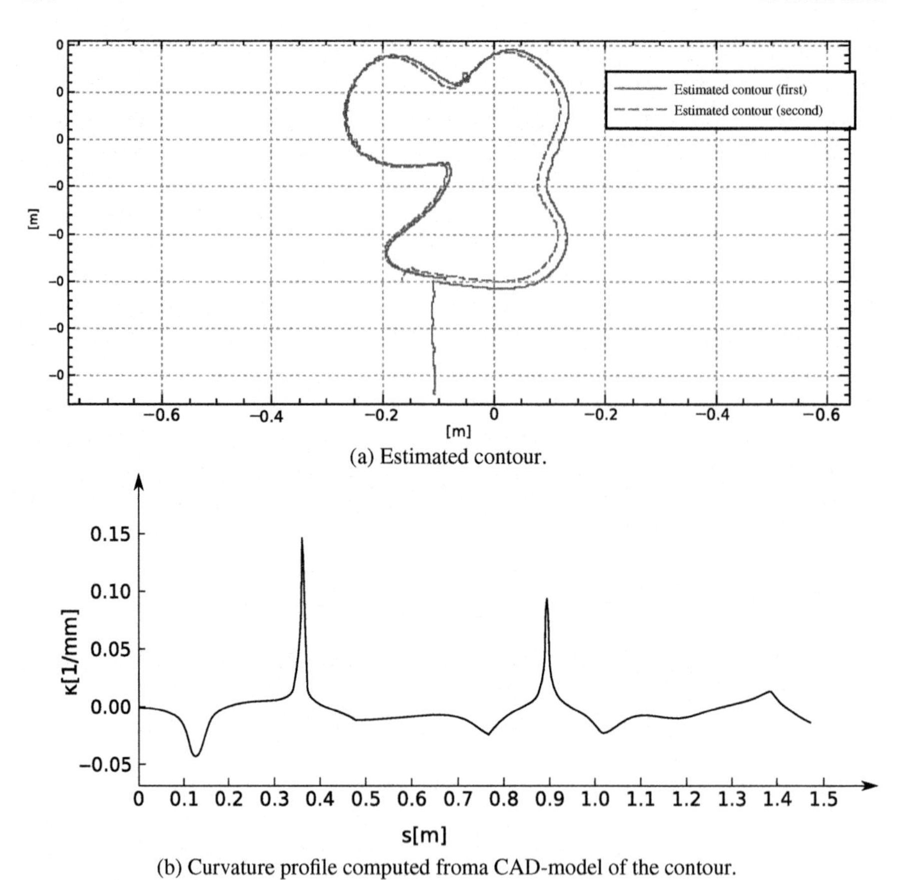

(a) Estimated contour.

(b) Curvature profile computed froma CAD-model of the contour.

Fig. 6 Contour and curvature profile of contour of workpiece

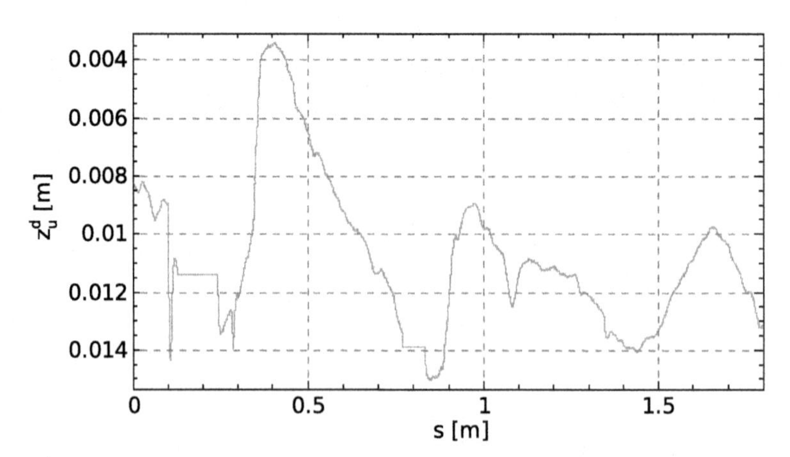

Fig. 7 z_u^d, estimated height of the frontplane of the workpiece

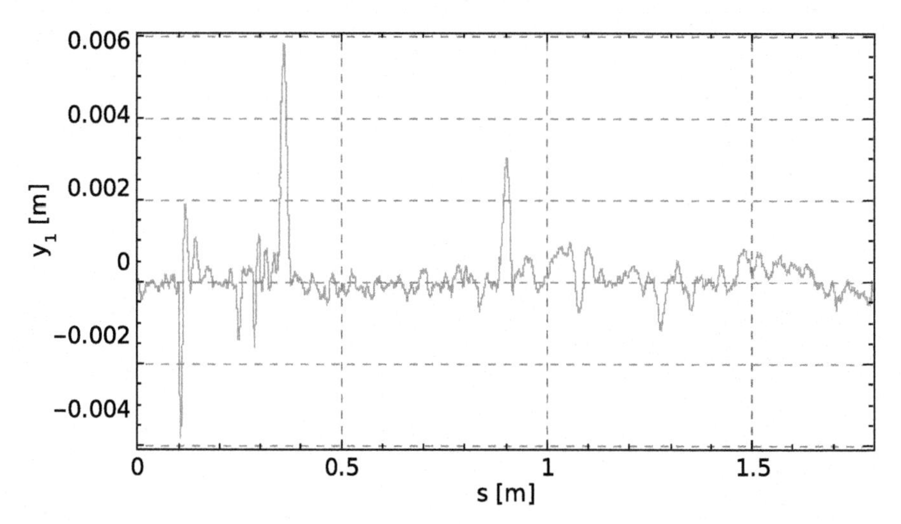

Fig. 8 y_1, output on the distance from the contact point to the frontplane of the workpiece (zero is starting position)

plane of the contour, y_4, (Fig. 10), is proportional to the curvature of the contour (Fig. 6(b)). This suggests that the tracking would benefit from feedforward of the curvature information, which can be obtained from the vision-based estimator (Sect. 6.2).

The closest person switches from one to the other when $s = 0.5m$ and $s = 0.9m$. This causes a sudden difference in the distance of the closest person (y_7) (Fig. 9) and of position of the closest person in the image (y_8 and y_9) for the first experiment (Fig. 11). This sudden error is also reflected in y_3 and y_4 (Fig. 12, Fig. 10).

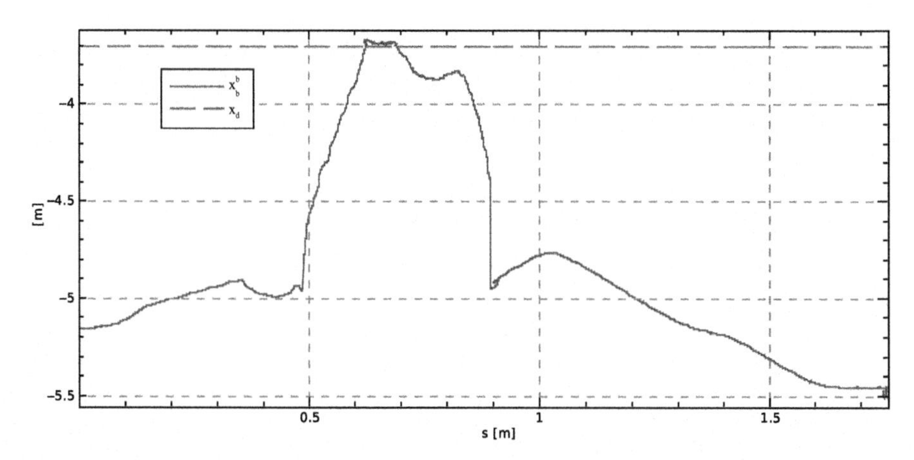

Fig. 9 y_7, distance to closest person

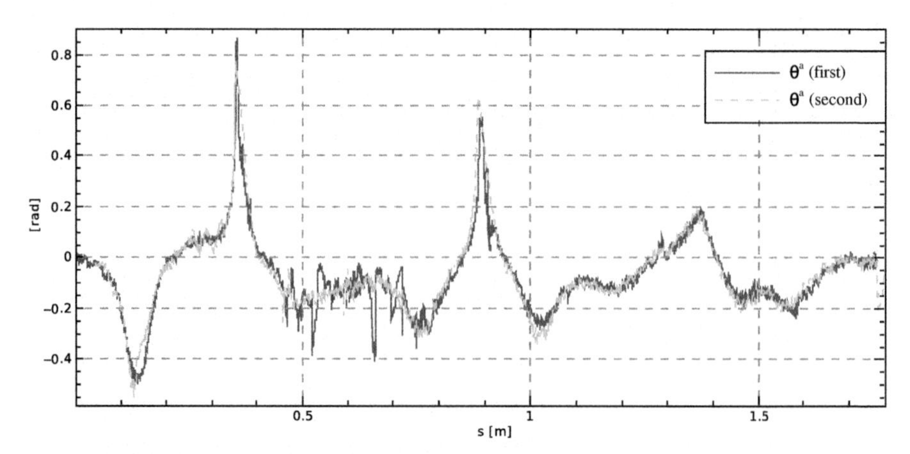

Fig. 10 y_4, orientation of probe with respect to the contour

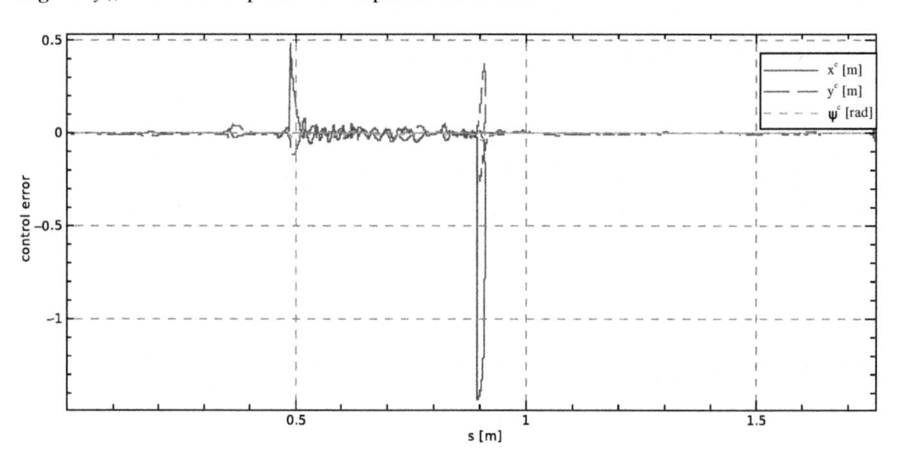

Fig. 11 y_8, y_9 and y_{10}, position and orientation of closest person in camera image

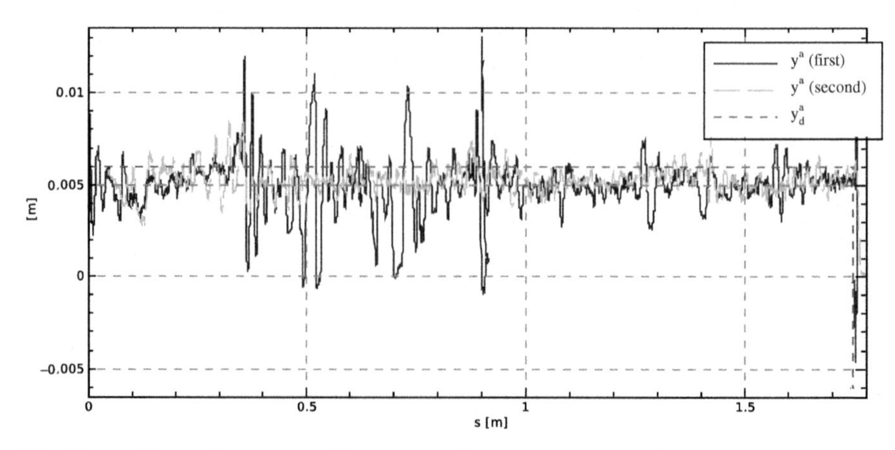

Fig. 12 y_3, deformation corresponding to contact force

Remark that the realized outputs are influenced by the imperfectly known kinematics, dynamics and relative position of the two robots, as well as the relatively stiff contact (we did not add any flexibility between contour and probe!).

Movies of different experiments are available at http://people.mech.kuleuven.be/~ orocos/

8 Conclusions and Future Work

Using an example task, this paper showed that iTASC is a powerful tool for multi-sensor integration in robot manipulation. The example task includes multiple sensors: encoders, a force sensor, cameras, a laser distance sensor and a laser scanner. The paper detailed the systematic modeling procedure, derived control and estimation equations for the task involving ten primary constraints, seven uncertainty coordinates, six scalar measurements and twelve secondary constraints, presented a laser scanner based people tracker and presented experimental results for the example task.

Future work includes the development of a user friendly interface to support the task specification to enable high-level task programming. We will also perform experiments with an acceleration-based control scheme including dynamic models of the robots, and compare the results with the velocity-based scheme.

Acknowledgments This work was supported by the K.U.Leuven's Concerted Research Action GOA/05/10. Tinne De Laet is a Doctoral Fellow of the Fund for Scientific Research–Flanders (F.W.O.) in Belgium.

References

1. J. De Schutter, T. De Laet, J. Rutgeerts, W. Decré, R. Smits, E. Aertbeliën, K. Claes, and H. Bruyninckx, "Constraint-based task specification and estimation for sensor-based robot systems in the presence of geometric uncertainty," *Int. J. Rob. Res.*, vol. 26, no. 5, pp. 433–455, 2007.
2. J. De Schutter and H. Van Brussel, "Compliant Motion I, II," *Int. J. Rob. Res.*, vol. 7, no. 4, pp. 3–33, August 1988.
3. N. Hogan, "Impedance control: an approach to manipulation. Parts I-III," *Trans. ASME J. Dyn. Syst. Meas. Control*, vol. 107, pp. 1–24, 1985.
4. N. Hosan, "Stable execution of contact tasks using impedance control," in *Int. Conf. Rob. Autom.*, Raleigh, NC, 1987, pp. 1047–1054.
5. H. Kazerooni, "On the robot compliant motion control," *Trans. ASME J. Dyn. Syst. Meas. Control*, vol. 111, pp. 416–425, 1989.
6. J. Baeten, H. Bruyninckx, and J. De Schutter, "Integrated vision/force robotics servoing in the task frame formalism," *Int. J. Rob. Res.*, vol. 22, no. 10, pp. 941–954, 2003.
7. M. T. Mason, "Compliance and force control for computer controlled manipulators," *IEEE Trans. Syst. Man, Cybern.*, vol. SMC-11, no. 6, pp. 418–432, 1981.
8. H. Bruyninckx and J. De Schutter, "Specification of force-controlled actions in the "Task Frame Formalism": a survey," *IEEE Trans. Rob. Autom.*, vol. 12, no. 5, pp. 581–589, 1996.

9. A. P. Ambler and R. J. Popplestone, "Inferring the positions of bodies from specified spatial relationships," *Artif. Intell.*, vol. 6, pp. 157–174, 1975.
10. C. Samson, M. Le Borgne, and B. Espiau, *Robot Control, the Task Function Approach*. Oxford, England: Clarendon Press, 1991.
11. J. De Schutter, J. Rutgeerts, E. Aertbelien, F. De Groote, T. De Laet, T. Lefebvre, W. Verdonck, and H. Bruyninckx, "Unified constraint-based task specification for complex sensor-based robot systems," in *Int. Conf. Rob. Autom.* Barcelona, Spain, 2005, pp. 3618–3623.
12. T. De Laet, W. Decré, J. Rutgeerts, H. Bruyninckx, and J. De Schutter, "An application of constraint-based task specification and estimation for sensor-based robot systems," in *Proc. IEEE/RSJ Int. Conf. Int. Rob. and Syst.*, San Diego, California, 2007, pp. 1658–1664.
13. W. Decré, T. De Laet, J. Rutgeerts, H. Bruyninckx, and J. De Schutter, "Application of a generic constraint-based programming approach to an industrial relevant robot task with uncertain geometry," in *IEEE Int. Conf. Comp. Tool*, Warsaw, Poland, September 2007, pp. 2620–2626.
14. T. De Laet and J. De Schutter, "Control schemes for constraint-based task specification in the presence of geometric uncertainty using auxiliary coordinates," Dept. Mech. Eng., Katholieke Univ. Leuven, Belgium, Internal report 07RP001, 2007.
15. K. L. Doty, C. Melchiorri, and C. Bonivento, "A theory of generalized inverses applied to robotics," *Int. J. Rob. Res.*, vol. 12, no. 1, pp. 1–19, 1993.
16. Y. Nakamura, *Advanced Robotics: Redundancy and Optimization*. Reading, MA: Addison-Wesley, 1991.
17. A. Ben-Israel and T. N. E. Greville, *Generalized Inverses: Theory and Applications*, reprinted ed. Huntington, NY: Robert E. Krieger Publishing Company, 1980.
18. C. M. Bishop, *Pattern Recognition and Machine Learning*. New York: Springer, 2006.
19. I. J. Cox, "A review of statistical data association techniques for motion correspondence," *Int. J. Comput. Vis.*, vol. 10, no. 1, pp. 53–667, 1993.
20. D. Schulz, W. Burgard, and D. Fox, "People tracking with mobile robots using sample-based joint probabilistic data association filters," *Int. J. Rob. Res.*, vol. 22, no. 2, pp. 99–116, 2003.
21. M. Isard and A. Blake, "CONDENSATION—conditional density propagation for visual tracking," *Int. J. Comp. Vis.*, vol. 29, no. 1, pp. 5–28, 1998.
22. J. Shen and S. Castan, "An optimal linear operator for step edge detection," *Comp. Vis. Graph. Image Proc.: Graph. Models Underst.*, vol. 54, no.2, pp. 112–133, 1992.

Behavioral Programming with Hierarchy and Parallelism in the DARPA Urban Challenge and RoboCup

Jesse G. Hurdus and Dennis W. Hong

Abstract Research in mobile robotics, unmanned systems, and autonomous man-portable vehicles has grown rapidly over the last decade. This push has taken the problems of robot cognition and behavioral control out of the lab and into the field. In such situations, completing complex, sophisticated tasks in a dynamic, partially observable and unpredictable environment is necessary. The use of a Hierarchical State Machine (HSM) for the construction, organization, and selection of behaviors can give a robot the ability to exhibit *contextual intelligence*. Such ability is important for maintaining situational awareness while pursuing important goals, sub-goals, and sub-sub goals. Using the approach presented in this paper, an assemblage of behaviors is activated with the possibility of competing behaviors being selected. Competing behaviors are then combined using known mechanisms to produce the appropriate *emergent behavior*. By combining *hierarchy* with *parallelism* we present an approach to behavior design that balances complexity and scalability with the practical demands of developing behavioral systems for use in the real-world. The effectiveness of merging our hierarchical arbitration scheme with parallel fusion mechanisms has been verified in two very important landmark challenges, the DARPA Urban Challenge autonomous vehicle race and the International RoboCup robot soccer competition.

Keywords Action Selection · Hybrid Architecture · DARPA Urban Challenge · RoboCup

1 Introduction

The problem of high-level behavioral programming is defined primarily by its position within a greater Hybrid Deliberative-Reactive control architecture such as [1–6]. Traditionally, behavior-based software agents are responsible for low-level

J.G. Hurdus (✉)
TORC Technologies, LLC 2200 Kraft Dr. Suite 1325, Blacksburg, Va 24060, USA
e-mail: hurdus@torctech.com

S. Lee et al. (eds.), *Multisensor Fusion and Integration for Intelligent Systems*,
Lecture Notes in Electrical Engineering 35, DOI 10.1007/978-3-540-89859-7_18,

Fig. 1 Inputs and outputs for a behavioral software module

reflexes and direct actuator control while deliberative agents are used for more cognitive, high-level functions. With the rapid growth of computing technology, however, there has been a re-emergence of deliberative methods for low-level motion planning [6–8]. Such methods provide the important traits of predictability and optimality, which are extremely useful from an engineering point of view. This trend, along with the need for robots capable of handling more and more complex problems, has resulted in a shift in scope and responsibility for behavior-based software agents within Hybrid control architectures .

The need now exists for a behavioral control component capable of bridging the gap between high-level mission planning and low-level motion control. This behavioral module must be capable of abstract decision making in order to complete complex, multi-faceted, temporal problems. This reactive, behavior-based software agent receives perception information about the world through *virtual sensors* and dictates desired high-level action through *virtual actuators*. Virtual sensors use sensor independent perception messages to provide a filtered view of the world and virtual actuators specify abstract motion commands to a deliberative motion planner.

The responsibility of this behavioral module is to provide two important aspects of embodied A.I., *contextual intelligence* and *emergent behavior*. Contextual intelligence provides the robot with a mechanism for understanding the current situation. This situation is dependent on both the current goals of the robot, as defined by the mission planner, as well as the current environment, as defined by the relevant objects present in the world model. Such insight is important for performance monitoring, self-awareness, and the ability to balance multiple goals and sub-goals. Emergent behavior is a very important trait of biological intelligence which is understood to be necessary for the success of living organisms in the real world. It allows for the emergence of complex behavior from the combination of simpler behaviors, which is important not only for individual intelligence, but cooperative intelligence in multi-agent systems as well.

This paper presents a novel formulation of a Hierarchical State Machine (HSM) for providing *contextual intelligence* within a behavioral agent. A generalized description of the behavioral HSM is described here for use on mobile robots with complex applications. This behavioral HSM allows for a subset of behaviors of varying levels of abstraction be activated and deactivated in real-time. Once a context dependent set of behaviors are activated, it is expected that conflicting behavioral outputs be resolved in a manner most appropriate for the specific robot application.

2 Background

2.1 The Action Selection Problem

The central focus of behavioral programming is determining at any given moment what type of action should be performed. Jim Albus, of the National Institute for Standards and Technology, defines mobile robot intelligence as the ability to "act appropriately in an uncertain environment, where appropriate action is that which increases the probability of success, and success is the achievement of *behavioral* goals" [9].

The process of deducing the most "appropriate" action is known as the *Action Selection Problem* (ASP). Unfortunately, the ability to evaluate "appropriateness" is a very complex problem and one that causes even many humans trouble. While choosing the absolutely rational, or optimal action is often impossible without seeing into the future, we can hope to select "good enough" or *satisficing* actions, as defined in [10]. According to Maes, the following requirements are needed of any Action Selection Mechanism (ASM) to produce "good enough" behavior [11].

- **Goal-orientedness** – the favoring of actions that contribute to one or several goals
- **Situatedness** – the favoring of actions that are relevant to the current situation
- **Persistence** – the favoring of actions that contribute to the ongoing goal
- **Planning** – the ability to avoid hazardous situations by looking ahead
- **Robustness** – the ability to degrade gracefully
- **Reactivity** – the ability to provide fast, timely response to surprise

In [10], the following requirements for an ASM capable of producing satisficing behavior were added.

- **Compromise** – the favoring of actions that are best for a collection of behaviors, rather than for individual behaviors
- **Opportunism** – the favoring of actions that interrupt the ongoing goal and pursue a new one

From our own experiences developing ASMs for both the Urban Challenge and RoboCup, a capable ASM should also take into account:

- **Temporal Sequencing** – the ability to define a necessary order for tasks and sub-tasks
- **Uncertainty Handling** – the ability to not react poorly to perception noise

It is very important to note that some of these many requirements conflict with each other. For example, persistence can be in conflict with opportunism and situatedness. Similarly, planning is in conflict with reactivity. It is therefore impossible to create an ASM which meets *all* of these requirements equally. Instead an ASM must attempt to *trade-off* between these requirements in a way that best fits the given application.

2.2 Existing Action Selection Mechanisms

Taxonomies of existing ASMs are seen in [12–14]. Of these taxonomies, the most complete and comprehensive is by Pirjanian in [14]. Pirjanian breaks down all ASMs as being either in the *arbitration* or *command fusion* class.

Arbitration ASMs allow "one or a set of behaviors at a time to take control for a period of time until another set of behaviors is activated" [14]. Arbitration ASMs are therefore most concerned with determining what behaviors are appropriate given the current situation. Once this has been determined it is guaranteed that there will be no conflict in outputs between the running behaviors and so no method of combination or integration is needed. ASMs within the arbitration category are further broken down into priority-based, state-based, or Winner-take-all subclasses.

Command fusion ASMs, on the other hand, "allow multiple behaviors to contribute to the final control of the robot" [14]. Rather than being concerned with selecting appropriate behaviors, command fusion ASMs let all behaviors run concurrently, then rely on a fusion scheme to filter out insignificant behavioral outputs. Command fusion ASMs are therefore typically described of as being *flat*. Since multiple behaviors can end up desiring the same control, these ASMs present novel methods of collaboration amongst behaviors. This *cooperative* approach, rather than *competitive*, can be extremely useful in situations with multiple, concurrent objectives. For example, in the robot navigation domain, command fusion ASMs are useful for both avoiding an obstacle and proceeding towards a goal at the same time. An arbitration ASM would be constrained to doing one or the other. ASMs within the command fusion category are further broken down into Voting, Superposition, Fuzzy, or Multiple Objective subclasses.

Arbitration mechanisms, on the other hand, are more efficient in their use of system resources. By selecting only one behavior from a group of competing behaviors, processing power and sensor focus can be wholly dedicated to one thing. In a flat, command fusion ASM, all behaviors must be operating at all times in order to vote for the action they prefer. As the complexity of the robot application grows, the number of behaviors needed grows, and so does the *necessary resources* in a command fusion ASM. In a hierarchical arbitration ASM, however, the library of behaviors can grow as much as it wants, but only a subset of those behaviors will ever be needed at any given moment.

Well known examples of arbitration ASMs include the Subsumption Architecture [15] and Activation Networks [11]. Popular examples of command fusion ASMs include Potential Fields [16], Motor Schemas [17], Distributed Architecture for Mobile Navigation (DAMN) [18], and Fuzzy DAMN [19].

In this paper, a method of merging these two different classes of ASMs is presented. In doing so, the strengths of *both* arbitration and command fusion mechanisms hope to be preserved. **This is possible by placing an arbitration ASM in sequence with a command fusion ASM.** The result, in essence, is the ability to select a subset of behaviors given the current situation. Then, if multiple behaviors competing for the same output are activated, they can still be cooperatively combined using a method of command fusion. Specifically, a state-based, hierarchical, arbitration ASM is used for behavior coordination. This method utilizes a hierarchical network of Finite State Automata (FSA), which can be referred to as a Hierarchical State Machine (HSM). To integrate the outputs of the activated behaviors, almost any known method of command fusion may be used. However, the chosen method should exhibit the qualities most conducive to the specific robotic application.

2.3 Hierarchy with Parallelism

It has been shown in [20] that the major bottleneck in developing behavioral intelligence is not selecting the best approach or architecture, but developing the *correct version* of this approach. While complexity is needed for multi-faceted problems, reducing complexity is important for making the robot designer's job simpler. It is not enough that a behavioral system be able to do a lot of things, it is equally important that they do all those things right, and at the right times.

In real-world applications with major repercussions for incorrect behavior, performance predictability can be paramount. The ability to hand code behaviors and ignore certain perceptual triggers at certain times is extremely useful and important for goal-orientedness. Hierarchy takes advantage of *selective attention* to make this hand-coding of behaviors possible and practical. Yet at the same time, complex combinations of behaviors are important for developing higher level intelligence.

Combining hierarchy with parallelism in the method presented in this paper provides important flexibility to the behavioral programmer. Situations in need of predictability can be catered to, while other situations can still take advantage complex, parallel, combination schemes. This approach balances quantity and complexity with design practicality.

3 Behavioral HSMs

Using a hierarchical approach to behavior decomposition is a common practice in *ethology*. It allows for the differentiation of behaviors according to their level of abstraction. According to Minsky in the Society of Mind [21], intelligent beings consist of agents and agencies. All agents are organized in a hierarchy where abstract agents are built upon lower, less abstract agents. Each agent has an individual motive which it pursues by activating and deactivating lower, subordinate agents. Groups of related agents in the hierarchy are viewed as sub-systems, and the hierarchy as a whole is the overall system.

3.1 Hierarchical Structure

A very similar organization has been adapted here, except *agents refer to individual behaviors*. All behaviors are similarly organized in a hierarchy with more abstract behaviors higher in the tree, and more physical behaviors lower in the tree. At any given time a subset of the total number of behaviors in the hierarchy are *activated* and the rest are *deactivated*. The activated behaviors are considered to be along the *activation path*. Each behavior, or node, in the tree is responsible for determining which of their *sub-behaviors* should be activated. This is determined by each behavior's internal state and is not limited to only one sub-behavior. For example, given behavior *A* in state *X*, two parallel, sub-behaviors may be activated at the same time. The result is a branch in the activation path and can be seen in Fig. 2.

We can also see from Fig. 2 that all behaviors have implied relationships based off of their position within the hierarchy tree. Behaviors can have parent-child relationships or sibling relationships, but it is important to note that these relationships do not necessarily imply importance or priority. While some arbitration ASMs use hierarchy to determine the relevance of a behavioral output [15], this approach uses hierarchy solely as an abstraction method for task decomposition. Simply put, the primary function of the hierarchical tree is to determine what behaviors to run. Using a hierarchy allows us to logically break down a complex task into smaller, more manageable pieces.

Establishing the final output to each *virtual actuator* (VA) is therefore handled by a set of command fusion ASMs. As seen in Fig. 2, two sibling behaviors are collaborating/competing for control of VA_1. VA_2, on the other hand, has a parent-child pair producing command messages. It is also possible for a single behavior to produce more than one VA command if it requires explicit coordination between two or more VAs. However, it is not *required* for every behavior to produce a VA command. Some behaviors, especially higher-level, more abstract behaviors may be used solely as decision nodes in the hierarchy. The internal state of these behaviors is important in determining the activation path and subsequently what lower-level behaviors will run, but do not necessarily request specific action themselves. These behaviors are seen in Fig. 2 as activated, but not having a specific texture.

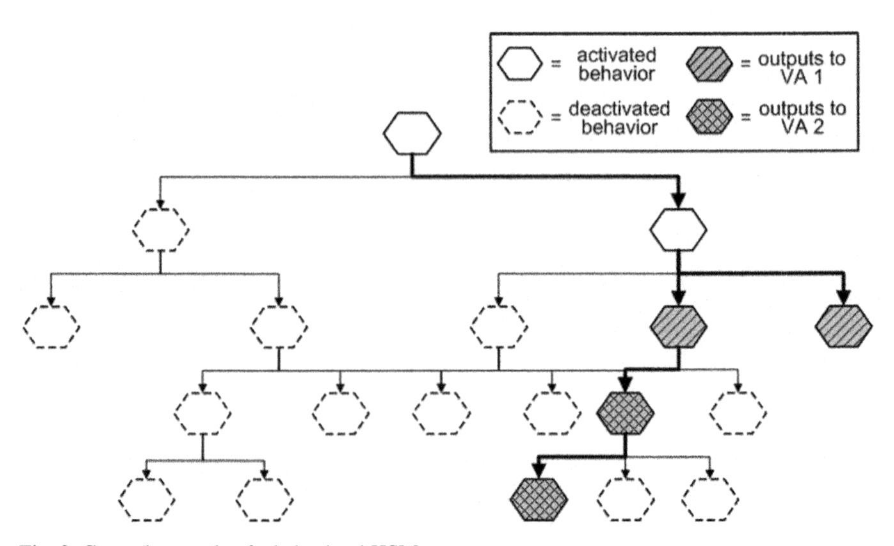

Fig. 2 General example of a behavioral HSM

Any behavior which produces one or more VA commands is classified as a *command behavior*. Any behavior which results in the activation of lower sub-behaviors (i.e., not a leaf node) is classified as a *decision behavior*. These classifications are not mutually exclusive, so it is possible for a behavior to be both a command and decision behavior.

3.2 Behaviors as Finite State Automata

Every behavior is modeled as an individual state machine, or finite state automata (FSA). Individual behaviors can therefore be formally described as consisting of a set of controls states $cs_i \in CS$. Each control state encodes a control policy, π_{va}, which is a function of the robot's internal state and its beliefs about the world (virtual sensor inputs). This policy, π_{va}, determines what action with respect to a specific VA to take when in control state cs_i. All behaviors have available to them the same list of virtual actuators $va_i \in VA$. Furthermore, each control state has hard-coded what sub-behaviors $sb_i \in SB$ to activate when in that state.

Transitions between control states occur as a function of the robot's perceptual beliefs, in the form of virtual sensors, or built-in events, such as an internal timer. While each behavior may have a "begin" and "end" state corresponding to the start and completion of a specific task, a single behavior, or state machine, cannot terminate itself. The higher, calling behavior always specifies what sub-behaviors should be running. Should a sub-behavior complete its state sequence and have nothing to do, it will remain in an idle state and not compete for control of any VA.

A simple example of an abstract behavior used for robot soccer is shown Fig. 3. The *Field Player – Attacker* behavior shown here is just one behavior within the

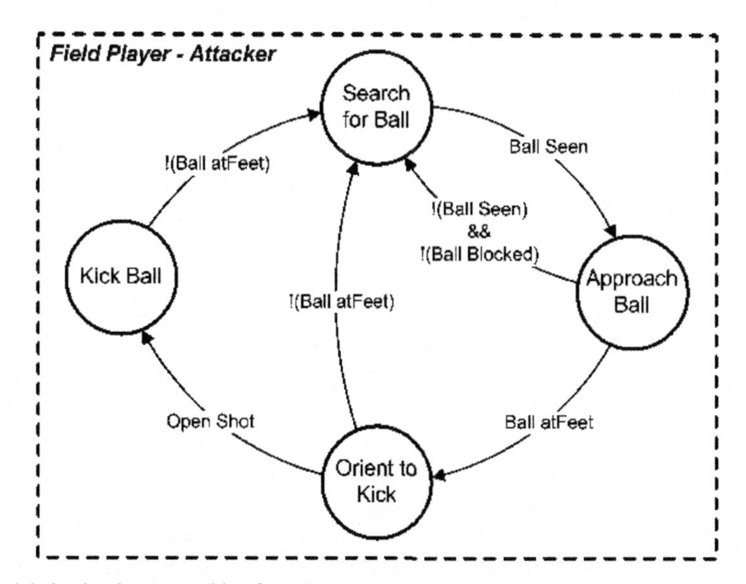

Fig. 3 A behavioral state machine for robot soccer

overall behavior hierarchy needed for a generic soccer playing robot. It is a *decision behavior* with four control states and a multitude of transitions for moving between these control states. While all transitions in this example are based off of perceptual occurrences, some may require a combination of virtual sensor inputs before being evaluated to true. For example, *Open Shot* may require perceiving the goal as being in front of the robot as well as perceiving the presence of no other robots before triggering.

Of course this individual behavior is only one within a hierarchy of other more, and less, abstract behaviors. A higher-level behavior might determine the role of the robot based off of the game situation or user inputs. For example, if the team is winning significantly it might be desired to have attacking players transition to a defender role, at which point the behavior shown in Fig. 3 might no longer be called. On the other side, each control state shown has a selection of sub-behaviors which are activated when in that control state. Let the *Field Player – Attacker* behavior be in $cs_{ApproachBall}$, it is possible then that $sb_{WalkToBall}$, $sb_{TrackBall}$, and $sb_{AvoidObstacle}$ are activated, each with their own state machine and corresponding sub-behaviors. Since the behavior shown here is a decision behavior and not a cmand behavior, $cs_{ApproachBall}$ has no control policy with respect to a virtual actuator. Instead, the primary function of this behavior is to determine what sub-behaviors to run given the current situation.

From these examples we see how a HSM, and particularly the current activation path within that hierarchy, are representative of the robot's current situation. This situation is a function of the robot's environment, the goals of the robot, and the *internal states* of the robot. In total, proper construction of the HSM will result in providing *contextual intelligence* to the robot. Producing *emergent behavior*, however, is left to the Command Fusion mechanism.

3.3 Application Specific Command Fusion

As stated earlier, the hierarchical relationship between behaviors has no relevance to the likelihood of that behavior's effect on a specific VA. Once all the behaviors along the *activation path* have been defined by the arbitration mechanism described previously, their hierarchy is thrown out and they are put in a 'flat' structure. Their individual outputs are then combined by a series of command fusion ASMs, with each instance corresponding to a single virtual actuator. The specific mechanism used for command fusion is not specified in this approach, and instead should be determined by the designer according to the robot application and specific virtual actuator. It is therefore possible to have one command fusion method for VA_1 of robot *X*, and a separate command fusion method for VA_2 and VA_3 of the same robot. This general approach to command fusion is seen in Fig. 4.

Returning to the robot soccer example presented in the previous section, let VA_1 be a vector which defines the direction and speed of a walking gait. Based on the current activation path in the HSM, the *walkToBall* behavior and the *avoidObstacle*

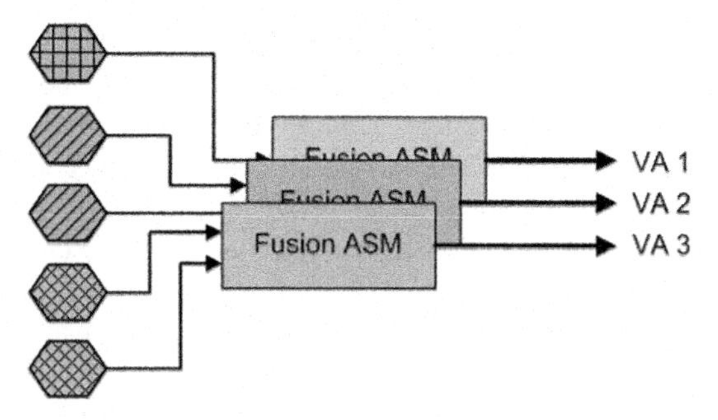

Fig. 4 Layered command fusion mechanisms

behavior are outputting desired gait vectors. It therefore makes sense, in this robot navigation example, to use a superposition mechanism of command fusion, such as *potential fields* or *motor schemas*. This would be the simplest way of producing the desired emergent behavior of approaching the ball while avoiding other robots along the way. Take now the situation where the robot is attempting to kick the ball into the opposing goal. Let VA_2 be a set of discrete kick types, *leftFoot_forward*, *leftFoot_backward*, *rightFoot_forward*, *rightFoot_backward*, etc. Just the fact that there are only a set number of discrete kick types makes a superposition-based ASM inappropriate. Instead a voting-based ASM would be much more applicable, where each behavior would vote for one type of kick, and the kick with the most votes would be selected. Taking yet another, further example, examine the behavior needed to select lanes when driving down in urban street in an autonomous vehicle. In this situation, one behavior desiring to stay in the right lane for an upcoming turn is running concurrently with a behavior desiring to pass a slow moving vehicle by moving to the left lane. Let the VA be the desired lane, and again we see that a superposition ASM is not appropriate. In this robot application, driving in between two lanes is unacceptable. Instead, a single lane should be chosen, either the left *or* the right.

We see from these examples the result of selecting different fusion ASMs. Depending on the exact mechanism chosen, completely different emergent behavior can result. This provides the robot designer with the flexibility to pick and choose the most appropriate method for the desired emergent behavior.

4 Real-World Application

The ultimate goal of action selection and behavior-based decision making research within mobile robotics is to build a physically embedded system that can exist autonomously in the real world. Action selection mechanisms that work in virtual environments are often unsatisfactory when transported to agents that must deal with real world uncertainty. It is therefore desirable to inspect the performance of any approach to behavioral programming on real robots performing real tasks.

The use of behavioral HSMs as described in this paper has been verified in two very important examples, the DARPA Urban Challenge and the International RoboCup soccer competition. At first glance, these two real-world robotic applications are extremely different. The DARPA Urban Challenge is concerned with building a full-sized autonomous ground vehicle capable of driving in an urban environment. RoboCup, on the other hand, is focused on creating a team of fully-autonomous humanoid robots capable of playing soccer. Across these two applications, the base platform is drastically different; from a 1.8 ton, 4-wheel, differentially steered vehicle to a bi-pedal, 2 foot tall humanoid robot. The goals of each robot are significantly different as well, from urban driving to goal scoring. In both of these landmark challenges, however, the core problem of a behavioral control structure is the same. Both robots must somehow balance dynamically changing desires while trying to achieve mission objectives in a real and unpredictable environment.

4.1 DARPA Urban Challenge – Team VictorTango

In November 2007, the Defense Advanced Research Projects Agency (DARPA) hosted the Urban Challenge, an autonomous ground vehicle race through an urban environment. In order to complete the course, the fully autonomous vehicle had to traverse 60 miles under 6 h while negotiating traffic (both human and robotic), through roads, intersections, and parking lots. Out of an original field of hundreds of teams from across the globe, only 35 were invited to the National Qualifying Event (NQE) in Victorville, California. After rigorous testing, only 11 teams were selected to participate in the Urban Challenge Event (UCE). Of these 11, only 6 teams managed to finish the course, with the top three places going to Carnegie Mellon University, Stanford University, and Team VictorTango of Virginia Tech.

In order to complete the challenge, vehicles had to contend with complex situations in crowded, unpredictable environments. A behavioral system capable of obeying California state driving laws in merging situations, stop sign intersections, multi-lane roads, and parking lots was needed. While a vehicle did not need to actively sense signs or signals such as traffic lights, right-of-way rules had to be followed as well as precedence-order at predefined intersections. This required the sensing, classification, and tracking of both static and dynamic obstacles at speeds up to 30 mph. To be successful, the vehicle had to balance goals of dynamically changing importance, traversing the course as quickly as possible while remaining a safe and "defensive" driver. The software module utilized by Team VictorTango to attack this problem employed a behavioral HSM for arbitration and a voting-based method for conflict resolution.

This implementation was able to produce an excellent performance at the Urban Challenge Final Event. Team VictorTango placed third overall, completing the course and all of its rigorous tests well within the 6 h limit and only minutes behind the leaders. After post-processing all the recorded data from the final race and exam-

Fig. 5 Odin, Team Victor Tango's entry in the DARPA Urban Challenge (Credit: Dr. Al Wicks, Mechanical Engineering Department, Virginia Tech)

ining hours of video, it was determined that the behavioral component made no incorrect decisions throughout the entire course of the race.

4.2 RoboCup – Team VT DARwIn

The landmark challenge presented by RoboCup is to develop a team of fully autonomous humanoid robots that can win against the human world soccer champion team by the year 2050. While it is unlikely that this will be accomplished in any near term, the idea of soccer as a standard arena for mobile robots has been widely accepted. It is estimated that more than 500 teams consisting of 3,000 scientists from 40 countries will participate in RoboCup 2008 in Suzhou, China, making it the largest competition in the project's history.

The Robotics and Mechanisms Laboratory (RoMeLa) of Virginia Tech has developed a team of fully autonomous humanoid robots for entry in the kid-size humanoid division [23, 24]. In this division a team of 3 fully autonomous humanoid robots must play the game of soccer against another team of robots. All sensing and processing must be performed on-board, and wireless transmission may be used only for communication amongst individual players. All sensing must be roughly equivalent to the capabilities of a human, prohibiting the use of active sensors that emit light, sound, or electromagnetic waves. In order to qualify for competition, robots must be able to localize an unknown ball position, walk to the ball while maintaining stability, localize a goal and position around the ball for kicking, kick the ball towards the goal, and autonomously detect and recover from a fall. To perform well in competition, robots must also be able to defend against other teams attacks, dive to block kicks if designated as a goalie, avoid contact with other robots, and work strategically as a team.

Fig. 6 DARwIn IIa and IIb competing in RoboCup 2007 (Credit: Dr. Dennis Hong, RoMeLa, Virginia Tech <www.me.vt.edu/romela/RoMeLa/Meda.html>)

Like the Urban Challenge, each individual robot must be able to handle complex situations in an unpredictable and noisy environment. A behavioral system is needed that can balance dynamic goals such as scoring, defending, and maneuvering. Therefore, a method for providing contextual intelligence and the ability to produce emergent behavior are again required for successful operation. For RoboCup, a software module built around a behavioral HSM was used. By developing this implementation and comparing it with the Urban Challenge implementation, the portability of high-level behavioral programming across drastically different platforms and functionality requirements can be seen.

5 Discussion and Conclusion

The arbitration ASM presented in this paper is a novel variant of existing state-based ASMs and utilizes a Hierarchical State Machine for task decomposition and behavior selection. The mechanism proposed in this paper provides the robot with *contextual intelligence* by maintaining a subset of activated behaviors with internal states that represent the robot's *current situation*. With environmental changes or the completion of sub-tasks and sub-sub-tasks, the activation path within the behavioral HSM will reflect the new situation.

In the case of multiple behaviors competing for control of a virtual actuator, the specific command fusion ASM is not specified and should be chosen based on the robot application. The organization of ASMs in this approach allows many typical and well known command fusion ASMs to be implemented. The selection and implementation of these command fusion mechanisms will result in the selected subset of behaviors producing the appropriate *emergent behavior*. In total, the use of behavioral HSMs addresses many important problems with existing ASMs, but like any solution, there are some important benefits and drawbacks which should be identified.

5.1 Benefits

Task Decomposition – The organization of behaviors in a hierarchical tree according to their level of abstraction is extremely useful for breaking down a task into manageable sub-tasks, and sub-sub-tasks that can be solved as independent solutions. Due to the fact that robotic behaviors still need to be largely hand-coded, a logical method for decomposition is very helpful in this process.

Temporal Sequencing – Through the use of state machines in each behavior, the robot designer can easily imply when the order of tasks is important and when it is not. Every behavior uses a state machine to define which sub-behaviors are activated. This designer can therefore use state transitions to imply order in the completion of those lower sub-behaviors.

Behavior Reuse – By taking a "divide-and-conquer" approach to behavioral problem solving, it is possible to reuse lower-level behaviors for similar problems. A sub-behavior for control state i of behavior x, can also be a sub-behavior for control state j of behavior y.

Behavior Commonalities – In conventional state machines, there are many commonalities amongst different states. In the behavioral programming example, it is possible that many different behaviors would encode the same policy for a specific VA. By using a hierarchical state machine, encoding this policy in every behavior is unnecessary. Instead, a higher-level behavior allows us to define common policies only once.

Perception Requirements – From a systems engineering perspective, the use of state machines is very useful because state transitions define all perception and virtual sensor requirements. By building the behavioral HSM first, a robot designer is aware of what information needs to be pulled from the environment.

Uncertainty Handling – A unique property of state-based behaviors is that they can be made robust to perception noise. This is possible because state transitions are directional. The requirements for transitioning from control state A to control state B can be different then the requirements for transitioning from B to A. If there is noise in the perception data (which there usually is), defining these transitions properly can prevent flip-flopping between states.

5.2 Drawbacks

Preprogrammed vs. Learned – Individual behaviors and their relationships within the greater hierarchy must be hand-coded. As a result, determining the control policies and parameters built into each state of each behavior is a time consuming and error prone process. Testing, both in simulation and on the actual robot, is absolutely essential but not always possible. It is desirable to automatically generate or learn behaviors, or at least autonomously modify parameters and control policies based off of the robots actual experience. Such learning methods are not addressed in our approach but are being researched elsewhere [22].

Performance Measurement – There exists no formal method for measuring and comparing the performance of the presented approach against other existing approaches. While "good enough" behavior defines important functional requirements, there is no quantitative method of comparison for "goal-orientedness," for example. Qualitative observations are the only major source of comparison which is generally insufficient. Performance comparison of ASMs can be done in a standard simulation environment [10] or even better in real-world competitions such as the DARPA Urban Challenge. However, with non-standardized platforms, sensors, and technology, the overall performance of any team is not a good indication of the smaller behavioral programming problem. Furthermore, since the behavior hierarchy is hand-coded, different implementations of the same approach can have very different results. The overall performance, therefore, is still dependent more on the designer than the approach itself.

Acknowledgments This work was supported in part by the Defense Advanced Research Projects Agency (DARPA) through the Track A – Urban Challenge development grant, Science Applications International Corporation (SAIC), and the National Science Foundation (NSF) under grant no. OISE 0730206.

References

1. R. C. Arkin, E. M. Riseman, and A. Hansen, AuRA: an architecture for vision-based robot navigation, *Proceedings of the DARPA Image Understanding Workshop,* pp. 414–417, Los Angeles, CA, 1987.
2. R. Murphy and A. Mali, Lessons learned in integrating sensing into autonomous mobile robot architectures, *Journal of Experimental and Theoretical Artificial Intelligence special issue on Software Architectures for Hardware Agents*, 9(2), 191–209, 1997.
3. E. Gat, Three-layer architectures, in *Artificial Intelligence and Mobile Robots*, D. Kortenkamp, R. Bonasson, and R. Murphy, editors. Cambridge, MA: MIT Press, 1998.
4. R. Simmons, R. Goodwin, K. Haigh, S. Koenig, and J. O'Sullivan, A layered architecture for office delivery robots, *Proceedings Autonomous Agents 97*, Marina del Rey, CA: ACM pp. 245–252, 1997.
5. K. Konolige and K. Myers, The saphira architecture for autonomous mobile robots, in *Artificial Intelligence and Mobile Robots*, D. Kortenkamp, R. Bonasson, and R. Murphy, editors. Cambridge, MA: MIT Press, 1998.
6. A. Bacha et al., Odin: Team VictorTango's Entry in the DARPA Urban Challenge, *Journal of Field Robotics*, 25(8), 467–492, 2008.
7. S. Thrun, M. Montemerlo, et al., Stanley: the robot that won the DARPA Grand Challenge: research articles, *Journal of Field Robotics*, 23(9), 661–692, September 2006.
8. C. Urmson, et al., A robust approach to high-speed navigation for unrehearsed desert terrain, *Journal of Field Robotics*, 23(8), 467, August 2006.

9. J. S. Albus, Outline for a theory of intelligence, *IEEE Transactions On Systems, Man, and Cybernetics*, 21(3), May/June 1991.
10. H. A. Simon, *The New Science of Management Decision*. New York: Harper and Row, 1960.
11. P. Maes, How to do the right thing, *Technical Report NE-43-836*, Cambridge, MA: AI Laboratory, MIT, 1989.
12. D. Mackenzie, R. Arkin, and J. Cameron, Specification and execution of multiagent missions, *Autonomous Robots*, 4(1), 29–52, 1997.
13. A. Saffiotti, The uses of fuzzy logic in autonomous robot navigation: a catalogue raisonne, *Technical Report 2.1*, IRIDA, Universite Libre de Bruxelles, 50 av. F. Roosevelt, CP 194/6, B-1050 Brussels, Belgium, 1997.
14. P. Pirjanian, Behavior coordination mechanisms – state-of-the-art," *Technical Report IRIS-99-375*, Institute for Robotics and Intelligent Systems, University of Southern California, Los Angeles, CA, 1999.
15. R. A. Brooks, A robust layered control system for a mobile robot, *IEEE Journal of Robotics and Automation*, 2(1), 14–23, 1986.
16. O. Khatib, Real-time obstacle avoidance for manipulators and mobile robots, *The International Journal of Robotics Research*, 5(1), 90–98, 1986.
17. R. C. Arkin, Motor schema based navigation for a mobile robot: an approach to programming by behavior, in *IEEE International Conference on Robotics and Automation*, pp. 264–271, 1987.
18. J. Rosenblatt, DAMN: a distributed architecture for mobile navigation, in *AAAI Spring Symposium on Lessons Learned from Implemented Software Architectures for Physical Agents*, Menlo Park, CA: AAAI Press, 1995.
19. J. Yen and N. Pfluger, A fuzzy logic based extension to Payton and Rosenblatt's command fusion method for mobile robot navigation, *IEEE Transactions on Systems, Man, and Cybernetics*, 25(6), 971–978, 1995.
20. J. J. Bryson, Hierarchy and sequence vs. full parallelism in reactive action selection architectures, in *From Animals to Animats 6 (SAB00)*, pp. 147–156. Cambridge, MA: MIT Press, 2000.
21. M. Minsky. *The Society of Mind*. New York, NY: Simon and Schuster, 1985.
22. B. Argall, B. Browning, and M. Veloso, Learning to select state machines using expert advice on an autonomous robot, in *IEEE International Conference on Robotics and Automation*, pp. 2124–2129, 2007.
23. K. Muecke and D. W. Hong, The synergistic combination of research, education, and international robot competitions through the development of a humanoid robot, 32nd ASME Mechanisms and Robotics Conference, New York City, NY, August 2008.
24. K. Muecke and D. W. Hong, DARwIn's evolution: development of a humanoid robot, *IEEE International Conference on Intelligent Robotics and Systems*, October 29–November 2, 2007.

Simultaneous Estimation of Road Region and Ego-Motion with Multiple Road Models

Yoshiteru Matsushita and Jun Miura

Abstract This paper describes a method of estimating road region and ego-motion for outdoor mobile robots. In outdoor navigation, we have to cope with various road scenes where, for example, road boundary features such as white lines and curbs may not be always available. Integration of sensor data from multiple sensors is thus effective for realizing a robust road region estimation. Since sensor data are obtained as a robot moves, and since an odometry-based dead reckoning suffers from accumulating errors, we develop a method which simultaneously estimates road region and robot ego-motion. We implement the method using a particle filter. The method also has a mechanism of periodically generating new particles using multiple road models to cope with gradual road shape changes. The proposed method has been successfully applied to autonomous navigation in various road scenes.

1 Introduction

Research on ITS (Intelligent Transportation Systems) has recently been active. One of the objectives of ITS research is to realize safe driving by, for example, driver assistance systems like lane departure warning. Development of autonomous robots like guide robots has also been widely conducted. These systems require an ability to recognize traversable regions such as road regions.

GPS systems, combined with an accurate map, can provide location information. But for safe driving, local information on the road region such as curbs, road and lane boundary lines, and road shoulders, should be utilized, and such information can only be obtained on-site. It is, therefore, necessary to estimate road regions using external sensors such as vision and range finders.

Y. Matsushita (✉)
Department of Mechanical Engineering, Osaka University, Yamadaoka, Suita, Osaka, 565-0871, Japan
e-mail: matsushita@cv.mech.eng.osaka-u.ac.jp

S. Lee et al. (eds.), *Multisensor Fusion and Integration for Intelligent Systems*, Lecture Notes in Electrical Engineering 35, DOI 10.1007/978-3-540-89859-7_19,

Many works use vision for detecting road boundaries [1, 2], but such boundaries are not always easily detectable. Others use range finders to detect road boundaries from their shape information. If we use a 2D scanning range fingers, guardrails or clear curbs should exist [3–5].

Road region detection only from the latest observation is vulnerable to occasional sensing failures or missing of features, such as shadows in the image or discontinuity of curbs. It is, therefore, necessary to temporally integrate sensor data for reliable detection.

Wijesoma et al. [3] developed a method of detecting and estimating road boundary from a sequence of curb positions detected by a laser range finder looking slightly downwards. They estimated road boundary parameters using Kalman filter. This method assumes that curb positions on both sides are obtained clearly from the laser data and the vehicle motion is correct. Kirchner and Heinrich [4] proposed a method of estimating road boundary parameters using a sequence of horizontal laser-scanned data. They used a 3rd order polynomial boundary model as an approximation of clothoid curves, and estimated its parameters using Kalman filter. They estimated only road boundary parameters by assuming a correct vehicle motion. In addition, since they use horizontal scanned data, some objects (e.g., guardrails) should exist at the roadside along the road. Cramer and Wanielik [5] proposed a similar method of estimating road boundary parameters.

Since an accurate ego-motion estimation only from internal sensors (i.e., dead reckoning) is difficult, it is necessary to estimate the ego-motion as well as the road region using external sensors. There are various road scenes and, therefore, appropriate features to estimate the road region may be different from place to place. For example, curbs may be removed at the entrance of shops; lanes are sometimes almost erased in an old road; unpaved roads may be detected by using only color differences between the road region and the roadside region. We, therefore, use multiple sensors and features to estimate the road region robustly.

Some previous works have used multiple sensors for navigation. Langer and Jochem [6] performed a fusion of radar and vision data for detecting roadway obstacles. Miura et al. [7] developed a method of reliable free space detection by integrating an omni-directional stereo and a laser range finder. However, the purpose of these works is not road region detection but obstacle detection.

This paper deals with simultaneous estimation of road region and ego-motion using vision and laser range data. We implement the estimation method using a particle filter. To cope with a gradual change of road type, we prepare multiple road models and devise a technique for generating new particles corresponding to such multiple models.

The rest of the paper is organized as follows. Section 2 presents a system overview. Section 3 defines the road models and the state vector. Section 4 explains the processing of range and image data and the calculation of likelihood of a state. Section 5 explains the particle filter-based data integration and the model generation. Section 6 presents experimental results of the method for various road scenes. Section 7 summarizes the paper and discusses future work.

2 Method Overview

Figure 1 shows an overview of the proposed method. After each iteration, a set of particles is kept in the system. Each particle contains a robot pose and the parameters of a road model with respect to the current origin. The state transition step performs the following two operations. One is the coordinate transformation using an estimated ego-motion by odometry. The other is road model generation based on the prediction of possible road type changes. The observation prediction step calculates what observation will be obtained from the robot position and the road parameters. The likelihood calculation step processes range and image data to extract features and calculates the likelihood of each particle from such features and the predicted observation for the particle. The final step is the resampling.

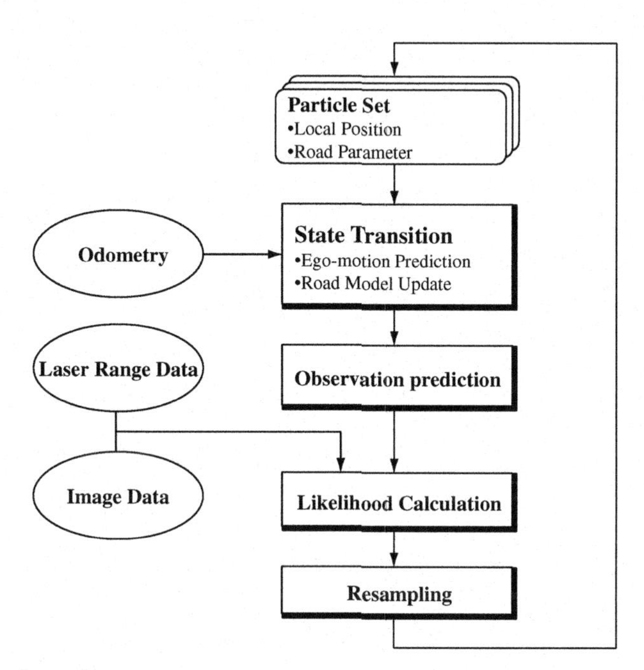

Fig. 1 Method overview

3 Road Models and State Vectors

Two-dimensional shape of the road can be classified roughly into straight lines and curves. Usually a clothoid curve, whose curvature changes smoothly, is used in the connected part of a straight line and a curve. Since the objective of the method is not estimating an accurate road shape but get an estimate which is sufficient for safe and efficient autonomous driving, we use only straight lines and circles as the models of road shape.

We connect two consecutive road models so that their tangents coincide with each other at the connection point. To avoid a frequent switch of road models due to errors in observations, we switch the road model, if necessary, only when the robot advances by a certain distance. Currently we set the distance to 10 m; this is the same as the maximum observation range. When the robot crosses a potential model-switching point, the new road models are generated (See Sect. 5 for the details).

For a simultaneous estimation of ego-motion and road region, the state vector includes parameters for both. The elements of the state vector are represented with respect to the local origin, which is defined by the pose at the previous time step. The origin is updated in the state transition step in Fig. 1.

The robot pose is represented by its 2D position, x, y and orientation, d. Concerning the road parameters, we use the gradient and the intercept for representing straight lines, and the center position and the radius for circles. Road parameters are divided into the front part before the switching point and the rear part beyond the point. We also use w and h as the road width and the distance to the switching point, respectively. Figure 2 shows five road models used in this paper. Road boundaries may sometimes be detected at different positions in range and image data. In such a

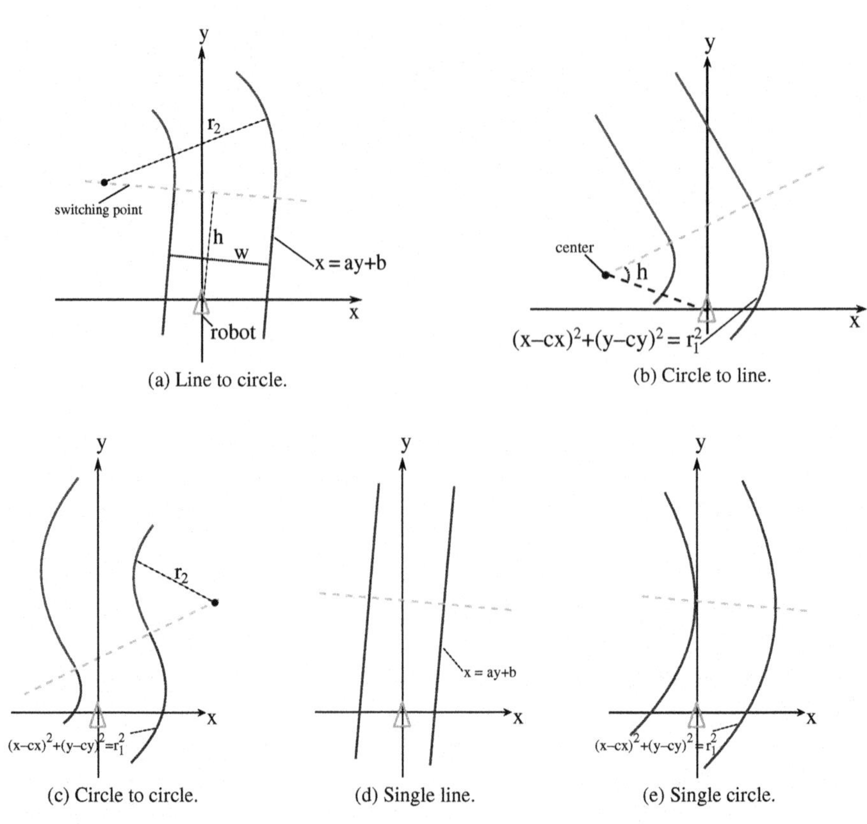

Fig. 2 Road models

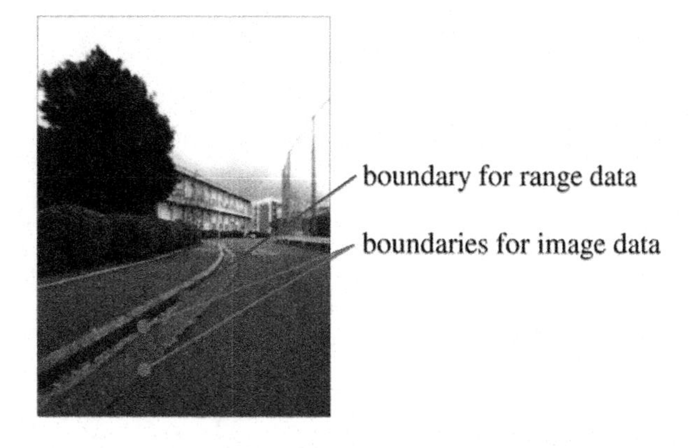

Fig. 3 Gap of road boundaries for range and image data

case, the boundary for range data usually exists outside the one for image data (see Fig. 3, for example). So we explicitly represent and estimate the gap between the boundaries, both on the right and the left side of the road.

4 Image and Range Data Processing for Likelihood Calculation

This section explains how to calculate the likelihood of each particle given a set of image and range data. We do not explicitly extract road boundaries but derive likelihood functions to be used for the calculation.

4.1 Range Data Processing

We use a SICK laser range finder (LRF). The LRF is set at the height of 0.45 m looking downward by 5° (see Fig. 4). Let h and ϕ be the height and the depression angle of LRF, and let l and α be the distance and the direction of a data point on the laser scanning plane. Then the position (x, y, z) of that point is given by:

$$x = l \sin \alpha, \tag{1}$$

$$y = l \cos \alpha \cos \phi, \tag{2}$$

$$z = h - l \cos \alpha \sin \phi. \tag{3}$$

At a curb position, a range data set has an L-shape but the data points are connected. The connectivity of data points are judged if the distance between a consecutive pair of data points is less than 1 m. A set of connected data points including the central one is analyzed to find L-shapes. We calculate the angle at each point from the two sets of the five neighboring points on both sides. A likelihood function of the angle, which assesses how likely a point is on the curb position, is then defined with a Gaussian; its mean and standard deviation are set to 90° and 30°, respectively.

Fig. 4 Placement of laser range finder

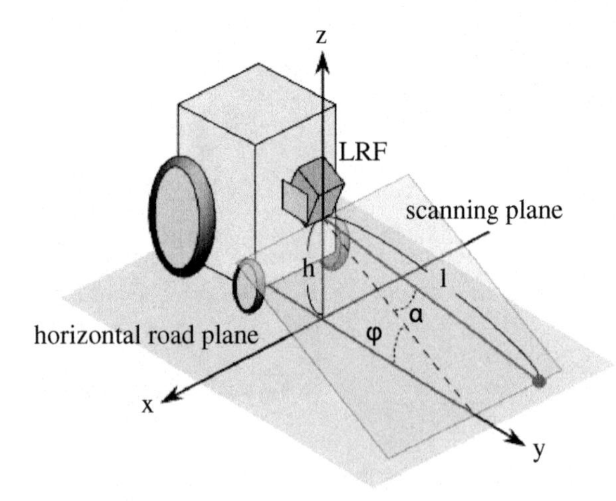

L-shape features may appear at places other than curb positions; for example, such a feature exists at the boundary of a pavement and a building wall. To exclude such spurious road boundaries, we use height information. We suppose that the robot is on a horizontal road plane, as shown in Fig. 4. In actual road scenes, however, since the gradient of the road can change, we consider the effect of such a change by considering the distribution of possible y positions (in the forward direction) of the road surface in the LRF data. The distribution is modeled as a Gaussian, whose mean and standard deviation are set to $h/\tan\phi = 5.1$ and 3 m, respectively.

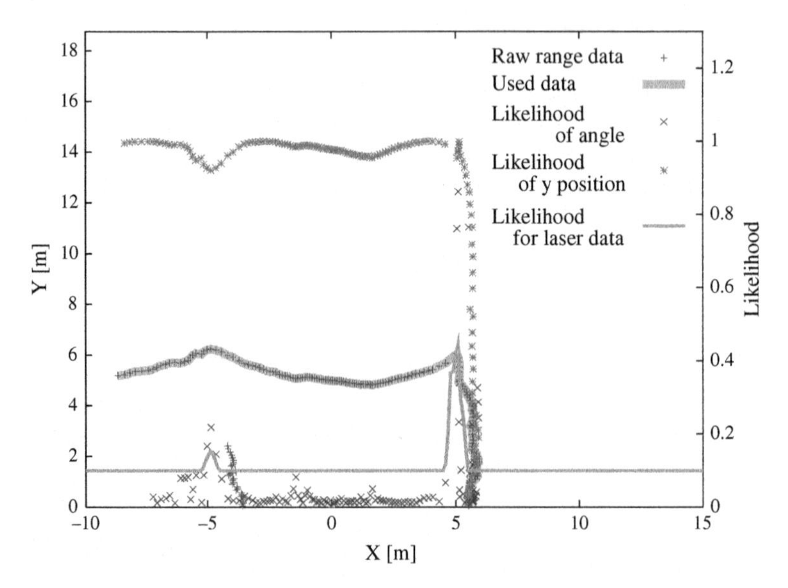

Fig. 5 LRF data and calculated likelihood values

We discretize the range of the horizontal position x with 0.1 m intervals. For each interval, we calculate the products of the two likelihood values for data points in the interval, and take the maximum product as the likelihood for that interval. For an interval with no data, we calculate the likelihood by linear interpolation. Figure 5 shows an example of likelihood calculation of the intervals.

We use this distribution of likelihood values to calculate the likelihood of each particle (i.e., each state vector). From the robot position and the road parameters of a particle, we can calculate the horizontal positions of road boundaries, which are actually the intersections of the laser scanning plane and the right and the left road boundary. The likelihood of the particle for laser data is then given by the product of the likelihood values at both horizontal positions.

4.2 Image Processing

We use a LadyBug2 (Pointgrey Research Inc.) omnidirectional camera system. This system has 5 CCD cameras, two of which are used in this research to cover the field of view of about 144°. We capture a pair of 512×384 images. We use two visual cues: road boundary lines and the boundaries between road and roadside regions. We use the intensity gradient images for the first cue and the color gradient images for the second one. The magnitude of gradient for each cue corresponds to the likelihood of the road boundary.

An intensity gradient image is obtained by applying a series of 3×3 median filter, a sobel filter, and a 11×11 smoothing filter. Figure 6(b) shows the magnitude-of-gradient image obtained from the input image shown in Fig. 6(a).

(a) Input image. (b) Intensity gradient image.

(c) Likelihood by color. (d) Color gradient image.

Fig. 6 Calculation of intensity and color gradient

Color gradient images are calculated as follows. We use the CIE L*a*b* color space, which fits well with the human perception. We model the color of a road surface with a Gaussian in the 2D color space and estimate it on-line. We sample color data in the estimated road region from the latest five frames, 100 samples from each frame. Using the estimated Gaussians, we make the image whose pixel values indicate how likely each pixel belongs to the road regions (see Fig. 6(c)). From this images, we calculate the color gradient of a pixel from the four regions of 15×7 pixels around that pixel (see Fig. 7). Let C_i be the averaged value of region i. At the left boundaries, since the gradient is rightward, we calculate the magnitude of gradient as:

$$\max(C_b - C_a, C_b - C_c, C_d - C_a, C_d - C_c). \tag{4}$$

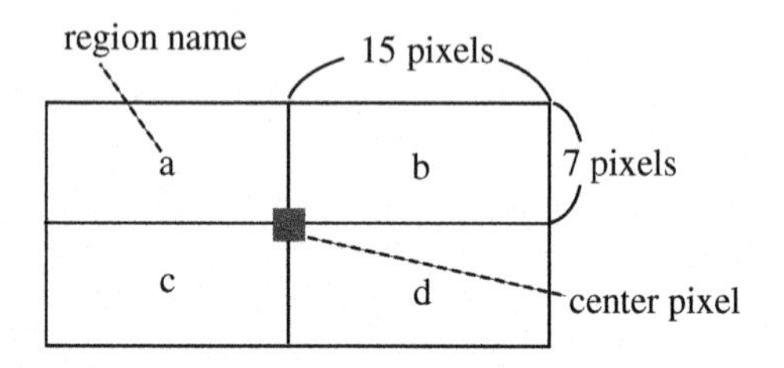

Fig. 7 Extraction of the color gradient

The magnitude of gradient for the right boundaries is calculated similarly. Figure 6(d) shows the gradient image of road boundaries (red: left, blue: right).

The likelihood of a particle for image data is calculated as follows. Road boundary points in the image coordinates can be calculated from the road parameters and the robot pose of the particle. The averaged value m of the magnitude on these points is then calculated and input to the following sigmoid function to calculate the likelihood:

$$l(m) = \frac{1}{1 + \exp(-k(m - m_c))}, \tag{5}$$

where k and m_c are experimentally determined parameters.

5 A Particle Filter-Based Estimation Algorithm

5.1 State Transition

5.1.1 Ego-Motion Prediction

This step performs the coordinate transformation (see Sect. 2) using odometry data and perturbs the robot pose part of each particle with an estimated odometry error. We currently use a very approximate error estimate; for a 1 m movement, we use the standard deviations of 0.1 m, 0.1 m, and 0.1 $[rad]$ for x, y, and d, respectively. These values are set to proportional to the moving distance.

5.1.2 Road Model Update

As we stated in Sect. 2, we switch the road model only when the robot advances by 10 m, and when the robot crosses a potential model-switching point, the new road models are generated. Each road model has two parts, the front and the rear part (see Sect. 3). When the robot crosses a model-switching point, the rear part of the current model is transferred to the front part of a new model and a new rear part is attached. Since we cannot know in advance what type of road appears in the rear part, we generate all of the following as the rear road type: the line and a set of circles. The curvatures of circles are limited to $1/r = 0.02, 0.04, 0.06, 0.08, 0.10\,[1/\mathrm{m}]$. For a particle which is judged to cross the switching point, multiple descendant particles are generated corresponding to various read road types. The weight w_{update} of a newly generated particle is set to $1/n_{new}$, the inverse of the number of newly generated particles. For a particle which is not judged to cross, the weight is one.

5.1.3 Road Parameter Prediction

In Sect. 3, we defined road parameters. Among them, a road width and a gap between boundaries from range and image data may change as the robot moves. In addition, since we generate circular road models at the rear road part with a limited set of curvatures, as described above, we need to gradually adjust the curvature to the observation. We therefore estimate on-line the width, the gap, and the curvature, and fix the other road parameters for each particle.

5.2 Likelihood Calculation

The weight of each particle is calculated by the weight determined by the state transition and the likelihood values for range and image data.

In some cases, however, the likelihood values for one of the sensors (range or image) on one side become very small due to, for example, a discontinuity of curb or strong cast shadows. In such a case, the weights of all particles become very small and, as a result, many promising particles might be deleted. To avoid this, if the maximum likelihood value for a sensor on one side is less than a threshold (currently, 0.3), the sensor is considered not to be effective on that side, and the likehood values for that combination of the sensor and the side are not used.

The final weight w_{model} is given by the product of the likelihood values l_{image} and l_{LRF} for image and range data and the weight w_{update} determined at the update step, that is,

$$w_{model} = (l_{image} \cdot l_{LRF}) \cdot w_{update}. \tag{6}$$

After calculating the weights, we perform a usual resampling step.

6 Experimental Result

This section describes the experimental results conducted at two locations in our campus (See Fig. 8). The first course is on a straight road partially with curbs. The second one is on a curved road with cast shadows.

Fig. 8 Courses of the experiments

(a) First experiment.

(b) Second experiment.

6.1 Straight Road

Figure 8(a) shows the course of the first experiment. We manually moved the robot along the road and obtained 130 sets of range, image, and odometry data. The data are processed off-line with the number of particles being 500. At first there are curbs on both sides, but as the robot moves, there appears a parking space on the left (see Fig. 9(a)). At the entrance of the space, there are no curbs and, in this case, LRF data are not effective for detecting the left road boundary (see Fig. 9(b)). The right road boundary is, on the other hand, clearly detectable. If we have information on curb position at least on one side, we can estimate the road parameters by using predicted road models. Image data are also effective in this case, as shown in Fig. 9(c) and (d). By integrating multiple information from both sensors, we can robustly estimate the road region. Figure 10 shows the estimation result when the data shown in Fig. 9 was obtained at step 55. The left figure superimposes the road boundaries obtained from the particles after resampling on the input image. We assign the three primary colors to represent the likelihood of each piece of information as follows:

- R: likelihood using color gradient,
- G: likelihood using intensity gradient,
- B: likelihood using range data.

So for example, a purple line indicate that information of color gradient and range data supports the line. Figure 10(b) shows a kind of certainty distribution of road

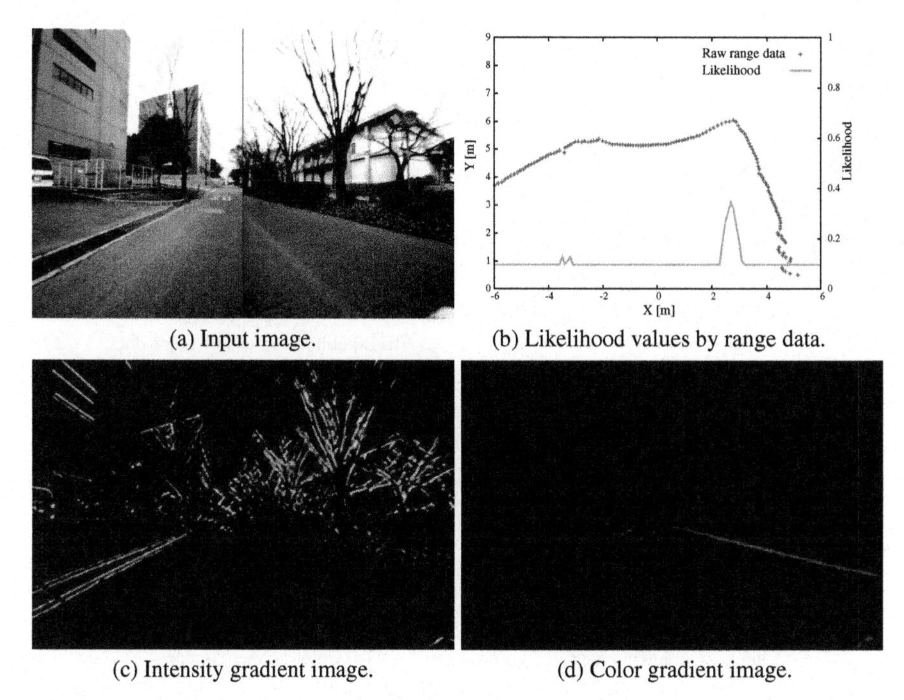

(a) Input image. (b) Likelihood values by range data.

(c) Intensity gradient image. (d) Color gradient image.

Fig. 9 Observation at step 55 in the first experiment

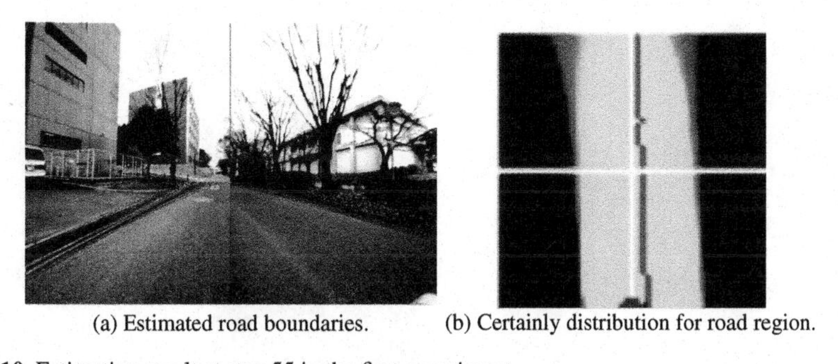

(a) Estimated road boundaries. (b) Certainly distribution for road region.

Fig. 10 Estimation result at step 55 in the first experiment

regions, obtained from the current set of particles, in the robot local coordindates (the green semicircle is the robot); brighter pixels indicate higher certainties. The center position of the road is also shown in red, which could be a guide for controlling the robot motion.

Figures 11 and 12 are the input image, range and gradient information, and the estimation result at step 40. Cues on the right side are undetectable due to a branch. On the left side, a curb gradually moves outside towards a parking space, so many left curves survive as the estimated road boundaries. A few steps later, however, a left boundary for the intensity gradient become clearly detectable, and only straight line road models survive (see Fig. 10).

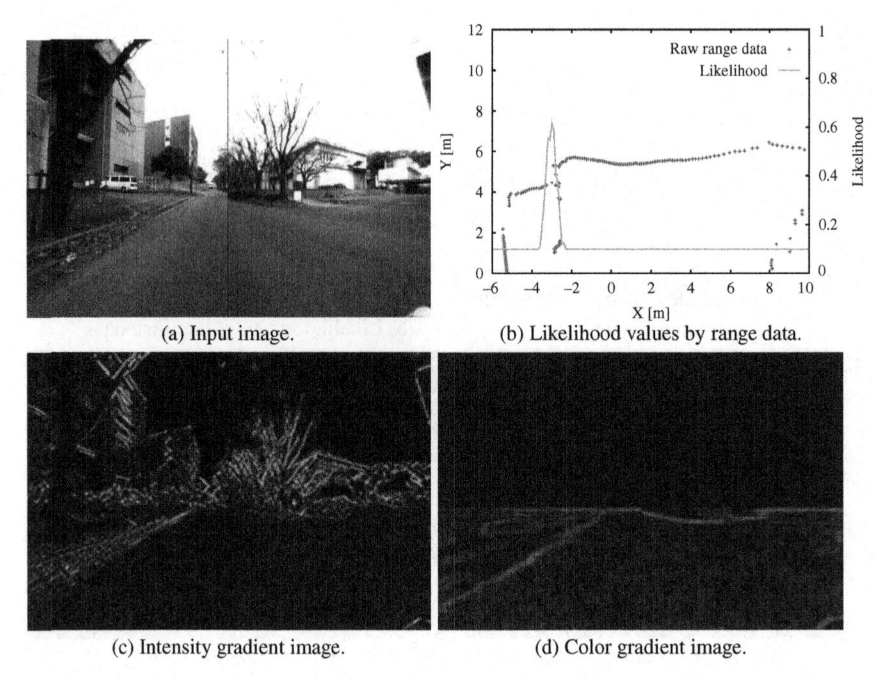

(a) Input image. (b) Likelihood values by range data.

(c) Intensity gradient image. (d) Color gradient image.

Fig. 11 Observation at step 40 in the first experiment

(a) Estimated road boundaries. (b) Certainty distribution for road region.

Fig. 12 Estimation result at step 40 in the first experiment

Each particle is split into a set of decendant particles at the road model update step, and a limited number of particles survive after the resampling. By tracing back from the current set of particles, we can obtain the global road shape and the motion history. Figure 13(a) shows the result obtained at step 55. The global shape of road boundaries and the robot motion histoties from the surviving particles are shown. Colors of road boundaries indicate road types (straight, left curve, and right curve). The history obtained from odometry is also shown for comparison. Figure 13(b)

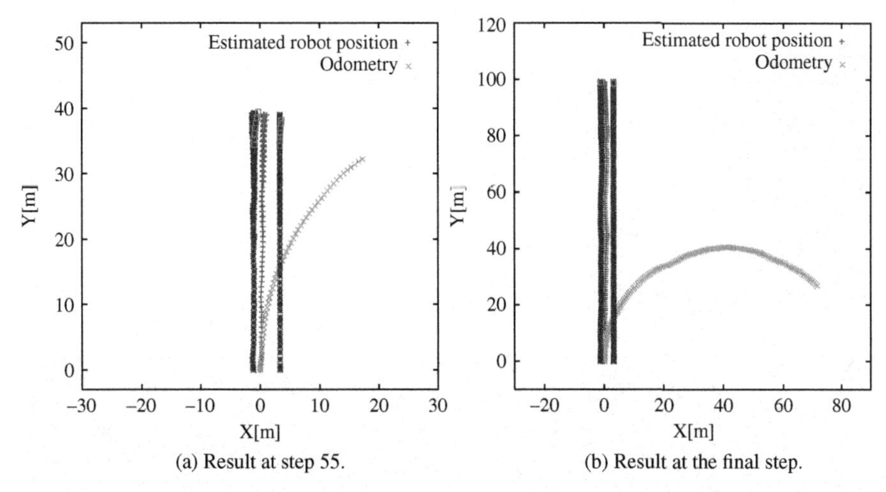

(a) Result at step 55.

(b) Result at the final step.

Fig. 13 Estimation results of global road shape in the first experiment

shows the result at the final step. Incorrect road models at step 55 have been eliminated and the straight shape of the road is clearly recovered.

6.2 Curves and Shadow

Figure 8(b) shows the course of the second experiment. The road includes curved parts and strong shadows are cast on many locations. We obtained 180 sets of data in this case. At step 38, strong shadows are cast on the road surface (see Fig. 14(a)); this makes it difficult to detect the right road boundary using colors (see Fig. 14(d)). The intensity gradient information is not very effective, either (see Fig. 14(c)). The curbs are clearly observable in the LRF data (see Fig. 14(b)), and this make it possible to correctly estimate the road region, as shown in Fig. 15.

Figures 16 and 17 are the input image, range and gradient information, and the estimation result at step 94. Again, a strong sunlight makes it difficult to detect road boundaries using color, but LRF data on both sides and the intensity gradient on the left are mainly used for road region estimation.

Figures 18 and 19 are the data at step 129. Because a branch exists on the left side and a right side curb is undetectable, there is no cue for the estimation. This results in a diffision of the estimated road boundaries (see Fig. 19). A few steps later, however, the range and image data on the left side become effective to make the boundaries converge.

Figure 20 shows the estimation results of the global road shape and the motion history at step 94 and at the final step. The approximate shape of the road is well recovered.

The number of particles certainly affect the estimation performance. We quantitatively examined their relationship for the two experimental situations. We ran the

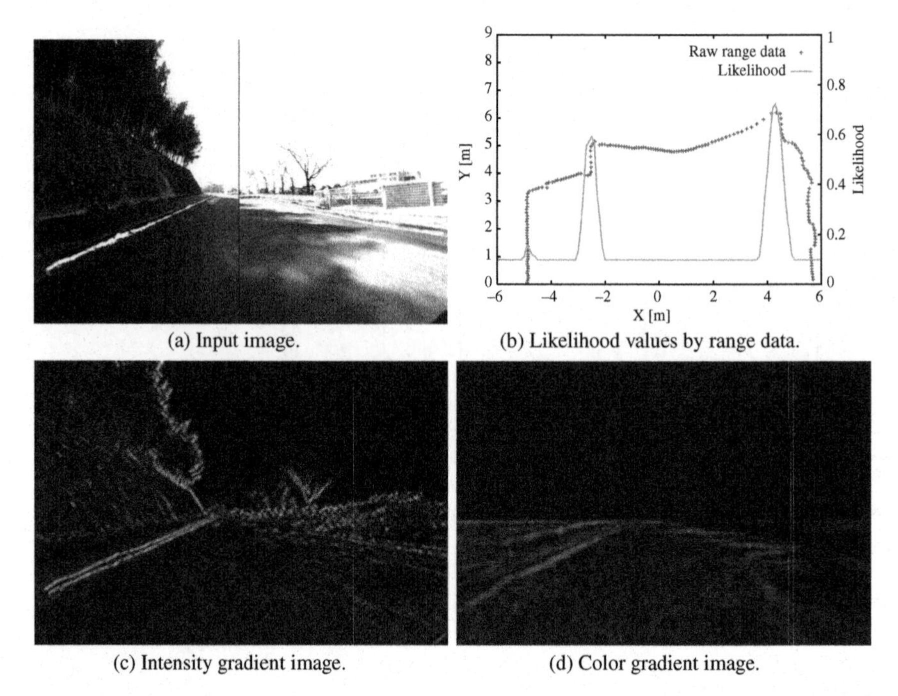

(a) Input image. (b) Likelihood values by range data.

(c) Intensity gradient image. (d) Color gradient image.

Fig. 14 Observation at step 38 in the second experiment

(a) Estimated road boundaries. (b) Certainty distribution for road region.

Fig. 15 Estimation result at step 38 in the second experiment

system 20 times for each number of particles and calculated the success rate. We judged if the estimation result is successful by visual inspection. Figure 21 shows the result. The computation time of the proposed estimation method is about 0.5 s per step using 500 particles. The number of particles depends on the accuracy of dead reckoning and a model variety. Increasing the accuracy by using, for example, gyroscope, would reduce the number of particles thus reducing the computation time.

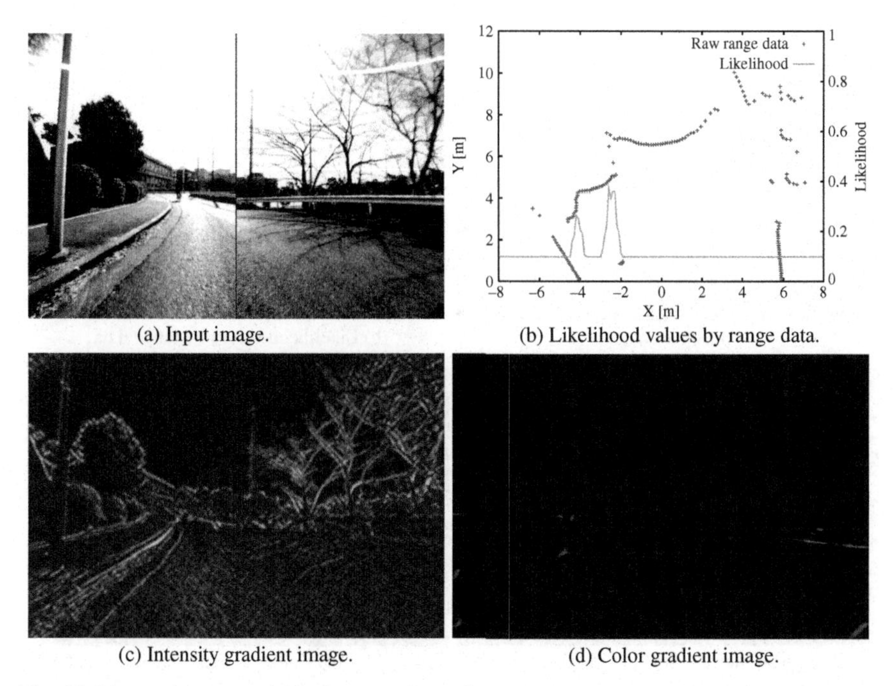

(a) Input image. (b) Likelihood values by range data.

(c) Intensity gradient image. (d) Color gradient image.

Fig. 16 Observation at step 94 in the second experiment

We combined the estimation method with a simple robot control procedure to perform the experiments of autonomous driving. Figure 22 shows some snapshots of the experiments. The robot was able to robustly move autonmously in various road environments.

(a) Estimated road boundaries. (b) Certainty distribution for road region.

Fig. 17 Estimation result at step 94 in the second experiment

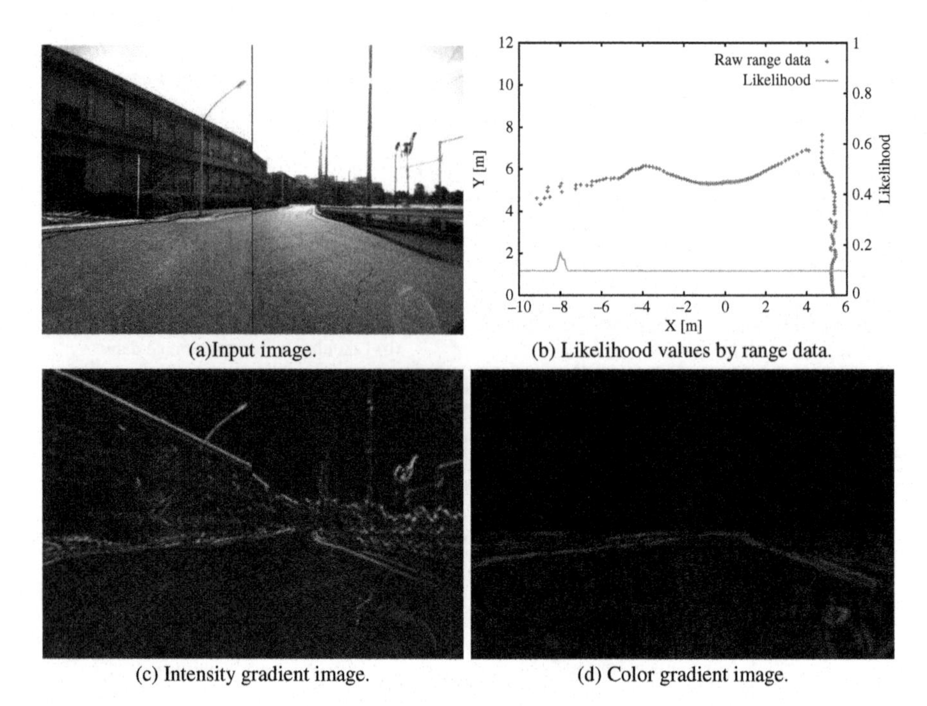

(a)Input image.

(b) Likelihood values by range data.

(c) Intensity gradient image.

(d) Color gradient image.

Fig. 18 Observation at step 129 in the second experiment

(a) Estimated road boundaries.

(b) Certainty distribution for road region.

Fig. 19 Estimation result at step 129 in the second experiment

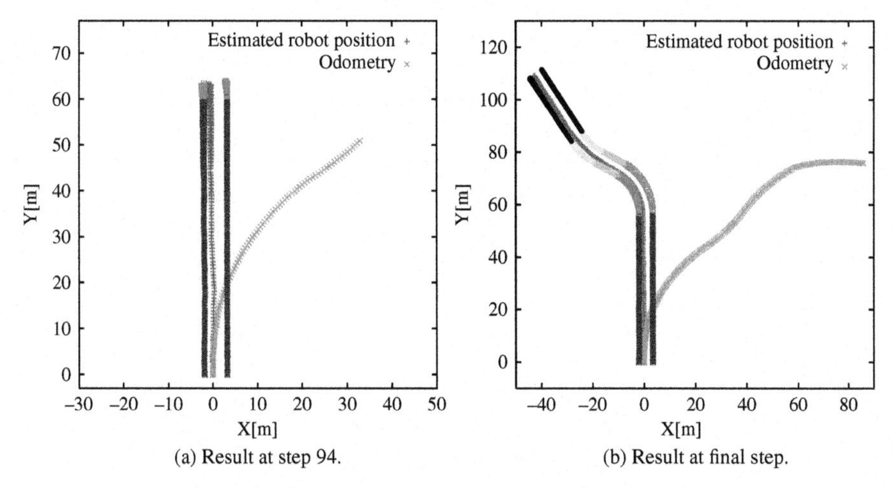

(a) Result at step 94. (b) Result at final step.

Fig. 20 Estimation results of global road shape in the second experiment

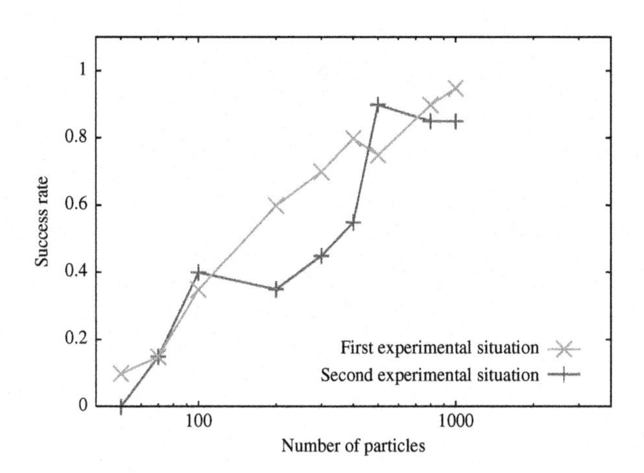

Fig. 21 Quantitative result

Fig. 22 Autonomous driving

7 Conclusion

This paper has described a new method of simultaneously estimating the road region and the robot ego-motion. The method effectively integrates two sources of information, vision and range finder, using a particle filter with new likelihood functions. It can also cope with the change of road types by devising a method of generating particles (hypotheses) corresponding possible road changes. The method has been tested in various real environments to show its effectiveness.

To cope with more various environments, we are planning to extend the method in the following two ways. One is to use more features to cope with various scenes. Human can recognize a road region even when no clear boundaries exist using some features which can separate road and non-road regions. Seeking and testing effective features are necessary. The other way is to add more various road models such as crossings and parking spaces where we cannot observe continuous road boundaries.

References

1. J.D. Crisman and C.E. Thorpe. SCARF: A Color Vision System that Tracks Roads and Intersections. *IEEE Trans. Rob. Autom.*, Vol. 9, No. 1, pp. 49–58, 1993.
2. H. Ishiguro, K. Nishikawa, and H. Mori. Mobile Robot Navigation by Visual Sign Patterns Existing in Outdoor Environments. In *Proc. IEEE/RSJ Int. Conf. on Intell. Rob. Syst.*, pp. 636–641, 1992.
3. W.S.Wijesoma, K.R.S. Kodagoda, and A.P. Balasuriya. Road Boundary Detection and Tracking Using Ladar Sensing. *IEEE Trans. Rob. Autom.*, Vol. 20, No. 3, pp. 456–464, 2004.
4. A. Kirchner and T. Heinrich. Model-Based Detection of Road Boundaries with a Laser Scanner. *In Proc. IEEE Int. Symp. Intell. Veh.*, pp. 93–98, 1998.
5. H. Cramer and G. Wanielik. Road Border Detection and Tracking in Non Cooperative Areas with a Laser Radar System. In *Proc. Ger. Radar Symp.*, 2002.
6. D. Langer and T. Jochem. Fusing Radar and Vision for Detecting, Classifying and Avoiding Roadway Obstacles. In *Proc. IEEE Int. Symp. on Intell. Veh.*, pp. 333–338, 1996.
7. J. Miura, Y. Negishi, and Y. Shirai. Mobile Robot Map Generation by Integrating Omnidirectional Stereo and Laser Range Finder. In *Proc. 2002 IEEE/RSJ Int. Conf. on Intell. Rob. Syst.*, pp. 250–255, 2002.

Model-Based Recognition of 3D Objects using Intersecting Lines

Hung Q. Truong, Sukhan Lee and Seok-Woo Jang

Abstract Exploiting geometric features, such as points, straight or curved lines and corners, plays an important role in object recognition. In this paper, we present a model-based recognition of 3D objects using intersecting lines. We concentrate on using perpendicular line pairs to test recognition of a parallelepiped model and represent the visible face of the object. From 2D images and point clouds, first, 3D line segments are extracted, and then intersecting lines are selected from them. By estimating the coverage ratio, we find the most accurate matching between detected perpendicular line pairs and the model database. Finally, the position and the pose of the object are determined. The experimental results show the performance of the proposed algorithm.

Keywords Line matching · Model-based recognition · Intersecting line

1 Introduction

Object recognition has been studied extensively in computer vision, and model-based object recognition is a well regarded method. Most approaches use the correspondence between model features and image features to estimate the pose of the object. With these approaches, the challenging problem is the selection of the most useful geometric features for matching.

Some researchers have exploited methods to recognize and estimate the shape of a 3D object using color features [1]; these algorithms may not be robust when objects lack texture or discriminating characteristics. Zhang and Faugeras [2] proposed a method to present line matching problems, but the resulting points are often not the midpoints of the corresponding line segments pairs. Guerra and Pascucci [3] presented an algorithm to match between two sets of 3D line segments

S. Lee (✉)
Intelligent Systems Research Center, Sungkyunkwan University, 300 Cheoncheon-dong, Jangan-gu, Suwon, Gyeonggi-do, 440-746, South Korea
e-mail: lsh@ece.skku.ac.kr

S. Lee et al. (eds.), *Multisensor Fusion and Integration for Intelligent Systems*, Lecture Notes in Electrical Engineering 35, DOI 10.1007/978-3-540-89859-7_20,

with unknown line correspondences. Some other matching methods [4–6] were issued which did not completely solve the general matching problem. Košecká and Zhang approached more complicated features such as rectangular structures [7] and Kamgar-Parsi [8], Polygonal Arc Matching. Many researchers are mentioned in the feature of 2D model line [9] as well as multiple features [10]. However, these methods have limitations with a large number of models. Lowe [11] determined the algorithm for SIFT features but this method seems to be ineffective when the object lacks texture or is occluded. Some approaches have focused on parallel line features and used that as an independent method to recognize and match up with the object. Here we have approached model-based recognition of 3D objects using intersecting lines with more advantages, for instance, almost the shape of the objects generally store the corners which are constructed by two or more intersecting line segments while parallel line segments do not appear popularly.

The main contribution of this paper is to demonstrate detection of the intersecting line segments in the scene, and then select the highest probability pairs which can fit the shape of one face of object. One proposed method is to represent one object's face using the selected intersecting line pairs and calculate the coverage ratios for each matching case. By comparison, we find the highest coverage probability to indicate the most accurate pose of the object in the scenario. In our algorithm, we assume the shape of the object which needs to be recognized contains perpendicular corners as a box in Fig. 1. Consequently, we extract perpendicular line segments and represent them as a rectangle for one face of the box.

The paper is organized as follows: Sect. 1 explains 3D line segments' representation from model and 3D line extraction from the scene. An algorithm which detects coplanar line segments and intersecting lines is described in Sect. 2. Interpretation and matching method between the model and the perpendicular line segment are explained in Sect. 3. Section 4 demonstrates that our experimental result can be implemented in the real environment. Finally, Sect. 5 summarizes our results and states our conclusions.

Fig. 1 The scene and box model

2 3D Line Extraction

2.1 3D Line Segments Representation from Model

In this implementation, we represents target object as a box in which its shape is like a parallelepiped. Besides its perpendicular corners, this model is also determined by three dimensions, the length, width and height of the parallelepiped, corresponding to Model_X, Model_Y and Model_Z. These dimensions are 3D line segments connecting the neighboring vertexes of the parallelepiped, and we assume that the length is known.

2.2 3D Line Extraction from the Scene

In this section, we first extract all lines from the 2D image, and these lines are converted to 3D lines through mapping the 3D points corresponding to the 2D lines, using the Hough transform for 2D line fitting. First, the edges are detected by the Canny edge algorithm. Then, the edges are categorized as horizontal, vertical and diagonal line segments, based on the connection of the edges. The 2D lines are found by connecting each line segment with adjoining line segments, based on the aliasing problem of lines in 2D. 3D lines can be obtained if there are corresponding 3D points at the pixels of the 2D line. In fact, the quality of data obtained by stereo imaging is not accurate, and each stereo point cloud has uncertainty. Therefore, we find a method to determine the error of the position or the range resolution (r) of the 3D point in space, described below.

Two images of the same object are taken from different viewpoints. The distance between the viewpoints is called the baseline b. The focal length of the lens is f. The horizontal distance from the image center to the object image is dl for the left image, and dr for the right image. The relationship of these parameters is:

$$r = \frac{b \cdot f}{d} \quad \text{where } d = dl - dr \tag{1}$$

$$\frac{\partial r}{\partial d} = -\frac{b \cdot f}{d^2} = -\frac{\left(\frac{b \cdot f}{d}\right)^2}{b \cdot f} = -\frac{r^2}{b \cdot f}$$

$$\Rightarrow \Delta r = \frac{r^2}{b \cdot f} \cdot \Delta d \tag{2}$$

The range resolution is a function of the range itself (1). At closer ranges, the resolution is much better than further away. Range resolution is represented by the equation (2), the range resolution Δr, is the smallest change in range that is discernable by the stereo geometry, given a change in disparity of Δd.

3 Detecting Intersecting 3D Line Segments

As mentioned above, we can obtain all 3D lines connected by two end points and each of point has range resolution Δr. In order to find the best perpendicular line pairs for model matching, we implemented some steps as follows:

3.1 Coplanar 3D Line Segments Inspection

The key point we concentrate on here is the distance between two 3D lines. In three dimensions, the two arbitrary lines generally may not intersect or intersect at a point, and they can be parallel and even though they may be coincident. If two 3D lines do not absolutely intersect at a point then the distance between them will be a line segment connected by two points lying on the each line. There is one and only one shortest line segment joins two lines in three dimensions. We assume two 3D line segments are P_1P_2 and P_3P_4, the shortest line segments that joins two 3D lines is P_aP_b as shown in Fig. 2. P_i can be expanded out in the (x,y,z) components where $i \in \{a, b, 1, 2, 3, 4\}$. In the analysis to follow, from at least two points we can define a line equation. A point P_a on line P_1P_2 defined by points P_1 and P_2 can be expressed by the equation,

$$P_a = P_1 + m_a(P_2 - P_1) \tag{3}$$

Similarly a point P_b on another line P_3P_4 defined by points P_3 and P_4 can be expressed by the equation,

$$P_b = P_3 + m_b(P_4 - P_3) \tag{4}$$

The line segment P_aP_b joins P_1P_2 and P_3P_4 and the values of m_a and m_b can be the arbitrary real number. Substituting the equations of each of lines gives

$$P_b - P_a = P_1 - P_3 + m_a(P_2 - P_1) - m_b(P_4 - P_3) \tag{5}$$

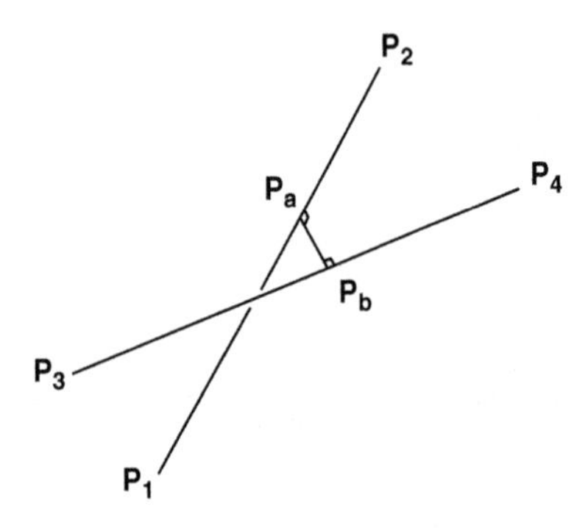

Fig. 2 The shortest line segment that joins two 3D lines

Since $P_a P_b$ is assumed to be like a shortest line segment, it must be perpendicular to either $P_1 P_2$ or $P_3 P_4$. We use the dot product equation to denote this property

$$(P_b - P_a).(P_2 - P_1) = 0 \tag{6}$$

$$(P_b - P_a).(P_4 - P_3) = 0 \tag{7}$$

From Eqs. (5–7), we obtain

$$[P_1 - P_3 + m_a(P_2 - P_1) - m_b(P_4 - P_3)](P_2 - P_1) = 0 \tag{8}$$

$$[P_1 - P_3 + m_a(P_2 - P_1) - m_b(P_4 - P_3)](P_4 - P_3) = 0 \tag{9}$$

Expanding these P_i in terms of the coordinates (*x*,*y*,*z*), we obtain m_a and m_b then P_a, P_b are determined by Eqs. (3) and (4). The length d of the line segment joining the two 3D lines is defined by (10)

$$d = |P_b - P_a| \tag{10}$$

P_a and P_b are points in three dimensional space, so they also have their range resolutions. Here we illustrate that the range resolution for a 3D point is a sphere with radius are Δr, with P_a and P_b corresponding to Δra and Δrb as shown in Fig. 3. Two 3D lines intersect if the shortest distance between them is equal to zero. In actual use, when each 3D point has its error in position in three dimensional space, we can not expect that the shortest distance joining two lines is equal to zero. Instead, a constraint is proposed to make the condition of two coplanar 3D lines be the most accurate

$$d \leq \Delta r_a + \Delta r_b \tag{11}$$

With the set of 3D line segments that are detected, we inspect the distance between every two lines of the set by Eq. (10). The pairs that satisfy the condition of Eq. (11) are considered to be intersected, and intersecting points are P_a, which

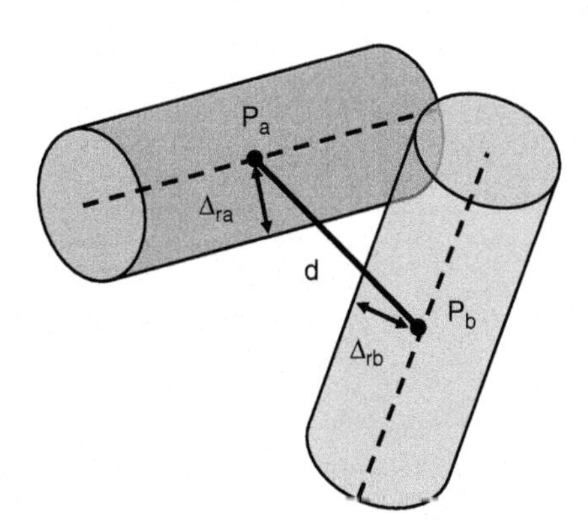

Fig. 3 Measure of the shortest distance between two 3D line segments

lies on the first line and P_b, which lies on the second line. If two lines absolutely intersect then P_a and P_b are coincident.

3.2 Candidate Intersecting Lines to Match with Model

Here we mention three steps to select perpendicular line segment pairs in order to interpret the rectangle face of the target object.

In the first step, we eliminate the cases where the intersecting point is interior to the line segment, as shown in Fig. 4 (a) and (b). We regard the other intersecting line cases if their intersection point is out of line segments or coincident with the vertexes of each line segment, as shown in Fig. 4 (c) and (d). These pairs have the ability to be represented by a rectangle.

In the second step, we consider the angle between two intersecting lines which satisfy the requirements of the first step. In this paper, since the target object is a parallelepiped, we therefore detect the pairs containing perpendicular angles. Each point has a range resolution, so its coordinate in three dimensions appears uncertain. Once a set of 3D line segments are connected by two points, it has a disparity in angle also. As Fig. 5 shows, the range of disparity in position of a line segment can be illustrated as a cylinder. We proposed an optimal measurement to decide the most accurate angle between two line segments described in Eq. (12)

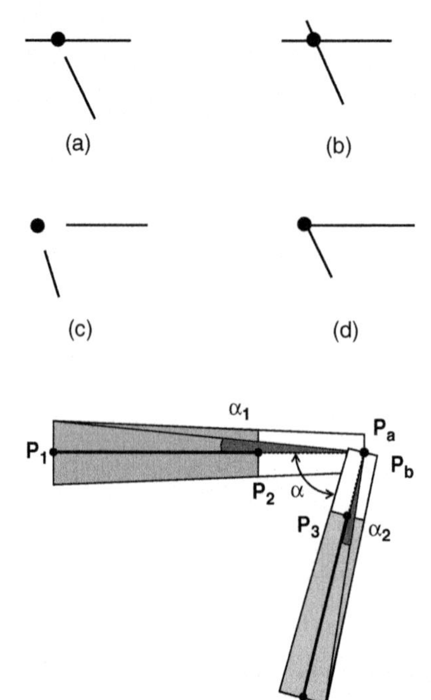

Fig. 4 Intersection cases of two arbitrary line segments

Fig. 5 Angle between two intersecting lines

$$\theta = \alpha_1 + \alpha + \alpha_2 \tag{12}$$

where θ is probably the angle of the two intersecting lines, as shown in Fig. 5, α is the angle calculated from the inner product between P_1P_2 and P_3P_4, and α_1 and α_2 are variant angles of P_1P_2, P_3P_4. They are defined by

$$\alpha_i = a \sin \frac{\Delta r_i}{l_i} \tag{13}$$

where $i = 1$ or 2. Δri is the range resolution of an end point on the line segment if the distance between this point and the intersecting point is larger than the distance between the remainder end point and intersecting point. l_i is the length of the line after it has been extended to the intersecting point. In the case where intersecting point is coincident with one of two end points of the line segment then l_i is length of that line segment. By the more accurate estimation, we issue a constraint to detect the perpendicular line segment pairs as follows:

$$\frac{\pi}{2} - (\alpha_1 + \alpha_2) \leq \alpha \leq \frac{\pi}{2} + (\alpha_1 + \alpha_2) \tag{14}$$

In the third step, we keep the perpendicular line pairs if the line length of each line is shorter than max(Model_X, Model_Y, Model_Z) and longer than 20% of min(Model_X, Model_Y, Model_Z).

After three steps, we obtain a set Ω of perpendicular 3D line segments which has a high probability to match with a rectangle face of model.

4 Matching with Model

4.1 Interpretation and Matching with Model

From the set Ω of perpendicular line pairs above, a method to represent a rectangle face of object is determined. We first assume that (P_1P_2, P_3P_4) is a pair in set Ω and A is its intersecting point. We describe this assumption as is shown in Fig. 6 (a) and our method is explained following these steps:

Consider the intersecting point A of $(P_1\underline{P}_2, P_3P_4)$ is a vertex of rectangle.

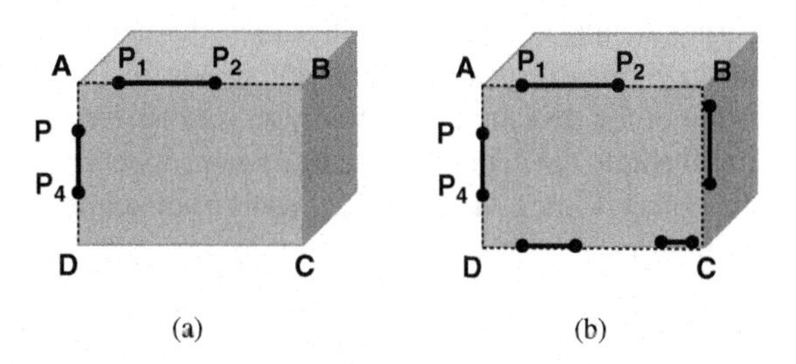

Fig. 6 Interpretation a rectangle from an intersecting line pair

If P_2A is less than $d1$ where $d_1 = \{\text{Model\_X, Model\_Y or Model\_Z}\}$, then extend P_2A so that its length is equal to d_1, and the new end point is vertex B.

Similarly, if P_4A is less than d_2 where $d_2 = \{\text{Model\_X, Model\_Y or Model\_Z}\}$ and $d_2 \neq d_1$, then extend P_4A so that its length is equal to d_2, the new end point is vertex D.

By the symmetric property, the remainder vertex C of rectangle is determined.

Each pair can represent one or more rectangles. With the set Ω, we can represent a set ξ of rectangles.

4.2 Calculation Coverage and Matching

In order to select the best interpretation of the set ξ, we select the cases that have the highest coverage ratio. This ratio is defined by the equation

$$w = \frac{N_{\text{matched\_line}}}{N_{\text{perimeter}}} \tag{15}$$

where w is the coverage ratio, each interpretation in set ξ has a calculated value for w, hence if ξ has n values then W also consists of n values of w. $N_{\text{perimeter}}$ is the perimeter of the rectangle. $N_{\text{matched\_line}}$ is the sum of the 3D line length on the margin of the rectangle or the boundary of the rectangle as shown in Fig. 6(b).

A line segment is considered to be on the margin of the rectangle if this line and the side of the rectangle it adjoins satisfy the coplanar condition, as mentioned in Sect. 3A and the angle between them is zero.

In the set W, we select one w which is the maximum value. From that w value, we get the corresponding rectangle's position, and this rectangle is predicted to be a visible portion of parallelepiped. Therefore we can indicate the position of target object.

5 Experimental Results

The proposed method has been tested with a box and we use a Bumblebee stereo camera to capture the sequence of images.

Fig. 7 shows the result of 3D line segment detection. After that, we select the perpendicular line pairs which satisfy the condition of set Ω. Interpretation of the rectangle based on these pairs are shown in Fig. 8. Figure 9 is the most accurate result of the matching. In some cases, the object is occluded, as shown in Fig. 10, so we illustrate the deficient rectangles as probability matching with one visible face of the object. Figure 11 shows the result of matching in the case of occluded objects.

Fig. 7 3D line segments detection result

Fig. 8 Interpretation the rectangles based on perpendicular line pairs in set Ω

Fig. 9 A matching result

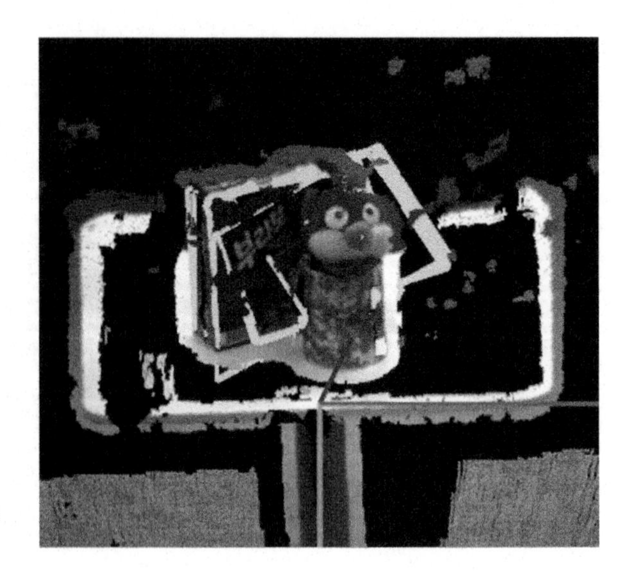

Fig. 10 Interpretation the rectangles based on perpendicular line pairs in set Ω (occlusion case)

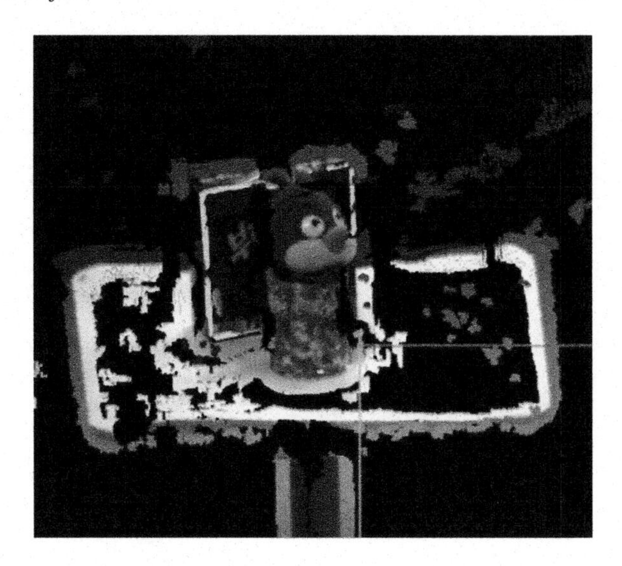

Fig. 11 A matching result in occlusion case

6 Conclusion

We have concentrated on the geometric feature to recognize a given object. Particularly, we exploit the perpendicular 3D line segment pair to match with the model. Through the experimental results, we found that intersecting line is necessary and important information as well as the algorithm to recognize the position of an object based on that property. Actually, there are many kinds of object with arbitrary shapes and their sides are not only straight lines but also curves or other shapes. This problem suggests to us an approach to a new challenge, recognizing a deformed object.

In our approach, even though the target object is occluded by other things, our algorithm can also detect a portion of the object and can still produce a reliable result in the face of that deficient information. However, in the weak light environment, the 3D line detection result is not clear. For our future studies, we will combine the geometric features, such as parallel line pairs or intersecting line pairs with arbitrary angles to improve the recognition algorithm.

References

1. M. Giessen and J. Schmidhuber, Fast color-based object recognition independent of position and orientation, ICANN 2005, LNCS 3696, pp. 469–474, 2005.
2. Z. Zhang and O. Faugeras, Determining motion from 3D line segment matches: A comparative study, *Image and Vision Computing*, 9(1), 10–19, 1991.
3. C. Guerra and V. Pascucci, On matching sets of 3D segments, Proceedings of SPIE Conference on Vision Geometry VIII, 3811, 157–167, 1999.
4. B. Kamgar-Parsi, Algorithm for matching 3D line sets, *IEEE Transactions on Pattern Analysis and Machine Intelligent*, 26(5), 582–593, 2004.
5. O. D Faugeras and M. Hebert, The representation, recognition, and locating of 3-D object, *International Journalof Robotics Research*, 5(3), 27–52, 1986.

6. B. Kamgar-Parsi and B. Kamgar-Parsi, An open problem in matching sets of 3D lines, *Proceedings of the IEEE Conference on Computer Vision and Pattern Recognition*, 1, 651–656, 2001.
7. J. Košecká and W. Zhang, Extraction, matching, and pose recovery based on dominant rectangular structures, *Computer Vision and Image Understanding*, 100(3), 274–293, 2005.
8. B. Kamgar-Parsi and B. Kamgar-Parsi, Matching sets of 3D line segments with application to polygonal arc matching, *IEEE Transactions on Pattern Analysis and Machine Intelligence*, 19, 1090–1099, 1997.
9. N. Ayache and O. Faugeras, A new approach for the recognition and positioning of two dimensional objects, *IEEE Transactions on Pattern Analysis Machine Intelligence*, 8(1), 44–54, 1986.
10. S. Lee, E. Y. Kim, and Y. C. Park, 3D object recognition using multiple feature for robotics manipulation, *IEEE International Conference on Robotics and Automation*, pp. 3768–3774, 2006.
11. D. Lowe, Object recognition from local scale invariant features, In *Proceedings of the Seventh International Conference on Computer Vision* (ICCV'99), pp. 116–128, 2001.

Visual SLAM in Indoor Environments Using Autonomous Detection and Registration of Objects

Yong-Ju Lee and Jae-Bok Song

Abstract For successful SLAM, landmarks for pose estimation should be continuously observed. This paper proposes autonomous detection of objects as visual landmarks for visual SLAM. Primitive features such as color and intensity, SIFT keypoints, and contour information are integrated to investigate environmental images and to distinguish objects from the background. Autonomous object detection can enable a robot to extract some objects without any prior information and it can help a vision system to cope with unknown environments. In addition, an adaptive weighting scheme and the use of a gradient of the gray scale are proposed to improve the performance of the proposed scheme. Using detected objects as landmarks, a robot can estimate its pose. A grid map of an unknown environment is built using an IR scanner and the detected objects are mapped in the grid map, which results in a hybrid grid/vision map. Visual SLAM using objects can have the less number of landmarks than other visual SLAM schemes using corners and lines. Various experiments show that the algorithm proposed in this paper can improve visual SLAM of a mobile robot.

Keywords SIFT · SLAM · Object recognition · Visual attention

1 Introduction

When a robot navigates in an unknown environment, both accurate pose estimation of the robot and map building of the environment are important issues. Therefore, SLAM (Simultaneous Localization and Mapping) has been one of the most fundamental and challenging issues in the field of mobile robotics in recent years.

Range sensors (i.e., laser scanners, sonar sensors, and IR scanners) and vision sensors (i.e., monocular and stereo cameras) are usually employed for SLAM.

Y.-J. Lee (✉)
Department of Mechanical Engineering, Korea University, Seoul, Korea
e-mail: yongju_lee@korea.ac.kr

S. Lee et al. (eds.), *Multisensor Fusion and Integration for Intelligent Systems*,
Lecture Notes in Electrical Engineering 35, DOI 10.1007/978-3-540-89859-7_21,

Early researchers preferred range sensors because they could provide the range information directly, which made feature extraction easier than vision sensors. However, features that can be extracted from the range information are limited to lines and corners. On the other hand, a vision sensor offers much more information than a range sensor. Although a vision sensor requires complicated image processing to extract visual features, recent SLAM approaches tend to employ vision sensors as a main sensor.

As both range- and vision-based schemes use features to estimate the robot pose, it is clear that observation of features is the most important factor of successful SLAM. Among various types of features for visual SLAM, objects can serve as a good visual landmark because some object recognition methods are relatively robust and invariant to scale and rotation. Objects enable data association simpler than corners or lines and objects are also found easily in real environment [1].

Two approaches have been mainly used in object recognition; model-based scheme and appearance-based scheme. While a model-based (top-down) approach uses the model of an object, an appearance-based (bottom-up) approach does not use any prior knowledge of an object. Obviously, the latter is more suitable for SLAM which deals with unknown environments.

Researchers proposed several appearance-based approaches for object recognition. The saliency-based region selection strategy extracts multi-scale image features to find salient objects within a complex natural scene [2, 3]. This scheme aims at searching objects as humans do and it can successfully extract objects from the background. However, it focused only on the image analysis and often extracts the objects that are too small or too easily movable (i.e., books and bags) to be used in navigation.

Another strategy for the appearance-based approach uses only SIFT keypoints or their clustering within an input image [4, 5]. The main idea of these approaches is to extract the SIFT keypoints or to use clustering of SIFT keypoints as point landmarks. The landmarks are used only to estimate a robot pose and they do not offer any information on the environment, which means that they are just scale invariant points rather than meaningful objects for human (i.e., sinks and beds). These schemes have some drawbacks of using too many point features in a relatively small environment because too many features can cause inefficiency of SLAM or an increase in computational complexity.

The contribution of this paper is to propose a novel approach to object recognition that is applicable to SLAM. We propose an approach which finds useful objects without any prior information and exploits them as natural landmarks to estimate the robot pose and build an accurate environment map. The proposed scheme consists of the extraction method of various primitive features for reliable outputs and several steps for not selecting too small objects such as books and bags or parts of objects. If some objects are determined to be suitable for navigation, these detected objects are separated from the source image and registered in the database. These registered objects are subsequently used as landmarks to estimate the robot pose. Figure 1 shows some useful objects for navigation of a mobile robot in real indoor environments.

Fig. 1 Objects useful for navigation in real environments

By both visual feature-based EKF SLAM and the proposed recognition algorithm in this paper, the robot autonomously models an unknown environment. In this research, both range and vision sensors are used for SLAM and the SLAM process can be implemented in real-time although it may take long to recognize objects as the number of objects in the database becomes larger.

The remainder of this paper is organized as follows. Section 2 presents an overall structure of the proposed scheme and Sect. 3 deals with extraction of various features from the camera image. Section 4 represents feature combination and object selection and Sect. 5 describes EKF-based SLAM using extracted objects. Section 6 describes Experimental Results. Finally, Sect. 7 presents Conclusions.

2 Overall Structure of Autonomous Registration of Objects

Figure 2 shows an overall structure of the proposed object recognition scheme. For successful performance, it is desirable to use various types of features that are not correlated with each other. The proposed scheme uses five types of features such as SIFT keypoints, object contours, hue, saturation, and intensity.

Among the five primitive features, the clustered region of SIFT keypoints and contours of objects are considered as object candidates. Inside the object candidates,

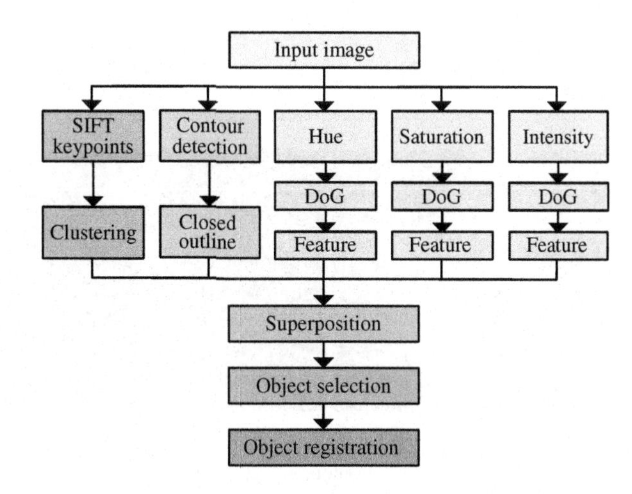

Fig. 2 Overall structure of the proposed scheme

color (hue, saturation, and intensity) information of the object candidates becomes a criterion to decide whether the object candidate can be selected as a useful object or not. After selecting an object, SIFT keypoints are exploited for object recognition. Object recognition offers the range and angle information of a recognized object because the stereo camera provides the position of the recognized object.

3 Feature Extraction from Camera Images

3.1 SIFT Keypoint Extraction and Contour Detection

SIFT (Scale Invariant Feature Transform) is one of the image recognition methods and extracts the feature points which are invariant to scale, rotation, and viewpoint [6]. The region where many SIFT keypoints exist is useful for navigation because the region is easily recognized by SIFT keypoints. The SIFT keypoints are clustered into several groups by the pixel distance in the image. Each group can be considered as a single object (although the keypoints of the group come from different physical objects) because SIFT keypoints tend to exist in the space whose pattern is obvious and remarkable (not flat and monotonous).

At the same time, the contour of an object is detected by the Canny edge algorithm [7]. The contour or outline of an object can help distinguish the object from the background or other objects. As objects in indoor environments are usually characterized by closed polygonal contours, it is useful to consider both the keypoints and contours together in selecting the objects. As an example, the detected contours and clustered regions of SIFT keypoints in the input image of Fig. 3 (a) are marked as rectangles in Fig. 3(b) and (c), respectively. If the size of object candidate is too small (smaller than 40 pixels in both width and length), this candidate is discarded because it is not likely to be reliably recognized due to its small size or the distance from the robot.

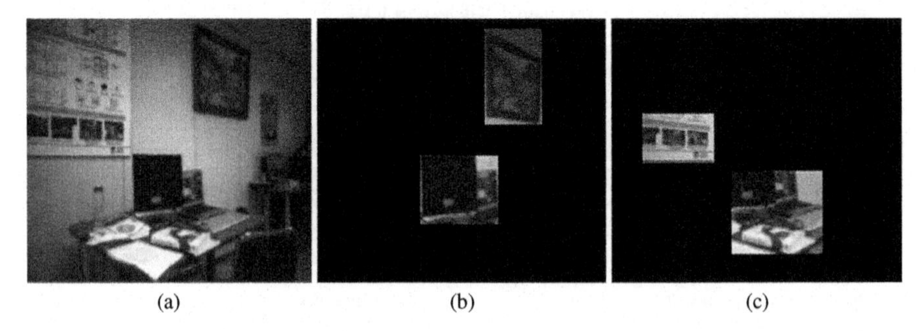

(a) (b) (c)

Fig. 3 (**a**) Input image, (**b**) contour detection, and (**c**) clustering of SIFT keypoints

3.2 Decomposition of Input Image

The input image in the form of RGB (i.e., red, green, and blue channels) is transformed into the form of HSI (i.e., three properties of color; hue, saturation, and intensity channels) through the procedure introduced in [8]. The HSI space is more intuitive and gives more information than the RGB space because the HSI space is similar to the human cognitive system and its three channels are not correlated. In the hue channel, all colors are represented as the values between 0 and 360. The saturation channel represents the degree of purity. For instance, dark blue and light blue are determined by adjusting the saturation channel. The intensity channel represents light information obtained by the conversion of the color image to the gray image.

3.3 Extraction of Features from Each Feature Image

Primitive features are extracted from the hue, saturation, and intensity channels. Gaussian convolution is conducted twice on each channel with variances of σ and 2σ. Boundaries become smooth through Gaussian convolution. Then, difference between the Gaussian convolution images represents the boundaries. The differences of Gaussian convolution images represent the complexity of patterns at hue (color), saturation (purity), and intensity (light) channels. The magnitudes of the features are represented as a gray scale image as shown in Fig. 4 and the feature images are obtained by Eq. (1).

$$I = |L(\sigma) - L(2\sigma)| \tag{1}$$

where $L(\sigma)$ and $L(2\sigma)$ are the Gaussian convolution images with masks whose variances are σ and $2\sigma\sigma$, respectively, I of Eq. (1) represents the primitive feature image. The magnitudes of the features are proportional to the gray scale of the I image.

A result of feature extraction is shown in Fig. 5. A camera image and feature images of hue, saturation, and intensity are shown in Fig. 5(a–d), respectively. The final combined image is made by summing them. The three feature images are normalized before they are combined because they represent their features with different ranges.

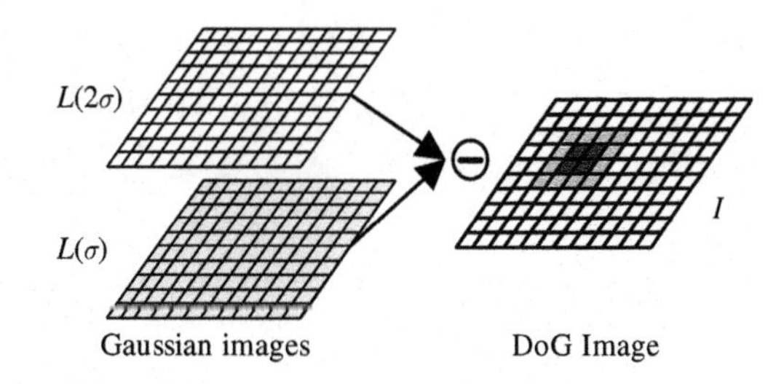

Fig. 4 DoG (Difference of Gaussian) image as a primitive feature image

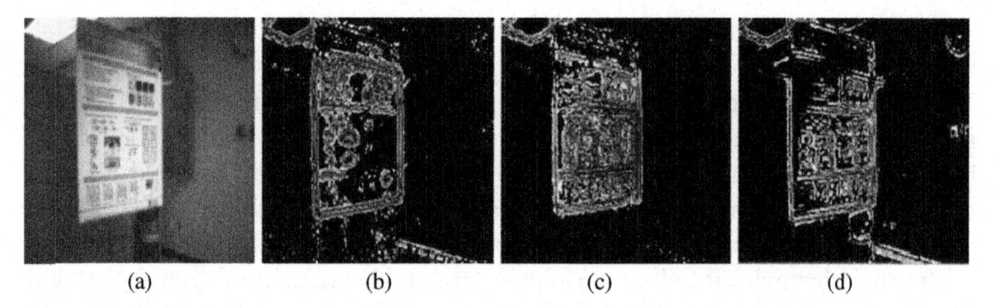

Fig. 5 Feature images; (a) Input camera image, (b) features extracted from hue image, (c) features extracted from saturation image, and (d) features extracted from intensity image

4 Combination of Primitive Features and Object Selection Mechanism

4.1 Combination of Primitive Features

The feature image extracted from hue, saturation, and intensity image inside an object candidate is the criterion for selecting suitable objects from the candidates. The gray scale of the feature images is related to salience. Salient objects or places look white in the feature images because the gray scale for the corresponding pixels is high.

Figure 6 shows examples of object candidates and their salience. Figure 6(a) is the input image. In Fig. 6(a), region A is the clustered region of SIFT keypoints and region B is the detected contour. The outer (yellow) rectangle in Fig. 6(a) represents a region of interest, which is used for prevention of the effect of insufficient information (i.e., the objects outside this region of interest are likely to be cut at the boundary of the camera image). Figure 6(b) represents the final combined image of Fig. 6(a). The three feature (hue, saturation, and intensity) images are combined with equal weights. Figure 6(c) shows the salience of regions A and B. In Fig. 6(c), all regions of Fig. 6(b) except A and B were discarded for better understanding. While region A is salient, region B is not salient, as shown in Fig. 6(c).

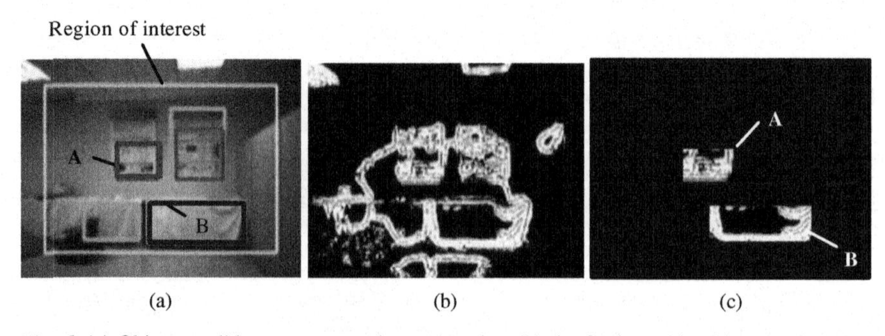

Fig. 6 (a) Object candidates represented as rectangles, (b) the final combined image where hue, saturation, and intensity feature images are combined with equal weights, and (c) HSI information of regions A and B in the final combined image

For successful results, we use an adaptive weighting strategy. The weights of feature images are determined according to the distribution of the gray scale values. That is, the weight is increased (or decreased) for the weight of dense (or sparse) features. It is described mathematically by

$$I_F = \omega_H I_H - \omega_S I_S - \omega_I I_I \tag{2}$$

$$\omega_H = \frac{\sigma_S^2 \sigma_I^2}{\sigma_H^2 \sigma_S^2 + \sigma_S^2 \sigma_I^2 + \sigma_I^2 \sigma_H^2} \tag{3}$$

$$\omega_S = \frac{\sigma_H^2 \sigma_I^2}{\sigma_H^2 \sigma_S^2 + \sigma_S^2 \sigma_I^2 + \sigma_I^2 \sigma_H^2} \tag{4}$$

$$\omega_I = \frac{\sigma_H^2 \sigma_S^2}{\sigma_H^2 \sigma_S^2 + \sigma_S^2 \sigma_I^2 + \sigma_I^2 \sigma_H^2} \tag{5}$$

where σ represents the distribution of the gray scale in the feature images, ω the weight of each image and I the feature image. The subscripts H, S, I, and F mean hue, saturation, intensity, and final combined feature. Whenever various scenes are captured, the weights change.

An example of adaptive weighting is illustrated in Fig. 7. A rectangle is extracted by the contour detection algorithm and the region inside the rectangle is assumed to be a candidate for an object. The variance of hue, saturation, and intensity inside the candidate is 2,944, 1,561, and 1,584, respectively. From Eqs. (3–5), the weight

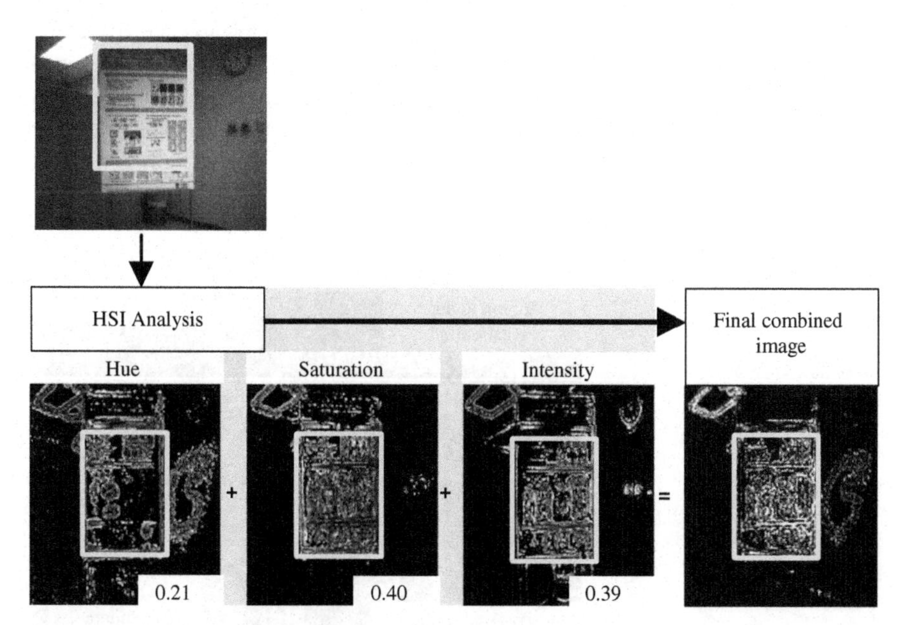

Fig. 7 Adaptive weighting; final combined image where hue, saturation, and intensity feature images are combined with different weights

becomes 0.21 for hue, 0.40 for saturation, and 0.39 for intensity. Features commonly extracted from all feature images become salient and they are represented in white, whereas non-salient areas in black.

4.2 Object Selection Mechanism

We also propose a filtering step using gray scale values in the final combined image for robust performance. The main idea is to investigate the average gray scale values along the boundary (10 pixels from the boundary) in four directions; left, right, top, and bottom. The scheme of investigating the average gray scale values to the left of the object candidate is shown in Fig. 8. As the object is distinguishable from the background in the final combined image, the area outside of the object candidate is not salient.

A more detailed explanation is described in Fig. 9. In Fig. 9(a), all the averages of the gray scale values outside the object candidate in four directions are low. On the other hand, in Fig. 9(b), the gray values outside of the object candidate are relatively high in all directions. Therefore, the object candidate of Fig. 9(a) is selected as an object, but that of Fig. 9(b) is discarded.

Figure 10 illustrates the recognized objects during navigation. By the proposed recognition method, a poster and part of a bookshelf were recognized, as shown in Fig. 10(a) and (b), respectively. In Fig. 10(c), another poster and a picture are matched from a quite long distance. In object recognition, the center of an object is selected as a point representing the object because the object has its own size at the input image. The red cross in the figures represents the center point of the recognized object. The affine transform, which calculates the geometrical relationship between the object recognized in the scene and that stored in the database, is used to extract the accurate center point with various viewpoints.

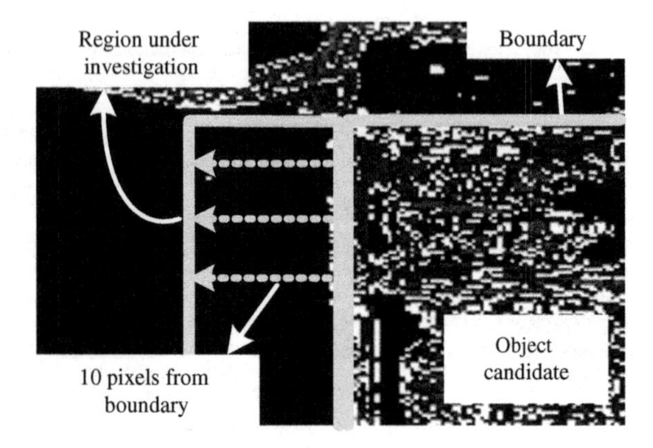

Fig. 8 Determination of the gradient from the boundary

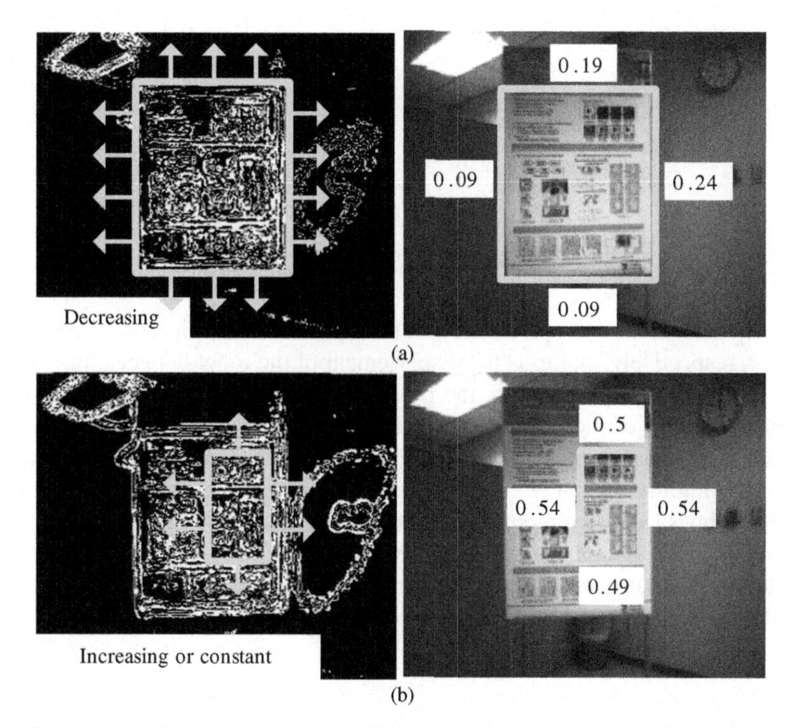

Fig. 9 Investigation of gradient of object candidates

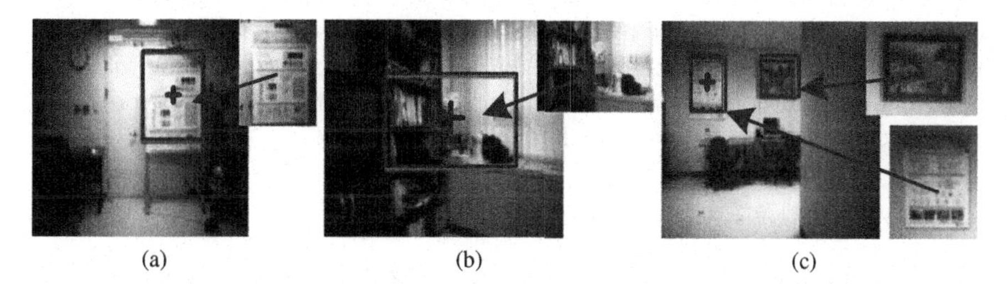

Fig. 10 Recognition of autonomously registered objects during navigation by the proposed method

5 EKF-Based SLAM

The EKF (Extended Kalman Filter) algorithm has proven to be the most appropriate framework in visual SLAM by much literature [9]. It compensates for the error accumulated due to both systematic and non-systematic errors during navigation. In EKF, the robot pose and landmark positions are stored in a state vector represented as X, and the position uncertainties of components of the state vector are stored in a covariance matrix denoted as P. The state vector and covariance matrix are updated recursively through sensor measurements.

5.1 Prediction

At the prediction stage, the state vector and its covariance matrix at time t are obtained as follows:

$$\hat{X}_t^- = f(\hat{X}_{t-1}, u_t, t) + w_t \tag{6}$$

$$P_t^- = F_x P_{t-1} F_x^T + F_u Q_t F_u^T \tag{7}$$

where $\hat{X}_t^-$ and P_t^- are the predictions of the state vector and its covariance matrix at time t, respectively, and u_t is the displacement of the robot between time t-1 and time t. The vector w_t represents the process noise with zero mean and Q is the covariance matrix of the process noise. The matrices $F_{\mathbf{x}}$ and $F_{\mathbf{u}}$ are the Jacobian matrices of the nonlinear motion model $f(\cdot)$ with respect to the state vector and the displacement u_t, respectively.

If the robot observes a feature, it compares this feature with the features in the state vector X. If it turns out to be a new feature, this feature is initialized and included in the state vector and its covariance matrix. If it is found to be one of the existing features, the EKF algorithm conducts the update stage.

5.2 Update

The state variables, the robot pose and landmark positions and the covariance matrix of the state vector are updated by the measurement of the sensor at the update stage. In this paper, the measurement is obtained from object recognition in the form of a relative range and orientation of the object from the robot. The state vector and its covariance matrix P at time t are updated as follows:

$$K_t = P_t^- H_t^T (H_t P_t^- H_t^T + R_t)^{-1} \tag{8}$$

$$\hat{X}_t = \hat{X}_t^- + K_t (Z_t - \hat{Z}_t) \tag{9}$$

$$P_t = (I - K_t H_t) P_t^- \tag{10}$$

where K_t represents the Kalman gain, and H_t is the Jacobian matrix of the sensor model with respect to the state vector. The error on the pose of the robot due to disturbances is compensated by the Kalman gain which is proportional to the difference between predictions and measurements. If none of landmarks are matched, the uncertainties of landmarks are kept unchanged. In this case, only the robot pose is calculated by the motion model and the uncertainty of the robot pose increases.

$$\hat{X}_t = \hat{X}_t^- \tag{11}$$

$$P_t = P_t^- \tag{12}$$

6 Experimental Results

Experiments were performed using a robot equipped with an IR scanner (Hokuyo PBS-03JN) and a stereo camera (Videre STH-MDI-C). The camera is used for object recognition and the IR scanner is used to build a grid map of the environment. The experimental environment consists of three rooms, as shown in the Fig. 11 (a). The total area of the experimental environment is $10\,\text{m} \times 10\,\text{m}$. Figure 11(b) shows the CAD data of the environment which will be compared with the map built by the proposed algorithm. The grid size of both the CAD data and the grid map built by SLAM is 10 cm.

Figure 12 illustrates the mapping process of the experimental environment using the proposed algorithm. In the experiment, pictures and bookshelves are selected as objects for estimating the robot pose. Since the cluttered environment such as chairs, table legs and small objects such as books are useless for localization, these objects are not selected. Figure 12(a) represents the initial state of the robot. In Fig. 12(b), the robot moves in the environment, builds the grid map and marks the objects in their own position. In room 2, it was difficult to detect some objects because of non-systematic errors generated by a carpet and slip of the wheels of the mobile robot. The map was distorted after navigating room 2, as shown in Fig. 12(c). However, the map was recovered from distortion by observing the registered object again, in Fig. 12(d). The recovery from the distortion is a result of data association, and object recognition can eliminate accumulated errors. It follows that object recognition makes data association easily compared to other features such as corners or lines.

Figure 13 shows the constructed map of the environment shown in Fig. 11 and the comparison of the trajectory estimated by the odometry (dotted line) with that by the proposed EKF-based SLAM (solid line) approach. The constructed map is referred to as a hybrid grid/vision map because it contains visual features as well as occupancy grids. Black objects or legs of tables cannot be represented in the map because an IR scanner cannot detect a light absorbing object and the object whose width is narrower than its angular resolution. The positional error of the resulting map is about ± 20 cm and orientation error is $5°$.

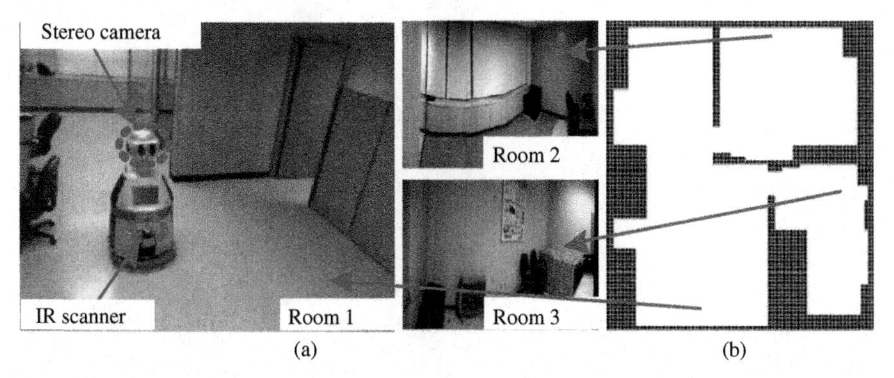

Fig. 11 Experimental environment; (**a**) mobile robot platform and experimental environment and (**b**) CAD data

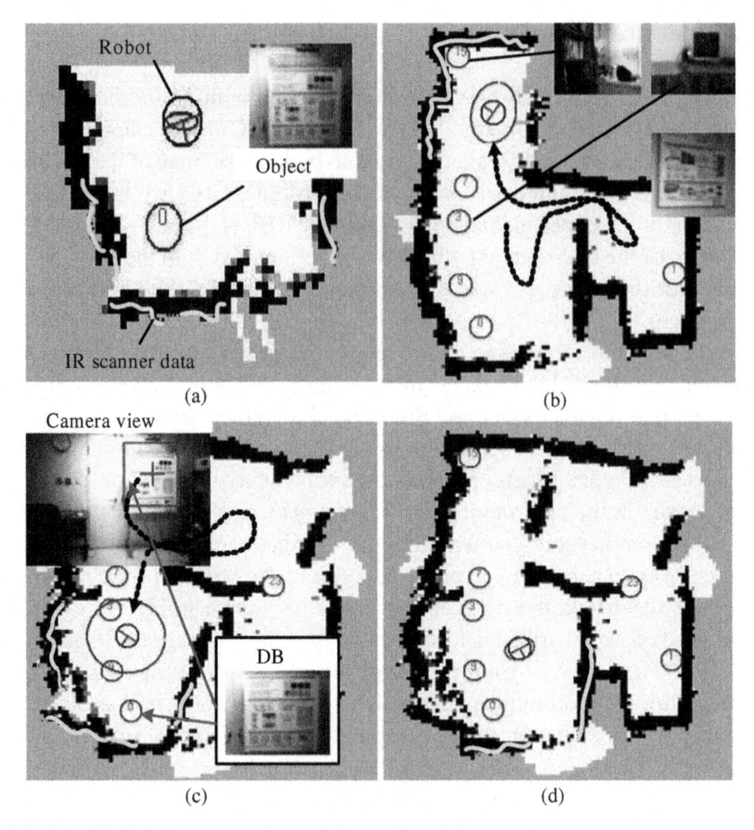

Fig. 12 Indoor SLAM with autonomous object registration

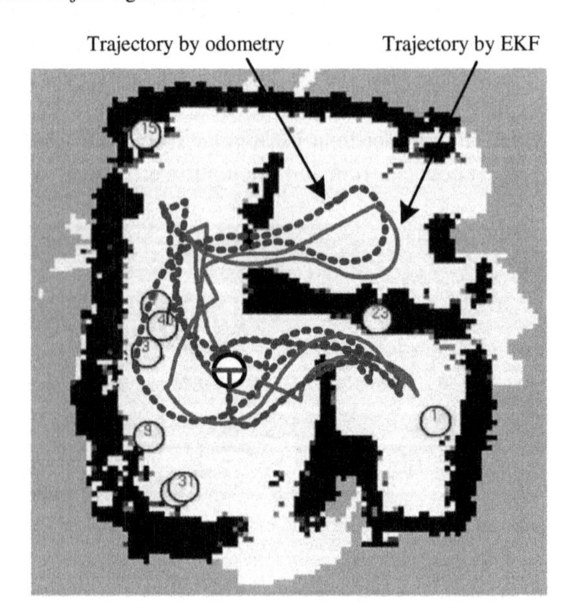

Fig. 13 Comparison of robot trajectory by odometry (dotted) with that by EKF-based SLAM (solid)

Fig. 14 Objects of different sizes from the same object

Several objects of different sizes can be registered from an identical object due to some factors such as lighting condition which affects the RGB image and thus HSI information. However, the hybrid grid/vision map is not much affected by the light condition because matching of the SIFT keypoints is relatively robust. For example, object B is hardly matched to object A in Fig. 14. Research on the consistent selection of an object under various lighting conditions is under way.

7 Conclusions

Object recognition is useful for navigation of a mobile robot because various objects exist in indoor environments. Current object recognition schemes require object information in the database, so objects which do not exist in the database cannot be recognized. However, the proposed scheme can recognize objects without any information. The experimental results of the proposed scheme and its application to SLAM are shown in the previous chapter. From the experiments, the following conclusions were drawn.

1. It is possible to autonomously select an object or a group of objects and register them as visual landmarks for SLAM without human interference.
2. The proposed scheme can solve data association or loop-closing problems relatively easily compared to that based on the range sensors alone because object recognition offers quite accurate feature matching results.

Acknowledgments This paper was performed for the Intelligent Robotics Development Program, one of the 21st Century Frontier R&D Programs funded by the Ministry of Commerce, Industry and Energy of Korea.

References

1. Newman, P., Cole, D., and Ho, K. (2006). Outdoor SLAM using visual appearance and laser ranging. In: *Proceedings of IEEE International Conference on Robotics and Automation*, pp. 1180–1187.

2. Itti, L., Koch, C., and Niebur, E. (1998). A model of saliency-based visual attention for rapid scene analysis. *IEEE Transactions on Pattern Analysis and Machine Intelligence*, **20**(11), 1254–1259.
3. Walther, D., Rutishauser, U., Koch, C., and Perona, P. (2004). On the usefulness of attention for object recognition. In: *2nd Workshop on Attention and Performance in Computational Vision*, pp. 96–103.
4. Sim, R., and Little, J. J. (2006). Autonomous vision-based exploration and mapping using hybrid maps and Rao-Blackwellised particle filters. In: *Proceedings of IEEE/RSJ International Conference on Intelligent Robots and Systems*, pp. 2082–2089.
5. Luke, R. H., Keller, J. M., Skubic, M., and Senger, S. (2005). Acquiring and maintaining abstract landmark chunks for cognitive robot navigation. In: *Proceedings of IEEE/RSJ International Conference on Intelligent Robots and Systems*, pp. 2566–2571.
6. Lowe, D. G. (2004). Distinctive image features from scale invariant keypoints. *International Journal of Computer Vision*, **60**(2), 91–110.
7. Bao, P., Zhang, L., and Xiaolin, W. (2005). Canny edge detection enhancement by scale multiplication. *IEEE Transactions on Pattern Analysis and Machine Intelligence*, **27**(9), 1485–1490.
8. Gonzalez, R. C., and Woods, R. E. (1992). *Digital Image Processing.* 3rd ed. Reading, MA: Addison-Wesley Publishing Company.
9. Dissanayake, G., Newman, P., Clark, S., Durrant-Whyte, H., and Csobra, M. (2001). A solution to the simultaneous localization and map building problem. *IEEE Transactions on Robotics and Automation*, **17**(3), 229–241.

People Detection using Double Layered Multiple Laser Range Finders by a Companion Robot

Alexander Carballo, Akihisa Ohya and Shin'ichi Yuta

Abstract Successful detection and tracking of people is a basic requirement to achieve a robot symbiosis in people daily life. Specifically, a mobile robot designed to follow people needs to keep track of people position through time, for it defines the robot's position and trajectory.

This work proposes a new method people detection and position estimation from a mobile robot by fusion of multiple Laser Range Finders arranged in two layers. Sensors facing opposite directions in a single row (layer) are combined to produce a 360° representation of robot's surroundings, then data from every layer is further fused to create a 3D model of people and from there their position.

The main problem of our research is an autonomous mobile robot acting as member of a people group moving in public areas, simple but accurate people detection and tracking is an important requirement. We present experimental results of fusion steps and people detection in an indoor environment.

Keywords People Detection · Feature Extraction · Multilayer LRF

1 Introduction

Companion robots are becoming part of daily life and are designed to directly interact with people, from a pet robot to aid the development process in children as to provide company to lonely elder people, to complex humanoid robots programmed with verbal interaction and providing services like guiding, entertainment and company. One necessary subsystem for such robots is detection, recognition and tracking of people as well as obstacles in the environment.

Laser Range Finders (LRF), besides being used for obstacle detection are also an important part of tracking systems, with important advantages over other sensing devices, like high accuracy, wide view angles, high scanning rates, robustness to

A. Carballo (✉)
Intelligent Robot Laboratory, Graduate School of Systems and Information Engineering, University of Tsukuba, 1-1-1 Tennoudai, Tsukuba City, Ibaraki Pref., 305-8573, Japan
e-mail: alexandr@roboken.esys.tsukuba.ac.jp

S. Lee et al. (eds.), *Multisensor Fusion and Integration for Intelligent Systems*,
Lecture Notes in Electrical Engineering 35, DOI 10.1007/978-3-540-89859-7_22,

changes in environment, usage simplicity and relatively less computing power to process data. However, $2D$ LRF data is not enough to solve important problems like occlusion in scan data, a set of multiple laser finders in different locations inside an area (for example [1, 2]) may reduce occlusion problems and effectively track multiple people, however detection is limited to the selected area and not suitable for tracking from a mobile robot.

Current approaches based on LRFs [3–6] place the sensors in the same height (single row), some 20 to 50 cm from the ground, to detect and track people's legs. However, detection of legs in cluttered environments is difficult especially if people are standing still. In Fod *et al.* [1] a row of several LRFs on different positions in a room were used for tracking moving objects, blobs (segments) are extracted and future positions estimated according to a motion model. Xavier *et al.* [4] focused on people detection using a fast method for line/arc detection but from a fixed position. Cui *et al.* [2] proposed a walking model to improve position prediction by including information about leg position, velocity and state. The later model was then used by Lee in [5] but this time from a mobile robot. Montemerlo *et al.* [7] also uses LRF from a mobile robot for people tracking and simultaneously robot localization by using conditional particle filters. Also in Zhao *et al.* [3] a mobile platform is used for a monitoring system based on a cart with two LRFs on a single row, to monitor people motion; their system also helps covering blind spots by moving the cart. Finally, Arras *et al.* [6] considers the problem of how to set the necessary threshold values and which features to use to successfully detect people from a mobile robot.

We propose in this work a new method for multiple people detection and position estimation by fusion of several LRF sensors, as other approaches, but installed in a multi-layered (multirow) arrangement. The idea is to obtain simultaneously different-but-complementary features to better detect people even in cluttered environments. This approach allows robust people detection even in presence of occlusion and is simple to implement using a mobile robot by placing sensor layers in the robot body at different heights from the ground depending on the features to detect.

Figure 1 represents our layered approach, every layer has two sensors facing opposite directions for 360° scanning (Fig. 1(a)), and two layers are used to extract features from upper and lower parts of a person's body (Fig. 1(b)).

Our method involves two fusion steps: fusion of sensors in a single layer and then fusion of layers. In the first step, sensors facing opposite directions in the same

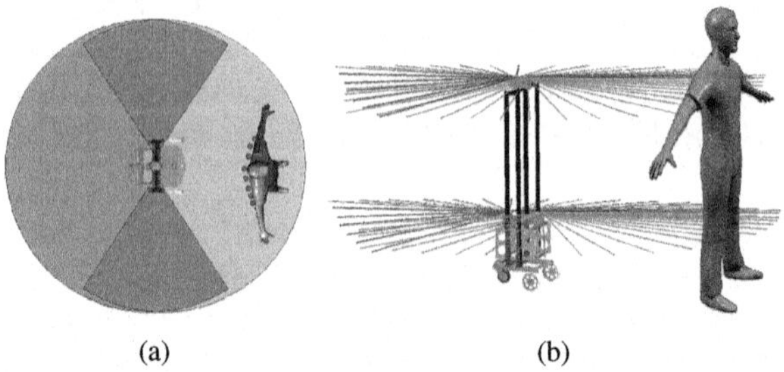

Fig. 1 Scanning from a double layered approach: (**a**) opposite facing sensors (top view) and (**b**) two layers of sensors (lateral view)

(a) (b)

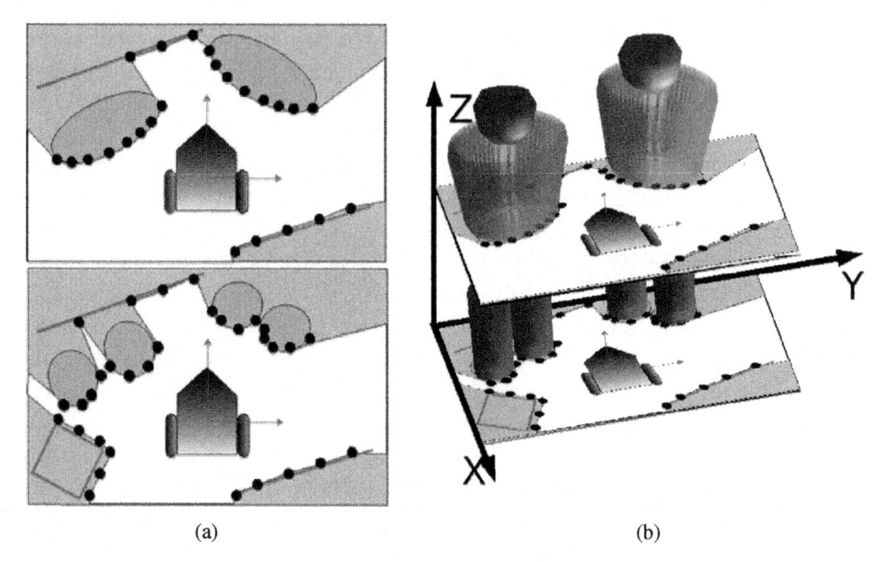

Fig. 2 (**a**) Projection of two 2D planes for a 3D representation, (**b**) a volumetric representation of people

layer are fused to produce a 360° representation of robot's surroundings. There is overlapping of scan data from both sensors (darker areas in Fig. 1(a)) so this fusion step must deal with data duplication. In the multiple layer fusion step, raw data from every layer is processed to extract features corresponding to people, then a $3D$ representation (a people model) is obtained allowing people detection and person position and direction estimation.

The main idea of our approach is depicted in Fig. 2, after fusion of sensors in every layer, geometrical features are extracted (Fig. 2(a)): in the upper one large elliptical shapes (chest areas) are seen while in the lower layer circular shapes (legs) are also visible. Then fusion of the extracted features from both layers define a 3D volume (Fig. 2(b)) enclosing every person detected.

A simple yet logical assumption is that an elliptical shape corresponding to a chest is always associated to one or two circular shapes corresponding to legs, and that the large elliptical shape (chest) is *always over* the set of small circles (legs). Figure 3 supports this concept, here we present a sequence of real scan images

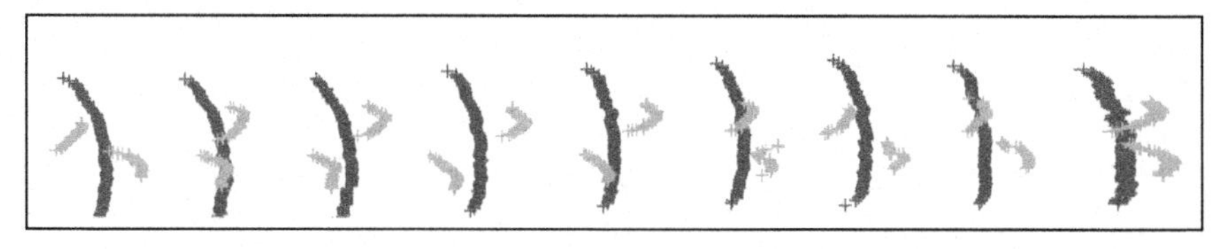

Fig. 3 A sequence of walking steps from scan data using sensors from both upper layer (darker points on large curve) and lower layer (smaller curves)

from a person walking (viewed from the top), both upper layer (large arc-like shape, chest) and lower layer (small arc-like shapes, legs) are visible. Fusion of extracted features allows creating a 3D volume (Fig. 2(b)) and from it the estimated person position is computed.

Our main research goal aims to develop a companion robot with the objective to study the relationship of an autonomous mobile robot and a group of multiple people in a complex environment like public areas, where the robot is to move and behave as another member of the group, while achieving navigation with obstacle avoidance. Some of the basic functions of such companion robot are depicted in Fig. 4, while the robot acts as another group member it has to detect, recognize and track the fellow human members (Fig. 4(a) and (c)) and also move in the environment like the rest of the members (Fig. 4(b) and (d)).

The robot used for our research is depicted in Fig. 5. The robot (Fig. 5(a)) is based on *Yamabico* robotic platform [8]. Two layers of LRF sensors are used, the lower layer is about 40 cm from the ground while the upper layer is about 120 cm. Every layer consists of 2 LRF sensors, one facing forwards and another facing backwards for a 360° coverage (Figs. 1 and 5). The sensors used in our system are the *URG-04LX* (Fig. 5(b)) range scanners ([9] provides a good description of the sensor's capabilities).

The rest of the paper is organized as follows. In Sect. 2 we describe the step of fusion of sensors in a single layer into a 360° representation. Section 3 presents our approach for fusion of multiple sensor layers, including feature extraction, people detection and position estimation. Section 4 presents experimental results for the different fusion steps and for people detection. Finally, conclusions and future work are left for Sect. 5.

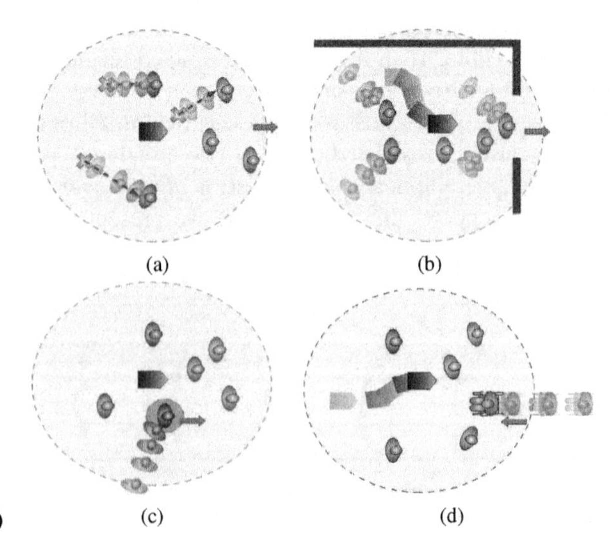

Fig. 4 Companion Robot with a group of people: motion tracking **(a)**, obstacle avoidance **(b)**, group members recognition **(c)**, mobile obstacle avoidance **(d)**

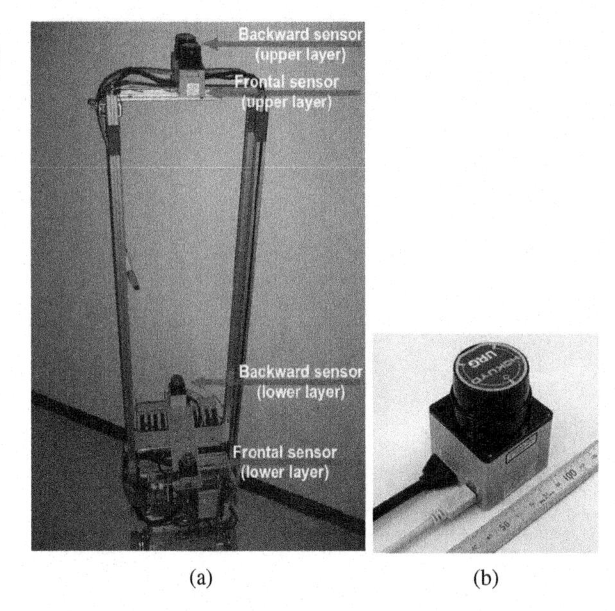

Fig. 5 Our robot system for multiple people detection and tracking **(a)**, four *URG-04LX* are used **(b)**

2 Combining LRF Sensors in Single Layer

The robot scans its 360° surroundings from both layers having two sensors in each layer, one facing forward and one facing backwards. A total of 4 LRFs are used, as presented in Fig. 5, raw scan data from every sensor is read, timestamped and integrated with odometry information. According to the top view representation in Fig. 1(a), real scan data obtained from both sensors in one layer is presented in Fig. 6 (data from the upper layer).

A previous step is to ensure sensor-to-sensor alignment which is to properly fix the pose of the two sensors so as the resulting combined scan data appears to be obtained from a single sensor. In other words scan data from areas where the sensors overlap should match and ideally be indistinguishable. A top view of real scan data is presented in Fig. 6, this overlap problem at left and right sides of the robot is also present.

Our application needs scan data from the complete surroundings of the robot, thus we need to combine readings of the two opposite facing sensors into a 360°. Ideally if the sensors we perfectly aligned data in the overlapping areas will be indistinguishable, however the different poses of the sensors provide different viewpoints with different problems, while a sensor's beam may suffer from reflection problems in one point the other sensor's beam (if both hitting the same point) may not. Our approach to fuse not-perfectly aligned scan data from two sensors in the same layer is described in next paragraphs.

A sensor s provides range scan points p_i^s from a set $\mathcal{P}^s$ consisting of range and direction $\left[r_i^s \; \theta_i^s \right]$ for every beam i, $i \in 1..N$ in a local coordinate system $\mathfrak{L}$ where direction is in the range $\{-120^{\circ}..120^{\circ}\}$. Data from sensors facing opposite directions

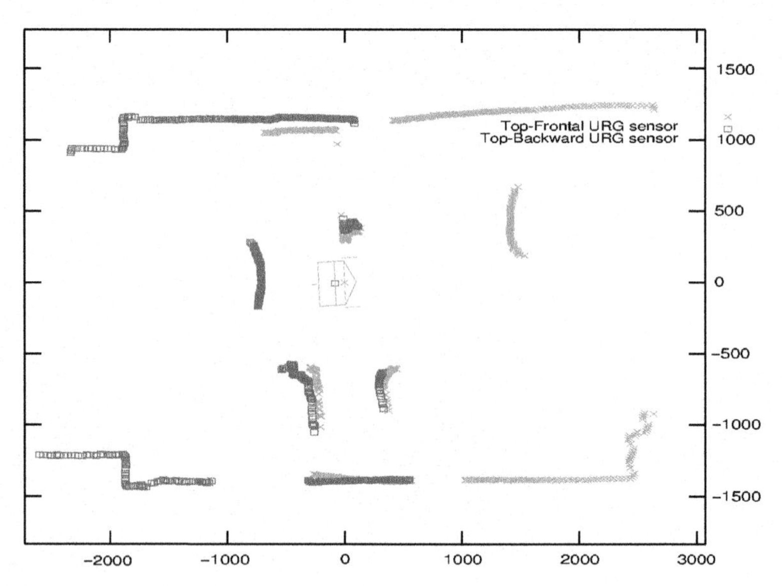

Fig. 6 Raw scan data from the upper layer

is transformed into a global system $\mathfrak{G}$ such that $\left[\, r_i^s \;\; \theta_i^s \,\right]$ is converted into $\left[\, R_k \;\; \phi_k \,\right]$ for each sensor.

Figure 7 represents this idea, the pair of sensors 1 and 2 facing opposite directions and separated in the vertical axis over a distance $d(1,2)$ are combined by transforming their scan data into $\mathfrak{G}$.

However sensors in this arrangement share scan points in *overlapping* areas (dark areas in Fig. 7(b) labeled as A and C, with points in range from -120° in one sensor to 120° in the opposite) thus a problem of duplication of data exists. *Non-overlapping* areas (B and D, right after the end of one overlap to 0° to the start of the next overlap, in both sensors) correspond to sensor's independent observations but those in the overlapping areas include both independent and shared observations, for the difference in pose of the sensors allows different points of view of the same object.

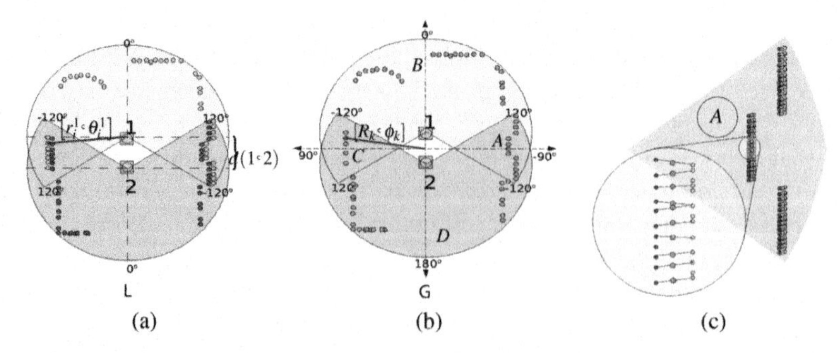

Fig. 7 Fusing scan data of two sensors in the same layer, (**a**) sensors in local coordinate system $\mathfrak{L}$, (**b**) converted to global system $\mathfrak{G}$ and fused, (**c**) averaging of matching points in overlapping areas A and C

To cope with this duplication of data in overlapping areas, a method was implemented to find for every scan data of one sensor the closest on the other sensor and obtain the average of their positions in $\mathfrak{G}$. Data in non-overlapping areas was left unchanged in $\mathfrak{G}$.

After joining every pair of sensors, scan data additionally transformed to include robot odometry information $\left[x \ y \ \varphi \right]^T$. The 2D representation of sensors in both layers (chest and legs) can be considered as two different XY planes in a 3D representation (Fig. 2), thus the Z component for every plane is the actual sensor height in the robot body (layers are parallel to the ground).

3 Fusion of double layered LRF sensors

Sensors in the same layer are facing opposite directions, individual scan data are combined into a 360° representation in the previous section. The next step is fusion of both sensor layers, here data will be divided into clusters with a segmentation function and then clusters will be classified according to their geometrical properties. Finally only those segments that match people features will be selected and joined into a 3D model from where people position is obtained.

3.1 Segmentation

Data clustering can be considered as the problem of breakpoint detection and finding breaking points in scan data can be considered as the problem of finding a threshold function $\mathcal{T}$ to measure separation of adjacent points. Every pair of neighboring points p_j and p_k are separated by an angle α which is proportional to the sensor's angular resolution (true for points of two adjacent scan steps) and by a distance $\mathcal{D}(p_j, p_k)$. Points are circularly ordered according to the scanning step of the sensor.

A cluster C_i, where $C_i = \{p_i, p_{i+1}, p_{i+2}, \cdots, p_m\}$, is defined according to a cluster membership function $\mathcal{M}$

$$\mathcal{M}(p_j, p_k) = (\theta_k - \theta_j) \leq \alpha \wedge \mathcal{D}(p_j, p_k) \leq \mathcal{T}(p_j, p_k) \tag{1}$$

such that for every pair $\langle p_j, p_k \rangle$ of adjacent points, the Euclidean distance $\mathcal{D}(p_j, p_k)$ between them is less than a given threshold function $\mathcal{T}(p_j, p_k)$ for p_j, p_k. A new point p_n is compared to the last known member p_m of a given cluster C_i as $\mathcal{M}(p_m, p_n)$.

Now, the threshold function $\mathcal{T}$ is defined for a pair of points, as in the work of Dietmayer [10], as:

$$\mathcal{T}(p_i, p_j) - C_0 + C_1 min(r_i, r_j) \tag{2}$$

with $C_1 = \sqrt{2(1 - cos(\alpha)}$. Dietmayer's work includes the constant C_0 to adjust the function to noise and overlapping. In our case C_0 is reemplaced by the radius $\mathcal{R}$ of the accuracy area for p_i as base point plus a fixed threshold value (10 cm in our case). $\mathcal{R}$ is defined according to the *URG-04LX* sensor specifications [9, 11] as:

$$\mathcal{R}(p_i) = \begin{cases} 10 & \text{if } 20\,\text{mm} \leq r_i \leq 1,000\,\text{mm} \\ 0.01 \times r_i & \text{otherwise} \end{cases} \quad (3)$$

The proposed threshold function $\mathcal{T}$ uses this accuracy information $\mathcal{R}$ when checking for break points, if two neighboring points have a large range value, it will be most probable that they form part of the same cluster for their bigger accuracy areas.

There is also a cluster filtering step that will drop segments very small to be considered of significance.

3.2 Feature Extraction

The idea of *feature extraction* is to match the sensor readings with one or more geometrical models representing expected behaviour of the data. For example if a LRF sensor is scanning a wall, then the *expected behaviour* of *a wall* scan data is *a straight line*. Also if the same sensor is to scan *a person* then the expected behaviour is a set of points forming *an arc*. So in order to identify walls a first requirement is to correctly associate the scan data with some straight line model, for people the same: associate a set of scan points to an arc-like shape (a circle or an ellipse).

Before applying any fitting method, it is important to have some information about the shape of the cluster that allows selecting the method. The information about clusters is extracted as a set of indicators like number of points, standard deviation, distances from previous and to next clusters, cluster curvature, etc.

One of the indicators is the cluster's *linearity*; our approach here is to classify the clusters into *long-and-thin* and those rather *short-and-thick*. The rationale behind this is that, straight line segments tend to be long and thin, round obstacles, irregular objects, etc., do not have this appearance.

Linearity is achieved by computing the covariance matrix Σ for the cluster $\mathcal{C}_i$ and then its eigenvalues λ_{max} and λ_{min} that define the scale and its eigenvectors v_1 and v_2 orientation (major and minor axes) of the dispersion of $\mathcal{C}$. The ratio $\ell = \lambda_{max}/\lambda_{min}$ defines the degree of longness/thinness of the cluster. We set threshold values for ratio $\mathcal{L}$ and for λ_{max}.

The *ellipticality* factor ε is computed as the standard deviation σ of the residuals of a ellipse fitting processes using the Fitzgibbon method [12]. The distance between a cluster point and an ellipse is computed using *Ramanujan's* approximation.

Only clusters with good ellipticality value are selected and segments passing the linearity criteria (that is lines) can be easily rejected since they do not belong to people.

Table 1 Example of indicators and their classifiers

Indicator	Classifier	Meaning
Width w	$w \leq W^{\Psi}_{max}$	A leg or a chest has a width no bigger than the threshold
Linearity ℓ	$\ell \leq \ell^{\Psi}_{max}$	Leg and chest features are not linear
Curvature $\bar{k}$	$\bar{k} \geq \bar{k}^{\Psi}_{min}$	Leg and chest features are curved
Ellipticality ε	$\varepsilon \leq \varepsilon^{\Psi}_{max}$	The fitting error of ellipse for chest under the threshold

We assign a weight value w to every indicator i and compute an scoring function $\mathbb{S}$ for every segment j in in layer Ψ, where $\Psi \in \{top, low\}$, as:

$$\mathbb{S}^j = \sum_{i}^{n} \mathrm{w}_i^{\Psi} \mathcal{H}_i^{\Psi}(I_i^j) \tag{4}$$

where $\mathcal{H}_i^{\Psi} : \mathbb{R} \rightarrow \{-1, 1\}$ is a binary classifier function for the i-th indicator which compares whether the given indicator is under some threshold value. Table 1 presents an example of indicators and their classifiers, the actual list of indicators is similar to that presented by Arras *et al.* in [6]. Weight values w_i and thresholds for every indicator i were manually defined after experimental validation.

3.3 People Model and Position Detection

As previously presented in Fig. 2, 3D projection of two planes of scan data from the layered sensors can be used to represent the position and direction of a person.

The set of geometrical features extracted from the former step are mostly ellipses and circles. If they belong to a person then another important criteria must be met: the large elliptical segment should come from the upper layer and the small circles from the lower layer. No large ellipses are possible for a person in the leg area. The small circles can not be over the large ellipse (the person height is restricted according to the height of the upper layer).

To properly establish the previous requirements, it is necessary to associate segments in the upper layer with those in the lower layer, this is to find the corresponding legs for a given chest. Even if data in both layers is aligned inside a 3D volume, as in Fig. 8, it is possible that legs lie inside or outside this volume according to the speed of motion and step length (length between feet when walking).

Latt *et al.* [13] present a study about how human motion, step length, walking speed, etc. are selected to optimize stability. Their study present data about different speeds people prefer when walking. If the average values of step length are used then it is possible to define the limits of motion of the legs with respect to the projected chest elliptical area. Figure 9 helps understanding this idea. The average leg height h is about 84 cm, and the height of the lower layer of sensors l is fixed at 40 cm. s is the step length which depends on the speed, for example 73 cm for an average speed of 1.2 m/s [13]. d is calculated as:

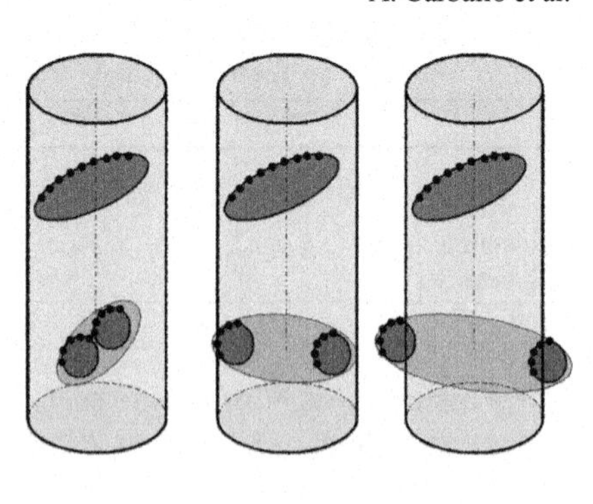

Fig. 8 Views of the walking of a person if both, the chest elliptical area and the leg circular areas (dark) are associated in the same 3D volume, step length is represented by dark color

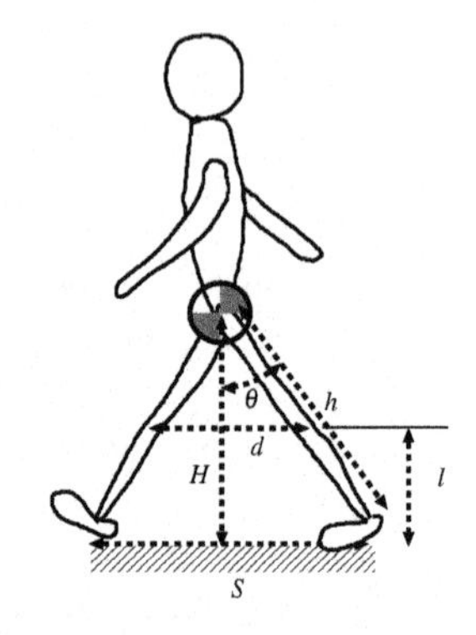

Fig. 9 Simple representation of human step to compute the distance d between leg segments while walking

Table 2 Step length according to speed and distance between leg segments d

Mode	Speed[a]	Step Length[b]	Distance between leg segments d[c]
Normal	1.2±0.04 m/s	73.0 ± 3 cm	34.40 cm
Very slow	0.5±0.05 m/s	47.0 ± 3 cm	22.29 cm
Very fast	2.1±0.1 m/s	86.0 ± 6 cm	39.64 cm

[a,b] Values according to Latt *et al.* [13].
[c] estimated from Eq. (5).

$$d = 2(H - l)\tan(\theta), \text{ where } \theta = \sin^{-1}\left(\frac{s}{2h}\right). \tag{5}$$

According to [13] the step length for three different walking speeds is presented in Table 2. In this table we include the parameter d from Fig. 9 about the distance between leg segments when walking at the different speeds.

Also from [13], the step width w (in Fig. 10) does not vary much with walking speed and is about 12.0 ± 1 cm.

With an estimation of the maximum value for d, the separation of legs at the lower layer height, we can set a search radius of $\frac{d}{2} \pm \xi$ at the center of the chest elliptical area projected into the lower layer to search for the corresponding legs

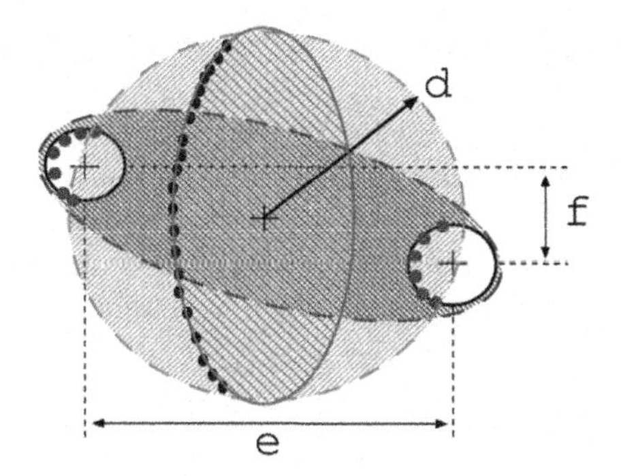

Fig. 10 Searching legs of a person in an area with radius $d/2 \pm \xi$ with center at chest ellipse in the 3D volume

for the chest. We use average walking step length from Latt *et al.* [13], at normal walking speed, to compute the value for d. This idea is represented in Fig. 10. Once associated, the position of the person will be finally set by the position of the chest ellipse center.

4 Experimental Results

The robot used for our research was presented in Fig. 5, the computer operating the robot is a Intel Pentium Core Duo based notebook running (Linux kernel 2.6.24) as operating system and robot control board is powered by a Hitachi SH-2 processor. The robot system uses *4 URG-04LX* range scanners from *Hokuyo Automatic Co., Ltd.* [11], small size (50 *x* 50 *x* 70 mm), maximum range 5.6 m, distance resolution of 10 mm and angular resolution of 0.36°, angular range of 240° operating at 10 Hz. Scan data from each sensor consists of 682 points circularly ordered according to scanning step.

Data from each sensor is read every 100 ms by a driver processes and registered in parallel into a shared memory system (*SSM* [14]) based on IPC messaging and multiple ring-buffers with automatic timestamping, one driver process per sensor. SSM also allows to record raw sensor data into log files and to play it back with the same rate as the sensor (10 Hz in this case).

Client processes read scan data from the ring-buffers according to sensor's pose (those in the top layer and those on the low layer), pairs of LRF sensors are processed in the fusion step, sensor layers are further fused and finally people position is computed.

The processing time for the two layers (four sensors), from single layer fusion to people position detection, was less than 50 ms, fast enough given the sensor's scanning speed.

In the following subsections we present the results of the different tasks involved in this method for people detection.

4.1 Fusion of LRF Sensors in Single Layer

Results of fusing opposite facing sensors are presented in Fig. 11. The fusion method joins correctly the data from both sensors, data from areas B and D (Fig. 7) is copied as it is, and data from areas A and C uses simple averaging of closest points from both sensors. Sensor data is joined and a 360° representation of the surrounding environment is possible.

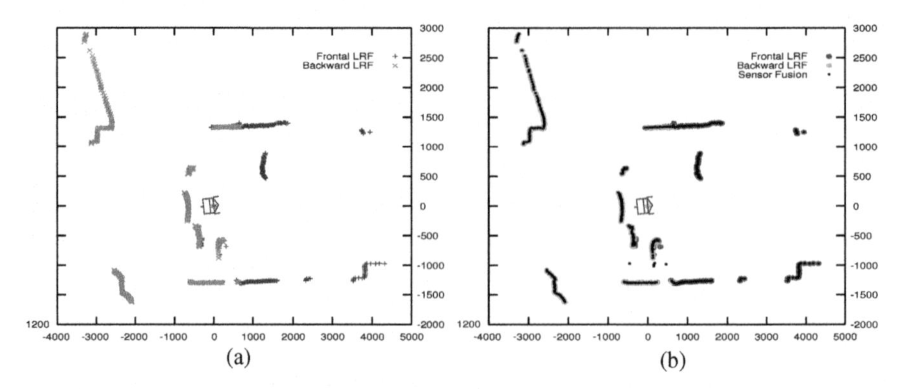

Fig. 11 Fusion of sensors in a single layer: **(a)** raw data (front sensor in red and backward in green points) **(b)** results of fusion (black points)

These results are important for the next steps since we obtained one set of points from the two sensors, from there the segmentation step in the fusion of double layers can extract clusters without having to consider the case of duplication and further merging of clusters.

4.2 Fusion of Double Layered LRF Sensors

We performed an experiment for people detection and position estimation from a mobile robot. In the experiment, five persons walked around the robot and additional person was taking the experiment video. Log data from each sensor was recorded, people position detection tests were performed off-line by playing back this log data using our SSM system. Figure 12 corresponds to the group of people surrounding the robot.

4.2.1 Segmentation and Feature Extraction

Figure 13 shows results of LRF data segmentation and feature extraction: raw data from each layer (top layer in Fig. 13(a)) is divided into clusters (Fig. 13(b)) and each cluster's indicators analyzed to extract those segments with human-like features and average sizes (Fig. 13(c)).

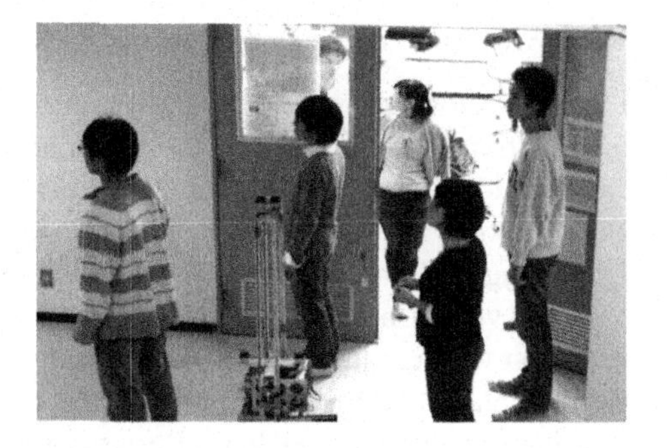

Fig. 12 An experiment for multiple people position estimation using the proposed method

In this figure, arrows in Fig. 13(a) represent the location of people in the environment, most of them were successfully detected in the results of feature extraction (Fig. 13(c)). However one of them has a height below the standard so top-level sensors were actually scanning his neck area, accordingly his chess ellipse is smaller than the allowed values, therefore was rejected. Another interesting case is the segment marked as "column" in Fig. 13(a), although its curvature and linearity indicators classify it as person, the boundary length and segment width were far bigger than the allowed values, reducing its scoring and marking it for rejection.

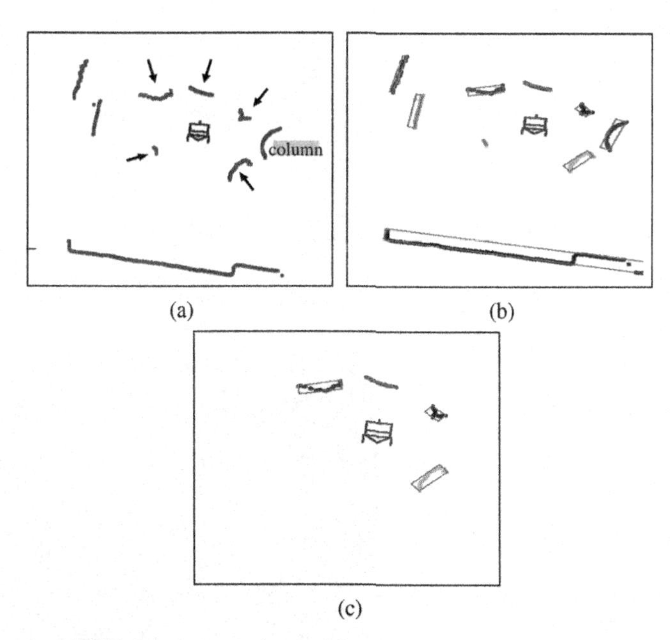

Fig. 13 Results of LRF data segmentation and feature extraction: raw data **(a)** is segmented **(b)** and then classified **(c)**

4.2.2 Multiple People Detection

Figure 14 shows the results of an experiment for people detection and position estimation from a mobile robot. In the experiment, five persons walked around the robot and additional person was taking the experiment video (Fig. 14(a)). Log data from each sensor was recorded, people position detection tests were performed off-line by playing back this log data using our SSM system.

A 3D visualization process was created to visualize and inspect how the people detection worked; in Fig. 14(b) chest ellipses and leg circular ellipses are detected then we place a 3D wooden doll, as a representation of a person, in the estimated position the person should have. Results were verified by human operator comparing the experiment video with results.

The members have varied body sizes, from broad and tall to thin and short. Some of the members have a height a little under the average, as result their chest ellipses were not correctly detected in the people detection step. As presented in Fig. 14(b), the person to the right of the robot (represented with blue line segments) is missing although circles from legs are present.

Additional snapshots of experimental results are presented in Fig. 15, the robot is represented in all cases as blue line segments. Figure 15(a) and (c) shows raw scan data from both layers (red for the upper layer and green for the lower one), and in Fig. 15(b) and (d) a 3D representation of the human detection and position estimation. In the cases of 3D representation, the raw scan data is plotted together with

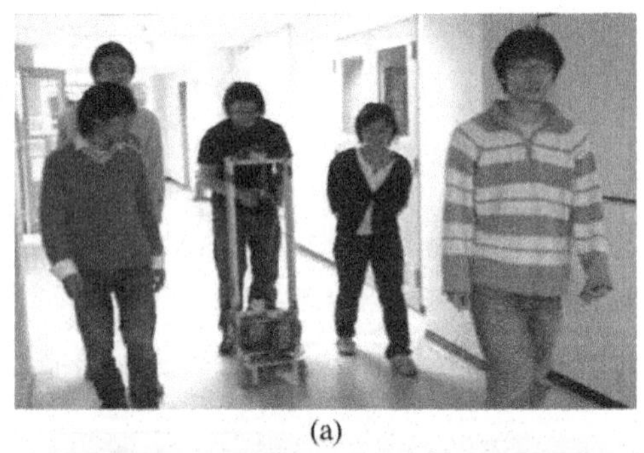

(a)

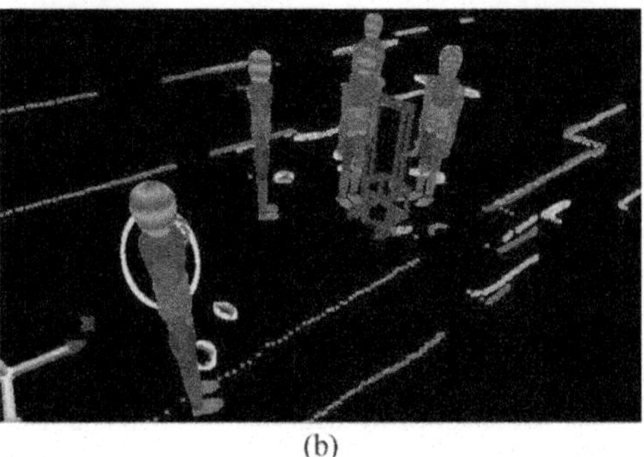

(b)

Fig. 14 Results of the people detection, (**a**) snapshot (**b**) 3D models in estimated positions of people in the experiment

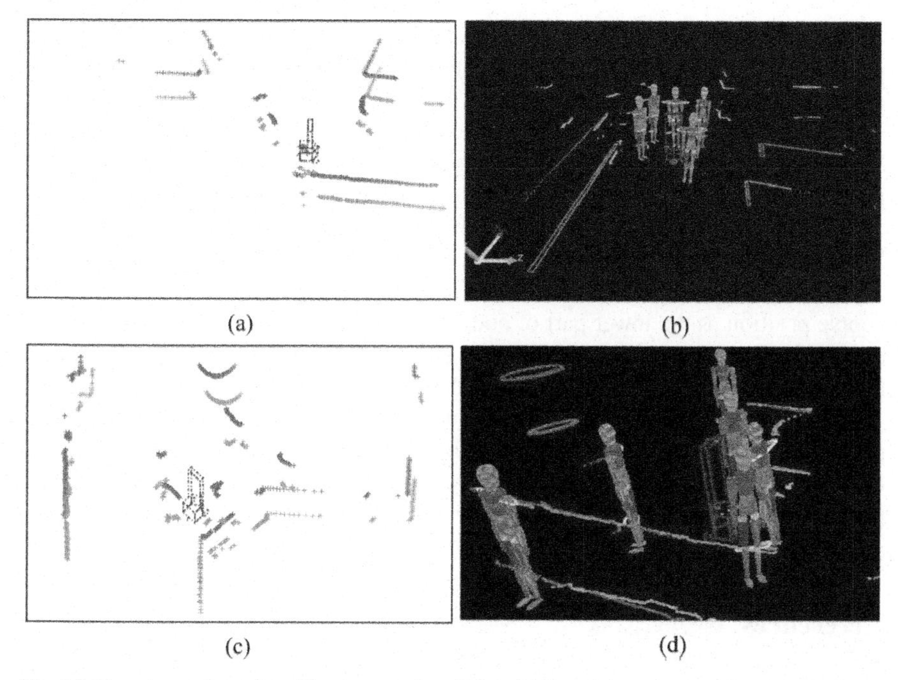

Fig. 15 Experimental results with raw scan data ((a) and (c)) and the corresponding people detection and position estimation ((b) and (d))

wooden dolls enclosed in the estimated people positions represented with elliptical shapes, a large one for the chest area and smaller ones for the extracted leg areas.

In Fig. 15(c) there are two rather large arc-like segments in the raw scan image and two large elliptical shapes in the 3D representation in Fig. 15(d), in both layers. That is the column inside the indoor environment, as already explained in Fig. 13(a). The people detection method discards this elliptical object because its dimensions are larger than the expected for people, those elliptical objects are represented with red color in this figure. Also we do not expect large elliptical objects from the lower layer so discarding this column as a non human object was simple.

5 Conclusions and Future Works

Fusion of multiple LRF sensors arranged in a double layer structure for multiple people position detection from a mobile robot studied in this paper. The proposed approach is simple and the double layer multiple LRF approach is practical enough to be implemented on a mobile robot. Instead of fusion of different sensors with complementary capabilities, we fused the same type but at different heights (layers), giving different perspectives which also helps solving simple cases of occlusion where one sensor is occluded and the other is not.

A simple method for fusion of opposite facing sensors in the same layer was presented, from it a simple 360° representation of the surrounding environment was possible and simplified the segmentation step in the fusion of both layers.

Fusion of double layers by segmentation of fused scan data, geometrical features extraction and association in a 3D volume for every detected person allows good position estimation and a measure of the possible direction the person is facing.

The method used here easily filters out non-people segments by analyzing the key indicators such as size, compactness, ellipticality and linearity. The addition of an extra layer of LRFs to detect chest elliptical areas improve the estimation of people position as the lower part of body (the legs) move faster and wider than the chest area. The combination of both areas creates a 3D volume which helps locating the position of the person more closely related to the center of this 3D volume.

As future work, estimation of the person direction from motion and multiple people tracking will be considered. Also the effectiveness of our method in cluttered environments will be studied. Future steps of our research include understanding people group motion and recognition of group members.

References

1. Fod A., Howard A., and Matarić, M. (2002) Laser-based people tracking. In: Proceedings of the IEEE International Conference on Robotics and Automation (ICRA), Washington DC, U.S.A, 3024–3029.
2. Cui J., Zha H., Zhao H., and Shibasaki R. (2006) Robust Tracking of Multiple People in Crowds Using Laser Range Scanners. In: Proceedings of the IEEE 18th International Conference on Pattern Recognition (ICPR), Hong Kong, China, 857–860.
3. Zhao H., Chen Y., Shao X., Katabira K., and Shibasaki R. (2007) Monitoring a Populated Environment Using Single-row Laser Range Scanners from a Mobile Platform. In: Proceedings of the IEEE International Conference on Robotics and Automation (ICRA), Roma, Italy, 4739–4745.
4. Xavier J., Pacheco M., Castro D., Ruano A., and Nunes U. (2005) Fast Line, Arc/Circle and Leg Detection from Laser Scan Data in a Player Driver. In: Proceedings of the IEEE International Conference in Robotics and Automation (ICRA), Barcelona, Spain, 3941–3946.
5. Lee J. H., Tsubouchi T., Yamamoto K., and Egawa S. (2006) People Tracking Using a Robot in Motion with Laser Range Finder. In: Proceedings of the IEEE/RSJ International Conference on Intelligent Robots and Systems (IROS), Beijing, China, 2936–2942.
6. Arras K. O, Martínez Mozos O., and Burgard W. (2007) Using Boosted Features for the Detection of People in 2D Range Data. In: Proceedings of the IEEE International Conference on Robotics and Automation (ICRA), Roma, Italy, 3402–3407.
7. Montemerlo M., Thrun S., and Whittaker W. (2002) Conditional particle filters for simultaneous mobile robot localization and people-tracking. In: Proceedings of the IEEE International Conference on Robotics and Automation (ICRA), Washington DC, U.S.A, 695–701.
8. Yuta S., Suzuki S., and Iida S. (1991) Implementation of a Small Size Experimental Self-contained Autonomous Robot - Sensors, Vehicle Control, and Description of Sensor Based Behavior. Lecture Notes in Control and Information Sciences: Second International Symposium on Experimental Robotics (ISER), Toulouse, France, Springer-Verlag, 190:344–358.
9. Kawata H., Kamimura S., Ohya A., Iijima J., and Yuta S. (2006) Advanced Functions of the Scanning Laser Range Sensor for Environment Recognition in Mobile Robots. In: Proceedings of the IEEE International Conference on Multisensor Fusion and Integration for Intelligent Systems, Heidelberg, Germany, 414–419.

10. Dietmayer K. C. J., Sparbert J., and Streller D. (2001) Model Based Classification and Object Tracking in Traffic Scenes from Range-Images. In: Proceedings of 4th IEEE Intelligent Vehicles Symposium, Tokyo, Japan.
11. Hokuyo Automatic Co., Ltd. http://www.hokuyo-aut.co.jp/
12. Pilu M., Fitzgibbon A. W., and Fisher R. B. (1996) Ellipse-specific Direct Least-square Fitting. In: Proceedings of the IEEE International Conference on Image Processing, Lausanne, Switzcrland, 3:16–19.
13. Latt M. D., Menz H. B., Fung V. S., and Lord S. R. (2008) Walking Speed, Cadence and Step Length are Selected to Optimize the Stability of Head and Pelvis Acceleration. Experimental Brain Research, Springer Berlin, 184(2):201–209.
14. Takeuchi E., Tsubouchi T., and Yuta S. (2003) Integration and Synchronization of External Sensor Data for a Mobile Robot. In: Proceedings of the Society of Instrument and Control Engineers (SICE) Annual Conference, Fukui, Japan, 1:332–337

Part III
Applications to Sensor Networks and Ubiquitous Computing Environments

Hernsoo Hahn

Sensor network is a concept of utilizing multiple sensors in an integrated way to achieve a single goal in various levels. Individual sensor nodes collect and process data for their own purpose and also may ask to other sensor nodes to send their raw or refined information. In this communication among the sensor nodes, each sensor node may assign belief measure to other sensor nodes individually. Therefore, the research topics in this field include communication problem among the sensor nodes, how to assign belief measure to individual sensor nodes, as well as network topologies. This concept can be applied to many real problems, such as path selection of robot and autonomous vehicle development. This chapter introduces 10 recent efforts on sensor networks and sensor applications dealing with the aforementioned issues.

1. "Path-selection Control of a Power Line Inspection Robot Using Sensor Fusion": A new wire detection algorithm for power line inspection by a mobile robot has been proposed in the paper. For the development of power line inspection robot, DECRO, the sensor fusion and fuzzy control algorithms are developed to detect the wire and slope of the wire.
2. "Intelligent Glasses: a Multimodal Interface for Data Communication to the Visually Impaired": This paper introduces the concept of bimodal visual-tactile interface, named Intelligent Glasses (IG), designed for assistance of the visually impaired in their daily tasks such as access to information and mobility. IG system can be an alternative solution that provides information on time-variant near space.
3. "Fourier Density Approximation for Belief Propagation in Wireless Sensor Networks": In order to make the algorithm efficient and accurate, messages which carry the belief information from to one node to the others should be formulated in an appropriate format. So this paper presents two belief propagation algorithms where non-linear and non-Gaussian beliefs are approximated by Fourier density approximations, which significantly reduces power consumptions in the belief computation and transmission.
4. "Sensor Node Localization Methods based on Local Observations of Distributed Natural Phenomena": This paper addresses the model-based localization of sensor networks based on local observations of a distributed phenomenon.

For localization process, the proposed method is rigorous exploitation of strong mathematical models of distributed phenomena. Also it introduces two approaches: first, the polynomial system localization method, and the other is the simultaneous reconstruction and localization method.

5. "Study on Spectral Transmission Characteristics of the Reflected and Self-emitted Radiations through the atmosphere": This paper develops a software that predicts spectral radiance from ground objects by considering spectral surface properties, that is to say, a software for analyzing the radiances through the path and from the objects in the scene which include the radiative intensities by self-emission and by reflection of incident radiation. Different material properties are considered in analyzing the thermal balance and the surface radiation.
6. "3D reflectivity Reconstruction by Means of Spatially Distributed Kalman filters": In this paper, a statistical approach is derived, which models the region of interest as probability density function (PDF) representing spatial reflectivity occurrences. To process the nonlinear measurements, the exact PDF is approximated by well-placed Extended Kalman Filters allowing for efficient and robust data processing.
7. "T-SLAM: Registering Topological and Geometric Maps for Robot Localization in Large Environments": This article reports on a map building method that integrates topological and geometric maps created independently using multiple sensors. The T-SLAM approach is mathematically formulated and applied to the localization problem within the Intelligent Robotic Porter System (IRPS) project, which is aimed at deploying mobile robots in large environments (e.g. airports).
8. "Map fusion based on a multi-map SLAM framework": This paper presents a method for fusing two maps of an environment: one estimated with an application of the simultaneous localization and mapping (SLAM) concept and the other one known a priori by a vehicle. Also, this paper shows how a priori knowledge available in the form of a map can be fused within an EKF-SLAM framework to obtain more accuracy on the vehicle poses and map estimates.
9. "Development of a Semi-Autonomous Vehicle Operable by the Visually Impaired": This paper presents the development of a system that will allow a visually-impaired person to safely operate a motor vehicle and specially to improve the independence of the visually-impaired by allowing them to travel at their convenience.

Path-Selection Control of a Power Line Inspection Robot Using Sensor Fusion

SunSin Han and JangMyung Lee

Abstract A new wire detection algorithm for power line inspection by a mobile robot has been proposed in this paper. There have been a lot of studies in order to support the high-quality electric power. For the high-quality power supply, it is necessary to investigate the power lines and insulators before the lines or insulators were disconnected or damaged. Although Korea Electric power Corp. has made many efforts for the quality improvement, it is not enough to inspect all the power lines by human inspectors. According to this problem, it is decided to replace the human operators by the power line inspection robot. When the robots are used for the inspection, there could be several advantages, for example, the working efficiency and the prevention of accident. And also the shortage of human power for dangerous jobs can be resolved. In this paper, as a part of the development of power line inspection robot, DICRO (DIstribution line Checking RObot), the sensor fusion and fuzzy control algorithms are developed to detect the wire and slope of the wire. The effectiveness of the proposed algorithms is proved by the real experiments with DICRO which is under development so far.

Keywords Power line inspection · Robot insulators

1 Introduction

High quality power supply is very important currently since most of the automated systems require the power without any perturbation.

To prevent any malfunctioning of the transmission line and equipment, the system need to be checked regularly and the fault prediction algorithms are necessary to be developed.

There are many electric poles and insulators to transmit the power to various areas. Processed wires are normally made of plastic coated aluminum–copper alloy

J.M. Lee (✉)
Pusan National University, KeumJung Gu, JangJeonDong, BUSAN, 609-735, South Korea
e-mail: jmlee@pusan.ac.kr

S. Lee et al. (eds.), *Multisensor Fusion and Integration for Intelligent Systems*, Lecture Notes in Electrical Engineering 35, DOI 10.1007/978-3-540-89859-7_23,

to transmit high voltage power, whose inner sides are filled by steel to keep the mechanical strength and to reduce the price. The processed wires are tied to the insulators on the pylon or on the electric pole.

Since the wires and insulators are used on the air long time, there are chemical reactions, aging, and thermal erosion which cause severe power loss through the transmission. For the conventional inspection, a human operator checks the wires and insulators with the aid of infrared cameras periodically. Some of the defects in hidden areas cannot be detected by the visual inspection [1].

There exists also accident possibility while the operators are working in the bucket truck to inspect the active power lines, which lowers the safety factor of the power transmission. Currently even though many researches on the power-line inspection in England, Japan, Poland, *etc.*, have been done, there is no commercial product yet. As a candidate tool for the laborious inspection task, a mobile robot is proposed in this research, which carries CCD and IR cameras, IR sensors, and microphones to detect noises from the insulators.

The power line inspection robot may reduce the accidents of human workers and may save money to raise and to maintain the expert for inspecting the power lines [2, 3]. This paper focuses on the algorithm to detect the existence of a neutral wire and to measure the tilt angle of the mobile robot against the neutral wire.

2 Sensor Fusion

2.1 Ultrasonic Sensor

Since ultrasonic sensors are cheap and simple to operate, there are many applications with high object detection reliability. By measuring the traveling time of the ultrasonic signal, the distance to the object can be measured since the traveling speed of ultrasonic is well defined. When the TOF (time of freight) is T, the distance to an object from the robot, L, can be determined as follows:

$$L = \frac{T}{2}u \tag{1}$$

where u represents the traveling speed of ultrasonic signal [4]. Generally, the accuracy becomes poor with the distance increase since the sensor has a certain range of radiation angle.

Even though it cannot provide either precise orientation of the object or the distance, it is very useful to determine the existence of the object with a rough distance value. The measuring is relatively robust against environmental condition such as illumination, temperature, humidity, *etc.* In this research, FW-H series ultrasonic sensors of KEYENCE Corp. have been used, which have the effective measuring range of 15~70 cm.

2.2 Infrared Sensor

Infrared sensor transmits the IR signal to an object and measures either the traveling time or the intensity of the reflected signal. Since the sensors based on the traveling time are expensive, the cheap IR sensors measure the intensity and calculate the distance to an object based on the intensity.

Compared to the ultrasonic sensor, the infrared sensor has a high directionality and accuracy. That is, it can only detect some objects which are on the path of the ray and measures the distance precisely. However it can neither detect a transparent object nor identify the color of an object. In this research, GP series IR sensors of SHARP Corp. have been used, which have the effective range of 10~80 cm.

2.3 Sensor Fusion Algorithm

When the mobile robot is moving along the neural wire of the power transmission lines, recognition of the electric pole and detection of the neural wire which is captured by the robot when it passes over the electric pole are the most important tasks. The mobile robot is vibrating by the wind while it passes over the electric pole, which makes it very difficult to detect the neural wire to hold. It also deteriorates the recognition accuracy of the electric pole, which may result unstable grip of the electric pole. Therefore detecting the neural wires by using only the infrared sensor may take too much time to apply for real situations where the location/orientation information on the wire are not available.

The distance between the robot and the neutral wire is about 23 cm which is within the effective range of the ultrasonic sensor. However when a high tilt angle between the robot and the neutral wire exists, the distance can be less than 15 cm that may not be detected by the ultrasonic sensor. The average distance measuring error of the ultrasonic sensor is about 3 cm. To improve the grasping accuracy of the electric pole and holding stability of the neural wire, ultrasonic sensor and infrared sensor data are heuristically fused.

That is, to measure the distance precisely with easy detection of the wires, ultrasonic sensors which have wide range detection capability are used with the infrared sensors which have precise distance measuring capability. In addition to this, the IR sensor data are sensitive to sun ray, while the ultrasonic sensor data are robust against sun ray.

On the top of the electric pole, there is a neutral wire to protect the power lines from lightning. The robot is holding this neutral wire and moves along the wire.

Figure 1 illustrates the arrangement of sensors on the robot to detect the distances to the neutral wire.

The sensor fusion process of robot is required when the ultrasonic sensors and infrared sensors are activated to detect the neural wires as shown in Fig. 1.

The wide region detection capability of ultrasonic sensor has been utilized first, and the precise distance data can be obtained by the infrared sensor later. In this

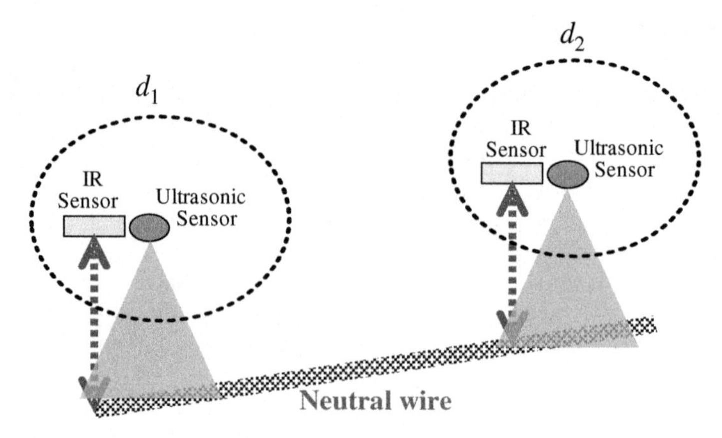

Fig. 1 Arrangement of sensors

process, there exists an error between the distances measured by the two sensors, which is represented as

$$E_d = d_{IR} - d_{sonic} \tag{2}$$

where d_{IR} is the distance to the neural wires measured by the infrared sensor, and d_{sonic} is that of the ultrasonic sensor. Generally, the distance value measured by the infrared sensor has high accuracy.

However it is very sensitive to the density of sun ray and is not so reliable. Therefore the sensor data fusion is applied for reducing the sun ray effects as follows:

$$\begin{cases} If\ |E_d| < \varepsilon,\ d_{out} = d_{sonic} + E_d (= d_{IR}) \\ otherwise\ d_{out} = d_{sonic} + \frac{\varepsilon}{2} \end{cases} \tag{3}$$

where d_{out} represents the determined distance from the sensor fusion, ε is the threshold value for the error. That is, when the error is less than ε, the distance measured by the infrared sensor is considered as a reliable one. Otherwise, the value is ignored and the output is determined by the reliable ultrasonic sensor data and the threshold value for the error. The threshold value, ε, is determined heuristically based on various experiments without disturbances.

3 Fuzzy Controller

Fuzzy controller became popular from 1970's by Prof. Mamdani [5]. It does not require plant equations for the system to be controlled. Therefore there are many applications for nonlinear systems where modeling of the plant is difficult or almost impossible [6]. Experts may provide some control rules as a mathematical model according to his own knowledge and experience, which are fuzzy rules.

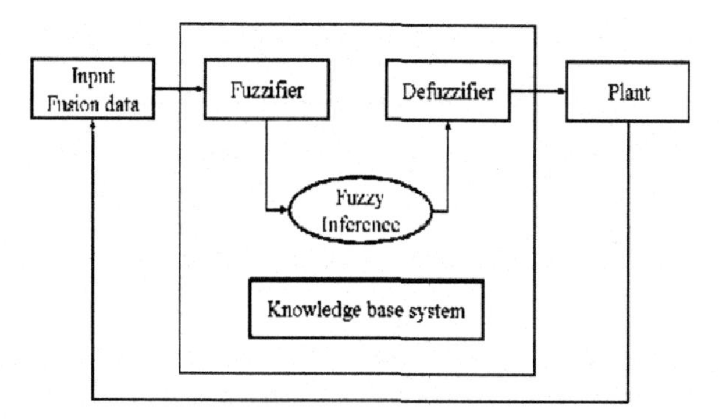

Fig. 2 Block diagram of a fuzzy controller

The fuzzy controller is composed of fuzzifier that transforms the measured values to fuzzy inputs, fuzzy inference part where the fuzzy outs are deduced from the fuzzy inputs, and defuzzifier that transforms the fuzzy outputs to applicable outputs. Fuzzy rules and membership functions can be derived from the knowledge base. A fuzzy controller is implemented in this research as shown in Fig. 2.

The inputs are the sensor outputs corresponding to the motor motion, which are normalized in the fuzzifier. The outputs of fuzzifier are passed through the fuzzy interference part and defuzzifier and fed to the inputs of plant as control signals.

3.1 Fuzzy Rules

The first step for implementation of a fuzzy controller is to set up fuzzy control rules. Initial fuzzy rules could be rough and those can be modified later to obtain better results. Therefore it is an important issue in the fuzzy controller design how to apply fuzzy rules to obtain desirable outputs as well as to set up control rules. The input variables of the power line inspection robot are the data from sensor fusion, which are represented by

$$\begin{aligned} e(k) &= (\text{reference}) - y(k) \\ ce(k) &= e(k) - e(k-1) \\ u(k) &= (\text{control input}) \end{aligned} \tag{4}$$

The control rules for fuzzy controller can be represented as [7, 8].

$$\begin{aligned} &R_1 : \text{If } e(k) = E_1 \text{ and } ce(k) = CE_1 \text{ then } u(k) = U_1 \\ &R_1 : \text{If } e(k) = E_2 \text{ and } ce(k) = CE_2 \text{ then } u(k) = U_2 \\ &R_1 : \text{If } e(k) = E_3 \text{ and } ce(k) = CE_3 \text{ then } u(k) = U_3 \\ &\qquad\vdots \\ &R_1 : \text{If } e(k) = E_n \text{ and } ce(k) = CE_n \text{ then } u(k) = U_n \end{aligned} \tag{5}$$

Table 1 Linguistic variables of fuzzy controller

Fuzzy Llabel	Meaning
PB	Positive Big
PS	Positive Small
ZE	Zero
NS	Negative Small
NB	Negative Big

Where $y(k)$ represents the system input, E, CE, and U are defined for the whole fuzzy space with fuzzy language labels of $e(k)$, $ce(k)$, and $u(k)$. For each input and output variables, fuzzy controllers can be represented by the fuzzy sets defined in Table 1.

Also fuzzy rules R_i represented in Eq. (5) can be expanded to the total spaces LD, RD, and U and represented as

$$R(E, CE, U) = R_1 U R_2 U \cdots U R_i U \cdots U R_n \quad (6)$$

Using membership function we can represent Eq. (6) as below.

$$\mu R(E, CE, U)(e, ce, u) = \max \left\{ \sum_{i=0}^{n} min\, [\mu E_i(e), \mu CE_i(ce), \mu U_i(u)] \right\} \quad (7)$$

where $\mu(\cdot)$ represents the membership function, n represents the number of fuzzy control rules, and $\mu E_i(e)$, $\mu CE_i(ce)$, and $\mu U_i(u)$represent the strength of fuzzy variables e, ce, and u to fuzzy functions E_i, CE_i, and U_i, respectively. And the fuzzy labels have the following set- relations:

$$E_i \in E, CE_i \in CE, U_i \in U \quad (8)$$

where E, CE and U are error, difference of error, and input space, respectively.

3.2 Fuzzification

To be used as inputs for fuzzy algorithm, membership functions are necessary for the measured variables. After applying the membership functions to the measured variables, the variables are fuzzy inputs. This process is fuzzification. The inputs are quantized and represented as discrete values before they are fed to membership functions that decide the strength of the inputs to the membership functions. There are various forms of membership functions. To make it easy, a discrete triangular form has been adopted in this paper.

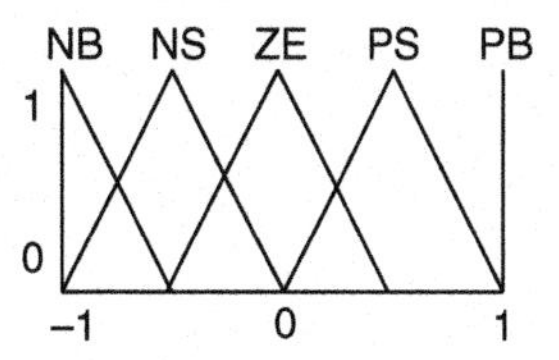

Fig. 3 Membership functions

Figure 3 illustrates the input membership function for this research. For the easy application of fuzzy logics and for the design independency of control rules, the membership functions are normalized.

3.3 Fuzzy Inference

To generate inputs for fuzzy controller, the measured inputs are applied to membership functions and properly inferred. The most popular min–max inference method is used, which is represented as

$$\mu U(u) = \max \left\{ \sum_{i=1}^{n} \min \left[\mu C_i(c), \mu CE_i(ce), \mu R_i(E, CE, U)(e, ce, u)\right] \right\} \tag{9}$$

3.4 Defuzzification

The fuzzified outputs of fuzzy inference cannot be applied to the controller directly. They need to be transformed back to non-fuzzy variables, which is defuzzification process. In this paper an average weight method is adopted, which is presented as

$$u_0(k) = \frac{\sum_{i=1}^{n} \mu U_i(u(k)) \cdot u^{*\,*}(k)}{\sum_{i=1}^{n} \mu U_i(u^*(k))} \tag{10}$$

where $u_0(k)$ is the control input after defuzzification, $u^*(k)$ represents a fuzzy control input that maximizes the following membership function by the ith rule, and $\mu U_i(u(k))$ is presented that membership function value of solving Eq. (9).

3.5 Knowledge Base

The knowledge base consists of data and rule bases. The data basis defines membership functions of input and output variables, rules for discrete representation, and normalization method. To determine the control rules, the experiences of expert are very important with some intelligent learning schemes.

4 Robot Navigation and Controller Design

4.1 Robot Navigation Principle

Robot navigation consists of four operations:

1. Detection of electric pole while it is moving along the neutral wire,
2. Grasping of the electric pole by the gripper,
3. Rotating the robot body to the new neutral wire to follow,
4. Grasping the neutral wire and following again.

To recognize the type of the electric poles, three ultrasonic sensors are attached at the bottom of the robot body. There are three types of electric poles as shown in Fig. 4. To move to the next neutral wire, after detecting the electric pole, the gripper needs to hold the electric pole. There two gripping modes (horizontal grip and vertical grip) for the three types of electric poles as shown in Fig. 5. To identify the stable grasp, photo sensors and limit switches are utilized.

While the gripper is holding the electric pole, the robot body is rotated to the new neutral wire to follow. To detect the neutral wire to hold by the robot body, two sets of ultrasonic and IR sensors are used to obtain a stable distance data. Figure 6 illustrates the cable detection and holding processes.

Fig. 4 Types of electric poles

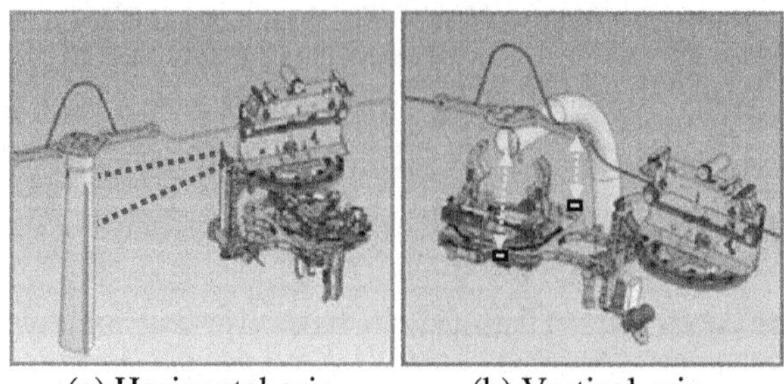

(a) Horizontal grip (b) Vertical grip

Fig. 5 Post detection and holding

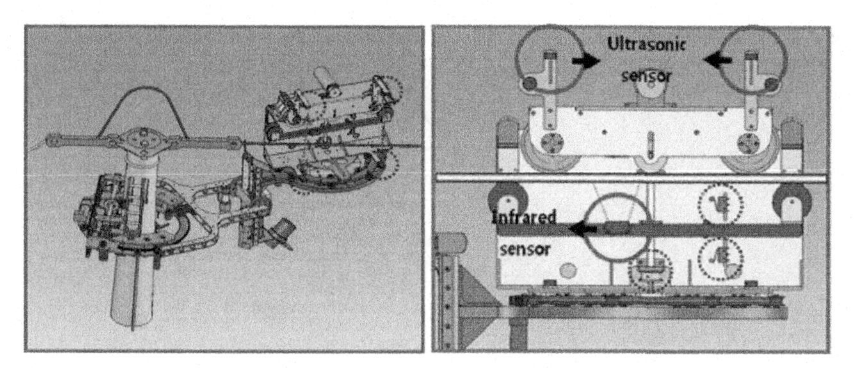

Fig. 6 Cable detection and holding

When the robot holds a new neutral wire, the gripper releases the electric pole and is folded back to the bottom of the robot body. This completes a sequence of electric pole passing operation. The inspection robot follows the neutral wire until it detects a new electric pole, while it is inspecting the power lines and insulators [9–12].

4.2 Controller Design

The control inputs are two distance data by the sensor fusion, which are used to detect the slope of the neutral wire. To make the robot parallel to the neutral wire to hold, the slope of the neutral wire *w.r.t* the robot is necessary. The control inputs are obtained as

$$\begin{aligned} d_1(k) &= \text{fused distance}(d_{out\_1}) \\ d_2(k) &= \text{fused distance}(d_{out\_2}) \\ \theta(k) &= \text{cable tilt}\left(\tan^{-1}\frac{|d_1 - d_2|}{d_L}\right) \end{aligned} \tag{11}$$

where d_1 and d_2 represent the distances between the robot and the neutral wire, and θ represents the slope of the neutral wire *w.r.t* the robot.

Figure 7 shows the membership functions for detecting tilt angle of the neutral wire. Upper two membership functions are defined for each set of the ultrasonic and IR sensors. And the lower membership function represents the membership function for the tilt angle of the neutral wire.

The 25 fuzzy control rules for inspection robot in a diverging point are summarized in Table 2, which controls the robot body to be parallel to the neutral wire to be held. To detect the existence of the neutral wire, NB (Negative Big), NS (Negative Small), ZE (Zero), PS (Positive Small), and PB (Positive Big), P (Positive), N (Negative), and Z (zero) are utilized as quantization variables. To make the robot body

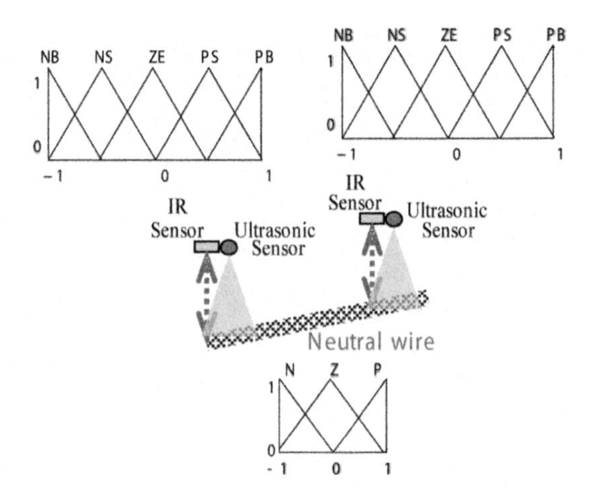

Fig. 7 Membership functions for tilt detecting of neutral wire

Table 2 Fuzzy control rules at a diverging point

	d_1	d_2	θ	action		d_1	d_2	θ	action
1	NB	NB	Z	BU	14	ZE	PS	P	TD
2	NB	NS	P	BU	15	ZE	PB	P	TL
3	NB	ZE	P	BU	16	PS	NB	N	TI
4	NB	PS	P	BU	17	PS	NS	N	TI
5	NB	PB	P	TL	18	PS	ZE	N	TI
6	NS	NB	N	TI	19	PS	PS	Z	BD
7	NS	NS	Z	BU	20	PS	PB	P	TL
8	NS	ZE	P	TD	21	PB	NB	N	TR
9	NS	PS	P	TD	22	PB	NS	N	TR
10	NS	PB	P	TL	23	PB	ZE	N	TR
11	ZE	NB	N	TI	24	PB	PS	N	TR
12	ZE	NS	N	TI	25	PB	PB	Z	GS
13	ZE	ZE	Z	CL					

and the neutral wire parallel, TL (Turn Left), TR (Turn Right), GS (Go Straight), BU (Body Up), BD (Body Down), TU (Tilt Up), and TD (Tilt Down) are utilized. When the parallel status achieved, CL (Cable Lock) function decides the robot to hold the wire.

Using the 25 fuzzy rules, the inspection robot performs three steps of actions: (1) Detecting the neutral wire, (2) Control of the tilt angle to match the robot and the neutral wire to be held together, and (3) To combine the robot body with the neutral wire.

5 Experiments and Results

5.1 Experimental Environment and Equipments

The experimental system consists of a mobile inspection robot, the controller, and a remote controller. The controller based on dspic30fxx series processor has a main control, sensing, and motor control parts. The sensor part processes various sensor data from ultrasonic, IR, and tilt sensors, and transmits the data to the main processor. The main control part gathers the sensor data, limit-switch signals, and motor encoders, and checks the operation status of the inspection robot to generate and send the next operation command to the motor control part. The motor control part governs the motor operation based on the input torque values, and it sends the control results gathered by the encoder back to the main control part.

5.2 Experiments for Sensor Fusion

The sensor fusion experiments are performed for the holding operation of the neutral wire by the inspection robot, where the distance measures by the two sets of sensors should be the same to provide a suitable holding condition. It is shown by experiments that by the sensor fusion the distance measuring error could be reduced.

Figure 8 shows the distance measurement results by IR sensor and ultrasonic sensor for the three real distance values of 20, 26, and 30 cm.

Figure 9 illustrates the distance measurement error of each sensor. The IR sensor has less than 5 mm in the five measurements. However the ultrasonic sensor has relatively large error at the location where the neural line exists because of its outward-conic wave propagation property. When there is 21~25 cm gap between the mobile

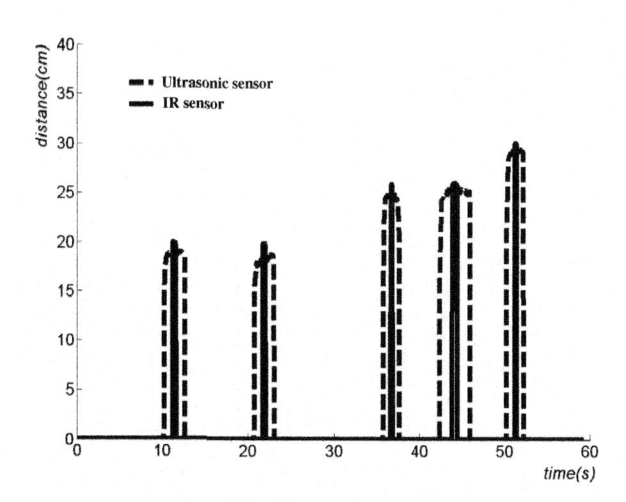

Fig. 8 Distance output of IR sensor and ultrasonic sensor

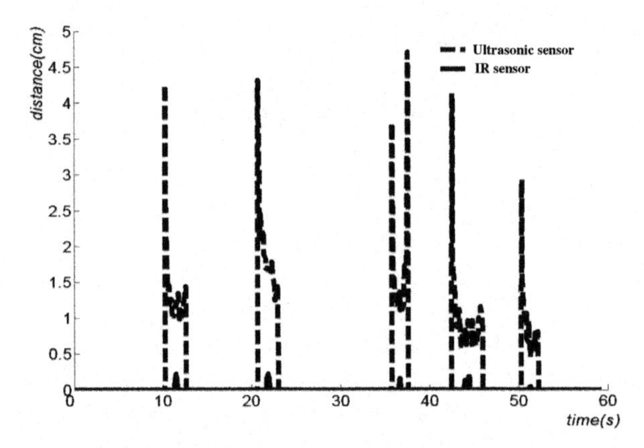

Fig. 9 Distance error

robot and the neutral wire, the maximum allowable error not to cause false operation was less than 2 cm in this research.

Figure 10 represents the finally determined distance value by the sensor fusion.

The error values are represented in Fig. 11, which are the differences to the real values for the five measurements. The maximum error could be kept less than 5 mm which can be acceptable for the inspection robot while it is passing over the electric poles.

5.3 Fuzzy Control Experiments for Sensor Fusion

To evaluate the performance of the proposed Fuzzy controller, the position tracking experiments are performed and compared to the PID controller. As it is shown in

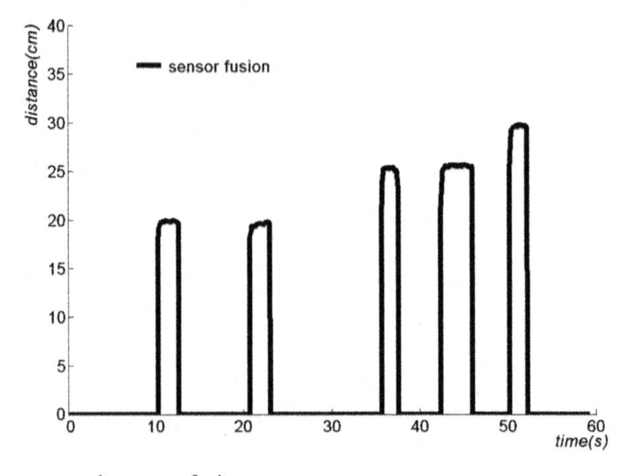

Fig. 10 Distance output by sensor fusion

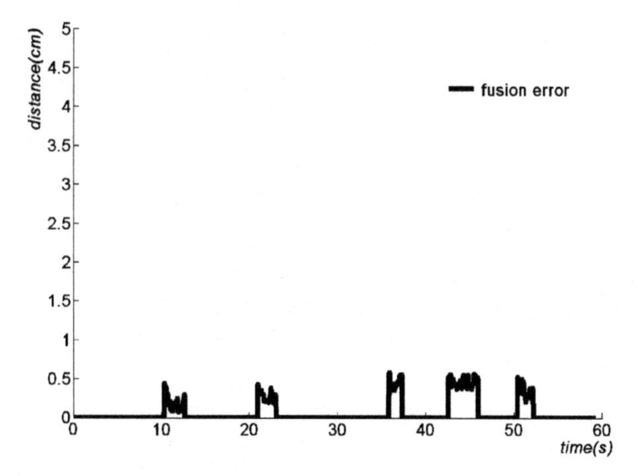

Fig. 11 Measurement error after sensor fusion

Fig. 9, when the left side distance, d_1, and right side distance d_2, become the same, the robot body becomes parallel to the neutral wire to be held.

Figure 12 illustrates the parallel matching (to make d_1 equal to d_2) time for each control algorithms of PID and the proposed Fuzzy controller. The necessary time to make the robot body to be parallel to the neutral wire is 1.5 s shorter than that of PID controller, the proposed Fuzzy controller.

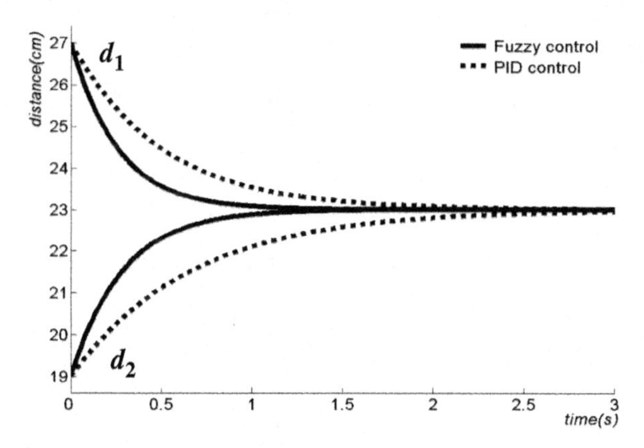

Fig. 12 Comparison of PID control with Fuzzy control

6 Conclusion

In this paper, a stable holding of the neutral wire by the power line inspection robot has been implemented by applying sensor fusion techniques that take advantages of both ultrasonic and IR sensors. The effectiveness of the proposed algorithm has been proven by real experiments to apply for the power line inspection robot. There are many researches to keep the power line stable, which is a vein for power industry. However there is not any specific commercial product yet.

The power line inspection robot is a new trial to save human workers and money to keep the stable power line. As a future research, a navigator with self-learning control algorithm is necessary to guarantee the robustness against environment.

References

1. Cho, Y. Statistic Prediction Model of Power Line Corona, *Journal of Korean Institute of Electrical Engineer*, 35(5), 290–294, 1986.
2. Cho, Y. Analysis of Radio Frequency Noises Caused by the Transmission Line Corona, *Journal of Korean Institute of Electrical Engineer*, 35(1), 5–10, 1986.
3. Kim, S.-D., and Lee, S.-H. Development of Deterioration Diagnosis System for Aged ACSR-OC, *Journal of the Korean Institute of Illuminating and Electrical Installation Engineers*, 14(6), 43–50, 2000.
4. Vazquez, J., and Malcolm, C. Fusion of Triangulated Sonar Plus Infrared Sensing for Localization and Mapping, *International Conference on Control and Automation*, Budapest, Hungary, June 2005.
5. Zadeh, L.A. Fuzzy Sets, *Information and control*, 8, 338–353, 1965.
6. Li, Y.F., and Lau, C.C. Development of Fuzzy Algorithms for Servo Systems, *IEEE International conference on Robotics and Automation*, April 1988.
7. Smith, S.M. and Comer, D.J. Automated Calibration of a Fuzzy Logic Controller Using a Cell State Space Algorithm, *IEEE Control Systems Magazine*, 2(5), 18–28, 1991.
8. Thongchai, S., and Kawamura, K. Application of Fuzzy Control to a Sonar-based Obstacle Avoidance Mobile Robot, *IEEE International Conference on Control Applications*, September 2000.
9. Sawada, J., Kusurnoto, K., Maikava, Y., Munakata, T., Ishikawa, Y. A Mobile Robot for Inspection of Power Transmission Lines, *IEEE Transactions on Power Delivery*, 6(1), 309–315, 1991.
10. Santamaria, A., Aracil, R., Tuduri, A., Martinez, P., Val, F., Penin, L.F., Ferre, M., Pinto, E., and Barrientos, A. Teleoperated Robots for Live Power Lines Maintenance(ROBTET), *14th International Conference and Exhibition on*, Vol. 3, pp. 311–315, 1997.
11. Souza, A., Moscato, L., Santos, M., Filho, W., Ferreia, G., Ventrella, A. Inspection Robot For High-Voltage Transmission Lines, *ABCM Symposium Series in Mechatronics*, Vol. 1, pp. 1–7, 2004.
12. Gu, J., Meng, M., Cook, A., and Liu, P.X. Sensor Fusion in Mobile Robot: Some Perspectives, *Proceedings of the 4th World Congress on Intelligent Control and Automation*, P. R. China, June 2002.

Intelligent Glasses: A Multimodal Interface for Data Communication to the Visually Impaired

Edwige Pissaloux, Ramiro Velázquez and Flavien Maingreaud

Abstract This paper introduces the concept of bimodal visuo-tactile interface, named Intelligent Glasses (IG), designed for assistance of the visually impaired in their daily tasks such as access to information and mobility. The IG architecture is outlined and its preliminary evaluation through original experiments with healthy blindfolded subjects is provided. The collected results show that the IG system can be used as a support for the considered tasks.

1 Introduction

Access to information, via reading, shape recognition and mobility are fundamental tasks for interacting with the external environment. Unfortunately, the neuro-cognitive processes which underlie these tasks are far from being identified neither well understood. Therefore, the existing technological assistances for the blind, the visually impaired and the elderly provide a very limited support for the effective execution of such tasks.

Indeed, the existing assistive systems offer some support but for separate tasks. Braille keyboards allow to communicate with PCs for one line data exchanges (not for window manipulation); Braille city maps (Fig. 1), scattered around towns are supposed to help the visually impaired to find their way but, in fact, they serve mainly as a good visual reference for the sighted. As with city maps for the sighted, maps for the visually impaired are arbitrary oriented and during navigation the user has no assistance for finding his/her current position in the city and in the map.

The oldest "assistances" for mobility of the visually impaired are the white cane and the guide dog; both have several limitations. The main one is the fact that they do not provide information on all obstacles (overhanging and dynamic, for example), neither some elementary data on the user's near but global environment. That is why

E. Pissaloux (✉)
Institut des Systèmes Intelligents et de Robotique (ISIR), Université Pierre et Marie Curie (France), Paris, France
e-mail: pissaloux@robot.jussieu.fr

S. Lee et al. (eds.), *Multisensor Fusion and Integration for Intelligent Systems*, 349
Lecture Notes in Electrical Engineering 35, DOI 10.1007/978-3-540-89859-7_24,

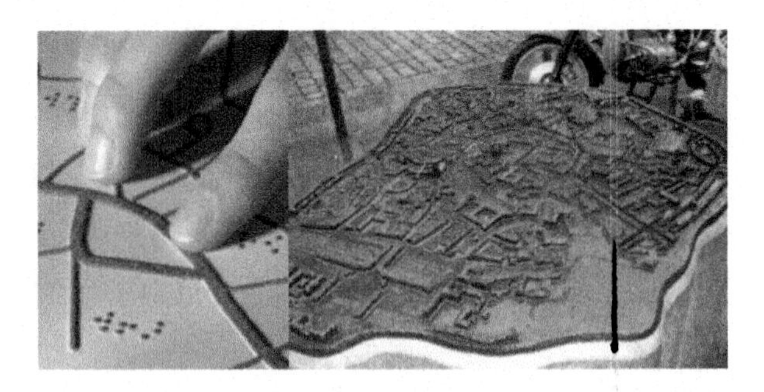

Fig. 1 Two examples of Braille maps

electronic assistances (ETA: electronic travel aids) using several sensors, such as Mowat's sensor, Borensteins' robotic guide cane, Farcy's TomPouce and Télétact, Nottingham Obstacle Detector, Key's SonicGuide or SonicGlasses [1–5] are not widely accepted by the targeted population. All of them provide point-wise feedback (tactile of audio) and require the environment's spatial scanning followed by cognitive (mental) integration to localise all obstacles in the nearby space. Furthermore, these assistances do not support the subject's nearest space global representation, which is time-variant in space, so they do not allow to understand the near local space and to establish the navigation strategy neither to anticipate dynamic events and the evolution of the environment while walking.

The Intelligent Glasses (IG) system is an alternative solution that provides information on the time-variant near space. We present a first bimodal device that explores vision and touch through a stereo vision perception system and a dedicated touch stimulating Braille-like surface.

The rest of the paper is organized as follows. Section 2 outlines the IG system design; Sect. 3 presents its preliminary evaluation while Sect. 4 concludes the paper with the IG future work perspectives.

2 Intelligent Glasses System Overview

The IG system consists of two synergetic elements: a visual perception system and a touch stimulating Braille-like surface (TactiPad) (Fig. 2). They will be presented and discussed in the following subsections once the IG neuro-cognitive bases are addressed.

2.1 The IG System Neurocognitive Basis

The IG system aims to provide a support for external world presentation useful for mobility of the visually impaired. The external world observation strategy elaborated by humans since their childhood induces a partition of the space in two zones: obstacles and obstacles-free zones (Fig. 3(a)). These two zones can be binary encoded and displayed on a 2D surface as shown on fig. 3(b). However, as this parti-

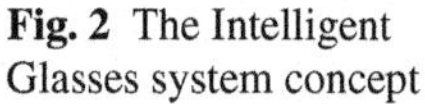

Fig. 2 The Intelligent Glasses system concept

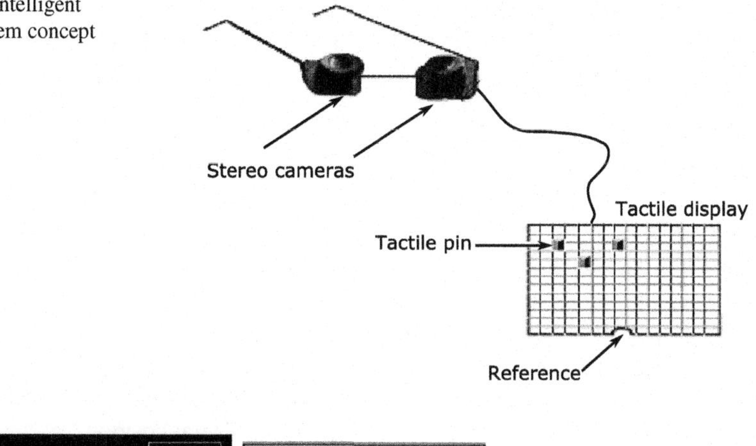

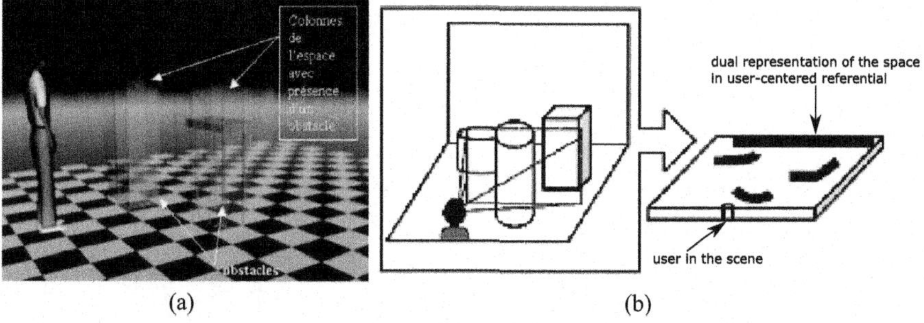

(a) (b)

Fig. 3 (**a**) Subject observing a 3D scene and (**b**) encoding of a 3D scene orthographic projection as nearest obstacles nearest boarders

tion depends on the observer's gaze direction and current position, it is necessary to materialize this observation point by using a notch on one of the boarders of the 2D surface (Fig. 3(b) left). This observation point is indeed the origin of the Euclidean reference frame of the 3D space orthogonally projected on 2D surface (cf. Fig. 7). Consequently, the IG system implements the whole process of such display: two images (convergent stereo) are acquired with a visual perception system and 3D scene orthogonal presentation (nearest obstacles nearest boarders) is displayed on a 2D surface.

Fig. 4 The Intelligent Glasses operation principle

The IG operation principle is shown in Fig. 4. This process exploits the fact that the mechanoreceptors on fingertips are sensitive to gradient of data, i.e. to edges or boarders (frames) of figures.

2.2 Visual Perception System

The choice of a convergent stereo vision system is mandatory; such stereo rig increases the size of the solid angle subtended by two cameras (Fig. 5 case 2, [6]).

A couple of portable stereoscopic epipolar cameras in conjunction with an inertial platform (Fig. 6) allow to acquire the orthogonal projection of the observed scene. The system permits to explore the global environment with different view points and its orthogonal projection is displayed on the tactile surface. This display varies with the head movements' and with user's spatial position.

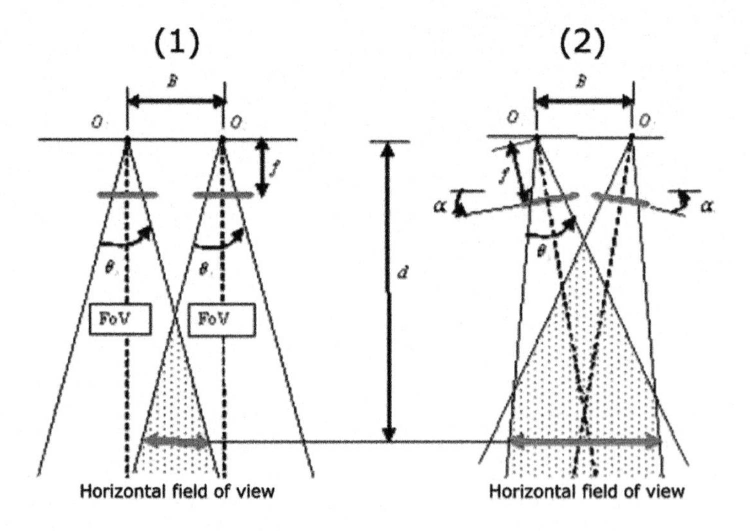

Fig. 5 Solid angle size induced by convergent and non-convergent stereo rig

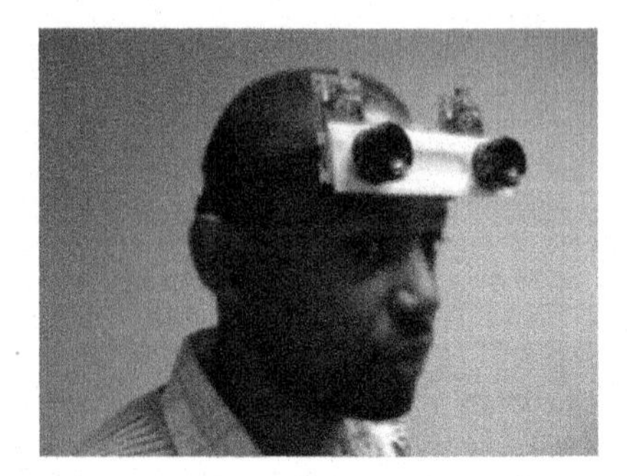

Fig. 6 The IG system visual perception portable subsystem

An obstacle in the peri-personal space can be localized via vision processing applied to stereovision epipolar rectified images. The system is initially calibrated by an original genetic algorithm [7] and then updated with the RANSAC algorithm [8]. Binocular vision principles allow the extraction of the distance to obstacles (in the cameras referential). Information given by the inertial platform allows to transform camera referenced data to gravity referenced information (human body usual referential). Therefore, it is possible to establish a navigation suitable space partition (Fig. 3(a)) which will be displayed on a touch stimulation surface.

2.3 Touch Stimulation Surface – TactiPad

A 2D matrix of taxels, ie. tactile elements or Braille points, could be a support for space partition dynamic display. An obstacle, for example a wall, is represented by a convenient taxel position raised ("1" coded), and its absence by taxel down position ("0" coded) (Fig. 7 the raised taxel (i,j) corresponds to an obstacle). Therefore, a binary code provided by a touch stimulating display could be manually explored for cognitive interpretation.

It should be noted that the proposed tactile interface provides simultaneously ego- and allo- centred representations of the user's nearest space (because of its reference point: a notch on the TactiPad boarder), thus allowing the estimation of both distance to obstacles (ego-centred) and distance between obstacles (allo-centred).

Since 1960, there have been many initiatives to develop an interface that could accurately represent the input information. Most relevant examples of these devices are the Optacon [9], the TVSS [10] (Tactile Vision Substitution System), the Dot Matrix [11], the VDT TVSS [12] and the Itacti [13]. Typical displays involve arrays of upward/downward moveable pins or taxels as skin indentation mechanisms. Tactile exploration concerns slight motions of the fingertips or palm over the raised pins. Actuation techniques already explored include servomotors [14], piezoceramics [15], pneumatics [16], shape memory alloys (SMA) [17–19] and fluids

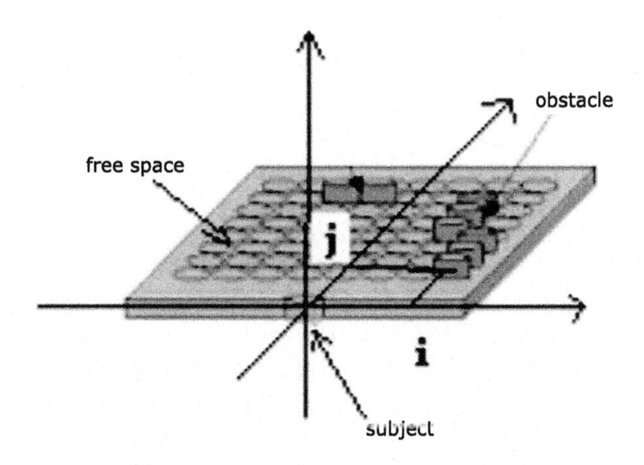

Fig. 7 TactiPad and its referential

technologies [20]. Each of these approaches has its own advantages and disadvantages, which finally determine the spatial and temporal resolution of the tactile stimuli. Recent progresses in the SMA actuation principle [19, 21], its simplicity, compactness, high power/weight ratio, clean and silent operation have allowed to design and realize actuators of appropriate dimensions to achieve a realistic sense of touch. Moreover, they offer high integration capacity in a small surface, the possibility of independent control, an extremely high fatigue resistance, market availability and low cost.

Fig. 8(c) shows the actuator integration in the TactiPad Braille-like surface designed at ISIR, Paris 6 University (UPMC). The TactiPad is a Braille matrix of 8 by 8 SMA-based taxels [19] having a refreshable rate suitable for reading, shape recognition and mobility of visually impaired.

The possibility and quality of the space integration from tactile snapshots has to be evaluated via dedicated virtual navigation experiments.

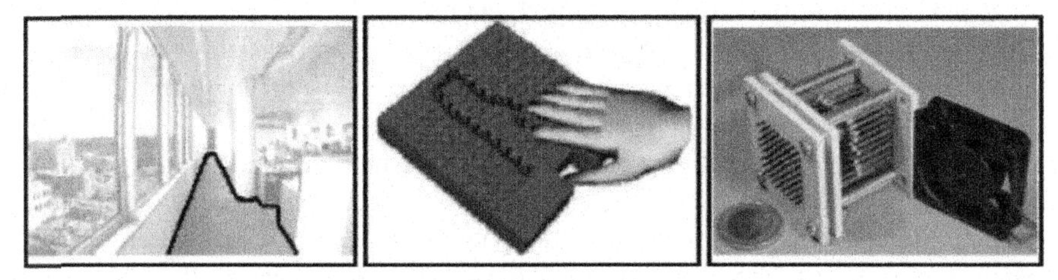

Fig. 8 (**a**) Information about the environment obtained with cameras and image processing, (**b**) the tactile representation of the environment, (**c**) the used tactile device

3 IG System Experimental Evaluation

Three experiments have been led with healthy sighted blindfolded voluntary subjects. In these experiments, we have used only the TactiPad as tactile percept formation is the basis of IG system. In Experiments 1 and 2 subjects have been seated in front of the TactiPad device, while they have carried the TactiPad in place closed to their gravity center in the Experiment 3.

3.1 Experiment 1: Static form Tactile Perception

The perception of static forms displayed on the tactile surface has been the goal of this experiment. We wanted to identify if there are (1) preferred geometric shapes in tactile perceptive modality (line, square, circle, and arrow), (2) preferred tactile representation (wired-frame (Fig. 9) or full); (3) the best shape size: large (6 taxels and more), medium (3–5 taxels) and small (1–2 taxels).

The collected results show that line segments and triangle have been recognized by almost all subjects; square and circle were frequently confused because

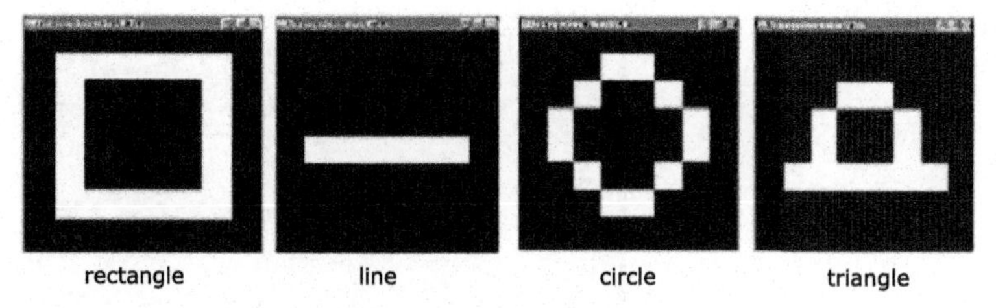

Fig. 9 Tested shapes for tactile perception: simple static geometric framed shapes

of TactiPad spatial resolution (intertaxel distance is of 2.54 mm). Indeed, many subjects complain about tactile surface too weak resolution. The recognition of filled forms is rather difficult. This confirms the fact that our fingertips mechanoreceptors react to gradient of data (boarders of shapes).

3.2 Experiment 2: Form and Moving Direction Tactile Perception

This experiment collected data in order to evaluate whether or not the displayed moving shape can induce the moving stimuli direction. Two shapes have been tested (Fig. 10): a line segment and an arrow (because its role for visual mobility). These shapes "moved" in orthogonal directions (NEWS: north, east, west and south).

Data collected during this experiment show that it is possible to recognize a direction, but the recognition process is shape insensitive. Almost all subjects have successfully recognized moving direction of the line segments, but very few of them recognized the moving direction of an arrow.

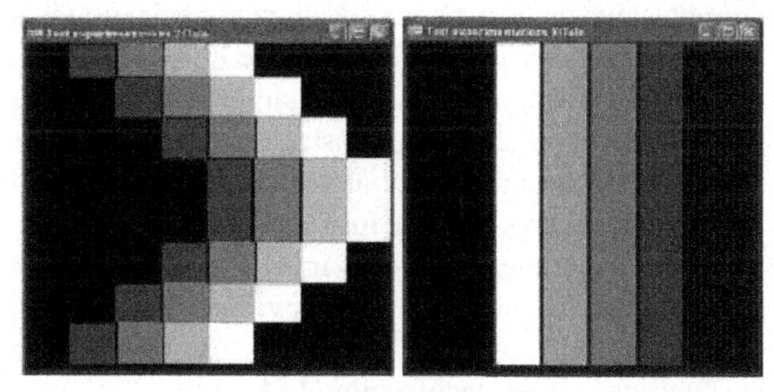

Fig. 10 Moving geometric shapes: arrow (left) and line segment (right)

3.3 Experiment 3: Navigation in 3D Space with Tactile Interface

Figure 11 shows navigation in the space using the IG device (natural visuo-tactile SLAM). The TactiPad is carried on by the end user at his height. The user navigates on the platform where several obstacles (paper boxes) were randomly placed. An

Fig. 11 Validation of the IG system for mobility

assistant (young women) supervises the navigation process and observes the obstacles avoidance procedure.

From the data collected during the Experiment 3, it is possible to conclude that (1) it is possible to perceive the space organization via its tactile representation; (2) obstacle edge representation can be appropriate for a space binary representation (obstacles - obstacle free space) ; (3) it is possible to integrate a space representation to a navigation tool.

4 Conclusion

This paper has introduced the concept of bimodal -visuo-tactile- interface and has briefly presented its implementation. The design of the IG system is based upon our present status on neuro-cognitive human navigation in space.

Future works should integrate more efficiently human navigation strategies and improve the TactiPad resolution. Larger classification of obstacles should be integrated as different classes of obstacles require different anticipation and avoidance strategies. A concept of natural visuo-tactile SLAM should be investigated in depth and new algorithms for data fusion built-in.

Future experiments have to involve visually impaired and seniors and should be performed in more complex environments. Association of vision system into experiments will validate the concept of visuo-tactile man-environment device and the concept to visuo-tactile natural SLAM.

Acknowledgments The authors would like to thank Prof. Alain Berthoz from Collège de France, the CNRS/France, Conacyt/Mexico and CEA/France for financial support of the IG project.

References

1. Maingreaud, F., Pissaloux, E., Velazquez, R., Hafez, M., and Fontaine, E. A Dynamic Tactile Map as a Tool for Space Organization Perception: Application to the Design of an Electronic Travel Aid for Visually Impaired and Blind People, IEEE EMBC (2005).

2. Maingreaud, F. Une contribution à la perception de l'espace et son intégration dans une aide pour la navigation des déficients visuels, PhD thesis, U. Paris 6 (2005).
3. Pissaloux, E., and Maingreaud, F. Déplacement dans l'espace: quelles aides technologiques, Journal de Basse Vision, Mars/Avril, pp. 22–26 (2007).
4. Velazquez, R., Maingreaud F. and Pissaloux, E. Intelligent Glasses: A New Man-Machine Interface Concept Integrating Computer Vision and Human Tactile Perception, in Proceedings of EuroHaptics 2003, Dublin, Ireland, pp. 456–460 (2003).
5. Velazquez, R. Contribution à la Conception et à la Réalisation d'Interfaces Tactiles Portables pour les Déficients Visuels, PhD thesis, U. Paris 6 (2006).
6. Bergame, F. Conception et Prototypage d'un Capteur de Vision Stéréoscopique Portable Pour le Recouvrement Précis de Profondeur d'une Scène 3D. Application au Systme "Lunettes Intelligentes", mémoire CNAM, 8 février (2008).
7. Maingreaud, F., and Pissaloux, E. Stereo Rig Calibration via a Genetic Algorithm, IEEE EUROCON2007 (International Conference on Computer as a tool), Varsovie, September 9–13, (CD) (2007).
8. Chen, Li. Conception et Mise en Oeuvre de l'Appariement Spatial d'Images Stéréo Par la Méthode de Ransac, Master Thesis, U. Paris 6 (2006).
9. Bliss, J. Dynamic Tactile Displays in Man-Machine Systems. IEEE Transactions on Man-Machine Systems. (Special issue: Tactile displays conference), Vol. 11, No. 1 (1970).
10. Bach-y-Rita, P., Collins, C., Saunders, F., White, F., and Scadden, L. Vision Substitution by Tactile Image Projection. Nature, Vol. 221, No. 963 (1969).
11. METEC GmbH 65a, D - 70176, Stuttgart, Germany. http://www.metec-ag.de
12. Thorsten, M., Karlheinz, M., and Johannes, S. The Heidelberg Tactile Vision Substitution System, in 7th International Conference on Computer Helping People with Special Needs, Karlsruhe, Germany 256–262 (2000).
13. EU project, IST-2001-32240 (2001–2003).
14. Wagner, C., Lederman, S., and Howe R. A Tactile Shape Display Using RC Servomotors, 10th Symposium on Haptic Interfaces for Virtual Environment and Teleoperator Systems, Orlando USA (2000).
15. Pasquero, J., and Hayward, V. STReSS: A Practical Tactile Display System with One Millimeter Spatial Resolution and 700 Hz Refresh Rate, EuroHaptics 2003, Dublin, Ireland, pp. 94–110 (2003).
16. Moy, G., Wagner, C., and Fearing, R. A Compliant Tactile Display for Teletaction, IEEE International Conference on Robotics and Automation, San Francisco, USA, April (2000).
17. Ikuta, K. Micro/Miniature Shape Memory Alloy Actuator, in Proceedings of IEEE International Conference on Robotics and Automation, Cincinnati, USA, pp. 2156–2161 (1990).
18. Taylor, P., Moser, A., and Creed, A. A 64 Element Tactile Display Using Shape Memory Alloy Wires, Elsevier Science B.V, pp. 163–168 (1998).
19. Velazquez, R., Pissaloux, E., Hafez, M., and Szewczyk, J. Tactile Rendering with Shape Memory Alloy Pin-Matrix, IEEE Transactions on Instrumentation and Measurement, Vol. 57, No. 5, pp. 1051–1057 (2008).
20. Kelaney G., and Cutkosky, M. Electrorheological Fluid Based Robotic Fingers with Tactile Sensing. IEEE International Conference on Robotics and Automation, Scottsdale AR, pp. 132–136 (1989).
21. Velazquez, R., Pissaloux, E., Hafez, M., and Szewczyk, J. Miniature Shape Memory Alloy Actuator for Tactile Binary Information Display, IEEE ICRA (2005).

Fourier Density Approximation for Belief Propagation in Wireless Sensor Networks

Chongning Na, Hui Wang, Dragan Obradovic and Uwe D. Hanebeck

Abstract Many distributed inference problems in wireless sensor networks can be represented by probabilistic graphical models, where belief propagation, an iterative message passing algorithm provides a promising solution. In order to make the algorithm efficient and accurate, messages which carry the belief information from one node to the others should be formulated in an appropriate format. This paper presents two belief propagation algorithms where non-linear and non-Gaussian beliefs are approximated by Fourier density approximations, which significantly reduces power consumptions in the belief computation and transmission. We use self-localization in wireless sensor networks as an example to illustrate the performance of this method.

Keywords Density approximation · Belief propagation · Distributed inference · Wireless sensor network

1 Introduction

DVANCES in sensor technology and telecommunications make wireless sensor network (WSN) an appropriate solution for a wide variety of applications [1, 2]. In a WSN, sensor nodes are spatially distributed to monitor the physical or environmental states. Information can be exchanged through the wireless channel so that the whole network works in a cooperative fashion. Many estimation problems in WSNs can be represented by probabilistic graphical models and solved by belief propagation methods. Belief propagation (BP) is an iterative message passing algorithm in which each node calculates its belief about other nodes and communicates with them to exchange their beliefs about each other. Compact messages that are transmitted between nodes carry the necessary information of the beliefs, based

C. Na (✉)
Siemens AG, Corporate Technology, Information and Communications, Otto-Hahn-Ring 6, 81739, Munich, Germany
e-mail: na.chongning.ext@siemens.com

S. Lee et al. (eds.), *Multisensor Fusion and Integration for Intelligent Systems*, Lecture Notes in Electrical Engineering 35, DOI 10.1007/978-3-540-89859-7_25,

on which the receiver can reconstruct the transmitter's belief about it. For discrete beliefs, messages can be a short vector of probabilities. For continuous beliefs with Gaussian distribution, it is enough to ensemble the mean and variance in the message. However, in many applications, beliefs have non-linear and non-Gaussian distributions so that belief calculation and transmission consumes a lot of power. That limits its application in WSNs which have strong power constraints. Hence, an appropriate representation of beliefs which reduces the complexity while keeping the accuracy is necessary but non-trivial.

Monte Carlo methods can be used where messages contain samples that are drawn from the distribution to represent the beliefs. Gibbs sampling is a popular method in this case. However, this is only possible for sufficiently small networks. Authors of [3] used non-parametric BP method where beliefs are represented by Gaussian mixtures. It generalizes particle filtering for inference in non-linear, non-Gaussian time series.

In this paper, we introduce Fourier density approximation (FDA) method to represent the beliefs. Fourier series were first employed to estimate probability densities in [4]. Recently, [5] and [6] ensured the non-negativity of Fourier series by approximating the square root of the density instead of the density itself. The usage of Fourier series in nonlinear Bayesian filtering is also derived in [5] and [6]. Using Fourier density approximation, the belief can be represented sufficiently by only a small number of Fourier coefficients. Hence, the transmission power and time between sensor nodes are significantly saved. Compared to other density representations like Gaussian mixture or Monte Carlo methods, the optimal number of coefficients under a required approximation error with respect to a density distance metric is more efficiently obtained. Furthermore, the sum-product operations in BP algorithms can be more effectively calculated in Fourier domain since some convolution-like integral operations are more easily calculated than in space domain. Since the Fourier series are orthogonal expansions, the coefficients are derived independently and effectively [5]. In practice, this is done by efficient Fast Fourier Transform (FFT).

In this paper, the self-localization in WSNs, a common practice of brief propagation, is used to evaluate the performance of Fourier density approximation. Two Fourier based algorithms are proposed, which are simplified transmission based on Fourier density approximation (ST-FDA) and simplified computation and transmission based on Fourier density approximation (SCT-FDA). ST-FDA reduces the size of the belief message to save radio transmission power, which is a critical factor for WSNs. SCT-FDA further simplifies the sum-product algorithm (SPA) to reduce computation power.

The paper is organized as follows. Section 2 presents BP as a general approach to the inference problems in WSNs. Fourier density approximation method will be introduced in Sect. 3. Section 4 uses a sensor self-calibration example to illustrate the use of Fourier density approximation for BP. ST-FDA and SCT-FDA algorithms are proposed. Their performances will be evaluated through simulation and the results will be shown in Sect. 5. Finally, Sect. 6 concludes the paper.

2 Belief Propagation in Wireless Sensor Networks

This chapter presents the general probabilistice inference problem in sensor networks and shows that belief propagation is a suitable solution.

2.1 Probabilistic Model of a Wireless Sensor Network

Lets consider a WSN with sensor nodes that are distributed in space. We use x_i to denote the physical state associated with sensor node i and use $\mathbf{x}$ to denote the collection of state variables at all sensor nodes. Each sensor makes a local noisy observation which we denote by y_i. In general, the following assumptions are valid:

1. Given the state variables, observations at different nodes are independent, i.e. $p(y_i, y_j | \mathbf{x}) = p(y_i | \mathbf{x}) p(y_j | \mathbf{x})$.
2. Observation at one node depends only on a subset of state variables, i.e. $p(y_i | \mathbf{x}) = p(y_i | \mathbf{x}_{Pa(y_i)})$ with $\{ \mathbf{x}_{Pa(y_i)} \} \subset \{ \mathbf{x} \}$.
3. Usually, local correlation exists between neighboring nodes. This indicates that the joint probability of state variables can be factorized into a product of local functions which present the correlation among the nodes in neighborhoods, i.e. $p(\mathbf{x}) = \prod_c p(\mathbf{x}_c)$.

Based on these assumptions and using the Bayes rule, the joint distribution of state variables and observations can be factorized in the following form:

$$\begin{aligned} p(\mathbf{x}, \mathbf{y}) &= p(\mathbf{y} | \mathbf{x}) p(\mathbf{x}) = \left(\prod_{i=1}^{N} p(y_i | \mathbf{x}) \right) p(\mathbf{x}) \\ &= \left(\prod_{i=1}^{N} p(y_i | \mathbf{x}_{Pa(y_i)}) \right) p(\mathbf{x}) = \left(\prod_{i=1}^{N} p(y_i | \mathbf{x}_{Pa(y_i)}) \right) \prod_{c} p(\mathbf{x}_c) \,. \end{aligned} \quad (1)$$

The conditional independences encoded in (1) can be presented by a graphical model, e.g. Markov random field [7]. A graphical model consists of a set of vertices which represent the variables. There exists an edge between two vertices which indicates the conditional dependence between them. So the whole graph represents the factorization of a joint distribution of all variables. The relationship between the graphical model and the joint distribution is given by the Hammersley-Clifford theorem [8], that is, a joint probability can be written as a product of potential functions which are defined on cliques (sub-graphs that are fully connected). In probabilistic inference in WSNs, we'd like to write this factorization as:

$$p(\mathbf{x},\ \mathbf{y}) = \prod_{i=1}^{N} \varphi_i(y_i|\ \mathbf{x}_{c_i}) \tag{2}$$

so that each factor in Eq. (2) can be associated with one sensor node. Such a factorization automatically provides the possibility to distribute the computation. Each node processes parts of the total computation and results are eventually disseminated through the communication between nodes.

Each potential function in Eq. (2) is obtained from Eq. (1). We first assign $p(y_i|\ \mathbf{x}_{Pa(y_i)})$ as a factor of $\varphi_i(y_i|\ \mathbf{x}_{c_i})$, then distribute each factors in $\prod_c p(\ \mathbf{x}_c)$ into one of the potential functions. In many applications, the distribution of $\prod_c p(\ \mathbf{x}_c)$ is not unique. For the assignment, we should also take factors such as computational complexity, communication connectivity and transmission power into consideration. Authors of [9] have introduced a method that first constructs a spanning tree and then assigns factors to the nodes of the tree. Such an assignment eventually results in a junction tree that can be solved by message passing algorithms [10]. In some other applications, the final graphical model is a graph with loops.

2.2 Belief Propagation in Wireless Sensor Networks

Inference of the variables defined on a graphical model has been intensively studied. For a graph without loops, this can be solved by junction tree algorithm. Exact inference on a graph with loops is generally an N-P hard problem. Approximate methods, such as loopy BP [11] have produced convictive results in many applications. BP is an iterative message passing algorithm in which each node calculates its belief about other nodes and communicates with them to exchange their beliefs about each other. Each node updates its beliefs when it receives messages from other nodes. Updated beliefs will be sent in messages to other nodes. This procedure repeats for a number of iterations or until a defined convergence criterion has been met.

In WSN applications, we are interested in the posterior probability of $p(x_i|\ \mathbf{y})$ for each state variable x_i. Such an inference problem on graphical models can be solved by using sum-product algorithm, which is a common practice [12].

Having defined the local potentials for each node like in (2), we can write the analytic formula for the belief updating in SPA at each sensor node. We define $m_{ij}^t(\ \mathbf{x}_{c_i} \cap\ \mathbf{x}_{c_i})$ to be the message sent from node i to node j in the tth iteration. Having received messages from all neighboring nodes in the tth iteration, node i calculates the message to be sent to node j for the $t+1^{\text{th}}$ iteration by:

$$m_{ij}^{t+1}(\ \mathbf{x}_{c_i} \cap\ \mathbf{x}_{c_i}) = \alpha \cdot \int_{\{\ \mathbf{x}_{c_i}\}\setminus\{\ \mathbf{x}_{c_j}\}} \varphi_i(y_i|\ \mathbf{x}_{c_i}) \prod_{k\in N(i)\setminus j} m_{ki}^t(\ \mathbf{x}_{c_i} \cap\ \mathbf{x}_{c_k}) \tag{3}$$

where α is a constant value to normalize the message. $N(i)$ denotes the neighbors of node i. At node i, we can also conclude the marginal probability of the variables in $\varphi_i(y_i|\,\mathbf{x}_{c_i})$. This is done by combining all the incoming messages with its local potential:

$$\hat{p}^t(\mathbf{x}_{c_i}) = \alpha \cdot \varphi_i(y_i|\,\mathbf{x}_{c_i}) \prod_{k \in N(i)} m_{ki}^t(\mathbf{x}_{c_i} \cap \mathbf{x}_{c_k}) \tag{4}$$

2.3 *Form of the Messages*

The computation in SPA is relatively simple if the messages and potential functions involved in (3) are discrete or Gaussion. However, in many cases, the local potential functions have a very complex non-Gaussian distribution and there exist high non-linear relationships between the variables. Discretizing the continuous functions (uniform sampling method) would be too expensive for many inference problems. Other forms of representation of the belief functions are needed.

A particle-based method, called non-parametric BP (NBP) is presented in [3] to solve self-localization problem in WSNs. In NBP, messages are presented by Gaussian particles which are generated from the belief functions. This method enables the use of SPA. However, calculating products of Gaussian mixtures and generating proper samples is not a trivial task.

The following part introduces a novel implementation of messages in BP using FDA method.

3 Fourier Density Approximation

Brunn et al. [5,6] derived the basic operations using Fourier density approximation. Here some important equations related to BP are briefly described.

3.1 *Definition of Fourier Densities*

A d-dimensional density function can be approximated by a d-dimensional Fourier expansion as:

$$p(\mathbf{x}) = \sum_{\kappa \in \mathrm{K}} \gamma_\kappa e^{j\kappa^T \mathbf{x}} = \sum_{\kappa \in \mathrm{K}} (\alpha_\kappa + \beta_\kappa) e^{j\kappa^T \mathbf{x}} \tag{5}$$

where $\mathbf{x} = [x_1, x_2, \ldots x_d]^{\mathrm{T}} \in [-\pi, \pi]^d$ is a multidimensional variable. $\gamma_\kappa = \alpha_\kappa + \beta_\kappa$ is the coefficients of the Fourier series. $\kappa = [\kappa_1, \kappa_2, \ldots \kappa_d]^{\mathrm{T}} \in \mathrm{K}$ is an index vector, where $\mathrm{K} = \left\{-\kappa_1^o, -\kappa_1^o + 1, \ldots \kappa_1^o\right\} \times \ldots \times \left\{-\kappa_d^o, -\kappa_d^o + 1, \ldots \kappa_d^o\right\}$ denotes the set of all valid indices [6].

In practice, the coefficients are obtained by the efficient Fast Fourier Transform (FFT) which has a complexity of $O\,(n \log(n))$ where n denotes the number of sampling points.

3.2 Fourier Density Product

Given two densities $p^a(\mathbf{x})$ and $p^b(\mathbf{x})$, they are represented by the Fourier density approximation as:

$$p^a(\mathbf{x}) = \sum_{\kappa \in \mathrm{K}_a} \gamma_\kappa^a e^{j\kappa^T \mathbf{x}} \tag{6}$$

and

$$p^b(\mathbf{x}) = \sum_{\kappa \in \mathrm{K}_b} \gamma_\kappa^b e^{j\kappa^T \mathbf{x}}, \tag{7}$$

their product can be expressed as:

$$p^c(\mathbf{x}) = p^a(\mathbf{x}) p^b(\mathbf{x}) = \sum_{\kappa \in \mathrm{K}_c} \gamma_\kappa^c e^{j\kappa^T \mathbf{x}} \tag{8}$$

with

$$\gamma_\kappa^c = \sum_{\mu \in \kappa_c} \bar{\gamma}_\mu^a \bar{\gamma}_{\mu-\kappa}^b \tag{9}$$

where the bar denotes a valid index:

$$\bar{\gamma}_\mu^{(\cdot)} = \begin{cases} \bar{\gamma}_\mu^{(\cdot)} & \mu \in \mathrm{K}_{(\cdot)} \\ 0 & otherwise \end{cases} \tag{10}$$

of γ_κ^a and γ_κ^b. The order of $p^c(\mathbf{x})$ is $\prod_{l=1}^d (\kappa_l^{o,a} + \kappa_l^{o,b})$, i.e. as many other approximation approaches like Gaussian mixture, the number of coefficient is significantly higher after production. But we can show later that the coefficient reduction in FDA is much easier than its counterparts.

3.3 Generalized Convolution Integral

Considering the Fourier densities of

$$p^a(\mathbf{x}) = \sum_{\mu \in \mathrm{K}_\mathbf{x}} \gamma_\mu^a e^{j\mu^T \mathbf{x}} \tag{11}$$

and

$$p^b(\mathbf{y}, \mathbf{x}) = \sum_{\substack{\mu \in \mathrm{K}_\mathbf{x}, \\ \kappa \in \mathrm{K}_\mathbf{y}}} \gamma_{\mu,\kappa}^b e^{j\mu^T \mathbf{x} + j\kappa^T \mathbf{y}} \tag{12}$$

their generalized convolution integral is given by:

$$p^c(\mathbf{y}) = \int p^b(\mathbf{x}, \mathbf{y}) p^a(\mathbf{x}) d\mathbf{x} = \sum_{\kappa \in \mathrm{K}_\mathbf{x}} \gamma_\kappa^c e^{j\kappa^T \mathbf{y}} \tag{13}$$

with $\gamma_\kappa^c = \sum_{\mu \in \kappa_\mathbf{y}} \gamma_{-\mu}^a \gamma_{\mu,\kappa}^b$.

Note that the order of resulting density only depends on the order of function $p^b(\mathbf{y}, \mathbf{x})$ not $p^a(\mathbf{x})$, which limits the computational complexity.

In addition, if the function $p^b(\mathbf{y}, \mathbf{x})$ has a form:

$$p^b(\mathbf{x}, \mathbf{y}) = p^b(\mathbf{y} - \mathbf{x}) , \tag{14}$$

(13) becomes

$$\begin{aligned} p^c(\mathbf{y}) &= \int p^b(\mathbf{x}, \mathbf{y}) p^a(\mathbf{x}) d\mathbf{x} = \int p^b(\mathbf{y} - \mathbf{x}) p^a(\mathbf{x}) d\mathbf{x} \\ &= p^b(\mathbf{y}) * p^a(\mathbf{x}) \end{aligned} \tag{15}$$

which is actually a convolution. Thus the coefficients of $p^c(\mathbf{y})$ are the multiplication of the coefficients of $p^b(\mathbf{y})$ and $p^a(\mathbf{x})$. In this way, the computation is simplified by replacing a high dimensional function $p^b(\mathbf{x}, \mathbf{y})$ with a low-dimensional function $p^b(\mathbf{y})$.

3.4 Coefficient Reduction

For many density mixture approximation approaches like Gaussian mixture, Dirac mixture or Monte Carlo methods, the number of coefficients increased exponentially after the product operation. Keeping all coefficients are practically impossible. Determining how many coefficients and which ones are needed is challenging. [13] provides a progressive way to calculate the parameters of mixture densities optimally. But the computational requirement is relatively high.

The coefficient reduction in FDA is relatively more efficient. As well known, the signal power in space domain and Fourier domain are equal. The Fourier coefficients ordered by their squared magnitude reflect the order of their influences to the square error between true density function and its Fourier approximation. Therefore, coefficient reduction in FDA is just deleting the coefficients with minimal squared magnitudes under the required density square error.

3.5 Ensuring Non-Negativity

FDA with reduced coefficients is sometimes negative which brings problem for further calculation. [5] proposed to use the square root of density function instead of density function itself for calculation. In this way, the final approximated density is ensured to be non-negative.

3.6 Computational Complexity

Table 1 lists the comparison of computational complexities for density product and generalized convolution between FDA and uniform sampling method where m denotes the number of coefficients used by Fourier density approximation. n is the number of uniform distributed samples. From this table, we see that the computation power is saved for the generalized convolution given the same number of m and n. By reducing the Fourier coefficients, both operations can be more efficient.

Table 1 Comparison of computational complexity

	FDA	Uniform sampling
Generalized convolution	$O(m)$	$O(n^2)$
Product	$O\,(m\log m)$	$O(n)$

4 Sensor Localization Example

In this chapter, we will use a sensor localization example to illustrate the BP method we proposed. Sensor localization is obtained by combining absolute positioning information (e.g. GPS) with relative distance information (e.g. time delay or power decay of the signal transmitted between sensors). In this paper, we restudy the self-calibration problem presented in [3] where each sensor has noisy measurements of its distances to neighboring nodes. The problem is formulated as a probabilistic inference problem that can be presented by probabilistic graphical model. BP algorithm is applied to exchange the calibration information between sensor nodes so that each sensor can obtain the MAP estimate of its location. Instead of Gaussian mixtures, FDA will be used to present the messages that are transmitted between nodes. Relative sensor geometry or the absolute sensor positions can be obtained depending on whether extra information about absolute positions is available at certain sensors.

4.1 System Model

Lets assume that we have a WSN with N sensors distributed in a planar space. The position of sensor i is denoted by $\mathbf{x}_i$. The measurement taken at sensor i about its distance to sensor j takes the form:

$$d_{ij} = \left\|\mathbf{x}_i - \mathbf{x}_j\right\| + \nu_{ij} \tag{16}$$

where d_{ij} denotes the observation, ν_{ij} is additive Gaussian noise with zero mean and standard deviation of σ. $\left\|\mathbf{x}_i - \mathbf{x}_j\right\|$ calculates the Euclidean distance between two points. d_{ij} is not always available since sensor i does not always detect its neighbor j. We use a binary random variable o_{ij} to indicate whether a distance measurement is available, i.e. $o_{ij} = 1$ when observation is made, $o_{ij} = 1$ otherwise.

According to [3], the probability that distance between sensor i and j is available with a probability of:

$$p(o_{ij} = 1|\mathbf{x}_i, \mathbf{x}_j) = \exp\left(-\frac{\|\mathbf{x}_i - \mathbf{x}_j\|^{\rho}}{R_1^{\rho}}\right) \tag{17}$$

Furthermore, each sensor has a prior knowledge about its position, which is given by a prior distribution $p(\mathbf{x}_i)$. The prior distribution is normally uninformative unless the sensor has obtained its position information from other resources, e.g. GPS signal. In this case, the prior distribution might look like a Dirac function.

4.2 Belief Propagation in Sensor Localization

Apparently, the assumptions mentioned in Sect. 2 are valid for this model. The joint distribution of the sensor locations $\{\mathbf{x}_i\}$ and the observations $\{d_{ij}\}$ and $\{o_{ij}\}$ can be factorized as:

$$p(\{\mathbf{x}_i\}, \{d_{ij}\}, \{o_{ij}\}) = \prod_{(i,j)} p(o_{ij}|\mathbf{x}_i, \mathbf{x}_j) \prod_{(i,j):o_{ij}=1} p_{\nu}(d_{ij}|\mathbf{x}_i, \mathbf{x}_j) \prod_{i} p(\mathbf{x}_i) \tag{18}$$

Based on Eq. (7), we can define the local potential for sensor i:

$$\phi_i(\mathbf{x}_1, \ldots \mathbf{x}_N) = p(\mathbf{x}_i) \cdot \prod_{j:j\neq i} p(o_{ij}|\mathbf{x}_i, \mathbf{x}_j) \prod_{j:j\neq i, o_{ij}=1} p_{\nu}(d_{ij}|\mathbf{x}_i, \mathbf{x}_j) \tag{19}$$

so that each sensor has now its local potential function.

Distributed inference can be done by using SPA. For sensor location problem, the message updating equation, obtained from Eqs. (3) and (19), takes the form:

$$m_{ij}^{t+1}(\mathbf{x}_1, \ldots \mathbf{x}_N) = \alpha \cdot \varphi_i(\mathbf{x}_1, \ldots \mathbf{x}_N) \prod_{k \in N(i)\backslash j} m_{ki}^{t}(\mathbf{x}_1, \ldots \mathbf{x}_N) \tag{20}$$

Each message in Eq. (20) involves N variables. The presentation of messages and the multiplication of messages will be too complicated that it makes the inference intractable. To simplify the problem, we define a message from node i to node j to be a function that only involves $\mathbf{x}_j$. In another word, message from node i to node j only contains a summary of sensor i's belief on the position of j. Position information about other sensor nodes are summed out. Based on this simplification, Eq. (20) will be revised to:

$$
\begin{aligned}
m_{ij}^{t+1}(\mathbf{x}_j) &= \alpha \int\limits_{\{\mathbf{x}_1,\ldots\mathbf{x}_N\}\backslash\mathbf{x}_j} \varphi_i(\mathbf{x}_1,\ldots\mathbf{x}_N) \prod_{k\in N(i)\backslash j} m_{ki}^t(\mathbf{x}_i) \\
&= \alpha \int\limits_{\{\mathbf{x}_1,\ldots\mathbf{x}_N\}\backslash\mathbf{x}_j} p(\mathbf{x}_i) \prod_{k:k\neq i} p(o_{ik}|\mathbf{x}_i,\mathbf{x}_k) \prod_{\substack{k:k\neq i\\ o_{kj}=1}} p_v(d_{ik}|\mathbf{x}_i,\mathbf{x}_k) \prod_{k\in N(i)\backslash j} m_{ki}^t(\mathbf{x}_i) \\
&= \alpha \int\limits_{\mathbf{x}_i} p(\mathbf{x}_i)\varphi_{ij}(\mathbf{x}_i,\mathbf{x}_j) \prod_{k\in N(i)\backslash j} m_{ki}^t(\mathbf{x}_i)
\end{aligned}
\tag{21}
$$

where $\varphi_{ij}(\mathbf{x}_i,\mathbf{x}_j)$ is defined as:

$$
\varphi_{ij}(\mathbf{x}_i,\mathbf{x}_j) = \begin{cases} p(o_{ij}=1|\mathbf{x}_i,\mathbf{x}_j)\cdot p(d_{ij}|\mathbf{x}_i,\mathbf{x}_j) \\ 1-p(o_{ij}=1|\mathbf{x}_i,\mathbf{x}_j) \end{cases}
\tag{22}
$$

The marginal probability of sensor location is given by:

$$
\hat{p}_i^t(\mathbf{x}_i) = \alpha\cdot p(\mathbf{x}_i)\prod_{k\in N(i)} m_{ki}^t(\mathbf{x}_i)
\tag{23}
$$

Although the complexity of messages has been greatly simplified in Eq. (21), calculation in Eqs. (21) and (23) is still complicated because of its non-linearity and the non-Gaussian distribution. To solve this problem, we use FDA method to approximate the density functions and present the messages as a collection of Fourier components and their coefficients.

4.3 Algorithm Description

Using FDA and the coefficient reduction method introduce in Sect. 3, the size of the messages are significantly reduced. This has brought benefits in two folds. On one side, it reduces the transmission power. On the other hand, it reduces the complexity of the SPA with a penalty of computing FFT.

We propose two algorithms. The ST-FDA algorithm, depicted in Table 2, uses FDA only to reduce the transmission power. SCT-FDA, depicted in Table 3, does all the calculation in the frequency domain thus reduces both the transmission power and the computational complexity.

5 Simulation Results

To verify the performance of the FDA based BP methods, we simulate the BP for self-localization problem in a WSN that is illustrated in Fig. 1. The positions of

Table 2 Description of ST-FDA algorithm

ST-FDA	
1.	Discretize the local potential functions.
2.	Initialize messages, e.g. a vector of ones.
3.	Calculate the outgoing message using Eq. (21). Since now the potential functions and the messages are discrete, we replace the integral in Eq. (21) with sum. Use FFT to transform the outgoing message into the frequency domain and use coefficient reduction method introduced in Sect. 3 to reduce the size of the messages.
4.	Once a new message (presented by Fourier coefficients) is received, an IFFT will be used to change the message to the 2D space domain for the SPA.
5.	Run SPA for a defined number of iterations.
6.	Posterior probability can be calculated by using Eq. (23).

Table 3 Description of SCT-FDA algorithm

SCT-FDA	
1.	Discretize the local potential functions.
2.	Initialize messages, e.g. a vector of ones.
3.	Use FFT to transform all messages and potential functions to frequency domain. Use coefficient reduction method (Sect. 3.4) to reduce the number of Fourier components. All messages stay in frequency domain until the end of the algorithm.
4.	The SPA of Eq. (21) in the frequency domain is implemented by using Eqs. (8) and (15). Coefficient reduction is done in each step.
5.	Run SPA for a defined number of iterations.
6.	Finally, use IFFT to convert the posterior probability from frequency domain into space domain.

sensor node 1, 2 and 3 are known as (0, 0), (1, 0) and (1, 1) respectively. Unknown sensor nodes 4 and 5 are located at $(-1, 0.4)$ and $(-0.2, 0.8)$. Note that although the Fourier densities are defined in $[-\pi, \pi]^d$ in Eq. (5), the definition in a large area can be also derived by a simple linear mapping. In this paper, we limit the area to $[-\pi, \pi]^2$ for simplicity.

The parameter ρ and R_1 in Eq. (17) are set to 2 and 3 m respectively. The standard deviation of distance measurements σ in Eq. (16) is set to 0.4 m. The BP is forced to stop after seven iterations. Figure 2 depicts the estimates of posterior distribution of sensor positions at node 4 (Fig. 2 (a2)–(a6)) and node 5 (Fig. 2 (a2)– (a6)) by SCT-FDA using different number of Fourier coefficients to represent a single potential function or a message and compare them with the true result generated by uniform sampling based method (Fig. 2 (a1) and (a2)). The sampling resolution is 65x65 for all experiments. From the results we can see with 100 Fourier coefficients, the approximation is already very close to the true value, whereas

Fig. 1 Sensor distribution

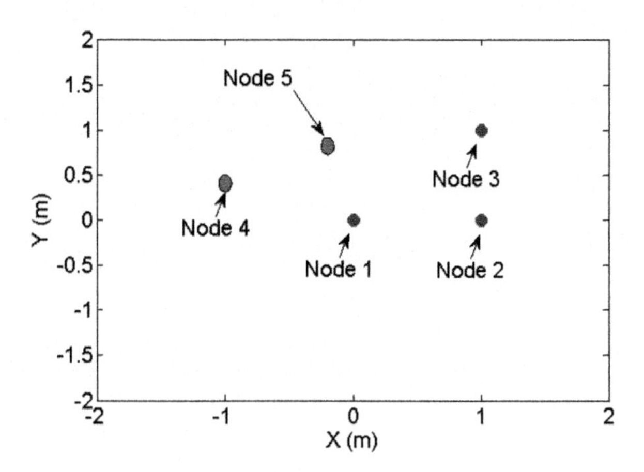

too few components can not fully characterize the very non-linear, non-Gaussian distribution.

Figure 3 shows the estimation results from SCT-FDA algorithm using different sampling resolutions. Sampling resolution of 15×15, 25×25, 35×35, 65×65 are applied to Fig. 3 (a1) to (a4) and Fig. 3 (b1) to (b4) respectively. The sampling resolution determines the precision of the estimate. According to *Nyquist Theorem*, original function can be recovered from its samples only if the sampling rate is greater than twice the maximum frequency of that function. Bad results can be observed from Fig. 3 (b3) and (b4) because the sampling rate is too low. Figure 4

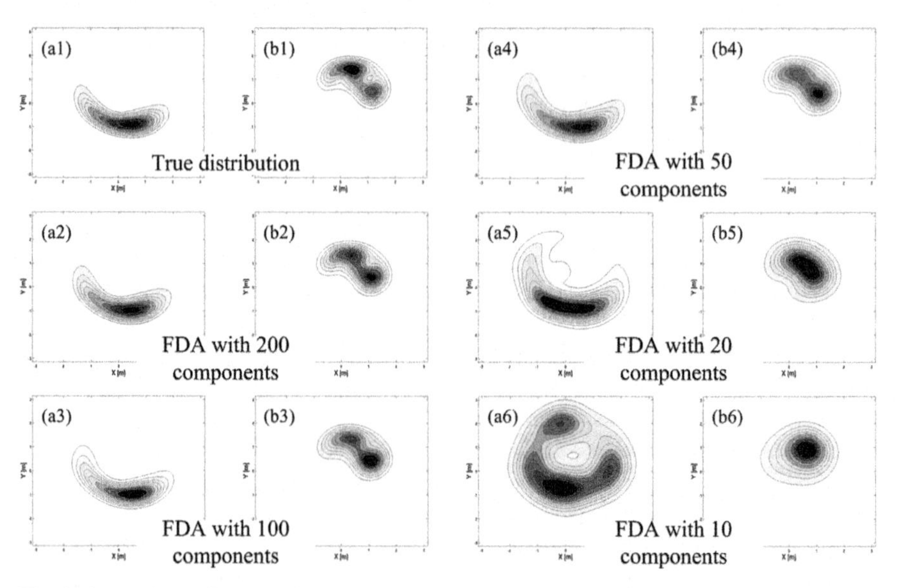

Fig. 2 Comparison of distribution estimates by SCT-FDA with different number of coefficients

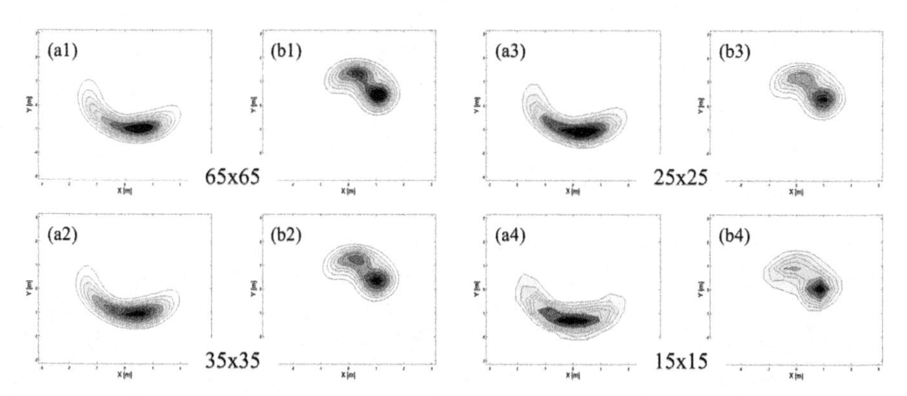

Fig. 3 Comparison of distribution estimate by SCT-FDA from different sample resolutions

plots the error of the position estimate of node 4 vs. number of Fourier coefficients curves of STFDA and SCT-FDA algorithms. It can be seen that, increasing number of coefficients results in a better performance for both methods. But compared to the result by uniform sampling which requires 4,225 samples to represent a message, a close result is achieved by much less Fourier coefficients. In the simulation, 119 messages are transmitted. If sample resolution is 65×65 and 50 Fourier coefficients are kept, ST-FDA and SCT-FDA methods transmit 23,800 components while uniform sampling based method has to transmit 502,775 components. Obviously, Fourier density approximation significantly reduces the transmission power.

ST-FDA method outperforms SCT-FDA methods in Fig. 4 because approximation is only made for the transmission in ST-FDA while SCT-FDA method also greatly simplifies the computation by using fewer coefficients in the SPA. Although SCT-FDA loses some accuracy, it saves the computation power and time. Furthermore, note that ST-FDA performs FFT and IFFT at the transmission and reception

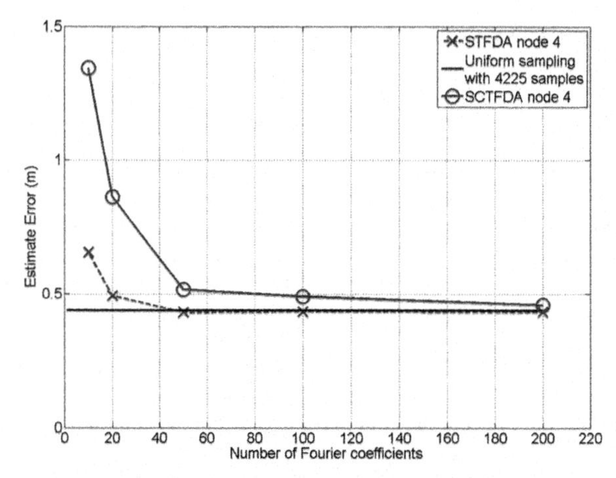

Fig. 4 Comparison of estimation errors in uniform sampling, STFDA and SCT-FDA

of each message while in SCT-FDA, FFT is only performed at the beginning and the end of BP, which further reduces the complexity.

6 Conclusions and Future Work

WSNs can be modeled by graphical models, where BP algorithm provides a promising solution. However, in WSNs, the computational ability and battery life of sensor nodes are limited. The intensive probability density computation and transmission between nodes required by BP make a big problem. This paper presents a method to use Fourier density approximation to represent belief densities. ST-FDA algorithm uses Fourier approximation to compress the complex non-Gaussian densities in order to reduce the radio transmission which is regarded as the most power consuming part in WSNs. Another algorithm SCT-FDA implements the SPA in Fourier domain so that it saves power consumptions not only in transmissions but also in belief calculations.

ST-FDA and SCT-FDA use a fixed number of Fourier coefficients. A more general algorithm with adaptive Fourier coefficient reduction can be investigated. In addition, other density representation like wavelet approximation could also be considered for the same application. The comparison between Fourier approximation and standard Gaussian mixture representation in [3] would also be interesting.

References

1. Gharavi, H., Kumar, S. (Eds.): Special issue on sensor networks and applications. In Proceedings of the IEEE, vol. 91, no. 8, Aug. 2003
2. Kumar, S., Zhao, F., Shepherd, D. (Eds.): Special issue on collaborative information processing. In IEEE Signal Processing Magazine, vol. 19, no. 2, Mar. 2002
3. Ihler, A., Fisher, J., Moses, R., Willsky, A.: Nonparametric belief propagation for self-calibration in sensor networks. In Proceedings of IPSN 2004
4. Kronmal, R., Tarter, M.: The estimation of probability densities and cumulatives by Fourier series methods. In Journal of the American Statistical Association, vol. 63, no. 323, pp. 925–952, Sep. 1968
5. Brunn, D., Sawo, F., Hanebeck, U.D.: Efficient nonlinear Bayesian estimation based on Fourier densities. In Proceedings of the 2006 IEEE International Conference on Multisensor Fusion and Integration for Intelligent Systems (MFI 2006), Germany 2006
6. Brunn, D., Sawo, F., Hanebeck, U.D.: Nonlinear multidimensional Bayesian estimation with Fourier densities. In Proceedings of the 2006 IEEE Conference on Decision and Control (CDC 2006), pp. 1303–1308, San Diego, California, Dec. 2006
7. Lauritzen, S.L.: Graphical Models. Oxford University Press, Oxford, 1996
8. Clifford, P.: Markov random fields in statistics. In Grimmett, G.R., Welsh, D.J.A. (Eds.) Disorder in Physical Systems, pp. 19C32. Oxford University Press, Oxford, 1990
9. Paskin, M., Guestrin, C.: A robust architecture for distributed inference in sensor networks. Intel Research, Technical Report IRB-TR-03-039, 2004.
10. Aji, S.M., McEliece, R.J.: The generalized distributive law. IEEE Transactions on Information Theory vol. 46, pp. 325–343, Mar. 2000

11. Murphy, K., Weiss, Y., Jordan, M.: Loopy-belief propagation for approximate inference: An empirical study. In Uncertainty in Artificial Intelligence vol. 15, pp. 467–475, Jul. 1999
12. Kschischang, F.R., Frey, B.J., Loeliger, H.A.: Factor graph and the sum-product algorithm. IEEE Transactions Information Theory, vol. 47, no. 2, pp. 498–518, Feb. 2001.
13. Hanebeck, U.D.: Progressive Bayesian estimation for nonlinear discrete-time systems: the measurement step. In Proceedings of the 2003 IEEE Conference on Multisensor Fusion and Integration for Intelligent Systems (MFI 2003), pp. 173–178, Tokyo, Japan, Jul. 2003.

Passive Localization Methods Exploiting Models of Distributed Natural Phenomena

Felix Sawo, Thomas C. Henderson, Christopher Sikorski and Uwe D. Hanebeck

Abstract This paper is devoted to methods for localizing individual sensor nodes connected in a network. The novelty of the proposed method is the model-based approach (i.e., rigorous exploitation of physical background knowledge) using local observations of a distributed phenomenon. By exploiting background phenomena, the individual sensor nodes can be localized by only locally measuring their surrounding without the necessity of heavy infrastructure. Two approaches are introduced: (a) the polynomial system localization method and (b) the simultaneous reconstruction and localization method. The first approach (PSL-method) is based on restating the mathematical model of the distributed phenomenon in terms of a polynomial system. Solving the system of polynomials for each individual sensor node directly leads to the desired locations. The second approach (SRL-method) regards the localization problem as a simultaneous state and parameter estimation problem within a Bayesian framework. By this means, the distributed phenomenon is reconstructed and the individual nodes are localized in a simultaneous fashion, while considering remaining stochastic uncertainties.

1 Introduction

The research work presented here is a modified version of [1], however explanations about the novel localization process are given in considerably extended way, with the focus on describing the actual estimation process with its different stages, i.e., identification/calibration stage and actual application stage. For more details about the used Bayesian estimator and its prospective applications, we refer to our previous research work [1–5].

F. Sawo (✉)
Intelligent Sensor-Actuator-Systems Laboratory (ISAS), Institute of Computer Science and Engineering, University of Karlsruhe, Germany
e-mail: sawo@ira.uka.de

S. Lee et al. (eds.), *Multisensor Fusion and Integration for Intelligent Systems*,
Lecture Notes in Electrical Engineering 35, DOI 10.1007/978-3-540-89859-7_26,

Recent developments in various areas dealing with sensor networks and the further miniaturization of individual nodes make it possible to apply wireless sensor networks for observing natural large-area physical phenomena [6]. Examples for such physical quantities are temperature distribution [4], chemical concentration [7], fluid flow, structural deflection or vibration in buildings, or the surface motion of a beating heart in minimally invasive surgery [8].

For the reconstruction of such distributed phenomena, the individual sensor nodes are densely deployed either inside the phenomenon or close to it. Then, by distributing local information to a global processing node, the phenomenon can be coöperatively reconstructed in an intelligent and autonomous manner [3, 9, 10]. In such scenarios, the sensor network can be exploited as a huge information field collecting data from its surrounding and then providing useful information both to mobile agents and to humans. Hence, the corresponding tasks are accomplished more efficiently, thanks to the extended perception provided by the sensor network. By this means, sensor networks can forecast or prevent dangerous situations, such as forest fires, seismic sea waves, or avalanches [11].

For most sensor network applications, the sensory data has only limited utility without location information. In particular for the accurate reconstruction of distributed phenomena, the locations of the individual sensor nodes are necessary. Manually measuring the location of every node in the network becomes infeasible, especially when the number of sensor nodes is large, the nodes are inaccessible or in the case of mobile sensor deployments. This makes the localization problem one of the most important issues to be considered in the area of sensor networks.

Classification of Localization Methods In general, the main goal of a localization system is to provide an estimate about the location of the individual nodes in the sensor network in the area of interest. There are several ways to classify the huge diversity of localization methods. In this work, they are classified into *active* localization methods and *passive* localization methods, depicted in Fig. 1.

- ***Active localization methods***: Active localization methods obtain an estimate of the sensor node location based on signals that are artificially stimulated and measured by the network itself or by a global positioning system. The stimuli

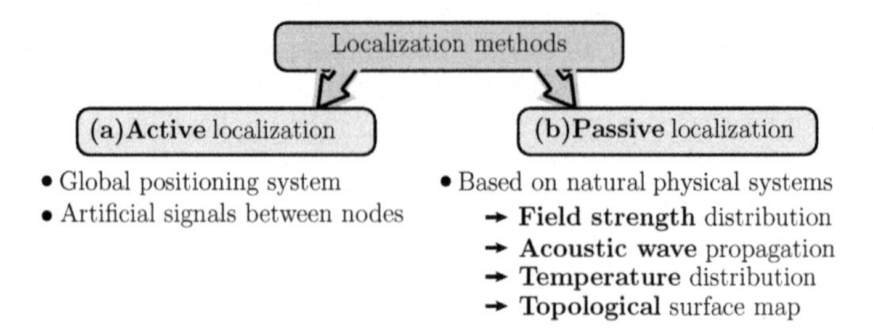

Fig. 1 Classification of localization methods: **(a)** Active localization, such as methods based on artificial signals between nodes and global positioning systems, and **(b)** passive localization, such as methods based on locally measuring a naturally existing distributed phenomenon

usually used in such scenarios consist of artificially generated acoustic events. It is obvious that the active localization process is performed in fairly controlled and well accessible environments. As it stands, these circumstances incur significant installation and maintenance costs. A comprehensive survey on active localization methods can be found in [12].

- ***Passive localization methods***: In the case of passive localization methods, which in contrary occur in a non-controlled and a possibly inaccessible environment, the stimuli necessary for the localization process occur naturally. In Fig. 1 **(b)** prospective examples of natural physical systems are stated, which can be used as stimuli for localizations. The clear advantage of using passive methods for the localization is that they do not need additional infrastructure. This certainly keeps the installation and maintenance costs at a very low level. In addition, these methods become particularly important for applications where global positioning systems are simply not available. This is for example the case of sensor networks for monitoring the snow cover [11], applications in deep sea, indoor localization [5, 13, 14], or robotic-based localization [15].

There are various techniques and methods that can be considered for localization systems using different kind of infrastructures in different scenarios. In general, for the estimation of a distributed phenomenon by a sensor network, the existing infrastructure could consist of both a number of sensor nodes with *known locations* and nodes with *unknown or uncertain locations*. For the minimization of the installation and maintenance costs, it is benefical to develop a method that requires no additional hardware such as a global positioning system or other heavy infrastructures. Moreover, there are various application scenarios without the possibility to access a global positioning system for the localization, such as the indoor localization of mobile phones [5, 16] or sensor networks deployed deep inside the snowpack for predicting snow avalanche risks, to name just a few. For that reason, a novel *passive process* is proposed that does not require such a global positioning system or the localization based on landmarks. It is important to emphasize that the passive localization technique proposed in this work can be employed in combination with other localization methods for further improving the location accuracy.

Key Idea of the Proposed Localization Method For the *passive localization* of sensor nodes, we present *model-based approaches* based on local observations. The novelty of the methods introduced in this work is the rigorous exploitation of a strong *mathematical model* of the distributed phenomenon for localizing individual sensor nodes. Furthermore, within this framework, the often remaining uncertainties in the sensor node locations can be considered during the reconstruction process of the distributed phenomenon [4]. The use of such a mathematical model for node localization was proposed in [11]. However, there was no consideration of uncertainties naturally occuring in the measurements and in the used model. The key idea of the proposed localization approach is depicted in Fig. 2. Roughly speaking, for localizing sensor nodes, the mathematical model and the resulting distribution of the spatially distributed phenomenon is exploited in an *inverse manner*. That means,

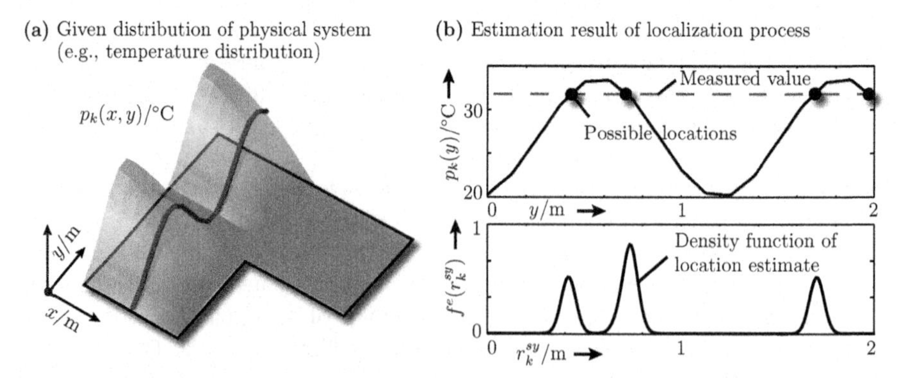

Fig. 2 Visualization of the key idea of the proposed novel localization method based on locally measuring a distributed phenomenon. **(a)** Possible distribution $p_k(x, y)$ of a physical system characterized by a strong mathematical model. **(b)** Sectional drawing of the system at a specific location in x-direction. Depicted are the possible locations (deterministic case) and the respective density function $f^e(r_k^{sy})$ (stochastic case)

locally measured physical quantities are used to obtain possible locations where the measured values could have been generated.

In this research work, we introduce two different methods for the model–based passive localization of sensor nodes based on local observations: (a) the polynomial system localization method, and (b) the simultaneous reconstruction and localization method. The *first approach* (***PSL-method***) is purely deterministic, meaning that neither uncertainties in the model description nor in the measurements are considered. This direct method is based on restating the model of the distributed phenomenon in terms of a polynomial system including the state of the physical system and the location to be identified. Then, solving a system of polynomial equations leads directly to the desired location of the sensor node. The *second approach* (***SRL-method***) considers uncertainties both in the mathematical model and the measurements during the localization process. It is shown that the localization problem can be regarded as a simultaneous state and parameter estimation problem, with node locations as the parameters to be identified. This leads to a high-dimensional nonlinear estimation problem, making the employment of special types of estimators necessary. By this means, the sensor nodes are localized and the distributed phenomenon is reconstructed in a simultaneous fashion. The improved knowledge can be exploited for other nodes to localize themselves.

2 Problem Formulation

The main goal is to design a novel localization method for sensor network applications, where individual nodes are able to locally measure a distributed phenomenon only. We assume to have a strong mathematical model of the phenomenon, i.e., with known model structure and model parameters. This model could possibly result

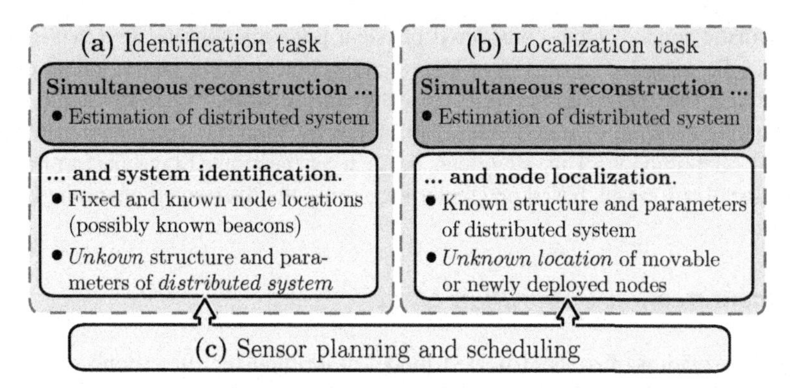

Fig. 3 Visualization of *two tasks* for the estimation of a distributed phenomenon. The individual tasks are managed by a planning and scheduling process (not considered in this research work)

from an earlier identification task, visualized in Fig. 3 (**a**). Based on this mathematical model and local measurements, newly deployed or movable sensor nodes can be efficiently localized without using a global positioning system, see Fig. 3 (**b**).

Considered Distributed Phenomenon Throughout this paper, we consider the localization based on the observation of a distributed phenomenon described by the one-dimensional diffusion equation

$$L\,(p(r,t)) = \frac{\partial p(r,t)}{\partial t} - \alpha \frac{\partial^2 p(r,t)}{\partial r^2} - s(r,t) = 0\,, \tag{1}$$

where $p(r,t)$ denotes the distributed state of the phenomenon at the spatial coordinate r and at the time t. The diffusion coefficient α can be varying in both time and space. Given an estimated solution $p(\cdot)$, the aim is the estimation of the location r_k^{si} of the individual sensor nodes based on local measurements $\hat{\underline{y}}_k$ of a realization of the distributed phenomenon $p(\cdot)$. In this work, we consider the worst-case scenario where the node locations are *completely unknown* and the phenomenon $p(\cdot)$ still contains some uncertainties. The same methods can be utilized for simply considering uncertainties in the locations during the reconstruction of distributed phenomena (i.e., without localizing sensor nodes).

3 Overview of the Passive Localization Method

The *model-based passive localization* method proposed in this research work can be considered as a ***two-stage technique***: The first stage is the so-called *identification/calibration stage*, which is responsible for building a sufficiently accurate probabilistic model of the considered physical phenomenon and its environment. This can be regarded as a system identification and training phase. Then, during the *localization stage*, the previously created and identified model is exploited to estimate the location of individual sensor nodes by local measurements of the distributed phenomenon. This stage can be seen as the usage stage performing the actual localization task based on locally measuring the distributed phenomenon.

3.1 Identification/Calibration Stage

For the derivation of a sophisticated model describing the underlying distributed phenomenon exploited for the localization, a series of calibration measurements is required. This can be performed by using a certain number of sensor nodes sensing the physical quantity at known locations. Here, these nodes are assumed to be responsible only for identifying the underlying phenomenon, however, not necessarily for the actual localization process. At each sensor node with the precisely known position $r_k^{\mathrm{BS}i}$, a realization of the distributed phenomenon $p_k(\cdot)$ is locally measured.

For physical phenomena distributed over a wide area, gathering the measurements can become tedious. However, the automation of this process can be achieved using mobile devices (with an accurate independent navigation system) moving in the area of interest in an autonomous and self-organized manner. Such a system was, for example, proposed in [17], where a mobile robot autonomously collects information about the signal strength for indoor localization purposes [13, 18].

The identification or calibration stage strongly differs in the way they actually make use of the measurements obtained. In this research work, the localization based on *static* as well as *dynamic* phenomena is of interest. In particular, depending on the type of the system, the description to be obtained during the identification stage is different. For static systems, a mathematical model only in terms of a probability density function is required, whereas for dynamic systems additional parameters describing the dynamic and distributed behavior need to be identified and calibrated.

Static Phenomena In the case of localizing sensor nodes based on a static distributed phenomenon, the identification stage consists only of finding an appropriate model description in terms of the conditional density function $f^e(p|r)$, as visualized in Fig. 4. This description characterizes the distribution of the considered physical quantity and its uncertainty in the area of interest. In this sense, for each position r a density function about the distributed phenomenon is obtained. There are several ways for the actual derivation of the model describing the distribution of the physical quantity. For example, this can be achieved by *data-driven approaches* [19], which use the calibration measurements to directly estimate the underlying density function $f^e(p|r)$ of the static distributed phenomenon. Another possibility is to use *probabilistic learning techniques*, such as the *simultaneous probabilistic localization and learning* method (*SPLL-method*) proposed in [20], which additionaly allows the simultaneous localization during the identification and calibration stage.

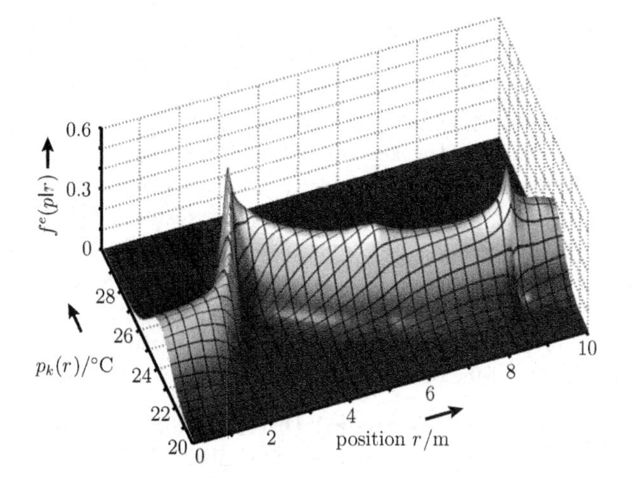

Fig. 4 Visualization of the model description for the localization of sensor nodes exploiting *static and distributed* physical phenomena. The model is given in terms of a conditional density function $f^e(p|r)$ over the position r and the distributed physical system p; here depicted for the one-dimensional case for simplicity purposes only

Dynamic Phenomena For dynamic distributed phenomena, it is not sufficient to derive a description only about the current spatial distribution of the physical quantity, rather additional parameters characterizing the dynamic behavior are necessary. The main advantage of exploiting dynamic phenomena for the localization is that additional information about the dynamics allows excluding specific values of the otherwise possibly ambiguous location estimates. However, this advantage is opposed by the more sophisticated and costly identification/calibration stage that must be accomplished before or simultaneous to the actual localization stage. That means, the precise identification of the structure and the parameters of the model description for the distributed phenomenon is required. This can be achieved by the **S**imultaneous **R**econstruction and **I**dentification method (*SRI-method*), see [3].

3.2 Localization Stage

In the localization stage, the individual sensor nodes with unknown location r_k^{si} measure the underlying distributed phenomenon locally, e.g., temperature distribution or signal strength distribution. The locations of the N sensor nodes to be identified are collected in the parameter vector $\underline{\eta}_k^M$, according to

$$\underline{\eta}_k^M := \left[r_k^{s1}, r_k^{s2}, \ldots, r_k^{sN}\right]^T \in \mathbb{R}^N .$$

In the following, two different approaches for the passive localization are introduced: **(a)** the polynomial system localization method (PSL-method) and **(b)** the simultaneous reconstruction and localization method (SRL-method).

4 Polynomial System Localization Method

This section is devoted to a deterministic approach for the localization of individual nodes in a sensor network based on local measurements of a distributed phenomenon. The key idea of the proposed direct method is to solve the partial differential equation (1) in terms of the unknown node locations. This leads to a straightforward solution as long as the resulting nonlinear equations can be readily solved. Solving these equations for all sensor locations is called the *Polynomial System Localization Method* (PSL-method). The PSL-method basically consists of two steps: (1) spatial and temporal discretization of the mathematical model, and (2) reformulating and finally solving the resulting system of polynomial equations in terms of the desired locations.

1. **Spatial and Temporal Discretization** The simplest method for the spatial and temporal discretization of distributed phenomena is the finite-difference method [10, 11]. In order to solve the partial differential equation (1), the derivatives need to be approximated with finite differences according to

$$\frac{\partial p(r,t)}{\partial t} = \frac{p_{k+1}^i - p_k^i}{\Delta t}\,, \qquad \frac{\partial^2 p(r,t)}{\partial r^2} = \frac{\frac{p_k^{i+1}-p_k^i}{r_k^{i+1}-r_k^i} - \frac{p_k^i-p_k^{i-1}}{r_k^i-r_k^{i-1}}}{\frac{1}{2}(r_k^{i+1} - r_k^{i-1})} \tag{2}$$

   where Δt is the sampling time. The superscript i and subscript k in p_k^i denote the value of the distributed phenomenon at the discretization node i and at the time step k. Plugging the finite differences (2) into the mathematical model of the distributed phenomenon (1), in general, leads to a system of polynomial equations of degree three. However, for the case of one unknown sensor node location, this reduces to a single quadratic equation, as shown in the next subsection.
2. **Solving Polynomial System Equations** Based on the spatial and temporal discretization, the partial differential equation (1) may be expressed as a finite difference equation and put in the following form at each discretization point, p_k^i, in the interval in question

$$\begin{aligned} 0 = {} & A_k^i(r_k^{i+1} - r_k^i)(r_k^i - r_k^{i-1})(r_k^{i+1} - r_k^{i-1}) \\ & - B_k^i(r_k^i - r_k^{i-1}) + C_k^i(r_k^{i+1} - r_k^i)\,, \end{aligned} \tag{3}$$

   where

$$A_k^i = \frac{p_{k+1}^i - p_k^i}{2\alpha\Delta t}\,, \quad B_k^i = p_k^{i+1} - p_k^i\,, \quad C_k^i = p_k^i - p_k^{i-1}\,.$$

At this point, it is important to mention that r_k^i represents the unknown location of the sensor node to be localized and r_k^{i+1} and r_k^{i-1} are the *known locations* of neighboring nodes. The derived system equation (3) can be simply regarded as an explicit relation between three positions on a line (two known endpoints and one unknown location between them), and four values of the measured phenomenon (all known and one at each location at time t and one at the unknown location at time $t+1$). In order to derive the unknown location r_k^i of sensor node i, the polynomial system of equations (3) needs to be solved and the root selected, which best fits the conditions (e.g., must be between the two known locations r_k^{i-1} and r_k^{i+1}).

The PSL-method assumes a densely deployed sensor network in which every node i communicates with its neighboring nodes $i-1$ and $i+1$. This means that measurements of the distributed phenomenon p_k^{i-1} and p_k^{i+1} need to be transmitted between adjacent nodes. It can be stated that the denser the sensor nodes are deployed, the more accurate the individual nodes in the network can be localized. The proposed localization approach involves neither errors in the mathematical model nor uncertainties in the measurements. However, it can be easily implemented and has low computational complexity.

5 Simultaneous Reconstruction and Localization Method

For the state reconstruction of distributed phenomena, the precise knowledge about the node locations is essential for deriving precise estimation results. However, using any kind of positioning system, some uncertainties in the location estimate remain. In order to obtain consistent and accurate reconstruction results, these uncertainties in the node location need to be systematically considered during the reconstruction process. Hence, the simultaneous method proposed in this section does not only allow (a) the *localization of sensor nodes*, but especially (b) the systematic consideration of uncertainties in the node locations during the *state reconstruction process*.

After the derivation of a finite-dimensional model for the node localization based on a system conversion, a method for the ***S***imultaneous ***R***econstruction of distributed phenomena and node ***L***ocalization (*SRL*-method) is introduced. There are four key features characterizing the novelties of the proposed method: (a) approach is based on local measurements only, (b) systematic consideration of uncertainties in the model description and the measurements, (c) derivation of an uncertainty measure for the estimated node location in terms of a density function, and (d) improvement of the estimation of distributed phenomena thanks to the simultaneous approach.

5.1 Conversion of Distributed Phenomena

The model–based state reconstruction of distributed systems based on a distributed–parameter description (1) is quite complex. The reason is that a Bayesian estimation method usually exploits a lumped–parameter system description. In order to cope with this problem, the system description has to be converted from a distributed–parameter form into a lumped–parameter form. In general, the conversion of the system description (1) can be achieved by methods for solving partial differential equations, such as modal analysis [8], the finite-difference method [10, 11], the finite-element method [4], and the finite-spectral method [21]. Basically, these methods consist of two steps, namely spatial decomposition and temporal decomposition.

1. **Spatial decomposition** By means of the spatial decomposition, the partial differential equation (1) is converted into a set of ordinary differential equations [4]. First, the solution domain $\Omega = [0, L]$ needs to be decomposed into N_x subdomains Ω^e. Then, the solution $p(r, t)$ in the entire domain Ω is represented by a piecewise approximation according to

$$p(r,t) = \sum_{i=1}^{N_x} \Psi^i(r)\, x^i(t)\,, \tag{4}$$

   where $\Psi^i(r)$ are analytic functions called *shape functions* and $x^i(t)$ their respective weighting coefficients. It is important to note that the individual shape functions $\Psi^i(r)$ are defined on the entire solution domain. The essence of all aforementioned conversion methods lies in the choice of the shape functions $\Psi^i(r)$, e.g., piecewise linear functions, orthogonal functions, or trigonometric functions [4].
2. **Temporal discretization** In order to derive a discrete-time system model the system of ordinary differential equations (derived from the spatial decomposition) needs to be discretized in time. The temporal discretization produces a linear system of equations for the state vector $\underline{x}_k$ containing the temporal discretized weighting factors x_k^i of the finite expansion (4). The resulting discrete-time lumped-parameter system represents the distributed system (1).

In the case of *linear* partial differential equations (1), the aforementioned methods for the spatial and temporal decomposition always result in a *linear* system of equations according to

$$\underline{x}_{k+1} = \mathbf{A}_k \underline{x}_k + \mathbf{B}_k \left(\underline{\hat{u}}_k + \underline{w}_k^x \right) . \tag{5}$$

The global state vector $\underline{x}_k$ characterizes the state of the distributed system and the vector $\underline{w}_k^x$ represents the system uncertainties. The structure of the system matrix $\mathbf{A}_k$ and the input matrix $\mathbf{B}_k$ merely depend on the applied conversion method [4].

5.2 Derivation of Measurement Model

In this section, we derive the *measurement model* for the purpose of *localizing* sensor nodes based on local observations of a physical phenomenon. The sensor nodes are assumed to measure directly a realization of the distributed phenomenon $p(r_k^{si}, t_k)$ at their individual locations r_k^{is}. Then, the measurement equation for the entire network is assembled from the individual shape functions $\Psi^j(\cdot)$ as follows

$$\underline{\hat{y}}_k = \underbrace{\begin{bmatrix} \Psi^1(r_k^{s1}) & \cdots & \Psi^{N_x}(r_k^{s1}) \\ \vdots & \ddots & \vdots \\ \Psi^1(r_k^{sN_s}) & \cdots & \Psi^{N_x}(\eta_k^{sN_s}) \end{bmatrix}}_{\mathbf{H}_k(\underline{\eta}_k^M)} \underline{x}_k + \underline{v}_k , \tag{6}$$

where $\underline{v}_k$ denotes the measurement uncertainty and N_s represents the number of sensor nodes used in the network. The measurement model (6) directly relates the measurements $\hat{y}_k^i$ to the state vector $\underline{x}_k$ characterizing the distributed phenomenon and to the location vector $\underline{\eta}_k^M$. The structure of the measurement matrix $\mathbf{H}_k$ for localizing sensor nodes in a network is shown in the following example:

Example of Measurement Model In this example, we visualize the structure of the measurement matrix $\mathbf{H}_k$ subject to *piecewise linear shape functions.* The entire solution domain Ω is represented by $N_x = 4$ shape functions $\Psi^i(\cdot)$. In addition, there are two sensor nodes located at r_k^{s1} and r_k^{s2} in the subdomains Ω^1 and Ω^2. Then, the measurement model is given as follows

$$\begin{bmatrix} \hat{y}_k^1 \\ \hat{y}_k^2 \end{bmatrix} = \begin{bmatrix} \overbrace{c_1^1 + c_2^1 r_k^{s1}}^{\Psi^1(r_k^{s1})} & \overbrace{c_3^1 + c_4^1 r_k^{s1}}^{\Psi^2(r_k^{s1})} & 0 & 0 \\ 0 & \underbrace{c_1^2 + c_2^2 r_k^{s2}}_{\Psi^2(r_k^{s2})} & \underbrace{c_3^2 + c_4^2 r_k^{s2}}_{\Psi^3(r_k^{s2})} & 0 \end{bmatrix} \begin{bmatrix} x_k^1 \\ x_k^2 \\ x_k^3 \\ x_k^4 \end{bmatrix} + \begin{bmatrix} v_k^1 \\ v_k^2 \end{bmatrix} ,$$

where the constants c_i^j arise from the definition of the piecewise linear shape functions in each subdomain and thus the geometry of the applied grid for the finite elements. The extension to orthogonal polynomials and trigonometric functions can be derived in a straightforward fashion [3,4].

From the previous example, it is obvious that the structure of the measurement matrix $\mathbf{H}_k$ merely depends on the location collected in the parameter vector $\underline{\eta}_k^M$ of the individual sensor nodes. That means, for the accurate reconstruction of the distributed phenomenon (1) based on a sensor network, the exact node locations are necessary. Due to this dependency, deviations of true locations from the modeled node locations lead to poor estimation results, as shown in our previous research work [1]. On the other hand, thanks to the dependency of the measurement matrix $\mathbf{H}_k$ on the node locations, the localization problem can be stated as a *simultaneous state and parameter estimation* problem. By this means, the distributed phenomenon can be reconstructed and the nodes can be localized in a simultaneous fashion.

5.3 Augmented System Description for Node Localization

For the simultaneous node localization and reconstruction of distributed phenomena, the unknown locations of the sensor nodes $\underline{\eta}_k^M$ are treated as additional state variables. By this means, conventional estimation techniques can be used to *simultaneously* estimate the location and the state of the distributed phenomenon. Hence, an *augmented state vector* $\underline{z}_k$ containing the system state $\underline{x}_k$ and the additional unknown node locations $\underline{\eta}_k^M$ is defined by $\underline{z}_k := [\underline{x}_k^T, \underline{\eta}_k^T]^T$.

The augmentation of the state vector with additional unknown parameters leads to the so-called *augmented system model*. In the case of localizing sensor nodes, the augmentation leads to the following augmented system model

$$\begin{bmatrix} \underline{x}_{k+1} \\ \underline{\eta}_{k+1}^M \end{bmatrix} = \begin{bmatrix} \mathbf{A}_k \underline{x}_k + \mathbf{B}_k \underline{\hat{u}}_k \\ \underline{a}_k(\underline{\eta}_k^M) \end{bmatrix} + \begin{bmatrix} \mathbf{B}_k \, \underline{w}_k^x \\ \underline{w}_k^\eta \end{bmatrix} , \tag{7}$$

and measurement model

$$\underline{\hat{y}}_k = \underbrace{\mathbf{H}_k(\underline{\eta}_k^M)\underline{x}_k}_{\underline{h}_k(\underline{x}_k, \underline{\eta}_k^M)} + \underline{v}_k , \tag{8}$$

where the nonlinear function $\underline{a}_k(\cdot)$ describes the dynamic behavior of the node locations contained in the vector $\underline{\eta}_k^M$ to be estimated.

The structure of the augmented system model (7) and (8) for the node localization is depicted in Fig. 5 (**a**). In this case, it is obvious that the augmented measure-

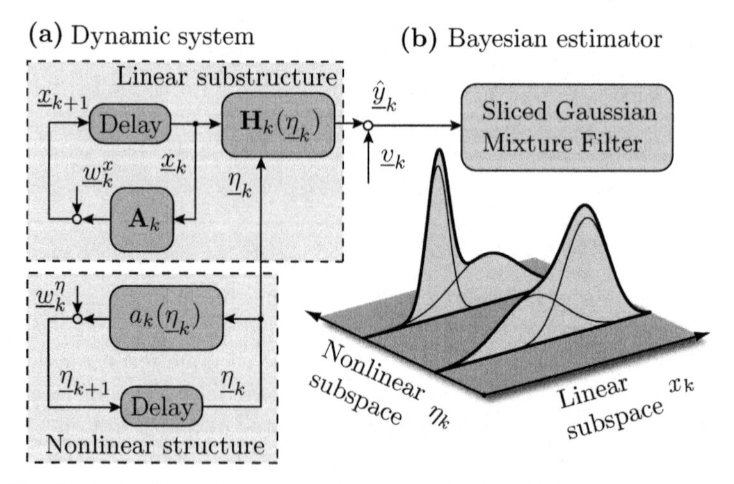

Fig. 5 Visualization of dynamic system and estimator for the node localization based on local observations. **(a)** The system description contains a high-dimensional linear substructure. The individual node locations r_k^{si} collected in the parameter vector $\underline{\eta}_k$ characterizes the measurement matrix $\mathbf{H}_k(\cdot)$, and thus, the individual measurements $\underline{\hat{y}}_k$. **(b)** The Bayesian estimator is based on *sliced Gaussian mixture densities* consisting of a Gaussian mixture and Dirac mixture

ment model is *nonlinear* in the augmented state vector $\underline{z}_k$ due to the multiplication of $\mathbf{H}_k(\underline{\eta}_k^M)$ and the system state $\underline{x}_k$. Since the parameter vector $\underline{\eta}_k^M$ characterizes the measurement matrix $\mathbf{H}_k$, it also has a direct influence on the actual measured values. It is important to emphasize that the measurement model (8) contains a high-dimensional linear substructure, which can be exploited by the application of a more efficient estimator. In the following section, we briefly describe a Bayesian estimator allowing the decomposition of the localization problem.

5.4 Estimation Based on Sliced Gaussian Mixture Densities

There are several methods to exploit the linear substructure in the combined linear/nonlinear system equation (7) and measurement equation (8). The marginalized particle filter [22] integrates over the linear subspace in order to reduce the dimensionality of the state-space. Based on this marginalization, the standard particle filter is extended by applying the Kalman filter to find the optimal estimate for the linear subspace (which is associated with the respective individual particles). The marginalized filter certainly improves the performance in comparison to the standard particle filter. However, some drawbacks still remain. For instance, special measures have to be taken in order to avoid effects like sample degeneration and impoverishment. More importantly, it does not provide a measure on how well the true joint density is represented by the estimated one.

For that reason, a more systematic Bayesian estimator is employed for the simultaneous reconstruction of distributed system and node localization. For the exploitation of linear substructures in general nonlinear systems, we introduced in our previous research work [2] a systematic estimator, the so-called ***S***liced ***G***aussian ***M***ixture ***F***ilter (SGMF). There are two key features leading to a significantly improved estimation result compared to other state of the art estimation approaches.

- **Novel density representation** The utilization of a special kind of density allows the decomposition of the general estimation problem into a linear and nonlinear problem. To be more specific, as a density representation the so-called *sliced Gaussian mixture density* is employed for the simultaneous reconstruction and localization of sensor nodes.
- **Systematic approximation** The systematic approximation of the density resulting from the estimation update leads to (close to) optimal approximation results. Thus, less parameters for the density representation are necessary and a measure for the approximation performance is provided.

Despite the high-dimensional nonlinear character, the systematic approach for the simultaneous reconstruction and localization for large-area distributed phenomena is feasible thanks to the decomposition based on the Sliced Gaussian Mixture Filter. Furthermore, the uncertainties occuring in the mathematical system description and arising from noisy measurements are considered by an integrated treatment. The systematic estimator exploiting linear substructures basically consists of three steps: the *decomposition* of the estimation problem, the utilizaton of an *efficient update*, and the *reapproximation* of the density representation [2, 3].

6 Simulation Results

In this section, the performance of the proposed localization methods is demonstrated by means of simulation results.

Assumption of Simulated Case Study In this simulation, we consider the localization based on the one-dimensional partial differential equation (1), with assumed initial condition and Dirichlet boundary conditions as considered in [1]. The nominal parameters for the system model (5) are given by

$$s(r,t) = 0\,, \quad \alpha = 1 \quad \Delta t = 0.2\,, \quad r^s_{\text{true}} = 16\,,$$

where r^s_{true} denotes the true node location. The aim is the localization of a sensor node with initially unknown location based on local observations only. The system noise term is $\mathbf{C}^w_l = \text{diag}\,\{20, \ldots, 20\}$, the noise term for the node location is given by $C^w_n = 0.02$, and for the local measurement of the node to be localized is assumed to be $C_v = 0.01$. Here, we compare different approaches for the passive localization based on local measurements: **(a)** PSL-method, **(b)** deterministic approach introduced in [11] (CSN-method), **(c)** SRL-method based on sliced Gaussian mixture filter (50 slices), **(d)** SRL-method based on marginalized particle filter (500 particles). These approaches are compared based on 100 independent simulation runs.

The simulation results for the PSL-method are depicted in Fig. 6. It is important to mention that this deterministic approach was simulated with perfect information, i.e., there is noise neither in the system nor in the measurements. Furthermore, we assume that the sensor node to be localized receives information about distributed

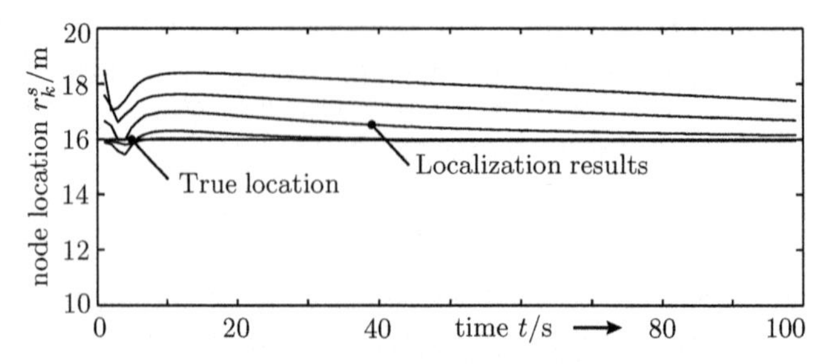

Fig. 6 Results of the polynomial systems localization method (PSL-method) for various neighboring nodes with known locations. The true location to be identified is $r^s_{\text{true}} = 16$

phenomenon and locations from neighboring nodes. Since the diffusion equation has derivatives involving Δt and Δx, the PSL-method is sensitive to the distance between the two adjacent known locations. Evidence of this effect is shown in Fig. 6 which plots the values found by the PSL-method for known points of varying distance from the unknown. It is obvious that the denser the nodes are deployed the more accurate the location can be identified.

The simulation results for the SRL-method with considering all the aforementioned uncertainties is shown in Fig. 7. Here, we assume the sensor network consists only of a single sensor node locally measuring the phenomenon. Furthermore, the sensor node has only very uncertain knowledge about the initial distributed phenomenon, see Fig. 7 (a).

Fig. 7 (c) depicts one specific simulation run for the estimation of the unknown node location η_k^S. It can be seen that after a certain transition time the SRL-method based on sliced Gaussian mixture filter (with 50 slices) offers a nearly exact location estimation, while the determinstic approach CSN-method strongly deviates (due to neglecting system and measurement noises). The root mean square error (rms) of all 100 simulation runs over time is depicted in Fig. 7 (d). It is obvious that in this example the SRL-method based on the Sliced Gaussian Mixture Filter (with 50 slices) outperforms both the deterministic approach (CSN-method) and the approach based on marginalized particle filter (with 500 particles); mainly due to the consideration of uncertainties and the systematic and deterministic approximation of the density.

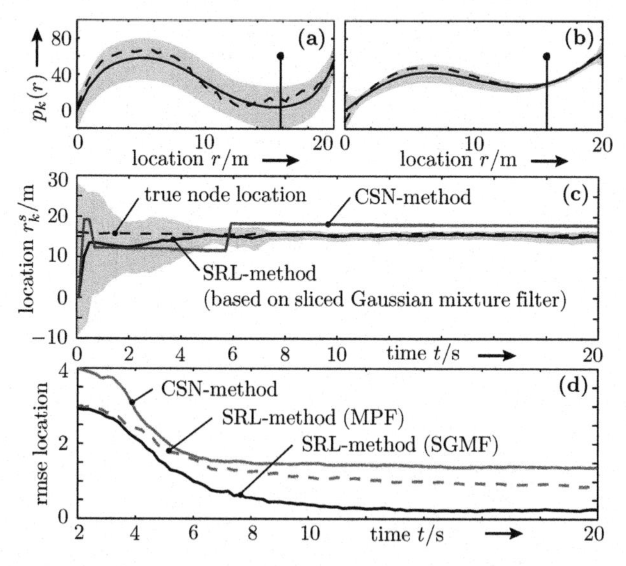

Fig. 7 Comparison of SRL-method based on SGMF, SRL-method based on MPF, and deterministic approach CSN-method. **(a)-(b)** Improvement of estimation of distributed phenomenon thanks to *simultaneous* approach. **(c)** Specific simulation run for the estimation of the node location r_k^s collected in the parameter vector $\underline{\eta}_k^M$. The true location is assumed to be $r_{\text{true}}^s = 16$. **(d)** Root mean square error (rmse) over time of 100 simulation runs

Comparing Fig. 7 (a) and (b), it is obvious that thanks to the *simultaneous property* of the SRL-method, not only can the sensor node be accurately localized, but also the estimation about the distributed phenomenon can be further improved. This can be exploited by other sensor nodes to localize themselves.

7 Conlusions and Future Work

In this paper, we introduce the methodology of two novel localization approaches for sensor nodes measuring locally only their surrounding. The *PSL-method* is a deterministic approach and is mainly based on restating the mathematical model in terms of the location. In the case of *no noise* in the model description and the measurement, this method leads to sufficient results for a *dense sensor network*. The stochastic SRL-method basically reformulates the localization problem as a simultaneous state and parameter estimation problem. This leads to a high-dimensional nonlinear estimation problem, which makes the employment of special types of estimators necessary. Here, the Sliced Gaussian Mixture Filter (SGMF) and the marginalized particle filter (MPF) are applied for the decomposition of this estimation problem. Thanks to the stochastic approach, the SRL-method leads to better estimation results for the location, even with noisy information. Furthermore, the simultaneous approach allows to improve the knowledge about the phenomenon, which then can be exploited by other nodes for the localization.

The application of the proposed localization methods (PSL-method and SRL-method) to sensor networks provides novel prospects. The network is able to localize the individual nodes without relying on a satellite positioning system (which is not always available) as long as a strong model of the surrounding is available.

For the PSL-method it is necessary to incorporate uncertainties into the mathematical model as well as the sensors, and to study the robustness of the method in the presence of noise. Another issue for future work is that if the locations of several nodes are unknown, they may be solved separately using the method described in this paper; however, we should compare it to the simultaneous solution of the system of degree three equations. So far, the model parameters and structure were assumed to be precisely known for the SRL-method. In many real world applications, the parameters contain uncertainties. The combination of the parameter identification of distributed phenomena and the node localization is left for future work. Finally, we intend to test the proposed localization methods on real sensor data.

Acknowledgments This work was partially supported by the German Research Foundation (DFG) within the Research Training Group GRK 1194 "Self-organizing Sensor-Actuator-Networks".

References

1. Felix Sawo, Thomas C. Henderson, Christopher Sikorski, , and Uwe D. Hanebeck. Sensor Node Localization Methods based on Local Observations of Distributed Natural Phenomena. In *Proceedings of the 2008 IEEE International Conference on Multisensor Fusion and Integration for Intelligent Systems (MFI 2008)*, Seoul, Republic of Korea, August 2008.
2. Vesa Klumpp, Felix Sawo, Uwe D. Hanebeck, and Dietrich Fränken. The Sliced Gaussian Mixture Filter for Efficient Nonlinear Estimation. In *Proceedings of the 11th International Conference on Information Fusion (Fusion 2008)*, Cologne, Germany, July 2008.
3. Felix Sawo, Vesa Klumpp, and Uwe D. Hanebeck. Simultaneous State and Parameter Estimation of Distributed-Parameter Physical Systems based on Sliced Gaussian Mixture Filter. In *Proceedings of the 11th International Conference on Information Fusion (Fusion 2008)*, Cologne, Germany, July 2008.
4. Felix Sawo, Kathrin Roberts, and Uwe D. Hanebeck. Bayesian Estimation of Distributed Phenomena using Discretized Representations of Partial Differential Equations. In *Proceedings of the 3rd International Conference on Informatics in Control, Automation and Robotics (ICINCO 2006)*, pages 16–23, Setúbal, Portugal, August 2006.
5. Hui Wang, Henning Lenz, Andrei Szabo, and Uwe D. Hanebeck. Fusion of Barometric Sensors, WLAN Signals and Building Information for 3–D Indoor/Campus Localization. In *Proceedings of the 2006 IEEE International Conference on Multisensor Fusion and Integration for Intelligent Systems (MFI 2006)*, Heidelberg, Germany, September 2006.
6. David Culler, Deborah Estrin, and Mani Srivastava. Overview of Sensor Networks. *IEEE Computer*, 37(8):41–49, 2004.
7. Tong Zhao and Arye Nehorai. Detecting and Estimating Biochemical Dispersion of a Moving Source in a Semi-Infinite Medium. *IEEE Transactions on Signal Processing*, 54(6): 2213–2225, June 2006.
8. Thomas Bader, Alexander Wiedemann, Kathrin Roberts, and Uwe D. Hanebeck. Model–based Motion Estimation of Elastic Surfaces for Minimally Invasive Cardiac Surgery. In *Proceedings of the 2007 IEEE International Conference on Robotics and Automation (ICRA 2007)*, pages 2261–2266, Rome, Italy, April 2007.
9. Aleksandar Jeremic and Arye Nehorai. Design of Chemical Sensor Arrays for Monitoring Disposal Sites on the Ocean Floor. *IEEE Journal of Oceanic Engineering*, 23:334–343, 1998.
10. Lorenzo A. Rossi, Bhaskar Krishnamachari, and C.-C.Jay. Kuo. Distributed Parameter Estimation for Monitoring Diffusion Phenomena Using Physical Models. In *First Annual IEEE Communications Society Conference on Sensor and Ad Hoc Communications and Networks (SECON 2006)*, pages 460–469, Los Angeles, USA, 2004.
11. Thomas C. Henderson, Christopher Sikorski, Edwart Grant, and Kyle Luthy. Computational Sensor Networks. In *Proceedings of the 2007 IEEE/RSJ International Conference on Intelligent Robots and Systems (IROS 2007)*, San Diego, USA, 2007.
12. Jeffrey Hightower and Gaetano Borriello. Location Systems for Ubiquitous Computing. *IEEE Computer*, 34(8):57–66, August 2001.
13. Andreas Rauh, Kai Briechle, Uwe D. Hanebeck, Joachim Bamberger, Clemens Hoffmann, and Marian Grigoras. Localization of DECT Mobile Phones Based on a New Nonlinear Filtering Technique. In *Proceedings of SPIE, Vol. 5084, AeroSense Symposium*, pages 39–50, Orlando, Florida, May 2003.
14. Ping Tao, Algis Rudys, Andrew M. Ladd, and Dan S. Wallach. Wireless LAN Location-sensing for Security Applications. In *Proceedings of the 2nd ACM Workshop on Wireless Security*, pages 11–20, 2003.
15. Andrew M. Ladd, Kostas E. Bekris, Algis Rudys, Guillaume Marceau, and Lydia E. Kavraki. Robotics-Based Location Sensing using Wireless Ethernet. In *Proceedings of the 8th Annual International Conference on Mobile Computing and Networking*, pages 227–238. ACM Press, 2002.

16. Hui Wang, Henning Lenz, Andrei Szabo, Joachim Bamberger, and Uwe D. Hanebeck. Enhancing the Map Usage for Indoor Location-Aware Systems. In *International Conference on Human-Computer Interaction (HCI 2007)*, Peking, China, July 2007.
17. Patrick Rößler, Uwe D. Hanebeck, Marian Grigoras, Paul T. Pilgram, Joachim Bamberger, and Clemens Hoffmann. Automatische Kartographierung der Signalcharakteristik in Funknetzwerken. October 2003.
18. Teemu Ross, Petri Myllymaki, Henry Tirri, Pauli Misikangas, and Juha Sievanen. A Probabilistic Approach to WLAN User Location Estimation. *International Journal of Wireless Information Networks*, 9(3), July 2002.
19. Marian Grigoras, Olga Feiermann, and Uwe D. Hanebeck. Data-Driven Modeling of Signal Strength Distributions for Localization in Cellular Radio Networks (Datengetriebene Modellierung von Feldstärkeverteilungen für die Ortung in zellulären Funknetzen). *at - Automatisierungstechnik – Automatisierungstechnik, Sonderheft: Datenfusion in der Automatisierungstechnik*, 53(7):314–321, July 2005.
20. Bruno Betoni Parodi, Andrei Szabo, Joachim Bamberger, and Joachim Horn. SPLL: Simultaneous Probabilistic Localization and Learning. In *Proceedings of the 17th IFAC World Congress (IFAC 2008)*, Seoul, Korea, 2008.
21. George E. Karniadakis and Spencer J. Sherwin. *Spectral/hp Element Methods for Computational Fluid Dynamics*. Oxford University Press, 2005.
22. Thomas Schön, Fredrik Gustafsson, and Per-Johan Nordlund. Marginalized Particle Filters for Nonlinear State-space Models. *Technical Report*, Linköping University, 2003.

Study on Spectral Transmission Characteristics of the Reflected and Self-emitted Radiations through the Atmosphere

Jun-Hyuk Choi and Tae-Kuk Kim

Abstract This paper is a part of developing a software that predicts spectral radiance from ground objects by considering spectral surface properties. The material surface properties are essential for determining the reflected radiance by solar energy and the self-emitted radiance from the object surface. We considered the composite heat transfer modes including conduction, convection and spectral solar irradiation for objects within a scene to calculate the surface temperature distribution. The software developed in this study could be used to model the thermal energy balance to obtain the temperature distribution over the object by considering the direct and diffuse solar irradiances and by assuming the conduction within the object as one-dimensional heat transfer into the depth. MODTRAN is used to model the spectral solar irradiation including the direct and diffuse solar energy components. Resulting spectral radiance in the MWIR ($3\sim5$ μm) region and LWIR ($8\sim12$ μm) regions arrived at the sensor are shown to be strongly dependent on the spectral surface properties of the objects geometric features, such as points, straight or curved lines and corners, plays an important role in object recognition. In this paper, we present a model-based recognition of 3D objects.

1 Introduction

The infrared (IR) signature is mainly affected by atmospheric effects, object temperature and surface properties etc. Models for calculating infrared signatures including the targets and background are useful assessment tools of IR signatures. We are encouraged to utilize the software to get the spectral images because it is almost impossible to obtain all of the desired data by measurement due to a large variety of objects, environments, and meteorological conditions [1]. A few of the developed countries possess the infrared image generation software which can deal with the

J.-H. Choi (✉)
Department of Mechanical Engineering, Chung-Ang University, 221 Huksuk-Dong, Dongjak-Ku, Seoul 156-756, South Korea
e-mail: miyasaki@wm.cau.ac.kr

S. Lee et al. (eds.), *Multisensor Fusion and Integration for Intelligent Systems*, Lecture Notes in Electrical Engineering 35, DOI 10.1007/978-3-540-89859-7_27,

targets within a certain designated region. These softwares can also be used to convert normal images to infrared images such as DIRSIG (Digital imaging and Remote Sensing Image Generation, USA, [2]), PRISM (Physically Reasonable Infrared Signature Model, USA, [3]), OKTAL (France, [4]), Ship IR/NTCS (Canada, [5]), SensorVision (Australia, [6]), CameoSim (CAMouflage Electro-Optic SIMulation, United Kingdom, [7]) and RadThermIR (USA, [8]) etc. Not all of these softwares are open to the public otherwise they are very expensive even with the limited applicability. These softwares can be used as the synthetic image generation tools which produce simulated imagery in the visible through thermal infrared region. They are designed to produce broad-band, multi-spectral and hyper-spectral imagery through the integration of a suite of first principles based on the radiation propagation sub models. These sub models are responsible for tasks ranging from the BRDF (Bi-directional Reflectance Distribution Function) predictions of a surface to the dynamic geometry scanning by a line scanning imaging instrument. The DIRSIG model is an integrated collection of independent first principles based models which work in conjunction to produce radiance field images with high radiometric fidelity in the 0.3∼30.0 μm region. The PRISM has been the US Army's standard tool for infrared signature and thermal modeling. The OKTAL has been utilized in the fields of virtual reality scene, synthetic environment design, and 3-D modeling. The ShipIR/NTCS includes generic models of an infrared seeker and IR flare countermeasure to simulate the IR signature of a ship, its flare deployments, and an infrared-guided missile. The SensorVision is a VEGA-based application that produces IR scenes in real time with a certain amount of simplifications in order to obtain the real time imagery. The CameoSim is an advanced IR program aiming at producing high fidelity physics based images originally applied to camouflage assessments. The RadThermIR is a three-dimensional heat transfer program that uses FDM (Finite Difference Methods) to predict the temperature distribution over a target which is used to predict the IR radiance.

The atmosphere radiative transmission introduces both multiplicative terms from absorption features and additive terms in the form of scattered and emitted energy. In the thermal region, reflected atmospheric emission and scattered radiance terms contribute to the target signature. Therefore, atmospheric propagation must be included to realistically synthesize an infrared scene. Atmospheric models, such as LOWTRAN [9], MODTRAN [10], and FASCODE [11] are used to simulate the effects of the atmosphere on radiation.

This study is aimed at the development of a software for analyzing the radiances through the path and from the objects in the scene which include the radiative intensities by self-emission and by reflection of incident radiation. Different material properties are considered in analyzing the thermal balance and the surface radiation. The spectral radiance received by a remote sensor is consisted of the self-emitted component directly from the object surface, the reflected component of the solar irradiance at the object surface, and the scattered component by the atmosphere without ever reaching the object surface. MODTRAN is used to model the scattered radiance by the atmosphere, and the solar radiation including the direct and the diffuse solar energy components.

2 Theoretical Backgrounds

2.1 Calculation of the Surface Temperature

The surface temperature of an object can be determined by considering the energy balance over a finite surface element as shown in Fig. 1. In this paper, the thermal model is based on an efficient one-dimensional, time dependent (unsteady state) equation for calculation of the surface temperature. The resulting equation is expressed as;

$$M_s C_{p,s}\left(\frac{dT_s}{dt}\right) = Q_{cond} + Q_{conv} + Q_{solar} + Q_{emiss} \tag{1}$$

where T_s is the surface temperature and t is time. M_s, $C_{p,s}$ are the mass and specific heat of the surface element. Q_{cond}, Q_{conv} are the conduction and convection heat transfer. Q_{emiss} is the surface emitted radiative energy. Q_{solar} is composed of a direct component ($q_{solar,direct}$ and a diffuse component ($q_{solar,diffuse}$), both depending on the azimuth and elevation angle of the sun. The radiation incident on an object surface depends on the scattering and absorbing processes of the atmosphere. The solar energy absorbed by the surface is determined by considering the surface absorptivity α as;

$$Q_{solar} = \alpha \cdot A_s \left(q_{solar,direct} + q_{solar,diffuse}\right) \tag{2}$$

where $q_{solar,direct}$ and $q_{solar,diffuse}$ are the total solar energy fluxes by direct and diffuse irradiations.

1. *Direct Solar Heat Flux:* The direct solar radiation is reached on an object surface after transmitting through the atmosphere and calculated by using the MODTRAN. The total direct solar radiation incident to a horizontal surface is calculated as;

$$q_{solar,direct} = \int_0^\infty I_{solar,direct}(\lambda)\cos\theta_s \, d\lambda \tag{3}$$

   where θ_s is the solar angle measured from the surface normal axis.
2. *Diffuse Solar Heat Flux:* The diffuse solar radiation is reached on an object surface from all directions exposed to the atmospheric environment through multiple scattering in the atmosphere and is calculated by considering the directional integration using the T_N quadrature [12]. The spectral diffuse solar radiation components obtained by using the MODTRAN are then used to obtain the hemispherical diffuse solar flux by using the T_N quadrature with 100 solid angles in hemisphere ($N = 5$). The total diffuse solar radiation incident to the object surface is then calculated as;

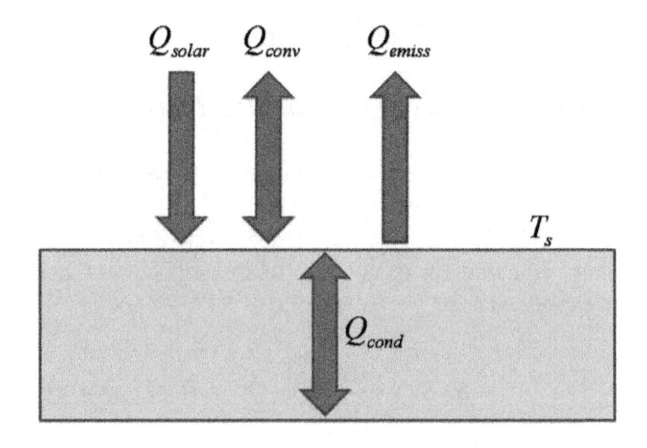

Fig. 1 Energy balance on a surface element

$$q_{solar,direct} = \int_0^\infty \int_{\hat{\omega}\cdot\hat{n}<0} I_{solar,diffuse}(\lambda,\omega)\cdot\hat{n}\,d\omega\,d\lambda \tag{4}$$

where $\hat{n}$ is a unit normal vector to the object surface and $d\omega$ is the solid angle.

2.2 Atmospheric Transmittance Model

In this study, the solar irradiation and path radiation are computed by using the MODTRAN. The MODTRAN is probably the most widely used and readily available of the propagation models. This model assumes that the atmosphere is divided into a number of homogeneous layers. The temperature of each layer can be determined from radiosonde data acquired at the time of data collection or from generic profiles stored in the MODTRAN model. The concentration of the permanent gases and water vapor can be estimated from radiosonde air pressure and relative humidity data as a function of altitude.

The MODTRAN models the atmosphere as many individual layers, each of which exhibits either pre-defined or user-specified meteorological conditions, atmospheric composition of gases, aerosol type and specific scattering phase function, as well as global position. The MODTRAN provides several standard atmospheric profiles if no detailed data are available from the time of the collection. The MODTRAN calculates atmospheric transmittance, atmospheric background radiance, single scattered solar and lunar radiance, direct solar irradiance, and multiple scattered and thermal radiance by using the supplied input data such as the perpendicular distributions of temperature, gas component etc through the atmosphere. The spectral resolution of the model used for the MODTRAN4 is $1\,\mathrm{cm}^{-1}$, but spectral bands of $20\,\mathrm{cm}^{-1}$ are considered for calculation of the spectral solar radiation and the spectral scattered radiance ranging from 0 to $50,000\,\mathrm{cm}^{-1}$. The spectral atmospheric solar transmittance and spectral direct solar radiation on July at mid-latitude region obtained by using the MODTRAN4 are depicted in Figs. 2 and 3.

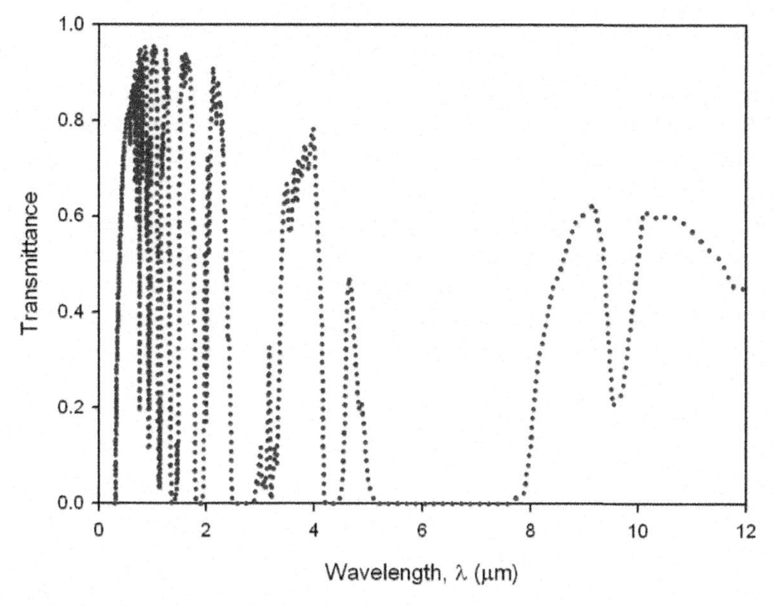

Fig. 2 Spectral atmospheric solar transmittance in July

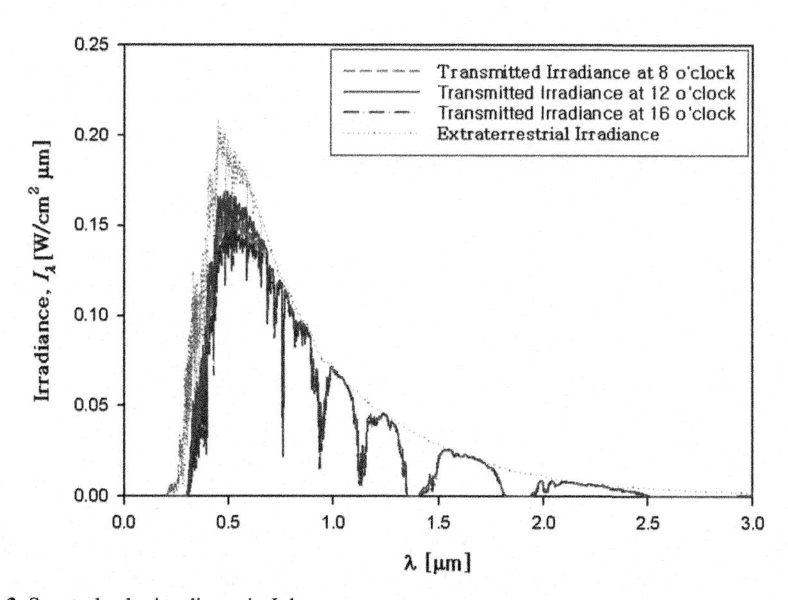

Fig. 3 Spectral solar irradiance in July

2.3 Spectral Radiance Received by a Remote Sensing

Infrared signatures by MWIR (Mid Wave Infra-Red, 3~5 μm) and LWIR (Long Wave Infra-Red, 8~12 μm) regions sensors are being increasingly used for a variety of remote sensing applications. In this paper, we analyzed the infrared signatures received by a remote sensor considering MWIR and LWIR radiances. The spectral radiance received by a remote sensor is consisted of the components emitted directly from the object surface, reflected at the object surface by the solar irradiance, and scattered by the atmosphere without ever reaching the object surface as shown in Fig. 4 [13–15]. The resulting equation for the radiance at wavelength λ_0 (wavelength band between λ_1 and λ_2) received by a remote sensor is expressed as;

$$I(\lambda_0) = I_{self,emitted}(\lambda_0) + I^r_{solar,direct}(\lambda_0) + I^r_{solar,diffuse}(\lambda_0) + I_{path}(\lambda_0) \quad (5)$$

where $I_{self,emitted}(\lambda_0)$ is the emitted radiance from the object surface. $I^r_{solar,direct}(\lambda_0)$ and $I^r_{solar,diffuse}(\lambda_0)$ are the reflected radiances from the ground object due to the direct and diffuse solar components. $I_{path}(\lambda_0)$ is the path-scattered radiance which is a combination of the path emission and the atmospheric scattering.

1. *Self-Emitted Radiance:* The self-emitted radiance from the object is calculated as;

$$I_{self,emitted}(\lambda_0) = \int_{\lambda_1}^{\lambda_2} \tau_a(\lambda) \cdot \varepsilon(\lambda) \cdot I_{\lambda b}(T_s)\, d\lambda \quad (6)$$

where $\tau_a(\lambda)$ is the spectral atmospheric transmittance between the object and the sensor. The blackbody intensity is obtained from the Plank's law as;

$$I_{\lambda b}(T_s) = \frac{C_1}{\lambda^5 \left(e^{C_2/\lambda T_s} - 1\right)} \quad (7)$$

where T_s is the absolute temperature of the surface. $\varepsilon(\lambda)$ is the spectral emissivity of the surface.

$$C_1 = 0.59552197 \times 10^8 \left[W \cdot \mu\text{m}^4/\text{m}^2 \cdot sr\right]$$
$$C_2 = 14,387.69 \left[\mu\text{m} \cdot \text{K}\right].$$

2. *Direct Solar Radiance:* The reflected spectral radiance of the direct solar radiation from the object surface that is directed to the remote sensor can be calculated as;

$$I^r_{solar,direct}(\lambda_0) = \int_{\lambda_1}^{\lambda_2} \tau_a(\lambda) \cdot \rho(\lambda) \cdot I_{solar,direct}(\lambda)\, d\lambda \quad (8)$$

where $I_{solar,direct}(\lambda)$ is the direct spectral solar intensity and $\rho(\lambda)$ is the spectral reflectivity of the object surface which is assumed to be diffuse by considering normal incidence.

3. *Diffuse Solar Radiance:* The reflected spectral radiance of the diffuse solar radiation from the hemisphere over a flat surface that is directed to the remote sensor can be calculated as;

$$I^r_{solar,diffuse}(\lambda_0) = \int_{\lambda_1}^{\lambda_2} \int_{2\pi} \tau_a(\lambda) \cdot \rho(\lambda) \cdot I_{solar,\,diffuse}(\lambda, \omega)\, d\omega\, d\lambda \quad (9)$$

where $I_{solar,diffuse}(\lambda, \omega)$ is the diffuse spectral solar intensity.

4. *Path-Scattered Radiance:* The path-scattered radiance is consisted of molecular rayleigh scattering for a clear atmosphere and mie scattering for an atmosphere with aerosols (water vapor) or particulates (dust, smoke). In MODTRAN the path-scattered radiance is obtained by summing up the radiance contribution of each layer segment. The path-scattered radiance between the sensor location and the ground object as depicted in Fig. 4 is given by [16]:

$$I_{path}(\lambda_0) = \int_0^S \tau_a\left(S^*, \lambda\right) J\left(S^*, \lambda\right) dS^* \quad (10)$$

where S^* and S are atmospheric optical depth from the sensor to the object along the LOS (Line-Of-Sight). J is the total source term including the path emitted and solar scattered components which are obtained by using the MODTRAN.

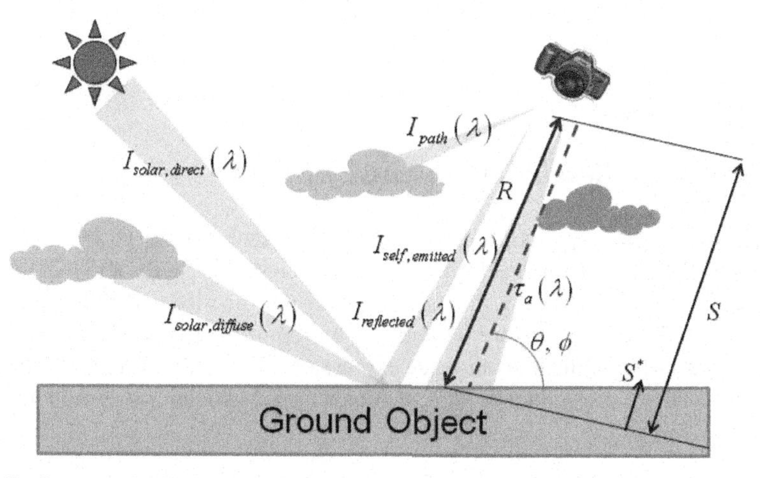

Fig. 4 Radiance received by a remote sensor

3 Numerical Demonstrations

3.1 Conditions Applied

To examine the capabilities of the software developed in this study, the system depicted in Fig. 4 is used to obtain the spectral radiance received by the remote sensor. The horizontal surface element is assumed to be located at the geographical location described in Table 1.

Table 1 Geographical information of the object considered

Location		Seoul, South Korea
Latitude		37.34N
Longitude		126.34E
Sensor angle and range	Zenith angle ($\theta\theta$)	15°
	Azimuth angle (ϕ)	0°
	Range (R)	1 km, 5 km, 20 km

Two different materials of asphalt and aluminum are considered to determine the infrared signatures to be received by the sensor. The thermodynamic properties of the materials considered are listed in Table 2. Figures 5 and 6 shows the spectral emissivities and reflectivities of asphalt and aluminum (ImageMapper, [17]). The solar absorptivities of the materials considered are determined by simply averaging the spectral emissivities over the whole spectrum and are given in Table 2. A uniform wind is assumed to be directed from south to north with the velocity of 2 m/s.

Table 2 Material properties considered

Material	Asphalt	Aluminum
Density	2,115 kg/m^3	7,870 kg/m^3
Specific heat	920.0 J/kg·K	447.7 J/kg·K
Thermal conductivity	0.062 W/m·K	71.965 W/m·K
Total absorptivity (α)	0.94	0.28

3.2 Numerical Results and Discussions

The IR signatures vary according to range (R) between the object and the sensor. The MODTRAN band models [10] assume that the line strengths and locations are statistical, and their accuracy increases as both the bin size and number of lines increase. Model parameters are determined by requiring agreement in the strong-line and weak-line of the transmittance expressions. In MODTRAN, the band averaged atmospheric transmittance (τ_a), at wavelength λ around a spectral band ($\Delta\lambda$) is given by [18]:

$$\tau_a(\lambda) = \left(\frac{2}{\Delta\lambda} \int_0^{\Delta\lambda/2} e^{-S\,u\,b(\lambda)}\, d\lambda \right)^n \tag{11}$$

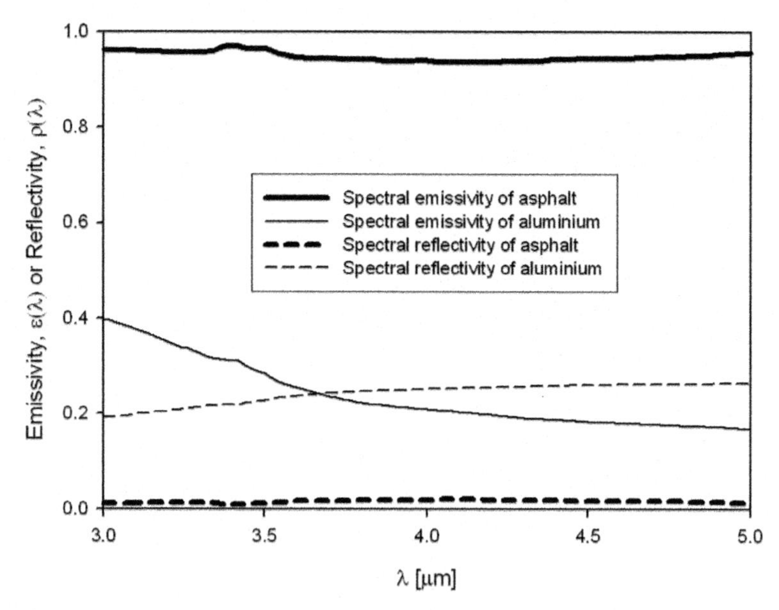

Fig. 5 Spectral emissivities and reflectivites in MWIR region

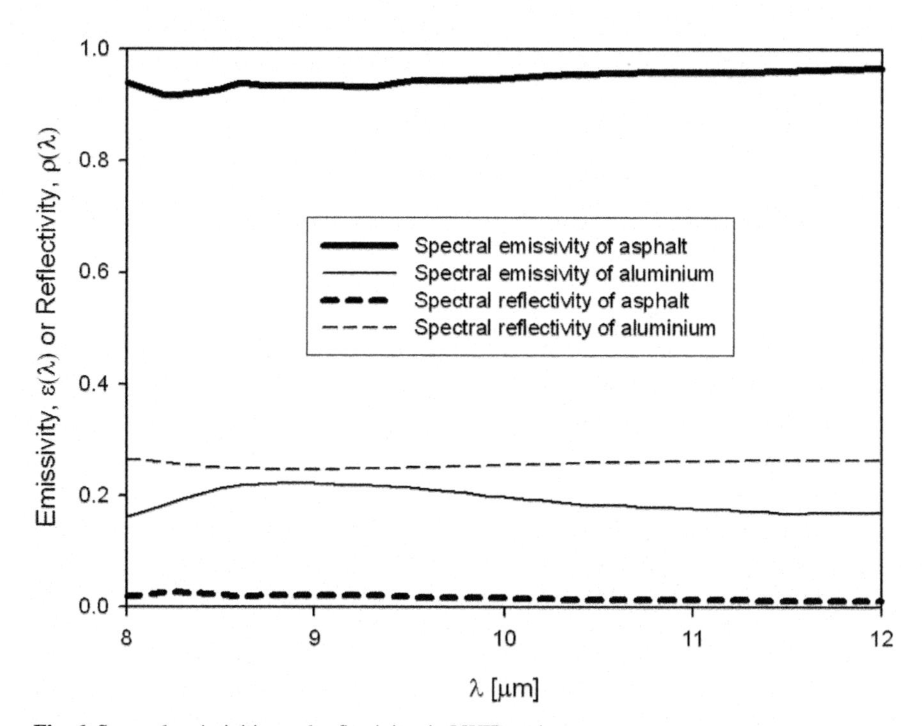

Fig. 6 Spectral emissivities and reflectivites in LWIR region

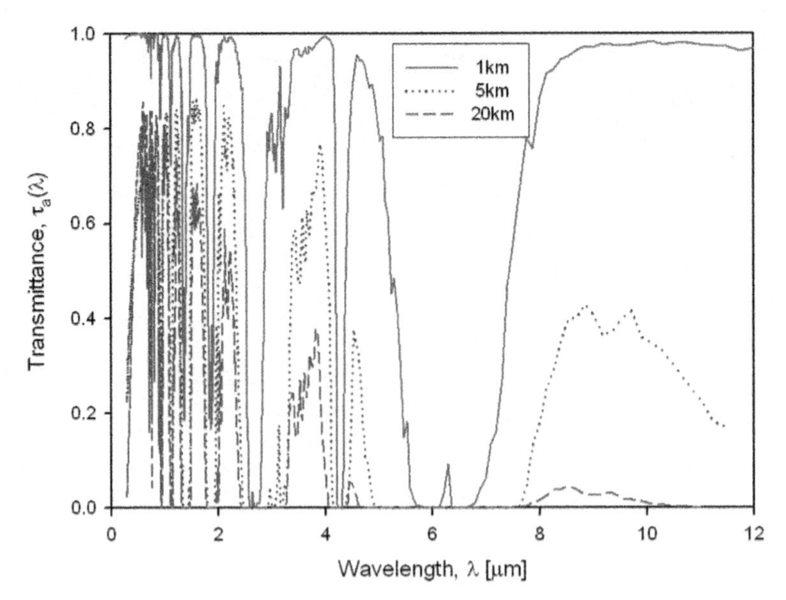

Fig. 7 Spectral atmospheric transmittance between the object and the sensor in July ($\theta = 15°$, $\phi = 0°$, $R = 1, 5, 20$ km)

where $b(\lambda)$ is the Voigt shape function, u is the absorber amount, and n is the average number of lines.

The spectral atmospheric transmittances between the object and the sensor in July obtained by using the MODTRAN with the data in Table 1 are depicted in Fig. 7.

The spectral radiance to be received by the remote sensor is computed for given conditions including the components emitted directly from the object surface, reflected at the object surface by the solar irradiation, and scattered by the atmosphere without ever reaching the object surface. The major spectral region used for remote sensing are the MWIR and LWIR regions because they contain relatively transparent atmospheric windows.

1. *Daily MWIR Results:* Figures 8 and 9 show the daily distributions of the radiances in MWIR from asphalt and aluminum for the three different distances considered between the sensor and the object. As shown in these figures the MWIR radiances from the asphalt are nearly 4 to 5 times higher than those from the aluminum. And the MWIR radiance signals transmitted through the atmosphere are strongly dependant on the range which many reduce the signal level up to 1/6 at 20 km range as compared to those at 1 km range.
2. *Daily LWIR Results:* The results obtained from this study show that the strength of LWIR signals are much stronger than the strength of MWIR signals. This is because the surface temperatures of the object considered in this study are relatively low temperature. Figures 10 and 11 show the daily distributions of the radiances in LWIR from asphalt and aluminum for the three different distances

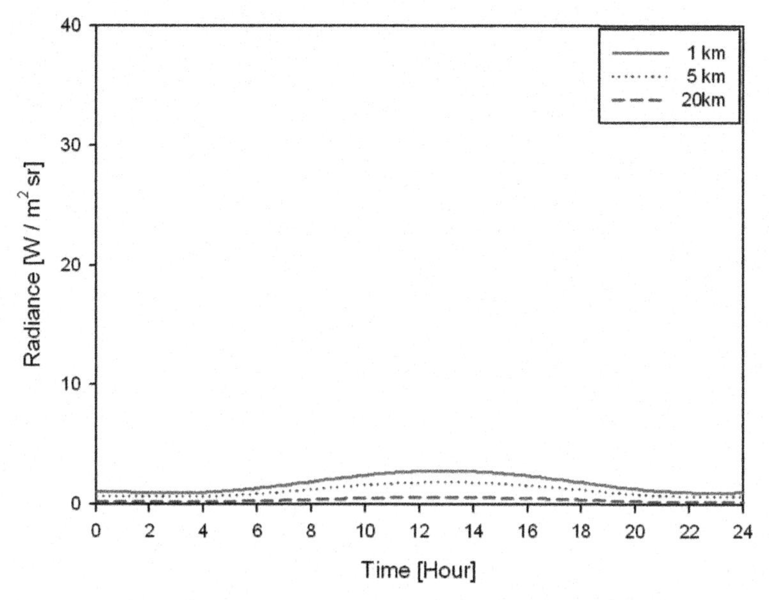

Fig. 8 Daily profiles of the radiances in MWIR region (July, Asphalt)

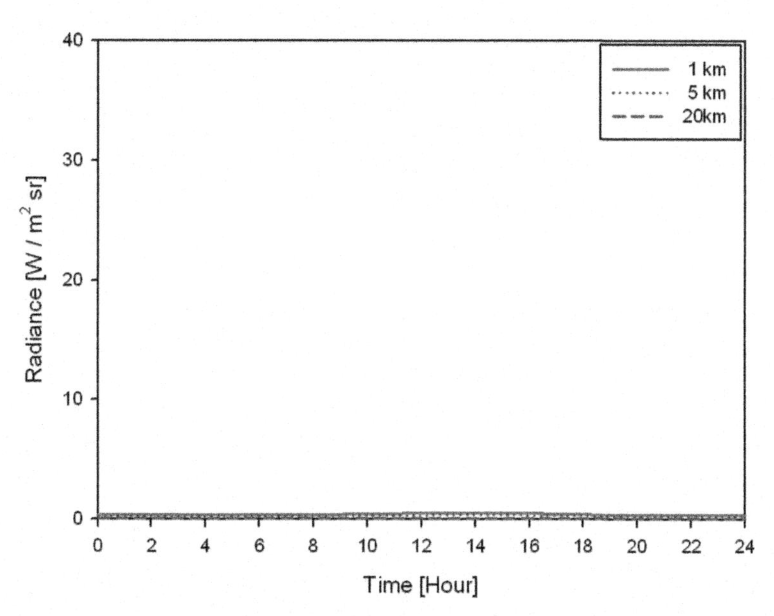

Fig. 9 Daily profiles of the radiances in MWIR region (July, Aluminum)

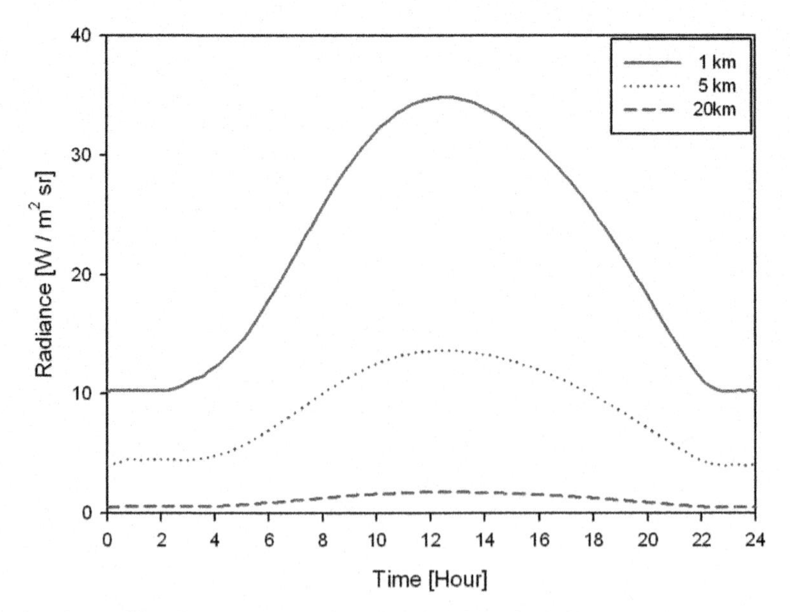

Fig. 10 Daily profiles of the radiances in LWIR region (July, Asphalt)

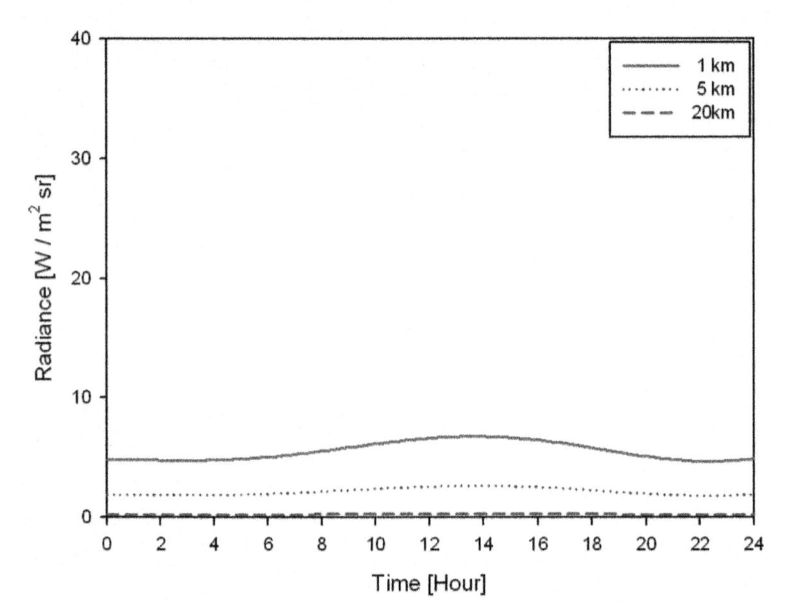

Fig. 11 Daily profiles of the radiances in LWIR region (July, Aluminum)

considered between the sensor and the object. As shown in these figures the LWIR radiances from the asphalt are approximately 6 times higher than those from the aluminum. And the LWIR radiance signals transmitted through the atmosphere are strongly dependant on the range which many reduce the signal level up to 1/17 at 20 km range as compared to those at 1 km range.

4 Conclusion

In this study, a computational tool to obtain the infrared radiance received by a remote sensor is developed and tested. The infrared radiance is obtained by considering the surface temperature, surface reflectivity and emissivity which are important parameters in estimating the upwelling radiance calculation. The resulting radiances received by the remote sensor are compared for different spectral radiative surface characteristics considered. The IR signature levels in LWIR show approximately 6 times those of the MWIR signals, and the IR signature level of an object is affected significantly by different surface characteristics; signals from asphalt show up to 4∼6 times stronger than those from aluminum. The IR signals are attenuated significantly due to the atmospheric transmission; the signal levels at 20 km show about 1/6 of those at 1 km. The radiance results obtained by using the software developed here show a reasonable trend expected and the software can be further developed for infrared image generation. Future improvements include covering the bi-directional reflectance characteristics and variable sensor characteristics.

Acknowledgments This research was supported by the Defense Acquisition Program Administration and Agency for Defense Development, Korea, through the Image Information Research Center at Korea Advanced Institute of Science & Technology under the contract UD070007AD.

References

1. Choi, J.-H., and Kim, T.-K. Study on the spectral transmission characteristics of MWIR through the atmosphere, *Proceeding of SPIE, Image and Signal Processing for the Remote Sensing XII*, vol. 6748, 2007.
2. Schott, J. R., Brown, S. D., Raqueno, R. V., Gross, H. N., and Robinson, G. An advanced synthetic image generation model and its application to multi/hyperspectral algorithm development, *Canadian Journal of Remote Sensing*, 25(2), June 1999, pp. 99–111.
3. Johnson, K. R. Implementation of Implicit Routine in PRISM, *MTU/KRC Technical Report for TACOM*, July 1992.
4. Le Goff, A., Kersaudy, P., Latger, J., Cathala, T., Stolte, N., and Barillot, P. Automatic temperature computation for realistic IR simulation, *Proceeding of SPIE, Targets and Backgrounds VI*, vol. 4029, 2000, pp.187–196.
5. Vaitekunas, D. A. Validation of ShipIR (v3.2): methodology and results,*Proceeding of SPIE, Targets and Backgrounds XII*, 2006.

6. Duong, N., and Wengener, M. SensorVision Validation: diurnal temperature variations in northern Australia, *Proceeding of SPIE, Technologies for Synthetic Environments: Hardware-in-the-Loop Testing V*, vol. 4027, 2000.
7. Kirk, A., Cowan, M., and Allen, R. CAMEO-SIM. An Ocean Model Extension to the physically accurate broadband EO scene generation system for the assessment of target vehicles within their natural environments, *Proceeding of SPIE, Targets and Backgrounds X: Characterisation and Representation*, vol. 5431, 2004.
8. Sanders, J. S. Ground target infrared signature modeling with the multi-service electro-optic signature (MuSES) code, *Proceeding of SPIE, Targets and Backgrounds VI*, 2000.
9. Neizys, F. X., Shette, E. P., Abreu, L. W., Chetwynd, J. H., Anderson, G. P., Gallery, W. O., Selby, J. E. A., and Clough, S. A. Users guide to LOWTRAN7, *Environmental Research Papers*. No. 1010 (1988).
10. Acharya, P. K., Berk, A., Anderson, G. P., Larsen, N. F., Tsay, S.-C. and Stammes, K. H. MODTRAN4: Mutiple scattering and bi-directional distribution function (BRDF) upgrades to MODTRAN, *Proceeding Of SPIE, Optical Spectroscopy Techniques and Instrumentation of Atmospheric and Space Research*, vol. 3756, 1999.
11. Clough, S. A., Kneizys, F. X., Shettle, E. P., and Anderson, G. P. Atmospheric radiance/transmission: FASCODE2, *Proceedings of the Sixth Annual Conference of Atmospheric Radiation*, 1986.
12. Thurgood, C. P., Pollard, A., and Becker, H. A. The TN quadrature set for the discrete ordinates methods, *Transaction of ASME* (*Journal of Heat Transfer*), 117, 1995, pp. 1068–1070.
13. Schowengerdt, R. A. *Remote Sensing: Models and Methods for Image Processing*, Academic Press, San Diego, CA, 1997.
14. Schott, J. R. *Remote Sensing: The Image Chain Approach*, Oxford University Press, New York, 1997.
15. Jacobs, P. A. *Thermal Infrared Characterization of Ground Targets and Background*, SPIE Press, Bellingham, Washington USA, 2006.
16. Acharya, P. K., Berk, A., Anderson, G. P., Larsen, N. F., Tsay, S.-C., and Stamnes, K. H. MODTRAN4: Multiple and bi-directional reflectance distribution function (BRDF) upgrades to MODTRAN, *Proceeding of SPIE, Optical Spectroscopic Techniques and Instrumentation for Atmospheric and Space Research III*, vol. 3756, 1999.
17. ImageMapper 2.0.5, http://www.surfaceoptics.com/
18. Acharya, P. K., Robertson, D. C., and Berk, A. Upgrade line-of-sight geometry package and band model parameters for MODTRAN, *Scientific Report No. 5*, 1993.

3D Reflectivity Reconstruction by Means of Spatially Distributed Kalman Filters

G.F. Schwarzenberg, U. Mayer, N.V. Ruiter and U.D. Hanebeck

Abstract In seismic, radar, and sonar imaging the exact determination of the reflectivity distribution is usually intractable so that approximations have to be applied. A method called synthetic aperture focusing technique (SAFT) is typically used for such applications as it provides a fast and simple method to reconstruct (3D) images. Nevertheless, this approach has several drawbacks such as causing image artifacts as well as offering no possibility to model system-specific uncertainties. In this paper, a statistical approach is derived, which models the region of interest as a probability density function (PDF) representing spatial reflectivity occurrences. To process the nonlinear measurements, the exact PDF is approximated by well-placed Extended Kalman Filters allowing for efficient and robust data processing. The performance of the proposed method is demonstrated for a 3D ultrasound computer tomograph and comparisons are carried out with the SAFT image reconstruction.

Keywords Data association · Extended Kalman filter · Synthetic aperture focusing technique · 3D image reconstruction

1 Introduction

The determination of the reflectivity distribution of a region of interest (ROI) addresses a wide area of applications. Application fields may be found in radar imaging of the earth [1], sonar imaging of the ocean bed [2], seismic imaging of the earth's crust [3] as well as in medical application based on ultrasound imaging systems [4].

The measurement setup for analyzing reflectivity distributions regarded in this paper consists of an arbitrarily distributed sensor network that acquires reflectivity information about a ROI. Additionally, unfocussed transmission of pulses is regarded as this leads to faster data acquisition especially for 3D applications. For each emitter and receiver one data set is acquired, which are then fused to achieve

G.F. Schwarzenberg (✉)
Institute for Data Processing and Electronics (IPE), Forschungszentrum Karlsruhe, Germany
e-mail: schwarzenberg@ipe.fzk.de

S. Lee et al. (eds.), *Multisensor Fusion and Integration for Intelligent Systems*, Lecture Notes in Electrical Engineering 35, DOI 10.1007/978-3-540-89859-7_28,

high resolution and high contrast images. By shifting the transducers to different positions, e.g., the movement of an airplane in synthetic aperture radar, the sensing aperture is increased.

The reconstruction of the unknown reflectivity from the measured data is called an inverse problem [5]. The inverse problem refers to the situation of knowing the incidence field of the emitter and the measured data of the receivers and trying to reconstruct the object causing the variation of the incidence field. This requires a precise knowledge of the characteristics of the transducers as well as the physics behind the propagation of the incidence wave and its interaction with the object under investigation. The mathematical solution of this inverse scattering problem is intractable, thus approximation schemes are applied to yield an analytic [6] or a numerical [7] solution. The solution with the smallest number of approximations is known as diffraction tomography [8], which is, however, computationally expensive as the solution has to be determined iteratively.

To overcome these difficulties, the 3D reflectivity distribution is described by a statistical approach. For this purpose the reflectivity in the ROI is modeled as a probability density function (PDF), which may be of arbitrary shape representing sharp peaks (point scatterers) and structural information about the object under study. The PDF is approximated by spatially well-placed Extended Kalman Filters, each of them estimating a local reflectivity. After the complete data set is processed, the estimates of all Kalman Filters are fused to construct a global reflectivity estimate in an efficient and robust way.

The paper is organized as follows: Sect. 2 gives a general overview on the problem of determining the reflectivity distribution measured by an arbitrarily placed sensor network and presents the general key idea. In Sect. 3, the proposed solution is introduced and explained. Section 4 presents an application of the proposed image reconstruction method on a 3D ultrasound computer tomograph and compares the results with the usually applied synthetic aperture focusing technique approach.

2 Problem Formulation

The problem addressed in this paper is an image reconstruction problem. As uncertainties in the overall system and measurement process cause this problem to be ill-posed [9], the solution for the inverse problem is intractable for the considered system setup. In order to render this nonlinear inversion problem tractable, the first-order Born approximation is employed, i.e., the incidence field at each scatterer is assumed to be the only source, neglecting the scattered fields from other scatterers. Furthermore, the refraction of the emitted pulse is ignored.

Each reflection acquired by a receiver is the integral of reflectivity (acoustic impedance, electrical permittivity) over a hypersurface, see Fig. 1.

By intersecting data sets of different receivers, the source can be located if just one reflector is present. The naïve solution would be to calculate the intersection of all ellipsoids. Both analytic and numerical approaches are very time-intensive. Additionally, in the presence of spatial noise or noise in the data preprocessing the

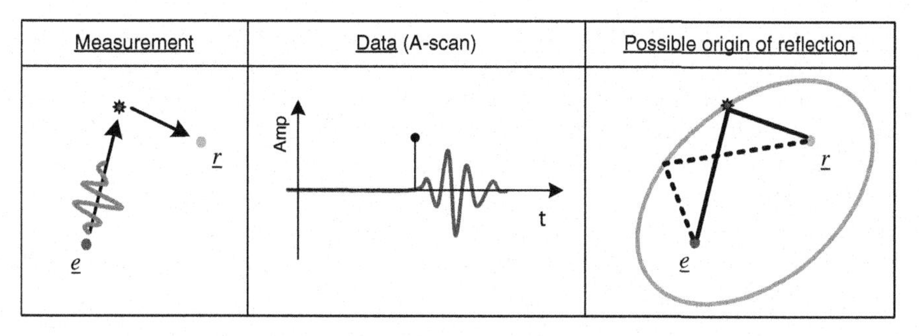

Fig. 1 Measurement and data interpretation: On the left an arbitrarily placed emitter $\underline{e}$ emits a pulse that is reflected by a scatterer and recorded by an arbitrarily placed receiver $\underline{r}$. The measurement of the receiver is plotted in the middle as an amplitude-over-time plot (A-scan). The value of interest is the time-of-arrival as shown by the flag. From only one measurement the exact position of the scatterer can not be derived, only a region containing the scatterer. Assuming a constant speed of propagation of the emitted pulse, this region becomes elliptical with emitter $\underline{e}$ and receiver $\underline{r}$ as focal points, as each position on this ellipse has the same summed distance to emitter and receiver (e.g., the solid and the dashed line). In 3D this elliptical region becomes a rotation symmetric ellipsoid, more precisely a prolate spheroid

intersections are not exact and statistical or heuristic measures have to be applied to compensate for deviations from the ideal intersection areas.

Since under the Born approximation each measurement restricts the position of the source on a hypersurface in 3D, only a few points on the surface are true scatterer positions. All others may be considered as false-positives and represent themselves as ghosts in the resulting reconstructed reflectivity image. The fusion of several measurements may therefore be interpreted as a data association problem. The complexity class of the corresponding optimal Bayesian solution is NP-hard and while further research in this area is still in progress, the data association problem is not explicitly addressed in this paper.

In addition, real objects do not consist of ideal point scatterers. Regarding non-isotropic scattering, damping, and the characteristics of the finite-sized transducers, reflections from the object are not present in all acquired measurements. This forbids solutions that fuse the measurements in a multiplicative manner, which would cause blindness for directive scatterers.

In order to avoid intersection calculations and regard realistic scattering behavior, the underlying space is sampled to create localized reflectivity estimates. The key idea is to keep track of all true- and false-positive reflections in the sampled volume and to estimate a probability density that denotes for each point the probability of being the source of a reflection. These local estimates enable the consideration of system-specific properties of the measurement process, e.g., regarding sensor characteristics with respect to the location of the estimator. After processing all measurements, these distributed samples are fused to an approximation of the global PDF, which is then used to create an image of reflectivity.

Here the question arises, how those local samples are represented and how the nonlinear measurements (hypersurfaces) are applied to update the PDF. Since the complete data set of the 3D sensor system is usually too large to be processed altogether, it is of interest to obtain a recursive update formulation. Additionally, due to the numerous error parameters interfering with each other, the overall measurement error is modeled as normal probability density, as stated by the central limit theorem. The Kalman Filter as an optimal recursive filter under the condition of normal distributions for the system model and noise is a good choice for representing the local samples. In order to handle the nonlinearity of the measurement data, the Extended Kalman Filter (EKF) is applied, which linearizes the model equations using a first order Taylor series approximation [10].

3 Reflectivity Probability Density

This section outlines the basic model setup derived to approximate the PDF of a 3D reflectivity distribution from the measurements of a distributed sensor network. At first, the approximation of the exact PDF is introduced, followed by the model equations of the EKF. Then an efficient measurement to filter assignment is presented and the update of the PDF is explained, concluded with the creation of an image based on the PDF.

3.1 PDF Approximation

The exact PDF is approximated by distributed Extended Kalman Filters that cover the ROI (Fig. 2). Each of them is a local estimator of reflectivity. The state vector of the filter is composed of the 3D position of the reflector and can be extended by any other parameter that may be extracted robustly from the raw input data, e.g., a frequency analysis by means of a short-time Fourier transform of the current echo. These parameters are an additional aid for the improvement of the reflectivity estimate of the object under study.

3.2 Extended Kalman Filter

In the following scenario, we assume a discrete-time dynamic system with linear system model, but nonlinear measurement model for the local estimates,

$$\begin{aligned} x_k &= \mathbf{A_{k-1}} x_{k-1} + w_{k-1}, \\ z_k &= h_k(x_k, v_k), \\ w_k &\sim \mathcal{N}(0, \mathbf{Q_k}), \\ v_k &\sim \mathcal{N}(0, \mathbf{R_k}), \\ \mathrm{e} v_k w_j^T &= 0 \qquad (\forall\, k \neq j). \end{aligned} \tag{1}$$

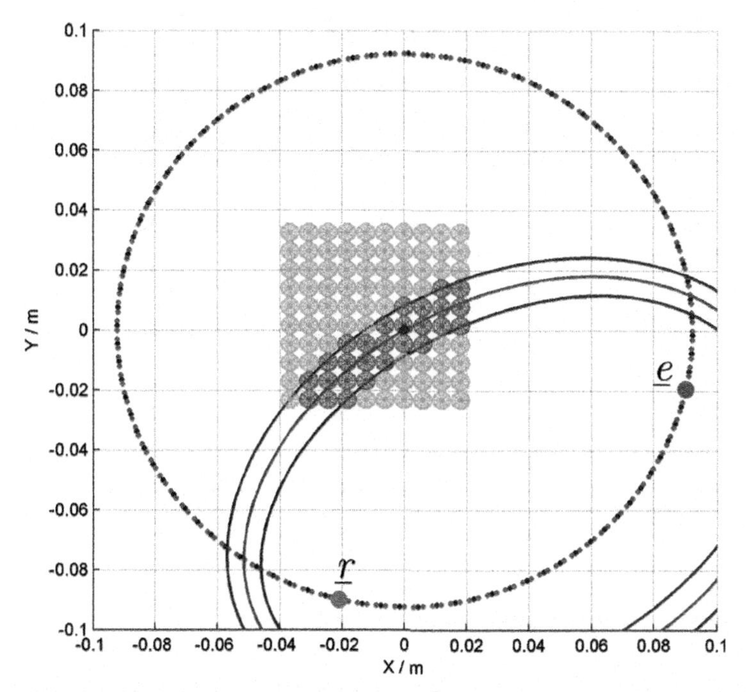

Fig. 2 Example for Kalman Filter placement and affected filters for one measurement. The large dashed circle represents the sensing aperture with one selected emitter $\underline{e}$ and receiver $\underline{r}$. One measurement (time-of-arrival) covers an elliptical region and the measurement noise determines the thickness of this region. The ROI is covered with Kalman Filters whose initial covariances are plotted as spheres with the standard deviation as radius. As shown, the current measurement does only affect a subset of all filters, which is exploited for an efficient measurement update

Here, x_k is a $n\times 1$ state vector at time step k, containing at least the $[x, y, z]^T$ position of a local reflection. $\mathbf{A_{k-1}}$ is the time-variant linear $n\times n$ system matrix, which relates the state vector at time step $k-1$ to time step k, disturbed by the system noise w_k drawn from a zero mean multivariate normal distribution with covariance $\mathbf{Q_k}$ of size $n\times n$. z_k is a $d_z\times 1$ measurement vector, which is a scalar if only the time-of-arrival (TOA) is taken into consideration. $h_k(x_k, v_k)$ is the time-variant, nonlinear measurement function, which returns a $d_z\times 1$ estimate of the next measurement, given x_k and the zero mean Gaussian white measurement noise v_k with covariance $\mathbf{R_k}$ of size $d_z\times d_z$.

The time update equations of the Kalman Filter are given as [11]

$$
\begin{aligned}
x_k^- &= \mathbf{A_{k-1}} x_{k-1}^+, \\
\mathbf{P_k^-} &= \mathbf{A_{k-1}} \mathbf{P_{k-1}^+} \mathbf{A_{k-1}}^T + \mathbf{Q_k}.
\end{aligned}
\tag{2}
$$

Based on the system model $\mathbf{A_{k-1}}$, the state x_k^- as well as the estimate covariance $\mathbf{P_k^-}$ is predicted for the current time step k.

The measurement update equations as defined by the Extended Kalman Filter are [10],

$$\begin{aligned}
\mathbf{S_k} &= \mathbf{H_k P_k^- H_k}^T + \mathbf{R_k}, \\
\mathbf{K_k} &= \mathbf{P_k^- H_k}^T \mathbf{S_k}^{-1}, \\
x_k^+ &= x_k^- + \mathbf{K_k}(z_k - h_k(x_k^-, 0)), \\
\mathbf{P_k^+} &= (\mathbf{I} - \mathbf{K_k H_k})\mathbf{P_k^-}.
\end{aligned} \tag{3}$$

The measurement prediction covariance $\mathbf{S_k}$ is the expected value of the innovation, calculated by means of the predicted covariance $\mathbf{P_k^-}$ and the linearization $\mathbf{H_k}$ of the measurement function h_k. The $d_z \times n$ Jacobian matrix $\mathbf{H_k}$ is dependent on the predicted state x_k^-,

$$\mathbf{H_k} = \frac{\partial h_k}{\partial x_k}(x_k^-, 0). \tag{4}$$

$\mathbf{K_k}$ is the $n \times d_z$ Kalman gain, which is used to update the state estimate to x_k^+ and its covariance $\mathbf{P_k^+}$.

3.3 Measurement to Filter Assignment

Each measurement of an emitter-receiver-combination is an amplitude-over-time signal (A-scan). For reflectivity imaging, parameters such as TOA, amplitude, frequency or phase information of each echo need to be extracted in a preprocessing step.

As demonstrated in Fig. 1, each preprocessed echo does only affect a specific volume (ellipsoidal shell) from where the possible scatterer(s) caused the reflection. The size of this volume increases with higher uncertainty of the preprocessing step. Therefore, a filter is only affected if it is close to this volume. This is exploited to achieve an efficient processing of the Kalman Filter updates as only a subset of all filters need to be updated for each measurement.

This is achieved by using one part of the Kalman Filter equations. The quadratic form of the innovation covariance $\mathbf{S_k}$ (Eq. (3)) may be regarded as a squared norm, weighted according to the filter covariance matrix $\mathbf{P_k^-}$ (Mahalanobis distance). This statistical distance is used to define a set $\mathcal{S}_k$ of Kalman Filters at the positions x_k that are updated by the current measurement z_k,

$$\begin{aligned}
\tilde{y}_k &= z_k - h_k(x_k^-, 0), \\
\mathcal{S}_k(\gamma) &= \left\{ x_k : \tilde{y}_k^T \mathbf{S_k^{-1}} \tilde{y}_k \le \gamma \right\}.
\end{aligned} \tag{5}$$

$\mathcal{S}_k(\gamma)$ is χ^2 distributed with d_z degrees of freedom. γ is selected beforehand and kept constant during the application.

An example for this gating procedure is shown in Fig. 2, where the possible origin of one measured reflection for the marked emitter and receiver is shown as elliptic region. The central ellipse represents the current measurement, the bounding ellipses represent the error interval defined by, e.g., measurement noise or the imprecise knowledge of the speed of propagation of the emitted pulse.

3.4 Independent Kalman Filters

Before processing the acquired data for reconstruction, Kalman Filters are placed throughout the ROI at a desired resolution. This initializes the state vector of each Kalman Filter.

As each TOA measurement is the integral of reflectivity along an ellipsoidal shell, multiple positions and thereby multiple Kalman Filters are affected by the same data. Nevertheless, in this paper the Kalman Filters are assumed to be independent of each other to avoid the large increase in complexity.

3.4.1 Update of the Kalman Filters

The nonlinear measurement function h_k (Eq.(1)) returns the summed travel time of the emitted pulse at the propagation speed v between x_k^- to the emitter $\underline{e}$ and x_k^- to the receiver $\underline{r}$,

$$h_k(x_k^-, 0) = \frac{\|x_k^- - \underline{e}\| + \|x_k^- - \underline{r}\|}{v}. \tag{6}$$

The Jacobian matrix $\mathbf{H_k}$ of this function as defined in equation (4) equals the normal vector of an ellipsoid through the position x_k^- with the focal points $\underline{e}$ and $\underline{r}$.

The update of the distributed Kalman Filters with a new measurement z_k is performed as follows: First, the gating procedure is applied to determine those Kalman Filters that have to be updated (Fig. 3(a)). The value of γ (Eq. (5)) is set accordingly so that it represents a reasonable amount of space, from where the measurement z_k could have originated. Then $\mathbf{H_k}$ and h_k are determined for each Kalman Filter state x_k^-. The update with z_k causes the filter state to be shifted along the local normal (defined by $\mathbf{H_k}$) on the ellipsoid that is defined by z_k and the emitter and receiver position (Fig. 3(b)). The update of the covariance matrices results in smaller eigenvalues along the local normal vector. During the processing of different spatial emitter-receiver-combinations, the eigenvalues of the covariance matrices are reduced along different normal vectors (Fig. 3(d)). For example all eigenvalues will become smaller at true scatterer positions if the object under study is surrounded by sensors.

3.5 Image Formation

After processing all measurements with the distributed Kalman Filters, an image has to be created, that shows high values at positions with a high probability of reflectivity. If a Kalman Filter has been placed at or is close to a true scatterer position, multiple measurements will have been used to update the covariance matrix,

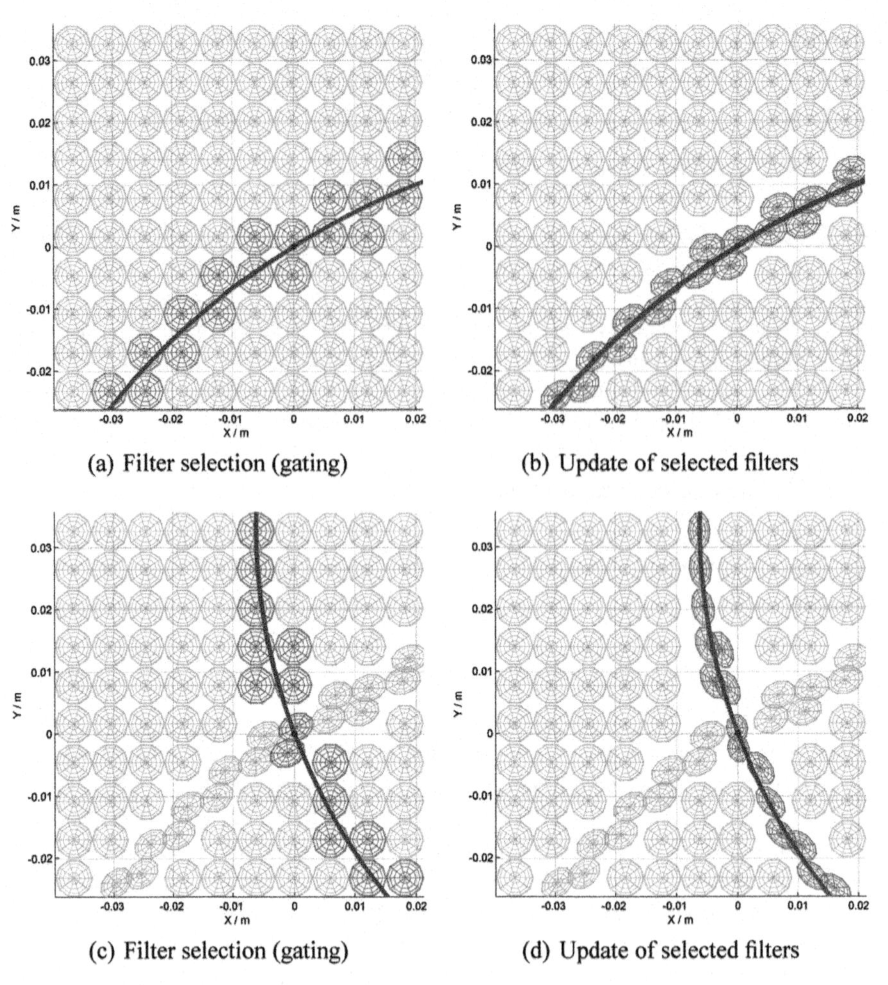

(a) Filter selection (gating)

(b) Update of selected filters

(c) Filter selection (gating)

(d) Update of selected filters

Fig. 3 Example of processing two TOA measurements for 100 distributed Kalman Filters. Their covariances are plotted as ellipsoids (initially spheres) centered at the positions of each filter. The point scatterer is shown as blue dot at the position $(x, y) = (0.0, 0.0)$. (a) shows the subset of filters (red) for the first measurement as determined by the gating procedure. The selected filters are updated, which shifts their positions towards the measurement and adapts their covariance matrices accordingly (**b**). The update procedure is shown for a second measurement ((**c**) and (**d**)), which causes the covariance of the filter closest to the scatterer to shrink the most

resulting in a denser probability mass around the filter position. Integrating this density over a predefined voxel grid with a desired resolution results in high image values at those positions where Kalman Filters with small eigenvalues in their covariance matrices are located.

One voxel of the final image is the sum of the integrals of all Kalman Filters over the volume of the regarded voxel. For speed-up purposes, only those Kalman Filters are regarded for a specific voxel that lie closer than four times the standard deviation corresponding to the covariance matrices.

4 Application: Ultrasound Computer Tomograph

The derived method is applied to reconstruct reflectivity images of a 3D ultrasound computer tomograph (USCT) [4], that has been built at the Institute for Data Processing and Electronics at Forschungszentrum Karlsruhe. This system has been developed for early breast cancer diagnosis and enables 3D imaging of a non-deformed breast with non-ionizing radiation. Fig. 4 shows the measurement setup.

The sensing aperture is cylindrical with a height of 15 cm and a diameter of 18 cm. It is equipped with 384 emitters and 1536 receivers grouped in 48 transducer array systems. The transducers have a size of $(1.4\,\text{mm})^2$, a center frequency of 2.4 MHz and an opening angle of ± 15 degree at -6 dB. A complete measurement results in a data set of approx. 600.000 A-scans (3 GB), which can be additionally increased by rotating the cylindrical aperture, thereby acquiring more information from different angles.

In this paper, a constant speed of sound is assumed. In future work, varying speed of sound can be introduced by including the speed of sound map of the ROI, which can be determined with the same data set [12].

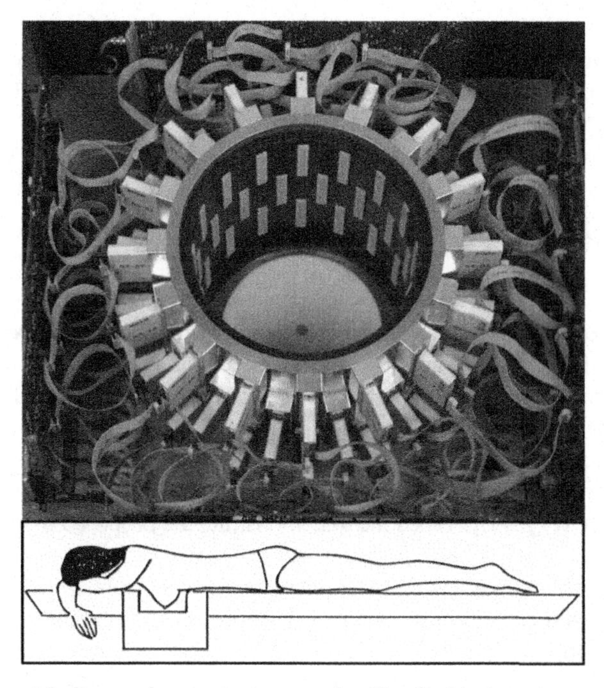

Fig. 4 Experimental ultrasound computer tomograph with cylindrical aperture and approx. 2000 transducers that are grouped into 48 transducer array systems (white blocks). The application is shown at the bottom, the woman lies in prone position on a bed while the breast is hanging in the measurement tank filled with water

4.1 Data Preprocessing

The TOA of each recorded echo is detected by means of a pulse detection method based on the wavelet transform [13]. Each TOA is used as a measurement for updating the distributed Kalman Filters. The ability of the pulse detection to separate two interfering echoes has been evaluated as 900 ns (center frequency of 2.4 MHz), which is used as basic magnitude of the measurement noise.

4.2 Kalman Filter Setup

For the following evaluation, the Kalman Filter equation are adapted as follows:

The system model $\mathbf{A_k}$ is set to identity as no prediction on the position of the state vector may be performed. The application of system noise is also neglected. The system noise would increase the eigenvalues of the covariance matrices and thereby unsharpen the image. The analysis of a useful application of system noise is part of future work.

The measurement noise is constant during the training process and has to be set depending on the amount of data and the initial values of the covariance matrices of the Kalman Filters.

The placement of the filters and their initialization is application-dependent. The choice of the number of Kalman Filters and their distribution is an empirical process so far and is based on the desired resolution in the final image or its desired quality with respect to structural information.

4.3 Evaluation

The proposed method is evaluated by means of two experiments. Image reconstructions with the PDF are compared to the currently applied image reconstruction, which is based on synthetic aperture focusing technique (SAFT) [14].

The first experiment consists of ten vertically spanned nylon threads (Fig. 5 left). This experiment was used for experimental resolution assessment in the horizontal plane. The reconstructed images of the nylon threads are compared by evaluating the contrast of each image. A second and more complex experiment with a clinical breast phantom (Fig. 5 right) is done to demonstrate the proposed method for breast imaging.

To reconstruct an image with SAFT, each recorded A-scan $A_{(i,j)}$ from an emitter and a receiver at the positions $\underline{e}_i$ and $\underline{r}_j$ is backprojected to the image position $\underline{x}$ of the image I using

$$I(\underline{x}) = \sum_{i=1}^{m} \sum_{j=1}^{n} A_{(i,j)} \left(\frac{\|\underline{e}_i - \underline{x}\| + \|\underline{r}_j - \underline{x}\|}{v} \right), \tag{7}$$

Fig. 5 Left: Ten vertically spanned nylon threads with a diameter of 0.2 mm, each spaced 2 mm apart for evaluating the horizontal resolution of the USCT. Right: Tissue mimicking triple biopsy breast phantom for evaluating the breast imaging capability of the USCT

where v is the speed of propagation of the emitted pulse, here assumed to be constant. The A-scans used for the image reconstruction with SAFT are created by convolving the TOA data from the preprocessing step with a Gaussian window of a temporal length of 1 μs. This compensates for the error induced by the preprocessing step and further errors caused by the imprecise knowledge of the speed of sound and the positioning of the transducers, respectively.

As measure for the contrast, the signal difference to noise ratio (SDNR) is evaluated, which is calculated by comparing the mean amplitude of the reconstructed object μ_{object} to the background artifacts. These are evaluated as the mean μ_{BG} and standard deviation σ_{BG},

$$SDNR = \frac{\mu_{object} - \mu_{BG}}{\sigma_{BG}}. \tag{8}$$

For evaluating the contrast of the nylon thread reconstructions, the background has to be segmented from the object. This is performed separately for each nylon thread by taking those pixel into account that have higher values than half of the local maximum.

4.3.1 Thread Experiment

The ten nylon threads have a diameter of 0.2 mm, each spaced 2 mm apart and are vertically spanned through the center of the USCT. Only the physically neighboring transducers closest to the slice image were used resulting in 16 sending and 64 receiving elements. This is sufficient for reconstructing the threads but also causes image artifacts due to the sparseness of the sensing aperture. This gives a good basis for comparing the two image reconstruction approaches via the contrast function. The ideal image reconstruction of these threads would result in ten distinct dots as shown at the top of Fig. 6.

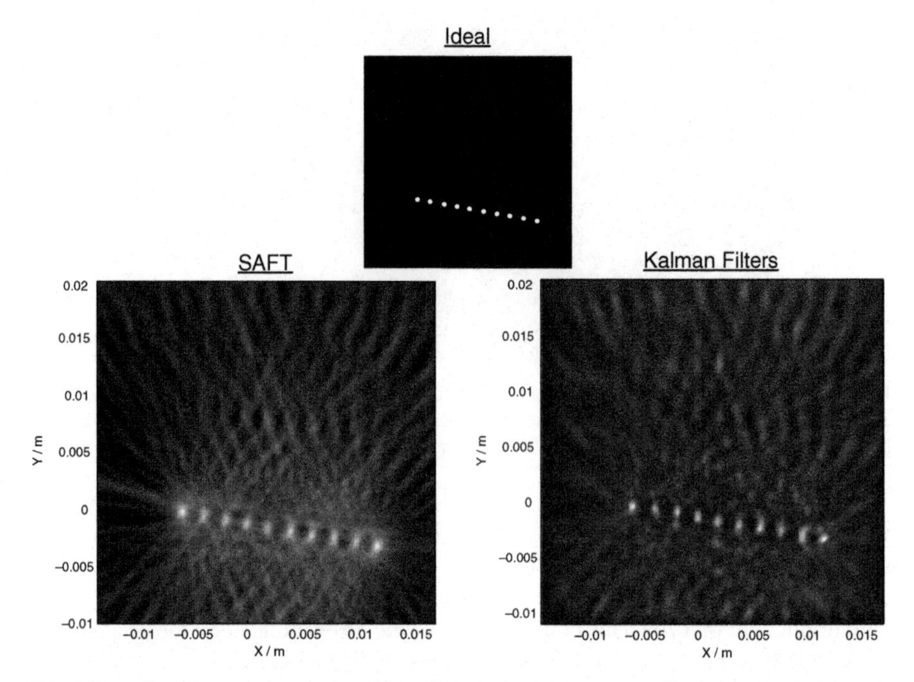

Fig. 6 The slice image reconstruction (407 × 407 pixel) of the ten vertically spanned nylon threads with SAFT shows ten clearly distinguishable points, see the ideal image at the top. Nevertheless, along the normal of the line connecting the ten threads, lots of artifacts are visible, which is due to the sparseness of our sensor aperture. The image reconstruction with 87 × 87 distributed Kalman Filters also clearly shows ten points. There are also reconstruction artifacts but less visible and more homogeneous. The SDNR value of this reconstruction is twice times higher compared to the SAFT image reconstruction

The reconstruction of the nylon threads with the SAFT approach (407 × 407 pixel) is shown on the left of Fig. 6. The ten threads are clearly visible, nevertheless, there are many artifacts in the proximity of the threads. The computed SDNR resulted in a value of 5.1.

The image reconstruction with 87 × 87 distributed Kalman Filters is shown on the right of Fig. 6. The threads are also imaged as distinctive points, but the reconstruction artifacts are significantly reduced. The evaluated contrast value of 10.5 is twice times higher than the value of the SDNR of the SAFT image reconstruction.

4.3.2 Breast Phantom Experiment

The clinical breast phantom is a triple modality test object for biopsy and can be imaged with X-ray, MRI, and ultrasound [15]. This breast phantom has several inclusions mimicking cysts and cancer structures. The average attenuation is 0.5 dB/MHz/cm. For comparison, a slice region was chosen that shows two cysts that have a strong directive scattering behavior and one cancerous structure that scatters more isotropic.

For this experiment the ground truth for ultrasonic reflectivity is unknown. In order to get an idea of the inner structure of the breast phantom, an MRI image of the same breast phantom has been acquired. The according slice region is shown at the top of Fig. 7. A high resolution reconstruction with SAFT (805 × 605 pixel) is shown on the left of Fig. 7. The boundaries of the cysts are not completely visible and the cancerous structure dominates the image, as most of the backprojections fell in this region. The image reconstruction with 73 × 55 Kalman Filters of the same region is shown on the right of Fig. 7. The boundaries of the cysts are more distinct and the cancerous structure does not dominate the image. The high amount of reflections from this region causes the covariance matrices to shrink resulting in a slightly visible grid. Compared to the boundaries of the cysts, which are formed by covariance matrices deformed along the tangent of the boundary, the value calculated from a single covariance matrix is not as high as the sum of largely overlapping Kalman Filters along the boundaries.

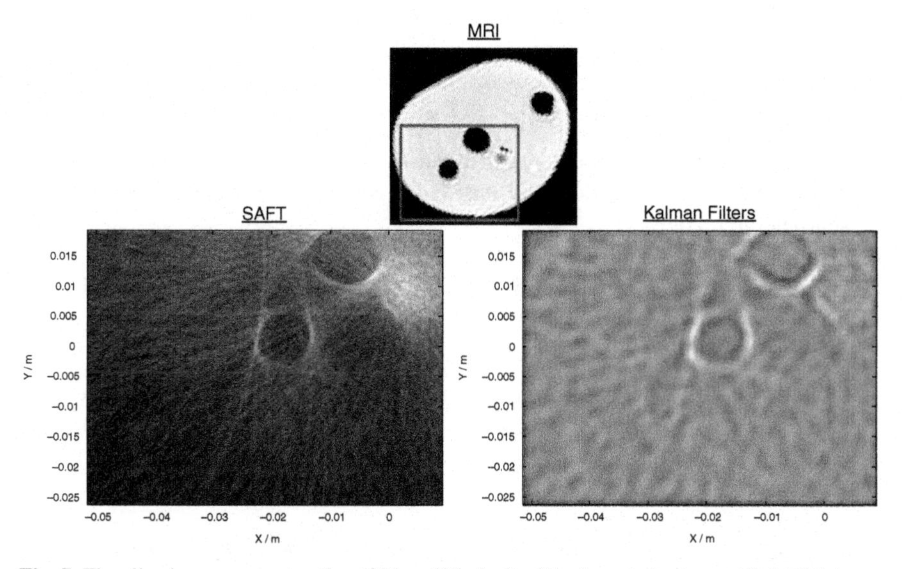

Fig. 7 The slice image reconstruction (805 × 605 pixel) of the breast phantom with SAFT shows two cyst mimicking structure that have a directive scattering behavior as well as another circular structure ("cancer") that has scatters more isotropic. For comparison, see the MRI image of the whole slice at the top with the marked region chosen for evaluation. The boundaries of the cysts are not completely visible and the strong scattering region dominates the image as a bright region. The image reconstruction with 73 × 55 distributed Kalman Filters shows the boundaries of the cysts more clearly. The cancerous region is displayed completely different compared to the SAFT image. The high amount of reflections from this area caused the covariance matrices to shrink, which is slightly visible as dot grid. Due to the reduced data set taken for this reconstruction, the skin of the breast is not visible in the ultrasound images, as the skin reflects the ultrasonic pulses to regions that are far below the regarded slice image

5 Conclusion and Future Work

A new approach was presented for the reconstruction of the 3D reflectivity distribution of a region of interest measured by an arbitrarily distributed sensor network. The imaging system regarded here is based on the synthetic aperture approach, which is widely applied in seismic, sonar, and radar imaging. The presented approach models the region of interest as a PDF representing spatial reflectivity occurrences. The data is processed in a recursive manner to update the distributed Extended Kalman Filters used to approximate the PDF. This allows to process the nonlinear measurements as well as fusing information of objects under study that are only partly available in the acquired measurements.

Experiments with a 3D ultrasound computer tomograph showed that the proposed method results in a higher image quality with less image artifacts and higher structural information. The run-time for reconstructing images with the distributed Kalman Filters is in the same order of magnitude as the SAFT approach. The results also showed that the amount of Kalman Filters does not have to be as high as the number of voxel used with the SAFT approach.

The image quality may be additionally improved by regarding system-specific parameters such as sensor characteristics, uncertainties in transducer positioning, and speed of propagation determination as well as object properties. With the proposed method, a basic framework is available for future work. More precisely, the following issues will be analyzed:

- The estimation of measurement noise during Kalman Filter training with a second estimation of the reflectivity may help reduce the artifacts further and also sharpen the image.
- The application of system noise to those Kalman Filters that were not affected by processing the TOA data of one A-scan could also decrease false-positives.
- The utilization of the information of non-occurring echoes at specific times could eventually be used to diminish directly artifacts in the image.

This work demonstrated a statistical approach for 3D image reconstruction, which is easily extendable with system-specific parameters and is able to consider uncertainties in system parameters and input data.

Acknowledgments The financial support of the German Research Foundation (DFG) is gratefully acknowledged. Thanks also to the reviewers for their comments.

References

1. Hovanessian, S.A.: *Introduction to Synthetic Array and Imaging Radar*. Artech House, Dedham, MA (1980).
2. Chatillon, J., Zakharia, M.: Synthetic aperture sonar for seabed imaging: relative merits of narrow-band and wideband approaches. *IEEE Journal of Oceanic Engineering* **17**(1), 95–105 (1992).

3. Claerbout, J.F.: Earth Soundings Analysis: Processing versus Inversion. Boston: Blackwell Scientific Publications, Inc. (1992).
4. Gemmeke, H., Ruiter, N.V.: 3D ultrasound computer tomography for medical imaging. *Nuclear Instruments and Methods in Physics* **580**(2), 1057–1065 (2007).
5. University of Alabama: Inverse problems. http://www.me.ua.edu/inverse/
6. Norton, S.J., Linzer, M.: Ultrasonic reflectivity imaging in three dimensions: Exact inverse scattering solutions for plane, cylindrical and spherical apertures. *IEEE Transactions on Biomedical Engineering* **28**, 202–220 (1981).
7. Norton, S.J., Linzer, M.: Ultrasonic reflectivity imaging in three dimensions: Reconstruction with spherical transducer arrays. *Ultrasonic Imaging* **1**, 210–231 (1979).
8. van Dongen, K.W.A., Wright, W.M.D.: A forward model and conjugate gradient inversion technique for low-frequency ultrasonic imaging. *Journal of the Acoustical Society of America* **120**(4), 2086–2095 (2006).
9. Lingvall, F.: Time-domain reconstruction methods for ultrasonic array imaging – a statistical approach. Ph.D. thesis, Uppsala University (2004).
10. Bar-Shalom, Y., Fortmann, T.E.: *Tracking and Data Association*. New York: Academic Press, Inc. (1988).
11. Kalman, R.E.: A new approach to linear filtering and prediction problems. *Transactions of the ASME, Journal of Basic Engineering* **580**(82), 35–45 (1960).
12. Ruiter, N.V., Schnell, R., Zapf, M., Kissel, J., Gemmeke, H.: Phase aberration correction for 3D ultrasound computer tomography images. *IEEE Ultrasonics Symposium*, pp. 1808–1811. New York, NY, USA (2007).
13. Schwarzenberg, G.F., Weber, M., Hopp, T., Ruiter, N.V.: Model-based pulse detection for 3D ultrasound computer tomography. In: *IEEE Ultrasonics Symposium*, pp. 1255–1258. New York, NY, USA (2007).
14. Mayer, K., Marklein, R., Langenberg, K.J., Kreutter, T.: Three-dimensional imaging system based on Fourier transform synthetic aperture focusing technique. *Ultrasonics* **28**, 241–255 (1990).
15. CIRS Incorporated: Tissue simulation & phantom technology. Norfolk, VA, USA. http:///www.cirsinc.com

T-SLAM: Registering Topological and Geometric Maps for Robot Localization

F. Ferreira, I. Amorim, R. Rocha and J. Dias

Abstract This article reports on a map building method that integrates topological and geometric maps created independently using multiple sensors. The procedure is termed T-SLAM to emphasize the integration of Topological and local Geometric maps that are created using a SLAM algorithm. The topological and metric representations are created independently, being local metric maps associated with topological places and registered at the topological level. The T-SLAM approach is mathematically formulated and applied to the localization problem within the Intelligent Robotic Porter System (IRPS) project, which is aimed at deploying mobile robots in large environments (e.g. airports). Some preliminary experimental results demonstrate the validity of the proposed method.

Keywords Topological maps · View-based localization · SLAM geometric maps · Robot localization.

1 Introduction

This article explores the use of combinations of topological and local geometric maps. There are a number of methods in the literature that attempt to exploit the perceived advantages of combined, hybrid or hierarchical maps for use in environment representation and mobile robot localization. There are some methods that are described in the literature that allude to Topological SLAM [1–13] in order to create an association with (geometric) Simultaneous Localization and Mapping (SLAM), a well-accepted means of building maps [14–16].

We could classify the methods that utilize both topological and metric information into hierarchical or hybrid methods. While the distinction is primarily semantic in nature, hierarchical methods could be seen as maintaining different

F. Ferreira (✉)
Institute of Systems and Robotics, Department of Electrical and Computer Engineering, University of Coimbra, Polo II, 3030 Coimbra (Portugal)
email: cfferreira@isr.uc.pt

S. Lee et al. (eds.), *Multisensor Fusion and Integration for Intelligent Systems*, Lecture Notes in Electrical Engineering 35, DOI 10.1007/978-3-540-89859-7_29,

Table 1 Glossary of terminology used in T-SLAM

K	total number of nodes in the topological map
k	an index, node occupied by the robot or position in the topological map
V^t_{obs}	the set of observations at time t
M	total number of distinct metric maps
m	an index, usually employed to denote a particular metric map
$\mathcal{X}^t_m$	representation of the metric map m at time t
α_i	the mixture component coefficient or component prior probability
$\mathcal{V}$	the FIM, or the complete set of views/vectors as collected during Environment-Familiarisation stage
π	the initial probability distribution over the hidden states of the Hidden Markov Model
a_{ij}	the probability of transiting from [hidden] state i to state j in the Hidden Markov Model
$b_i(n)$	the observation or emission probability for the symbol b_i at the place n within the Hidden Markov Model
j	an index, usually employed to denote a particular feature, j
k^*	the estimated place as obtained by applying the maximum criterion to the Belief over the indices of the Reference Sequence
M	the number of [hidden] states in the Hidden Markov Model
N	the number of [visible] observations/symbols in the Hidden Markov Model
Z	Hidden or incomplete data in a Mixture Mode
z_k	the vector from matrix Z corresponding to the view/vector V_k
λ	the parameter set, $\langle N, M, \{\pi_i\}, \{a_{ij}\}, \{b_i(n)\}\rangle$, of the Hidden Markov Model
Θ_i	a single component of the mixture model with the named features

representations in order to accomplish different purposes such as long-term planning in the topological map and precise motion control for navigating among obstacles using metric information.

One of the earliest and possibly one of the most well known approaches in this category is the Spatial Semantic Hierarchy or the SSH developed by Ben Kuipers [1,2]. The SSH is described as a model of knowledge of large-scale space consisting of multiple interacting representations, both qualitative and quantitative. The representation of the environment is maintained in the form of a hierarchy of maps – including metric and topological levels – each of which allows some abstraction of the perception and interaction of the robot with the environment. The advantages gained from using SSH or similar hierarchical model of representations is that incomplete or uncertainty in the information is handled in different forms depending on which particular localization or navigation problem is to be solved. Local metric maps help to perform place recognition, (middle-level) topological maps help create consistent maps in the face of challenges such as loop-closing problems, and the global metric maps maintain an overall consistency in the global position of the robot.

Hybrid approaches are usually employed to resolve specific disadvantages of one representation with regards to the other. In certain approaches that primarily depend on geometric maps, the loop closure problem has been resolved by simultaneously having locally precise geometrical information and globally consistent topological information about a (large) environment.

One such hybrid method is proposed by Choset and Nagatani [3] propose a SLAM method that exploits topology of the free space to localize the robot on a partial map. Low-level control laws are used to generate Voronoi graph (VG) and explore the unknown space. A graph matching process over the VG structure is used for robot localization, whereby the robot locates itself to one of the places of the VG, though the robot does not know its metric coordinates.

Thrun [17] builds a global metric (grid-based) map of the environment and then extracts a topological graph from this metric representation. Besides being not scalable to large environments, this method requires a globally consistent metric map, which is in general very difficult to obtain.

There are also attempts to utilize graph based approaches to solve particular problems that appear at the time of creation of metric maps. Methods such as [5], use graphical methods to maintain hypothesis for map expansion and closure, *i.e.* graph-like methods are used to maintain multiple map hypothesis of the main map which is geometrical. There are works that enhance the applicability of metric maps and the ability of users to interact with these such as representing individual objects. In [18], Limketkai and others store the representation of objects (some of which might also be used by persons) using a technique called Random Markov networks.

Tomatis *et al.* [13] developed a hybrid map representation wherein a global topological map connects local metric maps. The robot may switch between both representation when navigation conditions change (*e.g.* leaving a room and crossing a door). When doing this, the method requires a detectable metric feature in order to determine the transition point where the change from topological to metric has to be done and allows robust initialization of the metric localization (relocation and loop closure). The method was validated within office-like environments but its potential is unclear for different and larger environments.

In [19] Zimmer utilizes a clustering algorithm based on neural networks to cluster the local polar maps and ultimately register them in the global topological map. The experiments were performed on a small environment and the results are unclear regarding the scalability for larger environments. A similar idea is behind the procedure adopted by Zivkovic [20] where panoramic images are grouped and semantic information is associated with the groups. It is stated that this method bears semblance to the way animals represent their environment. As in the case of [4], a clustering approach is used to group images and represent places in the environment by using a typical set of images for that place. In [21], Thomas and Donikian hypothesize a hierarchical set of (topological) representations that represent the environment using similarity of places. The developers of these methods claim that such labeling of (similar-looking) places is in line with the spatial concepts that humans employ.

In the current work, we propose a generic method to integrate a global topological map with a set of two or more geometric maps. Some of the nodes of the topological map are associated with the individual metric maps, as depicted in Fig. 1. Our method tracks the global position of the robot only within the topological map. The localisation procedure in the topological map isolates features or properties of the

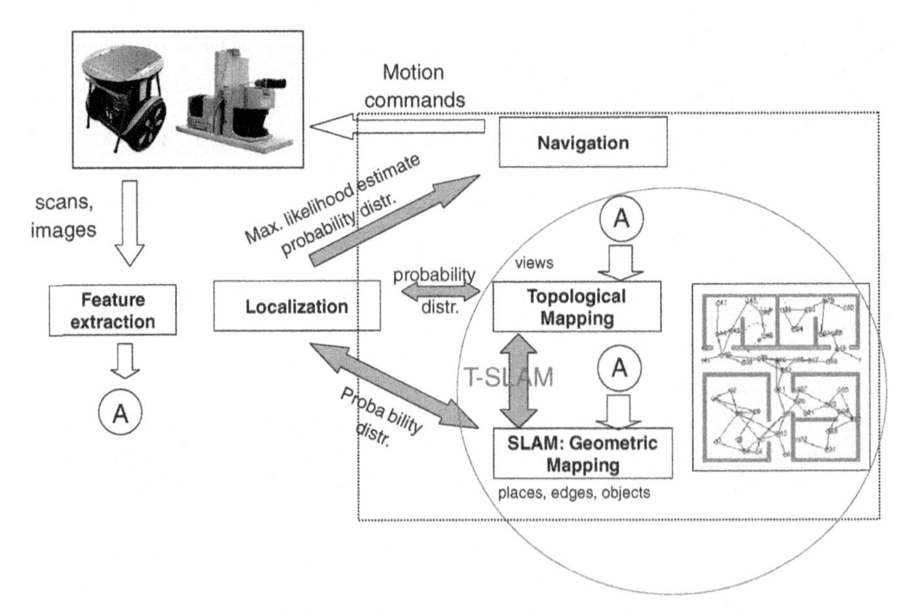

Fig. 1 Depiction of T-SLAM: the use of combined Topological and Geometric maps

environment into groups that are used to recover the node in the topological map that is currently occupied by the robot. By exploiting the associations between the nodes and the metric maps, we also maintain the local position of the robot enabling the precise geometric positioning of the robot. Localisation in each local metric map is performed independently and simultaneously. Map updating is performed simultaneously in these local metric maps as would be performed in a conventional SLAM algorithm.

In the next section, the method of localisation in the hierarchical representation is presented. In Sect. 2.2 a brief overview of the topological localization method is presented. In Sect. 2.3, a selected SLAM algorithm is used to create a geometric map. In Sect. 2, the combination of topological and geometric maps is described together with the localization system. In Sect. 3, the preliminary result from experiments using combined Topological and geometric maps.

2 Integrating Topological and Geometric Maps

Our representation of the environment is composed of a global topological map and a set of two or more local geometric maps. Let K be the total number of nodes in the topological map, these nodes indexed by the variable $k = 1, \ldots, K$. Let M be the total number of metric maps identified by $m = 1, \ldots, M$. Let $\mathcal{X}_m^t$ denote the representation of the robot in the topological map m at the discrete instance of time t. $\mathcal{X}_m^t$ varies depending on how the position of the robot is maintained in the metric map.

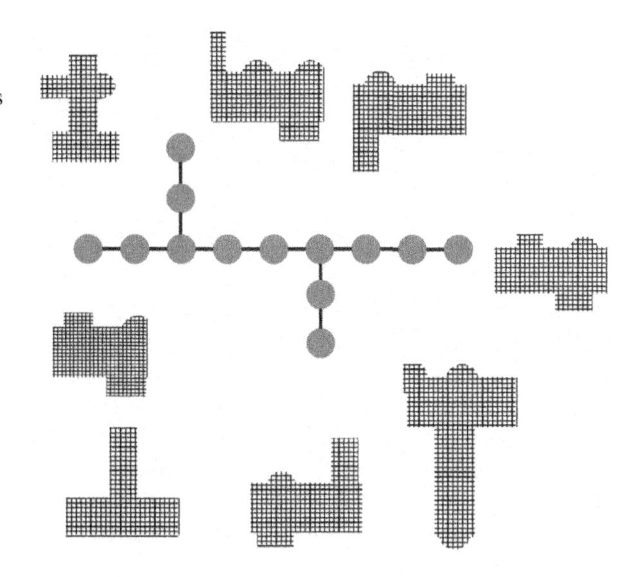

Fig. 2 Depiction of Independent creation of Topological and Metric maps

Since there exists a single global topological map and multiple geometric maps conditioned on this global topological map, as depicted in Figs. 2 and 3.

The probability of the robot being localised in both, the topological k and metric map m is given by $P(\mathcal{X}_m^t, k|V_{obs})$ in Eq. (1).

$$P(\mathcal{X}_m^t, k|V_{obs}) = P(\mathcal{X}_m^t|k, V_{obs}) \times P(k|V_{obs}) \tag{1}$$

The above expression conditions the probability of localisation on both maps on the probability of localisation on the global topological map. The term $P(k|V_{obs})$ in Eq. (1) denotes the localisation in the topological map. Without prejudice to the general case, the index indicating time has been removed from the remaining expressions.

$P(\mathcal{X}_m|k, V_{obs})$ represents the localisation in the metric map, conditioned on the robot being positioned at node k in the topological map and can be expanded as in Eq. (4).

Fig. 3 Depiction of Superimposition that actually exists between the topological map and the set of geometric maps

$$P(\mathcal{X}_m|k, V_{obs}) = \frac{P(\mathcal{X}_m, k, V_{obs})}{P(k, V_{obs})} \tag{2}$$

$$= \frac{P(V_{obs}|k, \mathcal{X}_m) \times P(\mathcal{X}_m|k) \times P(k)}{P(V_{obs}|k) \times P(k)} \tag{3}$$

$$= \frac{P(V_{obs}|k, \mathcal{X}_m) \times P(\mathcal{X}_m|k)}{P(V_{obs}|k)} \tag{4}$$

$P(\mathcal{X}_m|k)$ captures the association that the nodes of the topological map have with the individual metric maps. The exact nature of this association can vary depending on the features that are used with the topological and metric map and on the assumptions that are associated with the creation of each type of map. T-SLAM is an attempt to explore one type of association between a set of local geometric maps and a global topological map.

The real advantage of T-SLAM will emerge in scenarios in which the set of metric maps is registered to the global map only at certain places. For example, *some* nodes in the topological map might be associated with *way points* in the metric maps, through the use of artificial environment properties such as Radio Frequency Identification (RFID) tags [22] as depicted in Fig. 4.

The splitting of the environment, into many smaller regions or sections, that is described in this article is not new, see [6, 23–25] for a recent approach. The advantages of using the approach put forth in this article is that the knowledge of the position of the robot is conditioned on the nodes of the graph, the probability of which is valid globally, over the entire environment i.e. over all the sections of the environment.

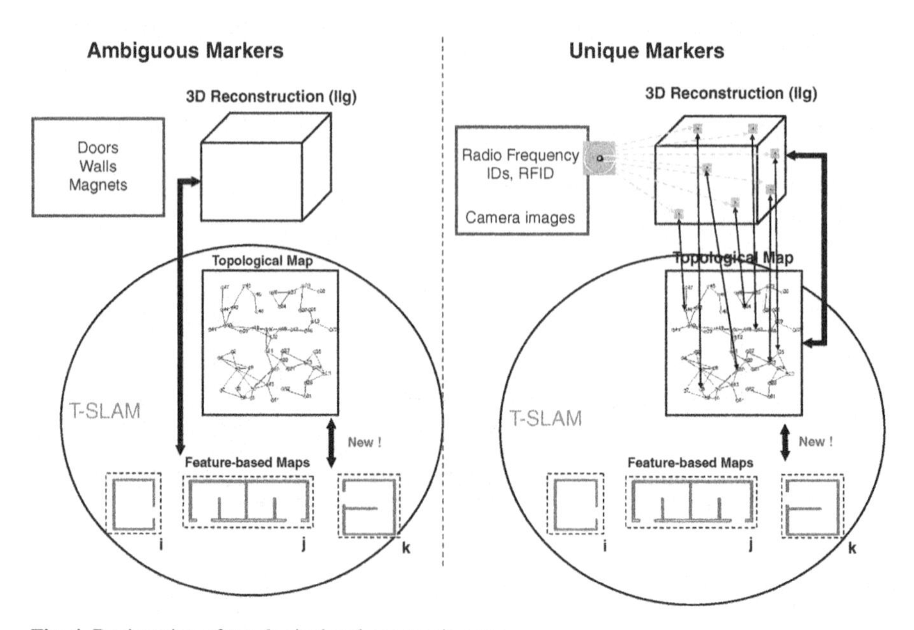

Fig. 4 Registration of topological and geometric maps

2.1 Current Problem formulation

In this article, the first version of T-SLAM is presented. Each node of the topological map is registered with every local geometric map. Each node k, in the topological map is associated with one or more geometric maps m by a human operator. This association is represented in the form of Node-Metric Map association matrix. Each element a_{mk} of this association matrix is assigned the value of 1 is the node is associated with the geometric map, 0 otherwise. Each line in the matrix corresponds to a particular metric map m and each column to a particular node k, leading to the expression (5). The Node-Metric map association matrix allows us to express the probability distribution associated to a map m^*, conditioned on the node k, $P(\mathcal{X}_{m^*}|k)$, by:

$$P(\mathcal{X}_{m^*}|k) = \frac{a_{m^*k}}{\sum_{m=1}^{M} a_{mk}} \tag{5}$$

The global probability of being at a particular position within the set of metric maps m is given by $\sum_{k=1}^{K} P(\mathcal{X}_m^t, k|V_{obs})$ and the location of the robot might be expressed as in (6) where $L(\mathcal{X}_m)$ is the Maximum Likelihood operator. The current observation is used to update the geometric map within which the robot is located.

Additionally, in the current method outlined in this article, we localise the robot in the topological and metric maps independently. This results in the simplification: $P(V_{obs}|k, \mathcal{X}_m) = P(V_{obs}|k) * P(V_{obs}|\mathcal{X}_m)$.

$$L(\mathcal{X}_m) = \mathcal{MLE}_k \left(\sum_{k=1}^{K} P(V_{obs}|\mathcal{X}_m) \times \frac{P(k|V_{obs})}{\sum_{m=1}^{M} a_{mk}} \right) \tag{6}$$

In the following Sect. 2.2, an expression will be developed for $P(k|V_{obs})$, where the topological map is built from a sequence of raw-image sequences. In Sect. 2.3, a well-known SLAM algorithm is utilised to create metric maps and localise the robot within them.

2.2 Topological Maps from Raw Sensor Data

In [26], a procedure was developed to localize a robot as it travelled along a path. During a first trip around the environment, the Environment Familiarization phase, depicted at left in Fig. 5, the robot samples the environment according to a sampling plan, collecting features by using its various sensors into the Reference Sequence.

A repetition of the motion performed during the place recognition should propel the robot along the path described by the Reference Sequence. Any maneuver other than the ones taken during the Environment Familiarization phase will take the robot to a place that was not sampled in the Environment Familiarization phase. The *Lost_Places*, in all a total of K in number, accommodate these possible views.

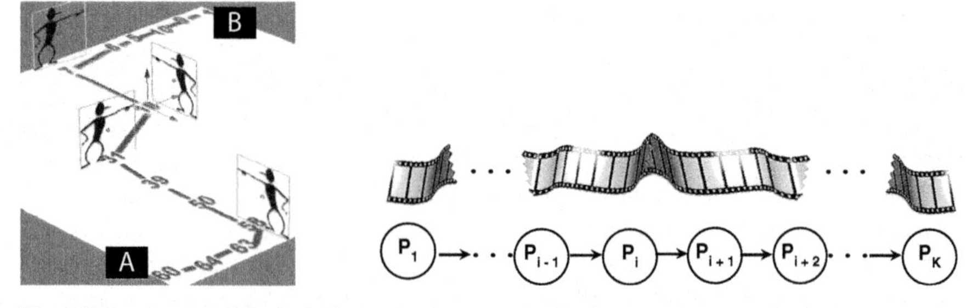

Fig. 5 The robot is led through the environment on the Environment Familiarization run to create the Reference Sequence (left). This Reference Sequence constitutes a left-to-right graph (right) composed of 'K' views, ordered as they were sampled during the Environment Familiarization

Thus, each *Lost_Place* takes into account the fact that the robot might be seeing views that were not seen in the Environment Familiarization phase.

The sampled views, which would normally be modelled as a left-to-right graph as at right in Fig. 5, are augmented by the insertion of '*Lost_Places*' as depicted in Fig. 6.

The sequence begins with P_{Lost_0} which indicates that the robot is completely lost or has never localized. Also, before every original place P_i, there is a P_{Lost_i}. By moving forward from one *Lost_Place* , the robot can transition from P_{Lost_i} to any node P_k where $k > i$. Similarly, from P_i the robot can transition to $P_k : k > i$ or to P_{Lost_i+1}. The graph does not allow a single-step transition from one P_{Lost_i} to another P_{Lost_j}.

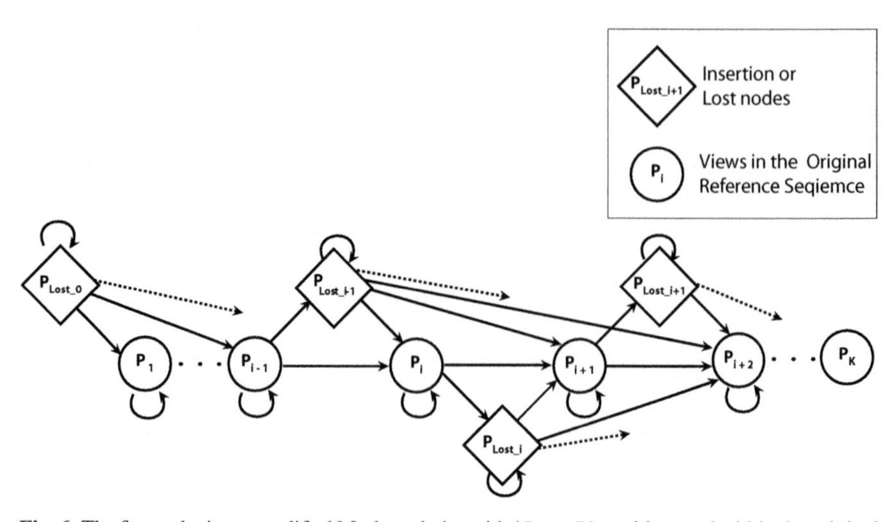

Fig. 6 The figure depicts a modified Markov chain, with '*Lost_Places*' inserted within the original Reference Sequence, to perform place-recognition. The dotted lines indicate the transitions to the Places in the original Reference Sequence and which have not been drawn to avoid cluttering the figure

When the robot needs to localize itself, it moves through the same environment, the current view is compared to the previously collected views and an inference is made of the current position of the robot. A Hidden Markov Model (HMM) is used to perform place recognition using the modified Markov Chain in Fig. 6 as a model for the transitions between the hidden states of the HMM. The Viterbi algorithm is commonly used in the context of HMMs to determine the most probable sequence of hidden states that gave rise to a particular sequence of observations. It is an inference tool that is associated with the process of making inferences in a HMM and is utilized to position the robot within the Reference Sequence by using the current sequence of observations. The HMM is specified in terms of its parameters λ, as in Eq. (7), where N corresponds to the number of states in the HMM, M the number of observations that will be used to make the inference, π represents the initial probability on the states, the a_{ij}s correspond to the transition probabilities between a pair of states i and j and $b_i(n)$ represents the probability of viewing symbol n at state i.

$$\lambda = \langle N, M, \{\pi_i\}, \{a_{ij}\}, \{b_i(n)\}\rangle \tag{7}$$

The transition between the states is influenced by the transition probabilities between a pair of places in the graph shown in Fig. 6. An elementary robot motion model is developed to evaluate the transition probability matrix. For each sequence of M observations, a simple distribution is used to model the transition probability distribution from each *Lost_Place* to the remaining original places in the Reference Sequence favoring places that lie closer in the Reference Sequence. The transition probability leading away from any of the original places in the Reference Sequence is uniformly split between the next original place (to the right) and to the corresponding *Lost_Place*. The one-step transition probability from one *Lost_Place* to another *Lost_Place* is zero.

The first hidden state is always matched to the first *Lost_Place*, P_{Lost_0}. This P_{Lost_0}, has a non-zero probability of reaching any place in the original Reference Sequence.

The observation model of the HMM is based on matching the current view with the views in the Reference Sequence. In the absence of any information regarding the view that will be visible at the '*Lost_Place*', we arbitrarily define the observation probability as an Uniform distribution over the K view in the original Reference Sequence. The features from each view in the Reference Sequence are converted into binary form and are represented within a Feature Incidence Matrix (FIM), $\mathcal{V}$. Due to the large dimensionality of the FIM, it is subsequently modelled as a Bernoulli Mixture Model (BMM). The parameters of the BMM are obtained by running the Expectation Maximization(EM) algorithm.

The Mixture parameters and the posterior probabilities over the components, the Z terms in Eq. (8), are used to evaluate the likelihood as depicted in Eq. (8), $P(V_k)$ representing the prior probabilities over each view k, in the Reference Sequence. As expressed in Eq. (9), the *Maximum Likelihood Estimation* is used to obtain the index k^* in V that best describes the object to be matched, V_{obs}.

$$P(k|V_{obs}) = \frac{\sum_{j=1}^{M} P(V_k) z_{kj} \alpha_j P(V_{obs}|\Theta_j)}{\sum_{k=1}^{K} \sum_{j=1}^{M} P(V_k) z_{kj} \alpha_j P(V_{obs}|\Theta_j)} \tag{8}$$

$$P(k^*|V_{obs}) = \underset{k}{\overset{K}{\operatorname{argmax}}}\, P(k|V_{obs}) \tag{9}$$

2.3 Creating and Updating Local Metric Maps Using SLAM

The incremental creation of Geometric maps from sensor data has been an area of much research over the last two decades. Simultaneous Localization and Mapping SLAM and Concurrent Mapping and Localization, CML, algorithms have been proposed by various researchers for the creation of different geometric maps. These algorithms have been very successful in the creation and utilisation of maps in indoor environments [14].

A couple of state of the art SLAM algorithms was used to create the local geometric maps. We experimented with the DP-SLAM [15] and the Fast-SLAM [27] algorithms. Both methods create grid-based metric maps using particle filters. The local geometric maps presented in this article were created using the Fast-SLAM algorithm.

A particle filter is a method of obtaining a description of a certain state space through partial observations of that space, which inevitably contain measurement errors. It maintains a weighted, and normalized, set of sampled states, $S = \{s_1, s_2, \ldots, s_p\}$, called particles. At each step, and given an observation vector z and a control vector u (in our context), the particle filter:

1. Samples m new states $S' = \{s'_1, s'_2, \ldots, s'_p\}$ from S, with replacement, using the probability density given by the weights of the elements in S.
2. Runs the state given by each particle through the corresponding motion model, using the previous states and u, obtaining in this way the new generation of particles.
3. Each new particle is then weighted, using the observation model together with the vector z.
4. Normalizes the weights of the new set of states.

The Fast-SLAM algorithm [27] is known for the speed at which the map is updated and for the relatively good quality of the geometric maps that are outputted. While the original Fast-SLAM algorithm [27] procedure was developed for metric maps using landmark, modifications and improvements were subsequently made including an adaptation to grid-based maps [16]. An implementation of this algorithm was obtained from the Open-SLAM web page [28]. In our current work we have create adopted a gird-based

3 Experiments and Results

Initial experiments have been performed on the localisation using a global topological map and a set of multiple metric maps. The topological representation of the environment was maintained in the form of a sequence of laser range scans and images gathered while leading the robot along one or more paths in the environment.

Our robot platform is equipped with two cameras and a Laser Range Finder, LRF, as seen in Fig. 7. The acquisition of data from the sensors and the control of the robot is performed within CARMEN. The two cameras, one facing forwards and the other facing onto one side, are capable of taking gray-scale 640×480 images. SIFT features [29] are utilised to perform matching between current observations and previously obtained images.

The forward-facing LRF provides a set of 361 range measurements through a 180 degree interval. Features from this sensor are used within the topological representation of the environment. The raw data from the sensor is used directly by the SLAM algorithm to build and maintain the topological map

The robot was first led along a path, depicted in Fig. 8, to create the topological and the set of geometric maps. The images from the cameras and the LRF were used to create the topological representation of the path, while raw laser range finder data and odometry were used to create the geometric maps. A new geometric maps was created after a specific amount of time of robot travel. A set of three geometric maps were created in all as shown in Fig. 9.

As stated in Sect. 2.1, the association between places that are represented in the topological map and the individual metric maps is represented in a Node-Metric maps association table. Excerpts of this map are shown in Table 2.

In a second experiment, the robot was driven along a path lined primarily by glass panes and pillars, Fig. 10. Typically, such an environment is difficult for SLAM applications given the absence of features in the direction lateral to the direction of robot travel. Excerpts of the Node-Metric Map association matrix are shown in Table 3. A few images from the set of 150 images that were used to construct the topological representation are presented in Fig. 11. As is seen in the above image, this environment, the robot is often surrounded by reflective and glazed surfaces, which make the SLAM difficult. The combined maps are depicted in Fig. 12.

Fig. 7 The sensor platform comprising of two laser range finders and two cameras is mounted on the Segway RMP 200

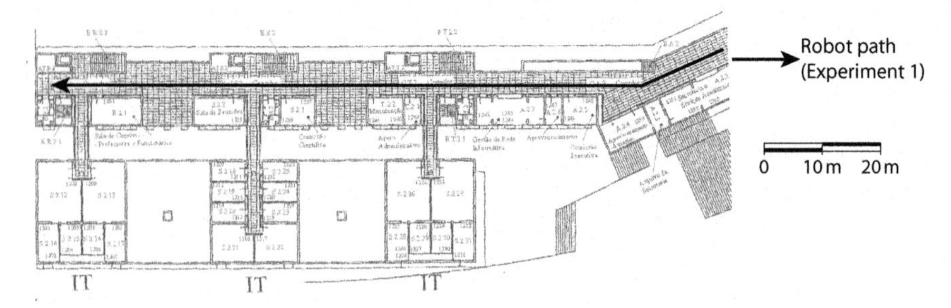

Fig. 8 Experiment 1: The robot was driven along a long hallway and map-building and localisation were performed to create independent topological and geometric representations

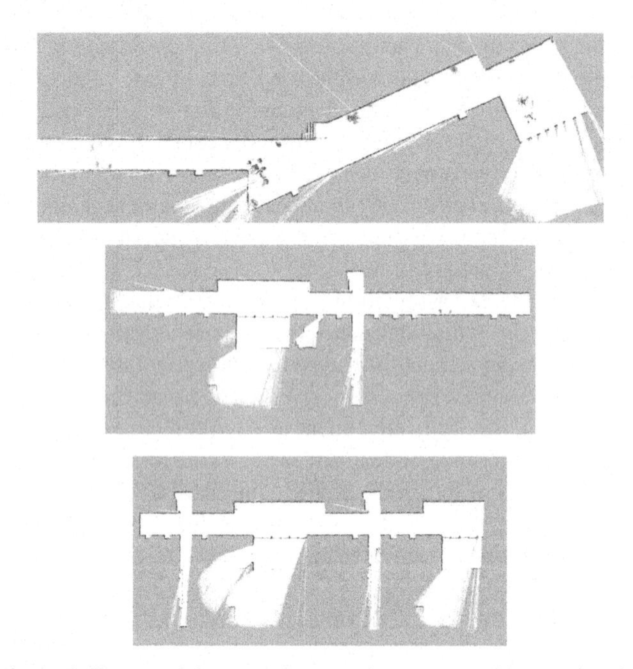

Fig. 9 Experiment 1: The set of three metric maps that are created by running the Fast-SLAM algorithm after the initial run through the environment in Experiment 1

Table 2 The Node-Metric map association matrix for Experiment 1

	1	2	...	53	54	55	...	94	95	96	...	143	144
$m = 1$	1	1	...	1	0	0	...	0	0	0	...	0	0
$m = 2$	0	0	...	1	1	1	...	1	0	0	...	0	0
$m = 3$	0	0	...	0	0	0	...	1	1	1	...	1	1

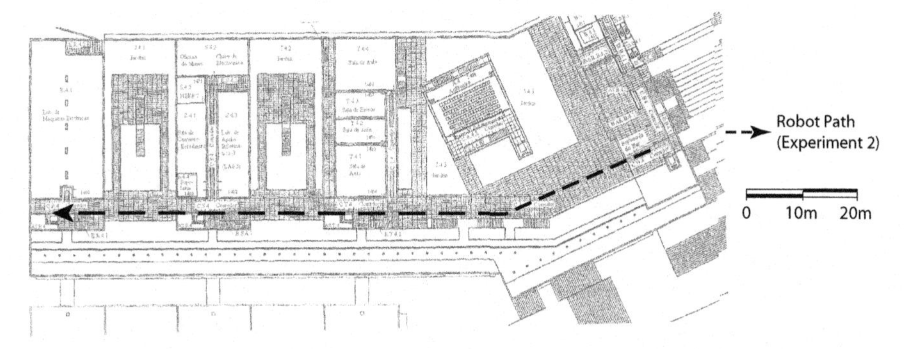

Fig. 10 Experiment 2: Mapping and localisation is performed in a second environment that comprises pillars and glass surfaces

Table 3 The Node-Metric map association matrix for Experiment 2

	1	2	...	54	55	55	...	114	115	116	...	149	150
$m = 1$	1	1	...	1	0	0	...	0	0	0	...	0	0
$m = 2$	0	0	...	1	1	1	...	1	0	0	...	0	0
$m = 3$	0	0	...	0	0	0	...	1	1	1	...	1	1

Fig. 11 Experiment 2: Typical images from a set of 150 images that comprise the topological representation of the path

There is some super position since the individual paths are created incrementally. Some of the larger amount of overlap that is present between the sections is removed during the process of merging topological paths. A small amount of overlap is maintained to allow transition between paths.

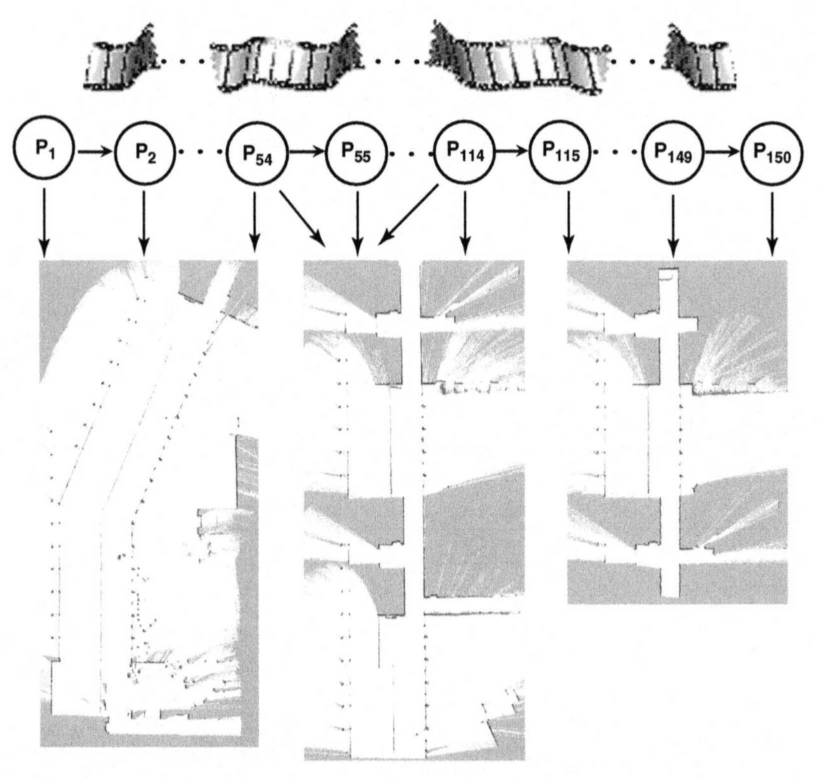

Fig. 12 Experiment 2: The association of the nodes of the topological map with the set of three metric maps. In the current version of T-SLAM, the registration of topological and metric maps is maintained in the form of the Node-Metric Map association matrix

4 Conclusions

Initial results were presented in this work on the simultaneous use of one global topological whose nodes are registered with multiple metric maps.

Current work includes the improved registration of the topological map with each metric map such that the uncertainty in the topological map can be transferred over to the metric maps and vice versa. We expect that this will lead to increased robustness in the localisation within the geometric maps and to reliable loop closing procedures in the topological map.

References

1. B. J. Kuipers. The Spatial Semantic Hierarchy. *Artificial Intelligence, Elsevier*, 119:191–233, 2000. http://www.cs.utexas.edu/users/qr/papers/Kuipers-aij-00.html.
2. B. Kuipers, J. Modayil, P. Beeson, M. MacMahon, and F. Savelli. Local metrical and global topological maps in the Hybrid Spatial Semantic Hierarchy. In *IEEE International Conference on Robotics and Automation (ICRA 2004)*, 2004.

3. H. Choset and K. Nagatani. Topological simultaneous localization and mapping (SLAM): toward exact localization without explicit localization. *IEEE Transactions on Robotics and Automation*, 17(2):125–137, 2001.
4. I. Posner, D. Schroeter, and P. Newman. Using scene similarity for place labelling. In *10th International Symposium on Experimental Robotics, ISER 2006*, 2006.
5. J. Folkesson and H. Christensen. Graphical slam - a self-correcting map. In *IEEE International Conference on Robotics and Automation, 2004*, 2004.
6. B. Lisien, D. Morales, D. Silver, G. Kantor, I. Rekleitis, and H. Choset. Hierarchical simultaneous localization and mapping. In *Proceedings of the IEEE/RSJ International Conference on Intelligent Robotic Systems, IROS*, pages 448–453, las Vegas, Nevada, October 2003.
7. A. Martinelli, V. Nguyen, N. Tomatis, and R. Siegwart. A relative map approach to slam based on shift and rotation invariants. *Robotics and Autonomous Systems*, 55(1):50–61, 2006.
8. P. Newman, D. Cole, and K. Ho. Outdoor slam using visual appearance and laser ranging. In *IEEE International Conference on Robotics and Automation, 2006*, 2006.
9. M. Pfingsthorn, B. Slamet, and A. Visser. A scalable hybrid multi-robot slam method for highly detailed maps. In *Proceedings of the 11th RoboCup International Symposium*, 2007.
10. A. Arleo, J. del R. Millán, and D. Floreano. Efficient learning of variable-resolution cognitive maps for autonomous indoor navigation. *IEEE Transactions on Robotics and Automation*, 15(6):990–1000, 1999.
11. J. Gasós and A. Saffiotti. Integrating fuzzy geometric maps and topological maps for robot navigation. In *Proceedings of 3rd International ICSC Symposium on Soft Computing (SOCO'99)*, pages 754–760, Genova, Italy, 1999.
12. E. Aguirre and A. González. Integrating fuzzy topological maps and fuzzy geometric maps for behavior-based robots. *International Journal of Intelligent Systems*, 17:333–368, 2002.
13. N. Tomatis, I. Nourbakhsh, and R. Siegwart. Hybrid simultaneous localization and map building: a natural integration of topological and metric. *Robotics and Autonomous Systems*, 44:3–14, 2003.
14. S. Thrun. Robotic mapping: A survey. In G. Lakemeyer and B. Nebel, editors, *Exploring Artificial Intelligence in the New Millenium*. Morgan Kaufmann, 2002.
15. A. Eliazar and R. Parr. Dp-slam: Fast, robust simultaneous localization and mapping without predetermined landmarks. In *Proceedings 18th International Joint Conference on Artificial Intelligence (IJCAI-03)*, pages 1135–1142, 2003.
16. C. Stachniss, D. Haehnel, W. Burgard, and G. Grisetti. On actively closing loops in grid-based fastslam. *The International Journal of the Robotics Society of Japan (RSJ)*, 19(10):1059–1080, 2005.
17. S. Thrun. Learning metric-topological maps for indoor mobile robot localization. *Artificial Intelligence*, 99(1):21–71, 1998.
18. B. Limketkai, L. Liao, and D. Fox. Relational object maps for mobile robots. In *International Joint Conference on Artificial Intelligence (IJCAI)*, 2005.
19. U. R. Zimmer. Embedding local metrical map patches in a globally consistent topological map. In *Symposium on Underwater Technology 2000, Tokyo, Japan, May 23-26 2000*, May 2000.
20. Z. Zivkovic, O. Booij, and B. Kröse. From images to rooms. *Robotics and Autonomous Systems*, 55(5):411–418, 2007.
21. R. Thomas and S. Donikian. A model of hierarchical cognitive map and human memory designed for reactive and planned navigation. *Technical report, l'Institut National de Recherche en Informatique et en Automatique, INRIA*, Project siames, 2000.
22. D. Haehnel, W. Burgard, D. Fox, K. Fishkin, and M. Philipose. Mapping and localization with rfid technology. In *IEEE International Conference on Robotics and Automation (ICRA), 2004*, 2004.
23. J.-L. Blanco, J.-A. Fernández-Madrigal, and J. González. Toward a unified bayesian approach to hybrid metric–topological slam. *IEEE Transactions on Robotics*, 24(2):259–270, 2008.

24. K. Rohanimanesh, G. Theocharous, and S. Mahadevan. Hierarchical map learning for robot navigation. In *In AIPS Workshop on Decision-Theoretic Planning*, 2000.
25. G. Theocharous, K. Murphy, and L. P. Kaelbling. Representing hierarchical pomdps as dbns for multi-scale robot localization. In *IEEE International Conference on Robotics and Automation, 2004*, 2004.
26. F. Ferreira, V. Santos, and J. Dias. A topological path layout for autonomous navigation of multi-sensor robots. *International Journal of Factory Automation, Robotics and Soft Computing*, 1:203–215, 2007.
27. M. Montemerlo, S. Thrun, D. Koller, and B. Wegbreit. FastSLAM: A factored solution to the simultaneous localization and mapping problem. In *Proceedings of the AAAI National Conference on Artificial Intelligence*, Edmonton, Canada, 2002. AAAI.
28. www.openslam.org
29. David G. Lowe. Distinctive Image Features-From Scale-Invariant Keypoints. *IJCV*, 60(2): 91–119, 2004.

Map Fusion Based on a Multi-Map SLAM Framework

François Chanier, Paul Checchin, Christophe Blanc and Laurent Trassoudaine

Abstract This paper presents a method for fusing two maps of an environment: one estimated with an application of the Simultaneous Localization and Mapping (SLAM) concept and the other one known a priori by a vehicle. The goal of such an application is double: first, to estimate the vehicle pose in this known map and, second, to constrain the map estimate with the known map using an implementation of the local maps fusion approach and a heterogeneous mapping of the environment. This article shows how a priori knowledge available in the form of a map can be fused within an EKF-SLAM framework to obtain more accuracy on the vehicle poses and map estimates. Simulation and experimental results are given to show these improvements.

Keywords EKF SLAM · Multi map fusion · Robotcentric local map approach

1 Introduction

The Simultaneous Localization and Mapping (SLAM) problem corresponds to the ability of a robot to build a map of its environment while localizing itself in that map. Since the seminal paper [1], the understanding of the Extended Kalman Filter (EKF) approach to the SLAM problem made real progress [2, 3]. Many researchers underlined the problem of filter consistency [4, 5] and improved it with multi-map solutions [6–8]. Other difficulties were addressed like data association [9] and computational burden [10].

Generally, in such applications, both the trajectory of the vehicle and the location of the landmarks are estimated online without the need for any a priori knowledge of location [3]. However, we choose to use absolute data in a form of an a priori map in a SLAM application. This information fusion is done to improve the EKF-SLAM estimates. Few SLAM applications use available information on the environment.

F. Chanier (✉)
Laboratoire des Sciences et Matériaux pour l'Electronique et d'Automatique, Université de Clermont-Ferrand, 24, avenue des Landais, 63177 Aubière cedex, France
e-mail: francois.chanier@lasmea.univ-bpclermont.fr

S. Lee et al. (eds.), *Multisensor Fusion and Integration for Intelligent Systems*, Lecture Notes in Electrical Engineering 35, DOI 10.1007/978-3-540-89859-7_30,

One example is described in Lee et al. [11] suggest a path-constrained SLAM solution using a priori information in the form of a digital road map.

Furthermore, such map fusion approaches can answer questions about solving the problem of observability in SLAM systems. This property is important, and is related to the problem of consistency. A lower limit on the estimated error can only be guaranteed if the designed state (vehicle and map) is completely observable. In [12], a demonstration proves that EKF-SLAM formulation is not fully observable without absolute information. In other words, SLAM approaches, with on board exteroceptive and proprioceptive sensors which provide relative measurements between the moving vehicle and the observed landmarks, give only partially observable estimates and fail to yield the absolute robot pose and feature positions in the global coordinates.

The method presented in this paper uses as absolute information an a priori known map of the environment as soon as the vehicle position is available in this known map. So, our method is composed of two parts and our paper contribution deals with the second part. First, Fig. 1, the vehicle uses a multi-map SLAM method [6] without constraint. The environment is divided into numerous local maps and the size of these maps depends on a maximal number of filter iterations. At an upper level, all these local maps are fused into a global one. Until the global map contains enough information, the vehicle position is estimated in the a priori known map. Then, the known map is used in an added EKF iteration to update the local and global maps, Fig. 1.

Finally, our method has three principal advantages: estimating the vehicle pose in a known map, answering to the observability problem and updating estimates with the known map. At the beginning of the experiment, initial vehicle position is not known in the a priori environment map. The system realizes the vehicle localization using matching between the estimate of the global map and the a priori known map. Since this position is estimated, the method solves the observability problem as stated in [12]. Besides, if the environment has changed, the a priori known map is updated using our constrained method.

The remainder of this paper is organized as follows: Sect. 2 presents the SLAM approach and the environment representation. In Sect. 3, details are given on the

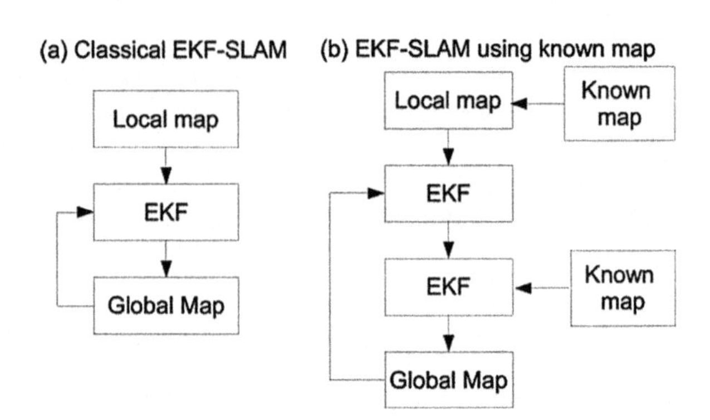

Fig. 1 Scheme of SLAM update step: classical and with known map

update stage with the known map and more especially the algorithm used to find vehicle position in the a priori map. In Sect. 4, simulation and experimental results underline the improvement with regard to the estimated vehicle path and mapped features. They also show the consistency of the fusion approach.

2 SLAM Approach Description

In this section, the SLAM part of our method is based on the *Robotcentric map joining* approach [6]. This SLAM approach is chosen because it is reliable, see the results presented in Castellanos et al. [6]. Indeed, if the vehicle pose in the a priori known map can not be estimated early in the experiment, the environment mapping has to remain consistent without being constrained, until the estimated map provides enough information to determine this vehicle pose, see Sect. 3.2.

2.1 Robotcentric Map Joining Approach

As reported in Castellanos et al. [6], the *Robotcentric map joining* approach consists in building a sequence of consecutive uncorrelated local maps and joining them together in a full correlated map. This technique limits the number of update steps for each local map, thus, limiting both feature location uncertainties and the effects of linearization errors [5]. This section introduces the different formulations of local and global maps used in the presented approach.

2.1.1 Local Maps

Each local map is represented with a *Robotcentric* approach. Features, $\mathbf{X}^R_{F_i}$ with $i = \{1, \cdots, n\}$, are referenced to a frame attached to the vehicle $\mathbf{R}$. The local reference $\mathbf{W}_{local}$ is included as a non-observable feature in the stochastic state vector, $\mathbf{X}_{local}$:

$$\mathbf{X}_{local} = \left[\mathbf{X}^R_{W_{local}}\ \mathbf{X}^R_{F_1} \cdots \mathbf{X}^R_{F_n}\right]^T \tag{1}$$

with the associated covariance matrix $\mathbf{C}_{local}$.

2.1.2 Global Map

Each local map is joined in a global map represented by the same *robotcentric* approach. Features, $\mathbf{X}^{W_{local}}_{G_i}$ with $i = \{1, \cdots, m\}$, are referenced to the current local frame $\mathbf{W}_{local}$ and the global reference is $\mathbf{W}_{global}$. The global vector state is written as:

$$\mathbf{X}_{global} = \left[\mathbf{X}^{W_{local}}_{W_{global}}\ \mathbf{X}^{W_{local}}_{G_1} \cdots \mathbf{X}^{W_{local}}_{G_m}\right]^T \tag{2}$$

with the associated covariance matrix $\mathbf{C}_{global}$.

To update the global map, the local map vector $\mathbf{X}_{local}$ is joined in $\mathbf{X}_{global}$. It is supposed that these local maps are not correlated. The vector state is written as:

$$\mathbf{X}_{system} = \left[\mathbf{X}_{global}\ \mathbf{X}_{local}\right]^T \tag{3}$$

with the associated covariance matrix:

$$\mathbf{C}_{system} = \begin{bmatrix} \mathbf{C}_{global} & 0 \\ 0 & \mathbf{C}_{local} \end{bmatrix} \tag{4}$$

Innovation calculation (distance between local and global associated features), used in EKF update step, is done in local map frame. The nonlinear observation function, $\mathbf{h}_k$ that formulates global feature coordinates in local map frame, is given by:

$$Z = \mathbf{h}_k(\mathbf{X}_{system}) = \ominus \mathbf{X}^R_{F_k} \oplus \mathbf{X}^R_{W_{local}} \oplus \mathbf{X}^{W_{local}}_{G_k} = 0 \tag{5}$$

The Jacobian matrix is given by:

$$\mathbf{H}_k = \left[0\ \mathbf{H}_{G_k}\ 0\ \mathbf{H}_R\ 0\ \mathbf{H}_{F_k}\ 0\right] \tag{6}$$

with $\mathbf{H}_{G_k}$ is the Jacobian of $\mathbf{h_k}$ with respect to $\mathbf{X}^R_{G_k}$, $\mathbf{H}_R$ is the Jacobian of $\mathbf{h_k}$ with respect to $\mathbf{X}^R_{R_g}$ and $\mathbf{H}_{F_k}$ is equal to identity matrix as covariance matrix associated to the measurement (local map feature) is null.

In the last step, named *composition*, vehicle and feature poses in the global map are updated with respect to the vector $\mathbf{X}^R_{W_{local}}$.

2.2 Heterogeneous Mapping

Heterogeneous representation allows a more exact mapping than in a one-feature based SLAM and, also, provides more information in order to estimate the vehicle position in the known map, Sect. 3.2. Two geometric feature types, line and circle, are used to represent the environment in two dimensions and are shown in Fig. 2.

2.2.1 Line

The SPmap model [13] is used to represent line features. The state vector is written as:

$$\mathbf{X}_{line} = \left[x_{line}\ y_{line}\ \theta_{line}\right]^T \tag{7}$$

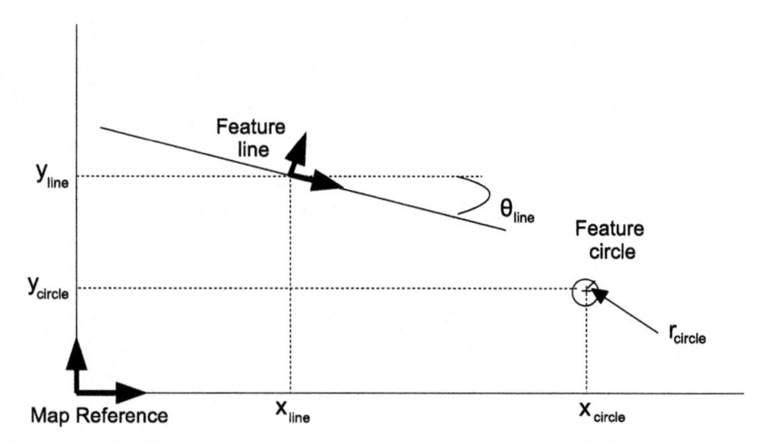

Fig. 2 Representation of circle and line states

The states, x_{line} and y_{line}, are the Cartesian coordinates of line frame and θ_{line} represents the orientation of this frame in the map reference, see Fig. 2. Only angular and lateral displacements are used in EKF update step. The other parameter x_{line} is used to improve data association step with ambiguous matching (parallel walls for example).

2.2.2 Circle

Circle features are represented by:

$$\mathbf{X}_{circle} = [x_{circle}\ y_{circle}\ r_{circle}]^T \tag{8}$$

The states, x_{circle} and y_{circle}, are the Cartesian coordinates of circle center and r_{circle} is the radius, see Fig. 2. The estimation of these parameters is based on a *Levenberg-Marquardt* algorithm, see Triggs et al. [14].

In this heterogeneous mapping concept, the general form of state vectors of the local and the global maps, respectively $\mathbf{X}_{local}$ and $\mathbf{X}_{global}$, is:

$$\mathbf{X}_{map} = [\mathbf{X}_{line}\ \mathbf{X}_{circle}]^T \tag{9}$$

3 Update SLAM with known Map

In this section, the part of our method where the EKF-SLAM estimates are constrained by the knowledge of an a priori map of the environment is presented, Fig. 1. It corresponds to the contribution to the approach presented in Sect. 2.1.

3.1 Known Map Description

Known map is composed of line and circle features, as presented in Sect. 2.2. This map is composed of l features. The state vector is written as:

$$\mathbf{X}_{known} = \left[\mathbf{X}_{E_1}^{W_{known}} \dots \mathbf{X}_{E_l}^{W_{known}} \right]^T \quad (10)$$

If this map is perfectly known, for example obtained from blueprints of building, map uncertainty is null. However, for the presented experiments in Sect. 4, feature poses are measured with a differential GPS sensor. The sensor standard deviation is around 5 cm. This uncertainty is translated in the known map by the matrix $\mathbf{C}_{known}$:

$$\mathbf{C}_{known} = \begin{bmatrix} \sigma^2 & & 0 \\ & \ddots & \\ 0 & & \sigma^2 \end{bmatrix} \quad (11)$$

with σ deviations on the feature states. Six different variables are defined:

- σ_{x_l}, σ_{y_l} and σ_{θ_l} for line type features,
- σ_{x_c}, σ_{y_c} and σ_{r_c} for circle type features.

3.2 Geometric Transformation Determination

Combined constraint data association (CCDA), presented in Bailey et al. [15], is used to determine geometric transformation between the references of the global map, W_{global} and the known map W_{known}, see Fig. 3. The vector $\mathbf{X}_{W_{known}}^{W_{global}}$, (12), represents the a priori map reference pose, W_{known}, in the global map.

$$\mathbf{X}_{W_{known}}^{W_{global}} = [x_{known}\ y_{known}\ \theta_{known}]^T \quad (12)$$

x_{known} et y_{known} are the Cartesian coordinates and θ_{known} is the orientation.

To calculate $\mathbf{X}_{W_{known}}^{W_{global}}$, two steps are necessary: matching global map with a priori map features and finding the position and the orientation of the a priori map reference, $\mathbf{W}_{known}$, in the global map.

First, CCDA algorithm performs batch-validation gating by constructing and searching a correspondence graph. The different features of a map are represented by a graph where each node is a feature and each edge is an invariant relationship between two features. Three different relationships are used by CCDA: between two lines, between two circles and between a circle and a line. These relationships only use center coordinates for circles and angular and lateral displacements for lines.

Having generated two feature graphs with the relationships, one for the global map and the other one for the known map, a correspondence graph is created. The

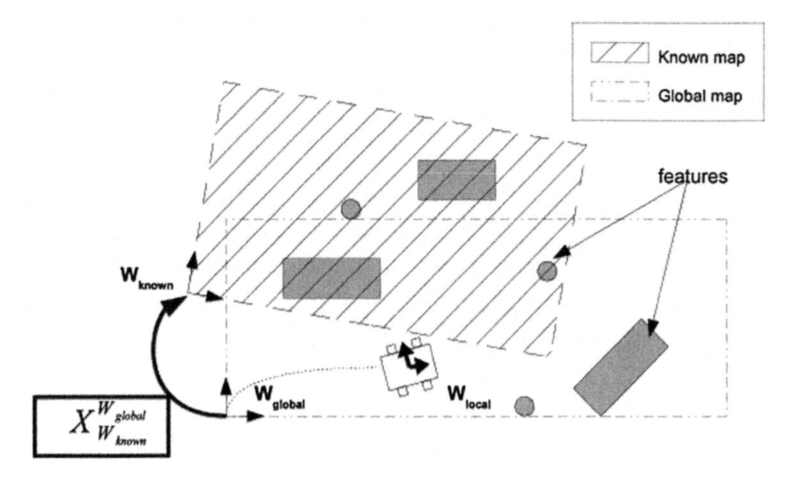

Fig. 3 Geometric transformation between the global map and the known map

nodes of this graph represent all the possible pairings of the different features of both maps. An algorithm, used in Bailey et al. [15], searches the maximum clique. If the clique is long enough, it means that there are enough feature pairings (more than three), the vector $\mathbf{X}^{W_{global}}_{W_{known}}$ and the covariance matrix $\mathbf{C}^{W_{global}}_{W_{known}}$ are estimated.

3.3 Update Step with known Map

3.3.1 Insertion of the Translation Vector in Local and Global Map Vectors

The vector $\mathbf{X}^{W_{global}}_{W_{known}}$ specified in Sect. 3.2, is now referenced to the frame attached to the vehicle $\mathbf{W_{local}}$:

$$\mathbf{X}^{W_{local}}_{W_{known}} = \mathbf{X}^{W_{local}}_{W_{global}} \oplus \mathbf{X}^{W_{global}}_{W_{known}} \tag{13}$$

$\mathbf{X}^{W_{local}}_{W_{known}}$ is included in the global map state vector (2) as a non observable feature. Its estimate can be improved during the update steps of the EKF filter:

$$\mathbf{X}_{global} = \left[\mathbf{X}^{W_{local}}_{W_{known}} \ \mathbf{X}^{W_{local}}_{W_{global}} \ \mathbf{X}^{W_{local}}_{F_1} \cdots \mathbf{X}^{W_{local}}_{F_m} \right]^T \tag{14}$$

The covariance matrix associated with the new vector $\mathbf{X}_{global}$ is given by:

$$\mathbf{C}_{global} = G \times \begin{bmatrix} \mathbf{C}^{W_{global}}_{W_{known}} & 0 \\ 0 & \mathbf{C}_{global} \end{bmatrix} \times G^T \tag{15}$$

with

$$\mathbf{G} = \begin{bmatrix} J_g & J_{X_{global}} \\ 0 & I \end{bmatrix} \tag{16}$$

and $\mathbf{J}_g$ is the Jacobian of equation (13) with respect to $\mathbf{X}_{W_{known}}^{W_{global}}$, $\mathbf{J}_{X_{global}}$ is the Jacobian of equation (13) with respect to $\mathbf{X}_{global}$ and I is equal to identity matrix.

The $\mathbf{X}_{W_{known}}^{W_{local}}$ is also included in each local map state vector. As these maps are initialized at the position of $\mathbf{X}_{W_{global}}^{W_{local}}$, the vector $\mathbf{X}_{W_{known}}^{R}$ can be initialized by the last value of $\mathbf{X}_{W_{known}}^{W_{local}}$. The vector $\mathbf{X}_{local}$ is now given by:

$$\mathbf{X}_{local} = \left[\mathbf{X}_{W_{known}}^{W_{local}} \ \mathbf{X}_{W_{local}}^{W_{local}}\right]^T \tag{17}$$

$\mathbf{X}_{W_{local}}^{W_{local}}$ is equal to zero. It is supposed that the vector $\mathbf{X}_{W_{known}}^{R}$ is uncorrelated with the initial vehicle pose in the local map. The initial values of the uncertainties on vehicle states are null. So, the associated covariance matrix $\mathbf{X}_{local}$ is initialized with:

$$\mathbf{C}_{local} = \begin{bmatrix} \mathbf{C}_{W_{known}}^{W_{local}} & 0 \\ 0 & 0 \end{bmatrix} \tag{18}$$

With this translation vector, states of local and global maps can be updated with the known map using equations presented in the next part. Moreover, the estimate of this vector is improved at each algorithm iteration.

3.3.2 Update of Local Maps and Global Map with known Map

This update is realized at every step after the update of local maps. To update the local map, the vector $\mathbf{X}_{known}$ is joined in the vector $\mathbf{X}_{local}$:

$$\mathbf{X}_{system} = [\mathbf{X}_{local} \ \mathbf{X}_{known}]^T \tag{19}$$

Feature positions are estimated, once the data association step is done, with the same formulation introduced in Castellanos et al. [6]. However, the observation equation (5) has to be changed to introduce the translation (13):

$$\mathbf{Z} = \mathbf{h}_k(\mathbf{X}_{system}) = \ominus \mathbf{X}_{F_k}^{W_{local}} \oplus \mathbf{X}_{W_{known}}^{W_{local}} \oplus \mathbf{X}_{E_k}^{W_{known}} = 0 \tag{20}$$

The Jacobian matrix is given by:

$$\mathbf{H}_k = \left[\mathbf{H}_{known} \ 0 \ \mathbf{H}_{F_k} \ 0 \ \mathbf{H}_{E_k} \ 0\right] \tag{21}$$

with $\mathbf{H}_{known}$ is the Jacobian of $\mathbf{h}_k$ with respect to $\mathbf{X}_{W_{known}}^{W_{local}}$, $\mathbf{H}_{F_k}$ is the Jacobian of $\mathbf{h}_k$ with respect to $\mathbf{X}_{F_k}^{W_{local}}$ and $\mathbf{H}_{E_k}$ is equal to identity matrix.

3.3.3 Update of the Translation Vector

Vector $\mathbf{X}_{W_{known}}^{W_{local}}$ is estimated using both the global and local map vectors. When a local map is joined to the global map, this vector is updated with the observation equation given by:

$$\mathbf{X}_{W_{known}}^{R} = \mathbf{h}_{known}(\mathbf{X}_{W_{local}}^{R}, \mathbf{X}_{W_{known}}^{W_{local}}) + \mathbf{w}_{known} \tag{22}$$

with $\mathbf{w}_{known}$, Gaussian noise and the observation function $\mathbf{h}_{known}$ defined by:

$$\mathbf{X}_{W_{known}}^{R} = \mathbf{X}_{W_{local}}^{R} \oplus \mathbf{X}_{W_{known}}^{W_{local}} \tag{23}$$

the corresponding Jacobian matrix is given by:

$$\mathbf{H}_{known} = \left[\mathbf{H}_{known}^{global}\ 0\ \mathbf{H}_{known}^{local}\ \mathbf{H}_{R}\ 0\right] \tag{24}$$

with $\mathbf{H}_{W_{known}}^{global}$ the Jacobian of $\mathbf{h}_{known}$ with respect to $\mathbf{X}_{W_{known}}^{W_{local}}$, $\mathbf{H}_R$ Jacobian of $\mathbf{h}_{known}$ with respect to $\mathbf{X}_{W_{local}}^{R}$ (vector equal to $\mathbf{X}_{W_{local}}^{R}$) and $\mathbf{H}_{W_{known}}^{local}$ the identity matrix.

3.3.4 SLAM Constrained by a known Map Algorithm

The method algorithm is presented in Algorithm 1. It summarizes the whole method and the conditions of switching between the SLAM approach and the constrained SLAM approach with an a priori known map.

Algorithm 1 : Constrained SLAM approach with an a priori known map algorithm

known ← false
hypothesis ← 0
loop
  if *known* is false **then**
    Building a local map without constraint (Sect. 2.1.1)
  else if *known* is true **then**
    Building a local map with constraint (Sect. 3.3)
  end if
  Joining local map in the global one (Sect. 2.1)
  if *known* is false **then**
    Matching global and known maps (Sect. 3.2)
    hypothesis ← number of validated matches
    if *hypothesis* > 3 **then**
      Estimate of $\mathbf{X}_{W_{known}}^{W_{global}}$ (Sect. 3.2)
      known ← true
    end if
  end if
  if *known* is true **then**
    Update of the translation vector (Sect. 3.3)
  end if
end loop

4 Simulation and Experimental Results

4.1 Consistency Definition

As ground-truth is available with simulation and experimental data (differential GPS) for the vehicle states, a statistical test for filter consistency can be carried out on the Normalized Estimation Error Squared (NEES):

$$d = (\mathbf{X}_{W_{local}}^{W_{global}} - \hat{\mathbf{X}}_{W_{local}}^{W_{global}})^T (\mathbf{P}_{W_{local}}^{W_{global}})^{-1} (\mathbf{X}_{W_{local}}^{W_{global}} - \hat{\mathbf{X}}_{W_{local}}^{W_{global}}) \tag{25}$$

As it is defined in Bar-Shalom et al. [16], the filter results are consistent if:

$$d \leq \chi^2_{dof,1-\alpha} \tag{26}$$

where $\chi^2_{dof,1-\alpha}$ is a threshold obtained from the χ^2 distribution with 3 degrees of freedom and a significance level equal to 0.05.

4.2 Simulation Results

The vehicle travels the map presented in Fig. 4 along a loop-trajectory of 600 meters, moving 0.25 m/step. The vehicle starts at $[-29, -11, \pi/3]$. It is equipped with a range-bearing sensor with a maximum range of 80 m, a 180 degrees frontal

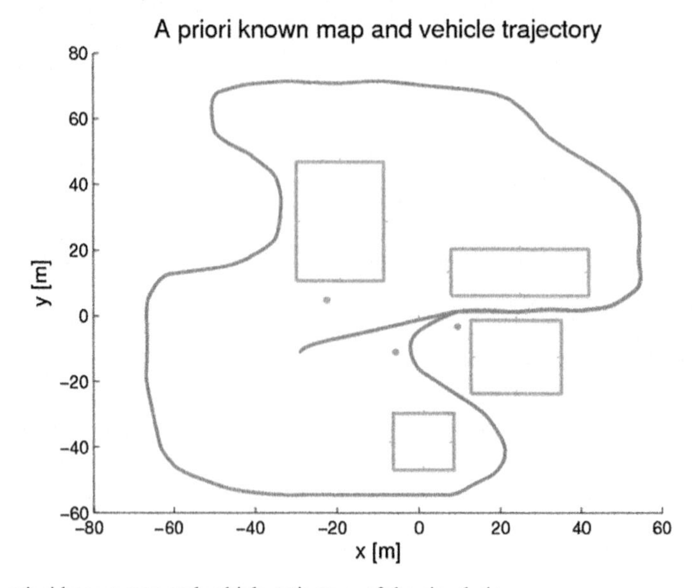

Fig. 4 A priori known map and vehicle trajectory of the simulation

Table 1 Range-bearing and control standard deviations

Sensor	Variable	units	Standard deviation value
Telemeter range	σ_ρ	m	0.05
Telemeter bearing	σ_β	rad	$\pi/720$
Vehicle velocity	σ_v	ms^{-1}	0.10
Vehicle heading angle	σ_φ	rad	$\pi/360$

field-of-view. The sensor frequency is set to 8 Hz. Gaussian-distributed errors are presented in Table 1.

The a priori known map of the simulated environment is shown in Fig. 5. The reference $\mathbf{W}_{known}$ is equal to [0,0,0] and the feature state deviations are fixed at 10 cm for coordinates, 10 cm for circle radiuses and 0.05 rad for line angles.

The evolution of errors and uncertainties (2 σ bounds) of the vehicle states is presented in Fig. 6. The error curves stay in the confidence bounds (dotted lines). These errors on the vehicle states are logical with the uncertainty values estimated in the associated covariance matrix. So the probability test of the filter consistency, defined by (26), is checked.

The application of our method brings some substantial improvement in the estimated vehicle path shown in Fig. 7. Errors on vehicle poses are maintained under 15 cm while errors with SLAM approach without constraint reach up to 50 cm in the environment part which is furthermost from the trajectory start. This accuracy improvement confirms the purpose to constrain a SLAM approach with an a priori known map.

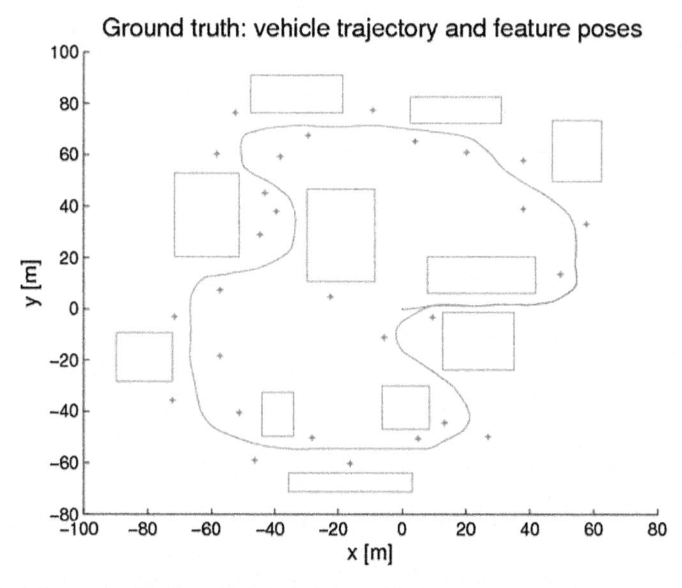

Fig. 5 Simulation ground truth: vehicle trajectory and feature poses

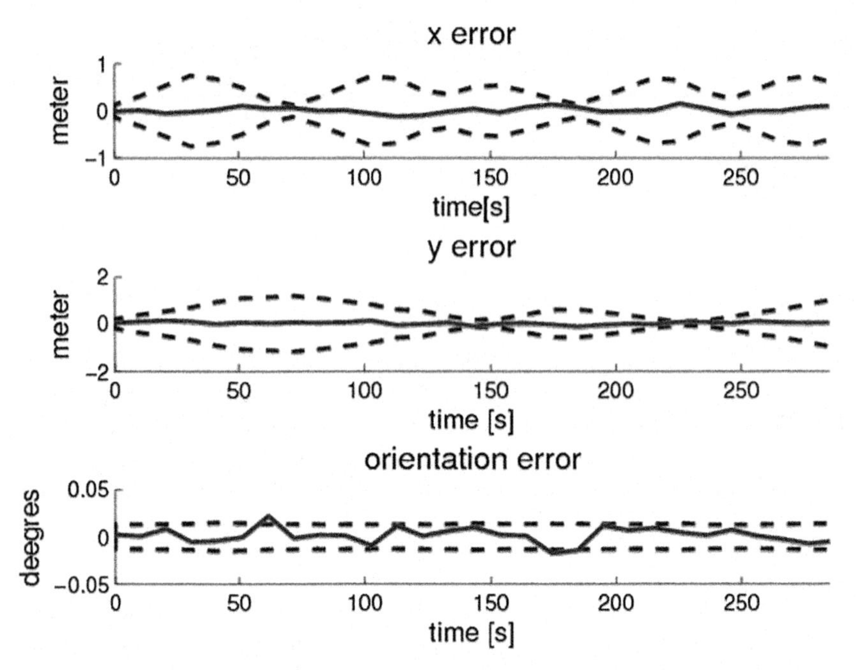

Fig. 6 Errors on the vehicle states bounded by the 2 σ confidence bounds (dotted curves)

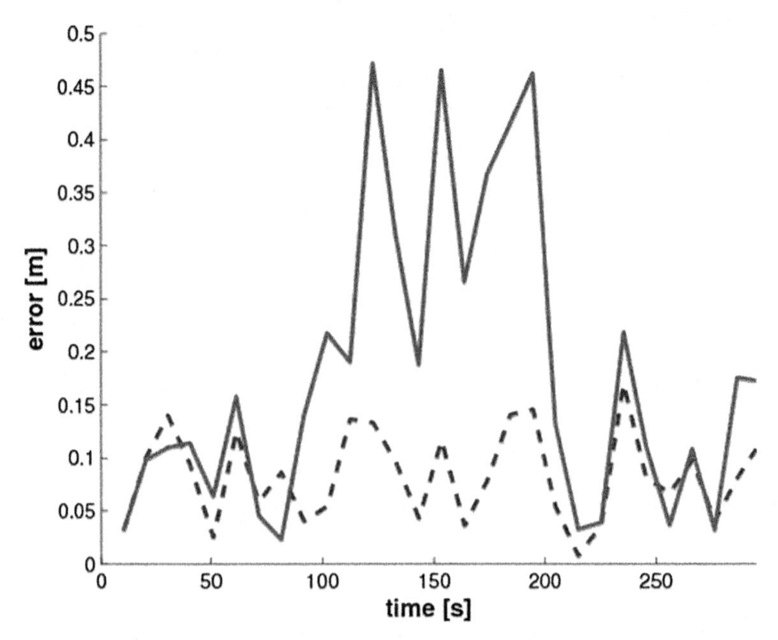

Fig. 7 Errors on the vehicle pose estimates: SLAM approach without map fusion (normal curve) and with map fusion (dotted curve)

The last result corresponds to the localization of the initial vehicle pose in the reference $\mathbf{W}_{known}$. The estimated vector $\mathbf{X}_{W_{known}}^{W_{global}}$ is $[-28.968, -10.980, 1.0472]$, it corresponds to a 0.14% error compared to the true pose equal to $[-29, -11, \pi/3]$. This error on the vehicle pose estimates proves that our method is accurate in a localization context.

4.3 Experimental results

This section shows experimental results of our fusion approach. The goal here is to confirm the results obtained using the simulated data in the previous section. A SICK LMS221 laser sensor and dead reckoning sensors equip one of our Automated Individual Public Vehicle (AIPV). The vehicle was hand driven along an outdoor path about 200 m, presented in Fig. 8 (gray line), at a speed of $1.2\,\text{ms}^{-1}$. This air-view image presents also the a priori known map of the experiment which is represented by circles and lines and is referenced in W, Fig. 8. A differential GPS was used to record the feature poses with an accuracy of 5 cm. All the other parameters of this experiment are the same than in the simulation part, Sect. 4.2 and Table 1.

A quantitative evaluation of the accuracy of our EKF-SLAM approach with or without map fusion is done studying the evolution of the error on the vehicle pose, Fig. 9. First, these results correspond to the simulation results and confirm the improvement carried by the map fusion. Errors with our approach are maintained at

Fig. 8 Air-view image of experimental environment: red draws (line, circle and reference W) represents a priori known map and green line shows real vehicle trajectory

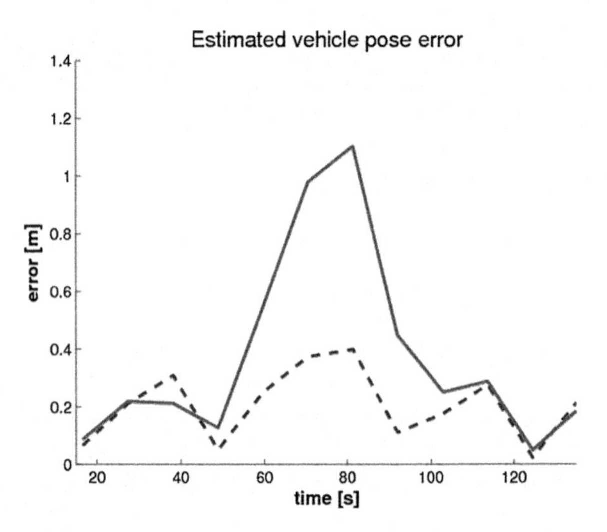

Fig. 9 Errors on the vehicle pose estimates: SLAM approach without constraint (normal curve) and with constraint (dotted curve)

a lower level than with the classical SLAM approach. The difference with simulation results on error amplitude can be explained by the fact that the floor is not flat in this environment.

The estimated map with the known map used in this experiment is presented in Fig. 10. An air-view of the final results is shown in Fig. 11. The mean error of 20 cm on the vehicle position is explained by the fact that the estimated map is not exactly overlaid with the known map. Two interesting things can be underlined. First, the loop closing is well done. The features observed at the beginning of the

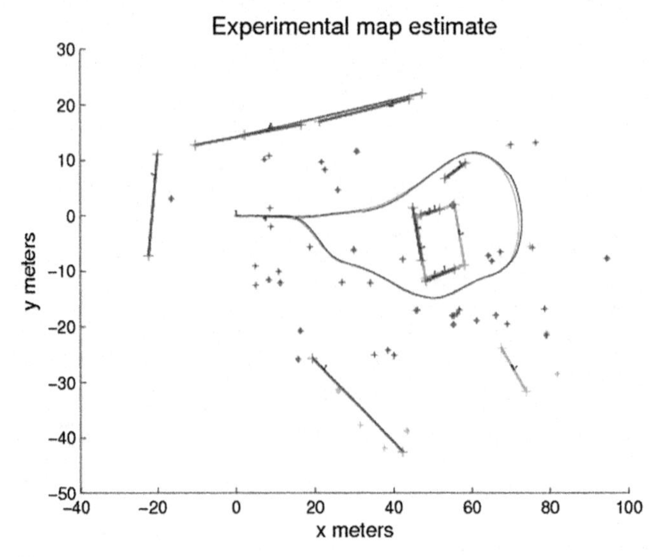

Fig. 10 Vehicle trajectory (black line) and map (red draws) estimates with presented approach, known map of the environment (blue draws) and real trajectory (green line)

Fig. 11 Air-view image of experimental mapping results: red draws (line and circle) represents the estimated map, magenta line represents vehicle trajectory estimate and green line shows real vehicle trajectory

experiment are matched at the end of the trajectory. Second, the known map of this environment is updated. The new and the old features compared with the known are clearly identified.

5 Conclusion

In this paper, a fusion between an a priori known map and an estimated map in a multi-map SLAM framework is presented. The simulation and experimental results prove that this fusion increases the estimate accuracy of vehicle pose and map compared to a classical SLAM approach. Our approach allows to use available map of the visited environment and to update these information. A direct application of our method is to detect changes on a known map of any place without GPS system.

References

1. Smith, R., Self, M., Cheeseman, P.: Estimating Uncertain Spatial Relationships in Robotics. In: *Proceedings, 2nd workshop on Uncertainty in Artificial Intelligence (AAAI)*, Philadelphia, U.S. pp. 1–21 (1986).
2. Durrant-Whyte, H., Bailey, T.: Simultaneous Localisation and Mapping (SLAM) : Part I The Essential Algorithms. *IEEE Robotics and Automation Magazine* **13**(2), 99–110 (2006).

3. Bailey, T., Durrant-Whyte, H.: Simultaneous Localisation and Mapping (SLAM) : Part II State of the Art. *IEEE Robotics and Automation Magazine* **13**(3), 108–117 (2006).
4. Julier, S., Uhlmann, K.: A Counter Example to the Theory of Simultaneous Localization and Map Building. In: *Proceedings, IEEE International Conference on Robotics and Automation*, vol. 4. Seoul, Korea (2001).
5. Bailey, T., Nieto, J., Guivant, J., Stevens, M., Nebot, E.: Consistency of the EKF-SLAM Algorithm. In: *Proceedings, IEEE/RSJ International Conference on Robots and Systems*, pp. 3562–3568. Beijing, China (2006).
6. Castellanos, J., Martinez-Cantin, R., Tardos, J., Neira, J.: Robocentric map joining: improving the consistency of EKF-SLAM. *Robotics and Autonomous Systems* **55**(1), 21–29 (2007).
7. Paz, L., Jensfelt, P., Tardos, J., Neira, J.: EKF SLAM Updates in O(n) with Divide and Conquer SLAM. In: *Proceedings, IEEE Conference on Robotics and Automation*, pp. 1657–1663. Roma, Italy (2007).
8. Rodriguez-Losada, D., Matia, F., Jimenez, A.: Local Maps Fusion for Real Time Multi-robot Indoor Simultaneous Localization and Mapping. In: *Proceedings, IEEE Conference on Robotics and Automation*, New Orleans, U.S., vol. 2, pp. 1308–1313 (2004).
9. Neira, J., Tardos, J.: Data Association in Stochastic Mapping Using the Joint Compatibility Test. *IEEE Transactions on Robotics and Automation* **17**(6), 890–897 (2001).
10. Guivant, J., Nebot, E.: Optimization of the Simultaneous Localization and Map-Building Algorithm for Real-Time Implementation. *IEEE Transactions on Robotics and Automation* **17**(3), 242–257 (2001).
11. Lee, K., Wijesoma, W., Guzman, J.: A Constrained SLAM Approach to Robust and Accurate Localisation of Autonomous Ground Vehicles. *Robotics and Autonomous Systems* **55**(7), 527–540 (2007).
12. Lee, K., Wijesoma, W., Guzman, J.: On the Observability and Observability Analysis of SLAM. In: *Proceedings, IEEE/RSJ International Conference on Intelligent Robots and Systems*, pp. 3569–3574. Beijing, China (2006).
13. Castellanos, J., Montiel, J., Neira, J., Tardos, J.: The SPmap: a probabilistic Framework for Simultaneous Localization and Map Building. *IEEE Transactions on Robotics and Automation* **15**(5), 948–953 (1999).
14. Triggs, B., McLauchlan, P., Hartley, R., Fitzgibbon, A.: Bundle Adjustment – A Modern Synthesis. In: W. Triggs, A. Zisserman, R. Szeliski (eds.) *Vision Algorithms: Theory and Practice, LNCS*, pp. 298–375. Springer Verlag, London (U.K.) (2000).
15. Bailey, T., Nebot, E., Rosenblatt, J., Durrant-Whyte, H.: Data Association for Mobile Robot Navigation: A Graph Theoretic Approach. In: *Proceedings, IEEE International Conference on Robotics and Automation*, vol. 3, pp. 2512–2517. Washington, US (2000).
16. Bar-Shalom, Y., Li, X., Kirubarajan, T.: *Estimation with Applications to Tracking and Navigation*. J. Wiley and Sons, Inc., New York (U.S.).

Development of a Semi-Autonomous Vehicle Operable by the Visually-Impaired

Dennis W. Hong, Shawn Kimmel, Rett Boehling, Nina Camoriano, Wes Cardwell, Greg Jannaman, Alex Purcell, Dan Ross and Eric Russel

Abstract This paper presents the development of a system that will allow a visually-impaired person to safely operate a motor vehicle. Named the Blind Driver Challenge, the purpose of the project is to improve the independence of the visually-impaired by allowing them to travel at their convenience under their own direction. The system is targeted to be deployed on Team Victor Tango's DARPA Urban Challenge vehicle, "Odin." DARPA stands for the Defense Advanced Research Projects Agency. The system uses tactile and audio interfaces to relay information to the driver about vehicle heading and speed. The driver then corrects his steering and speed using a joystick. The tactile interface is a modified massage chair, which directs the driver to accelerate or brake. The audio interface is a pair of headphones, which direct the driver where to turn. Testing software has been developed to evaluate the effectiveness of the system by tracking the users' ability to follow signals generated by the Blind Driver Challenge code. The system will then be improved based on test results and user feedback. Eventually, through a partnership with the National Federation of the Blind, a refined user interface and feedback system will be implemented on a small-scale vehicle platform and tested by visually impaired drivers.

1 Introduction

The National Federation of the Blind Jernigan Institute created the Blind Driver Challenge (BDC) with the goal of creating a semi-autonomous vehicle that can be driven by people with significant visual impairments. Currently, there are about 10 million blind and visually-impaired people in the United States, and 1.3 million of them are legally blind [1]. In today's world, people desire independence to live freely and lessen their reliance on others. New technologies are being created to help the blind attain this goal. The main focus of this project is to pioneer the development of non-visual interfaces as assistive driving technologies for the blind.

D.W. Hong (✉)
Department of Mechanical Engineering, Virginia Tech, Blacksburg, VA 24060, USA
e-mail: dhong@vt.edu

S. Lee et al. (eds.), *Multisensor Fusion and Integration for Intelligent Systems*, Lecture Notes in Electrical Engineering 35, DOI 10.1007/978-3-540-89859-7_31,

2 Background

The idea of allowing a blind person to drive a vehicle is viewed by many as overly ambitious. However, many driver assistance systems currently being researched directly relate to the BDC. One of these is the Honda Advanced Safety Vehicle. This vehicle has the ability to recognize imminent collisions and take precautions to minimize damage or avoid a collision altogether [2, 3]. This technology can be used in fully autonomous vehicles such as those found in the DARPA Urban Challenge. These intelligent vehicle systems can collect enough information to drive without human sight. Theoretically, this information can be provided to a blind person through other senses to allow him to drive.

Other senses that a person possesses include olfaction (smell), audition (sound), tactition (touch), and gestation (taste). Given the current state of technology, the most feasible senses with which to interface are audition and tactition. Purdue University is conducting research on a haptic back display using a chair outfitted with tactors. This research has discovered links between visual information and tactile cues, which can be used to increase the effectiveness of tactile feedback [4]. The University of Genova is investigating the use of 3D sound to provide information to the driver. This allows sound to be generated at any spatial coordinate or even along an arbitrary 3D trajectory [5].

In order to provide the driver with sufficient information to drive a vehicle, it is expected that the interface will need to include multiple senses. The Department of Veterans Affairs Medical Center has researched a wearable computer orientation system for the visually-impaired [6]. Three systems are being tested: a virtual sonic beacon, a speech output, and a shoulder-tapping system. Similar non-visual human interfaces may be used to supply necessary feedback to blind drivers regarding their surrounding environment.

3 Odin: The Autonomous Vehicle

Odin, pictured in Fig. 1, is the autonomous vehicle developed by Team Victor Tango as Virginia Tech's entry into the DARPA Urban Challenge. The BDC system was designed to be integrated into Odin for testing in actual driving scenarios. Odin has the ability to give the BDC system all the information it needs to allow a blind driver to operate the vehicle.

3.1 Odin's Sensors

Odin's ability to recognize the surrounding environment is crucial for safe, autonomous driving. To accomplish this task, Odin is equipped with the sensor array shown in Fig. 2. The sensors gather data regarding drivable area and potential static and dynamic obstacles.

Fig. 1 Odin. Team Victor Tango's DARPA Urban Challenge Vehicle

The sensor array consists of two IBEO Alasca XT LIDARs, two IEEE 1394 color cameras, four SICK LMS rangefinders, one IBEO Alasca A0 LIDAR, and a GPS/INS (Global Positioning System/Inertial Navigation System). The two IBEO Alasca XTs and the IBEO Alasca A0 are used to detect obstacles and the drivable area in front and behind Odin respectively. The two Alasca XTs are controlled by an external control unit (ECU). The ECU pre-processes data before sending it to Odin's computer. This feature creates a larger sensor resolution and was the factor for selecting the Alasca XTs as Odin's main forward sensors [7]. The four SICK LMS rangefinders monitor side blind spots and provide extended road detection. The two forward-facing IEEE 1394 color cameras are used as the primary means of road detection.

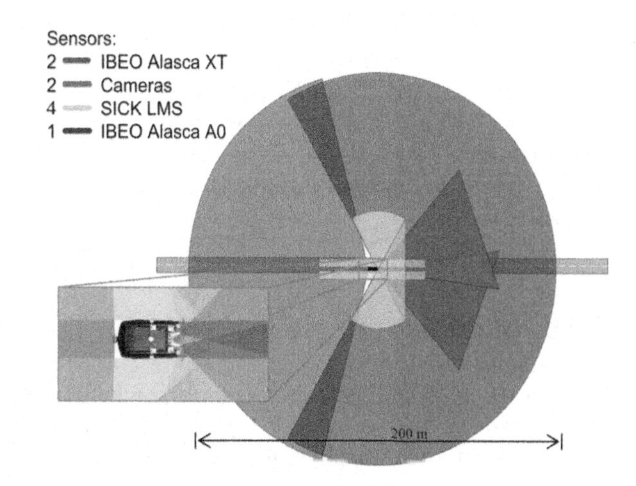

Fig. 2 Odin's sensor coverage. The colored areas indicate the maximum range of the sensor or the point at which the sensors scanning plane intersects the ground. Odin is facing towards the right

Odin uses the GPS/INS system in coordination with the RNDF (Route Network Definition File) and MDF (Mission Data File) for navigating to a final destination. The RNDF contains a GPS-based map of all drivable paths within the route network where Odin is operating. This file is used for creating the fastest route to Odin's destination. The MDF contains road checkpoints that Odin must pass through to reach the final destination.

3.2 Odin's Software

The sensor data must be translated into useful information that is understood by Odin's planning software. First, Odin's classification software interprets and converts the raw data into information such as drivable area, obstacles, and location. This refined data is sent to Odin's planning software, which processes the information further to determine the next move. Odin's planning hierarchy is constantly receiving updated information and reforming its plans to maintain its desired route. Each part of the software has its own tasks to complete which are critical to safely reaching the destination.

The Route Planner uses the RNDF and MDF to plan an overall route before Odin begins driving. This route is optimized to take Odin to the destination in the shortest possible time. Driving Behaviors handles the "rules of the road" such as changing lanes, passing, intersection negotiation, and parking [7]. Driving Behaviors also monitors obstacles that are detected by Odin's sensors. If Odin's path is blocked, Driving Behaviors instructs the Route Planner to plan a new set of directions. Motion Planning is responsible for the real-time motion of Odin. Motion Planning creates the Motion Profile that keeps Odin in the correct lane and allows it to avoid objects in the road. The Motion Profile is also essential to the BDC system, and will be discussed later in more detail. The final piece of the planning software is the Vehicle Interface. The Vehicle Interface software receives the motion profile from Motion Planning and translates it into commands to actuate the vehicle.

To effectively communicate between all of Odin's components, the JAUS (Joint Architecture for Unmanned Systems) message protocol is used. Originally developed for military use, Team Victor Tango was the first to implement JAUS with LabVIEW [7, 8]. JAUS allows software running on the same network to communicate using standard messages. Any systems that are JAUS-compatible can communicate between one-another regardless of their programming languages or affiliated hardware.

4 The Blind Driver System

The goal of creating a semi-autonomous vehicle for use by a blind driver required development of both hardware and software systems. Tactile and audio cues are provided through a physical interface to present the driver with necessary information, including desired heading and speed. The audio system gives the driver directional guidance while the tactile system gives speed guidance. A joystick was chosen as the driver input device due to its portability and low cost. Rationale for these design decisions, as well as details about each of the individual components, will be discussed in the following sections.

4.1 BDC Software Architecture

The BDC software architecture, as seen in Fig. 3, is designed to be highly modular. It combines audio, tactile, and user interface systems in parallel, independently running sub-functions. The front control panel uses multiple tabs to organize the code by component. Each tab holds the component's configuration settings. The BDC code receives two standardized cluster inputs: the Motion Profile and the Run Settings. The Motion Profile consists of all the information needed to navigate the vehicle: velocity, acceleration, curvature, and rate of change of curvature. The Run Settings cluster includes the configuration settings needed for each component. The BDC code only needs the Motion Profile and the Run Settings cluster to work. This creates modular software suitable for use on multiple platforms.

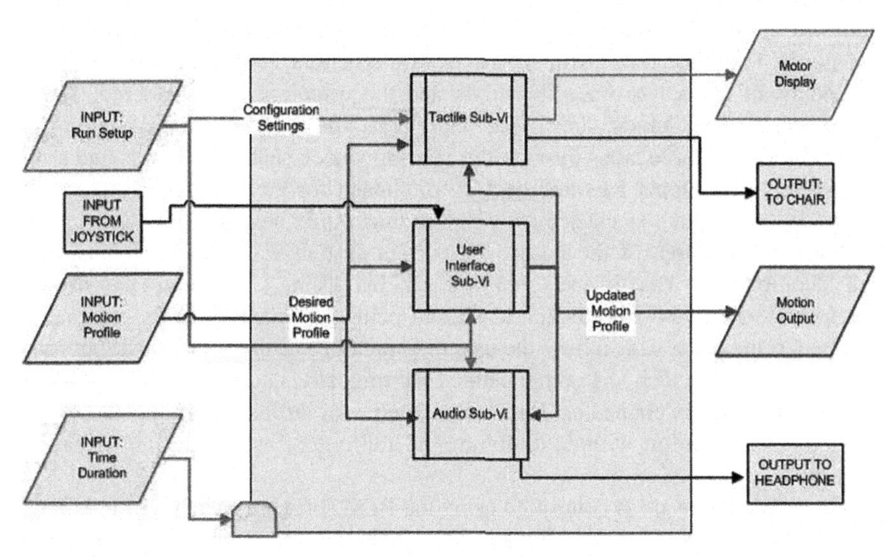

Fig. 3 BDC software architecture diagram. The tactile, audio, and user interface sub-functions are processed in parallel within the BDC code

4.2 The Tactile System

The BDC tactile system utilizes the driver's sense of touch to provide feedback regarding the speed of the vehicle under operation. The system uses a modified

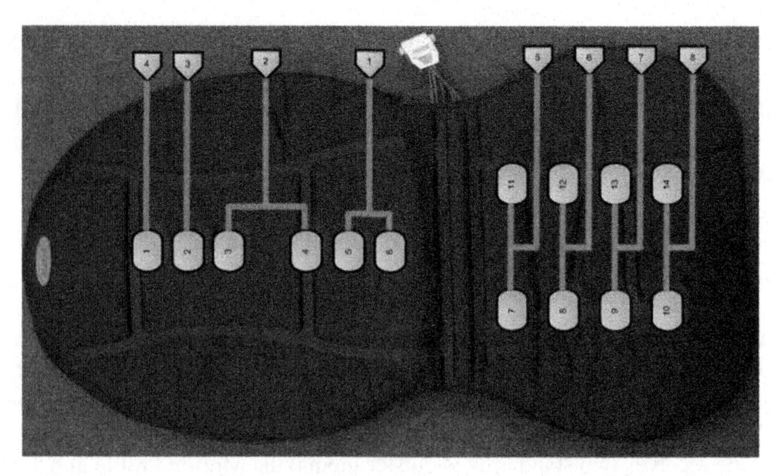

Fig. 4 Tactile feedback system. The vibrating chair is shown with a schematic depicting the motor wiring and configurations

vibrating massage chair, as seen in Fig. 4, to provide the user with tactile cues. These cues alert the driver to accelerate or decelerate by selectively varying the vibration intensity in the individual motors. The chair is controlled by a custom circuit board.

In order to choose how tactile messages would be relayed to the driver, all physical points of contact between the driver and the vehicle were considered. These points included the hands, chest, feet, upper-legs, and back. Feedback could have been provided to these areas through the steering wheel, seat belt, pedals, and seat, respectively. The upper-legs and back were chosen for the research discussed in this paper. A design was chosen for a device that would best accommodate tactile feedback on this area of the body. The current system was chosen after careful consideration of its ease of use, comfort, and cost. Using a vibrating chair allows the driver to receive tactile feedback without being burdened by bulky equipment attached to the body. This allows the user to enter and exit the vehicle without having to attach and detach any components. Electronically, the chair has one point of connection, a DB-25 connector. The chair is fitted with sufficient padding in order to maximize user comfort without inhibiting the ability to sense changes in vibration intensity.

The tactile hardware is controlled using the BDC software written in LabVIEW. The tactile component of the software takes in the desired vehicle speed from the Motion Profile, the actual vehicle speed, and the tactile configuration from the Run Setup. From the difference between the desired and actual speed of the vehicle, the software sends the appropriate signals to the circuit board. The signals activate a specified number of motors at either a low or high intensity. The motors are supplied either 9 or 15 V, causing this low or high intensity vibration. As the difference between the desired and actual speed increases, the system increases the number of vibrating motors and their intensity. According to the configuration used for testing, when the driver needs to accelerate, the chair vibrates on the driver's back, beginning at the top and traveling down the back as the required acceleration increases. Conversely, when the driver needs to decelerate, the chair vibrates on the back of the thighs, beginning at the front of the seat and traveling to the back of the seat as required deceleration increases.

4.3 The Audio System

An audio alert system was chosen to provide the directional guidance. The system relays information to the driver by using audible tones in the driver's ears which tell the driver both the direction and the magnitude of the desired turn, which is determined by curvature. The only hardware necessary for the audio system is a pair of headphones. The Motion Profile provides the code with a desired curvature, which is the curvature that the vehicle should be traveling, and a current curvature, which is the curvature at which the vehicle is actually traveling. The program then finds the difference and determines the degree to which the driver needs to correct his steering. The system then alerts the driver with a tone which indicates both the direction and magnitude of the required curvature. The change in "magnitude" of the tone reflects how much the driver should turn. This change can be in the form of increasing volume, increasing frequency, or some other change depending on the configuration chosen. The magnitude of the change is governed by Eq. (1).

$$M = V \cdot \frac{|C_d - C_c|}{C_{\max}} \tag{1}$$

where C_d is the desired curvature, C_c is the current curvature, $C_{\max}$ is the maximum curvature of the vehicle, and V is the maximum strength of the signal. V could be maximum volume, maximum frequency, or maximum beeping frequency depending on the configuration. The direction of the tone is governed by Eq. (2).

$$D = C_d - C_c \tag{2}$$

If D is positive, it means the driver is steering too far to the left, and if negative, the driver is steering too far to the right. D determines where the tone will be heard by the driver, depending on the configuration.

The driver may choose which configuration is most suitable from the options listed in Fig. 5. The configuration chosen for testing was the "center zero" setting. This setting tells the driver the direction of the desired turn by outputting a constant tone to the driver's ear corresponding to the desired direction. If the driver is steering in the correct direction, then there is no sound output, hence the name "center zero." This also saves the driver from hearing a tone for an extended period of time. Also, the magnitude of the turn is conveyed by changing the volume of the tone, where V is the maximum volume in Eq. (1). Encountering a sharper turn will result in a

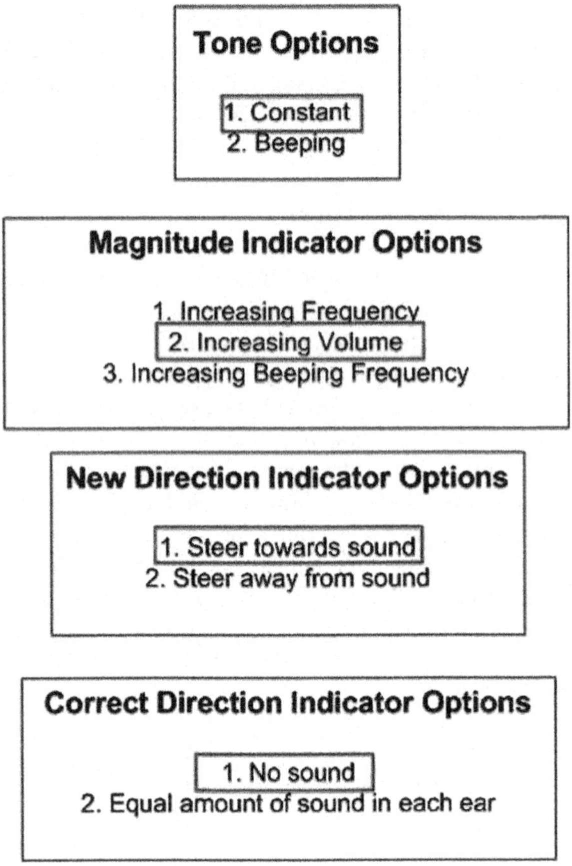

Fig. 5 Configuration options for audio system. The options for the "center zero" setting are shown with a box around them

Tone Options

1. Constant
2. Beeping

Magnitude Indicator Options

1. Increasing Frequency
2. Increasing Volume
3. Increasing Beeping Frequency

New Direction Indicator Options

1. Steer towards sound
2. Steer away from sound

Correct Direction Indicator Options

1. No sound
2. Equal amount of sound in each ear

louder tone. The audio code is source-independent, resulting in a modular and adaptive design. When the code is implemented in an autonomous vehicle, the desired curvature comes from the vehicle, and the current curvature is supplied from the steering wheel.

4.4 User Input

The user input device allows the driver to control the motion of the vehicle. It takes the place of the steering wheel and pedals in a normal automobile, allowing the user to drive the car in simulation. Currently, a joystick is used for this system. The y-axis of the joystick controls the acceleration of the vehicle. Forward (+y) creates a positive acceleration and backward (−y) creates a negative acceleration. The x-axis of the joystick controls the direction of the vehicle. Left (−x) steers the vehicle to the left and right (+x) steers the vehicle to the right. The trigger button may be used to simulate an emergency brake, creating a large deceleration when depressed.

5 Experimentation

A standardized test was created to benchmark the current system. The goal of this test is to provide objective information about the performance of the system in its current state. The test consists of the collection of both qualitative and quantitative data based on user opinions and performance, respectively. This data will help to indicate possible improvements to the current system. In addition, the data will be used as a point of comparison for future system developments.

5.1 Setup and Procedure

The testing software uses a sequence in LabVIEW to implement the BDC code. The testing software takes advantage of the BDC code's modularity because it is able to pass the Run Settings and a pre-defined Motion Profile to the BDC code without any modifications. Once the testing code passes a Motion Profile to the BDC code, the system functions as it would in the driving simulator or an autonomous vehicle.

The test begins with an orienting paragraph read by the test administrator. This paragraph outlines the test procedure for the test subject and assures confidentiality and safety. After this point, the test is run solely by the software. Using a text-to-speech module in LabVIEW, the test subject is advised that there will be an interface test. Using the buttons on the joystick, the user can choose to replay or continue to the next set of instructions. The interface test begins with instructions on how to use the buttons on the joystick. Next, the software verifies that the headphones work and are worn correctly. Following the headphone test, joystick functionality is verified by asking the test subject to tilt the joystick left. To complete the interface test, the system checks the vibration chair by alerting the user that he will feel vibrations in his legs and back. Then the program explains and demonstrates which vibrations indicate acceleration and which vibrations indicate braking.

Once the interface test is successfully completed, the system loads a pre-defined 1-minute driving course. The predefined function follows a figure-8 path to simulate the full range of motion experienced under normal driving conditions. The rate of change of speed and curvature vary throughout the driving schedule to test the user's reaction time. The test subject hears a 5 s countdown and then begins driving. The first driving test is broken into three sections. In the first section, the subject receives only audio feedback. The tactile system is disabled so that the subject will feel no vibrations during the audio test. In the second section, the same course is driven with the audio system disabled and the tactile system enabled so the subject only receives acceleration and braking instructions. The third section combines the first two so that the driver navigates the same pre-defined course while controlling both speed and direction. Between these sections, the computer explains which system(s) will be in use for the next section. The third section is repeated two more times in order to benchmark driver improvement. The computer simulation is followed by a brief survey.

5.2 Data Collection and Analysis

The software collects, processes, and stores test data to be used for evaluating the user's performance. The error between the desired and actual motion is continuously calculated and logged during the test at a rate of about 3 Hz. Upon completion of the test, the average error for speed and curvature is calculated. These statistics provide an easy way to track improvement between successive tests and compare performance for different tactile and audio configurations. They may also provide insight into performance trends between different user demographics. In order to maintain user anonymity, the data is logged under a testing code known only to that user. This allows future testing administrators to affiliate this data with any future data if the user returns for additional testing.

6 Discussion of Results

Ten Virginia Tech students with an average of 6 years of driving experience participated in the initial research of the BDC. It was predicted that the average error would decrease with subsequent runs of the combined audio and tactile tests. However, the average errors for curvature and speed showed no significant improvement through the three runs.

User feedback indicated some possible factors which may have contributed to this error. The lag in reaction time is one such contributor. This lag can be seen in Figs. 6 and 7 as a slight phase shift between the desired and actual curvature, and speed respectively. Another important parameter shown in these figures is the degree of oscillation in the user responses for curvature and speed. User feedback confirmed that oscillation was more prevalent in the audio system than the tactile system. One possible solution is to create a larger dead-band region for the center-zero setting of the audio system. This would decrease the amount of overshoot and oscillation because the user would have more leeway to obtain a satisfactory curvature.

Recall that the users were tested independently on the audio and tactile system, followed by three tests of the operation of both systems simultaneously. The average curvature error increased by 64% between the audio-only and the combined test. Similarly, the average speed error increased by 52% between the tactile-only and the combined test. This indicates that the increased mental load of controlling both systems decreases the driver's performance.

Although the increase in reaction time due to the combined tests appears substantial, keep in mind that the user is given only one minute to become familiar with each individual feedback and control system before beginning the combined tests. The overall reaction time would most likely decrease with sufficient training and practice.

In a real-world driving environment, reaction time is vital for the safe operation of a vehicle. The predefined desired path, depicted as a bold line in Figs. 6 and 7, is standardized for all tests, and is designed to simulate various real-world driving

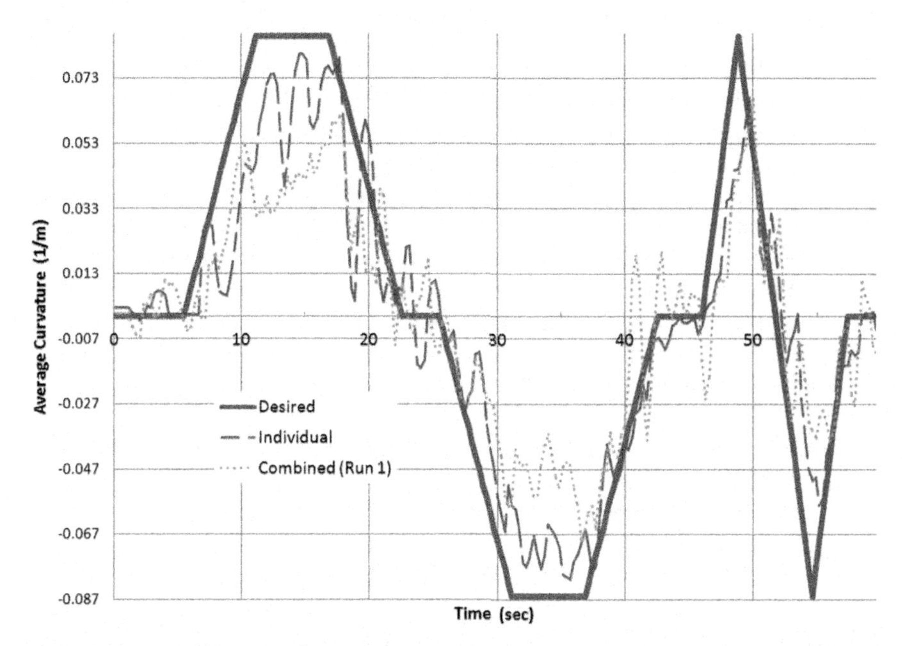

Fig. 6 Curvature vs. time plot. Shows the average time-response of curvature during the audio only and combined audio-tactile test. A positive curvature value indicates a right turn, and a negative curvature value indicates a left turn

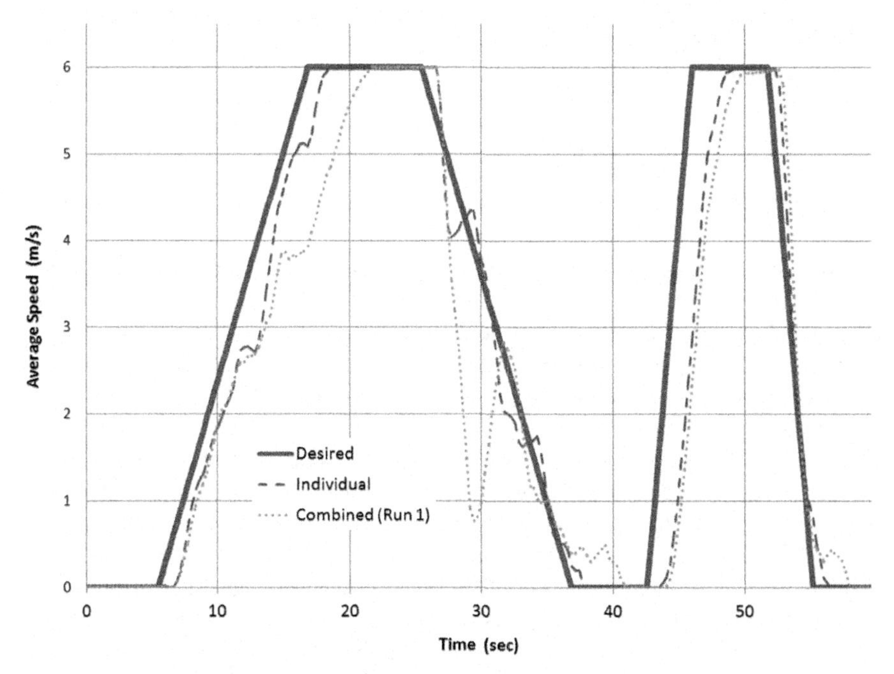

Fig. 7 Speed vs. time plot. Shows the average time-response of speed during the tactile only and combined audio-tactile test

conditions. The first 42 s of each test are designed to simulate normal driving conditions that may be encountered, for example, at a stop light or intersection. Whereas the final 18 s of each test are designed to simulate the sudden reactions required of a driver during critical situations like avoiding a potential collision. Of these reactions, the ability to decelerate and stop quickly is arguably the most important. To address this need, the BDC system is structured so that whenever the desired velocity suddenly jumps to zero, all motors in the chair that indicate deceleration will vibrate on full intensity, regardless of the difference between actual and desired velocity. Figure 7 shows that this feature successfully reduces the reaction time to an almost instantaneous compliance. Although it can be seen that not all driving situations resulted in such outstanding user performance, it is important to highlight that despite the lack of training, users were still able to react quickly in the event of a critical situation.

Many participants indicated that the tactile system would be more intuitive if the Run Settings were changed so that acceleration is indicated by vibration in the legs and deceleration in the back. Also, 85% of participants indicated that a steering wheel and pedals would be a more suitable and realistic form of user input. Making these configurations more intuitive would increase the reaction time, which is necessary for vehicle operation in real-world environments. Users also indicated the potential for numerous alternate applications of this audio and tactile feedback system. Some of these applications include increased situational awareness for sighted drivers, new driver education, aircraft pilots flying in instrument conditions, surgical operations, mining, biking, military operations, deep sea exploration, operation of construction equipment, adaptive automobile cruise control, obstacle proximity sensing for remote robotics applications, and video games.

7 Conclusions and Future Work

The BDC team designed, implemented, and tested non-visual interfaces to simulate a blind individual safely navigating a motor vehicle along a pre-determined course. Directional guidance is provided to the driver through audio cues via a set of headphones. Speed guidance is provided to the driver through tactile cues via a vibration chair. These systems are controlled by a modular software system which allowed them to be tested independently from a vehicle. Future integration with Odin and Team Victor Tango's software simulator is under consideration. The simulator was developed for testing software in a virtual environment [9]. Odin's on-board computer will create the Motion Profile required for integration with the BDC software.

The BDC system was tested and shown to be effective when the audio and tactile systems were independent of each other. However, when integrated, they proved to be overwhelming. It was not experimentally determined whether the feedback (audio and tactile) or input method (joystick) was the cause. Feedback from test subjects gave valuable insight on possible improvements to the user interface. The three main suggestions were to use a steering wheel and pedals in place of the joystick, reverse the tactile configuration, and increase the dead-band range in the audio settings. The current sample of test participants does not include any legally blind individuals. However, through a renewed partnership with the National Federation of the Blind, future testing will include blind individuals in order to authenticate our benchmarking analysis.

The BDC project has the potential to serve a dual purpose by supporting blind drivers and satisfying the demand for driving assistance in other demographics. The innovative non-visual interface technology being developed through this research can be used in a myriad of applications to provide vital feedback in ways rarely utilized in today's technological world.

Abbreviations

BDC	Blind Driver Challenge
DARPA	Defense Advanced Research Projects Agency
ECU	External Control Unit
GPJ/INS	Global Position System/Inertial Navigation System
JAUS	Joint Architecture for Unmanned Systems
MDF	Mission Data File
RNDF	Route Network Definition File

References

1. American Federation for the Blind, http://www.afb.org/section.asp?sectionid=15. Statistical Snapshots.
2. Takahashi, N., and Asanuma, A. (2008). Introduction of Honda Asv-2 (advanced safety vehicle-phase 2), *Proceedings of the IEEE Intelligent Vehicles Symposium*, pp. 694–701.
3. Labayrade, R., Royere, C., and Aubert, D. (2005). A collision mitigation system using laser scanner and stereovision fusion and its assessment, *Proceedings of the IEEE Intelligent Vehicles Symposium*, pp. 441–446.
4. Tan, H., Gray, R., Young, J., and Taylor, R. (2003). *A haptic back display for attentional and directional cueing: Purdue university*, Nissan Cambridge basic research, tech. rep., Haptic Interface Research Laboratory.
5. Bellotti, F., Berta, R., Gloria, A., and Margarone, M..(2002). Using 3D sound to improve the effectiveness of the advanced driver assistance systems, *Personal and Ubiquitous Computing*, 6(3), 155–163.
6. Ross, D., and Blasch, B. (2002). Development of a wearable computer orientation system, *Personal and Ubiquitous Computing*, 6(1), 49–63.
7. Reinholtz, C., Hong, D., Wicks. A. et al. (2008). Odin: team Victor Tango's entry in the DARPA Urban Challenge, *Journal of Field Robotics*, 25(8), 467–492.
8. Faruque, R. (2006). A JAUS toolkit for LabVIEW, and a series of implementation case studies with recommendations to the SAE AS-4 Standards Committee, *Master of Science Thesis*, Mechanical Engineering, Virginia Tech.
9. Reinholtz, C., Hong, D., Wicks, A. et al. (2007). DARPA Urban Challenge final report, DARPA

Process Diagnosis and Monitoring of Field Bus based Automation Systems using Self-Organizing Maps and Watershed Transformations

Christian W. Frey

Abstract A cost-effective operation of complex automation systems requires the continuous diagnosis of the asset functionality. The early detection of potential failures and malfunctions, the identification and localization of present or impending component failures and, in particular, the monitoring of the underlying physical process are of crucial importance for the efficient operation of complex process industry assets. With respect to these suppositions a software agent based diagnosis and monitoring concept has been developed, which allows an integrated and continuous diagnosis of the communication network and the underlying physical process behavior. The present paper outlines the architecture of the developed distributed diagnostic concept based on software agents and presents the functionality for the diagnosis of the unknown process behaviour of the underlying automation system based on machine learning methods.

Keywords Machine learning · Self organizing maps · Monitoring · Diagnosis · Software agents

1 Motivation

Technological progress in automation system engineering presents great challenges for system operators and especially for implementation and maintenance personnel. Functions which have hitherto been performed by mechanical or electromechanical devices are being increasingly replaced by software-based mechatronic systems. Functions that were previously visible and easily understood are now implemented using software and digital communications systems. Information that was previously transmitted, for example, as analogue electrical signals and was thereby simple to verify, is now transmitted as digital messages in fast data networks. Due to the decentralization of intelligence and the transition to distributed automation

C.W. Frey (✉)
Fraunhofer Institute for Information and Data Processing, Fraunhoferstrasse 1, 76131 Karlsruhe Germany
e-mail: christian.frey@iitb.fraunhofer.de

S. Lee et al. (eds.), *Multisensor Fusion and Integration for Intelligent Systems*, Lecture Notes in Electrical Engineering 35, DOI 10.1007/978-3-540-89859-7_32,

architectures, the engineering and implementation of systems is becoming ever more complex. The associated costs meanwhile exceed the basic hardware costs of the components by a large factor. It is here that effective diagnostic concepts offer great potential for cost savings. Verification that a system is functioning correctly, fault diagnosis, early detection of impending component failures and, in particular, monitoring of the process are of crucial importance for the cost-effective operation of complex automated processes.

The continuous diagnosis of the functionality, the early detection of potential failures and the continuous monitoring of the underlying physical process itself is essential for the cost-effective operation of complex automation systems. With respect to these suppositions a software agent based diagnosis and monitoring concept has been developed, which allows an integrated and continuous diagnosis of the communication network and the underlying physical process behavior. The present paper outlines the architecture of the developed distributed diagnostic concept based on software agents and explains the functionality for the diagnosis of the unknown process behaviour of the underlying automation system.

2 Diagnostic Concept Based on Software Agents

The field bus forms the "central nervous system" of distributed automation systems and represents both the object of the diagnosis as well as the access point for the diagnosis of the system components and the system functionality. The diagnosis of technical systems can be understood as a two-stage hierarchical process, in which quantitative information in the form of sensor signals is transformed into qualitative information, that is to say, diagnostic results [1].

In the first stage, the feature generation, the objective is to convert the measurable state variables of the system by transformation into a suitable compressed representation format, such that the possible diagnostic results are reliably reflected. The second step of the diagnostic process, the feature evaluation, represents a logical decision-making process in which the compressed quantitative knowledge in the form of features is transformed into qualitative diagnostic knowledge.

The developed diagnostic concept, illustrated in Fig. 1, is based on the idea of continuously analysing the messages transmitted on the field bus and extracting characteristic features, which describe in compressed form the behaviour of the respective field device and its interaction with the overall automation system. Features can be generated from the field bus messages concerning both the communication behaviour, for example, by analysing the response time of a field device, and the actual physical process behaviour, for example, by analysing a manipulated variable or a measurement value. The features represent a compressed description of the system functionality on the communications as well as on the process levels, which in turn can be linked together by a logical decision-making process in the higher-level feature-analysis stage to form an integrated diagnostic result.

The permanent analysis of the field bus data stream requires an effective compression of the data transmitted over the field bus, as otherwise the quantity of data

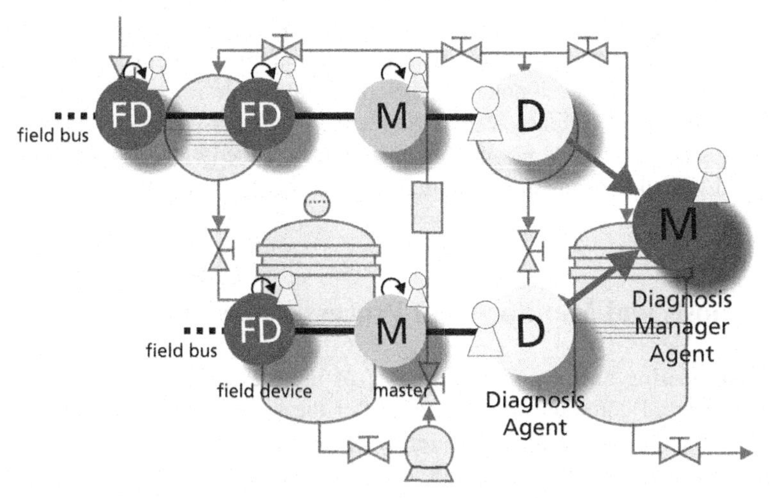

Fig. 1 Schematic overview of the developed software agent-based diagnostic concept for field bus based automation networks

produced would no longer be manageable. As an example of this, in a automation network with 12 Mbit/s transmission speed, as much as 1 MByte of data is transmitted per second – in Ethernet-based systems with transmission rates of up to 1 Gbit/s this can be over 100 MByte/s. It is clear that under these boundary conditions even a selective sampling of field bus data messages with subsequent off-line analysis would fail due to the non-availability of the requisite levels of computing capacity or storage media. The generation of meaningful features based on the immense data stream coming from the field bus places the highest demands on the efficiency and intelligence of the software design that is used. When considering the design of the diagnostic system, and especially in view of the heterogeneous communications networks or the different software platforms deployed on the control level in an automation system, particular attention must be paid to the effective coordination and cooperation of the distributed software components. Taking account of these requirements, a hierarchical diagnosis and monitoring concept for field bus based systems has been developed, based on software agents.

The diagnostic agents, in the lowest level of the hierarchy, are software components which as far as possible run autonomously on dedicated embedded systems, for example, so-called field bus gateways. The task of the various diagnostic agents is to analyse the field bus messages of the field devices assigned to the section of the field bus and to generate meaningful features. The higher-level diagnosis manager is also implemented as a software agent and contains the actual diagnostic functionality. Implemented as a separate executable application or as a component of the host system, the diagnosis manager cyclically retrieves the features generated in the diagnostic agents of the respective strands of the field bus and passes them to the feature analysis stage.

With regard to integration into existing automation systems, the newly developed concept provides additional software agents, for example, for linking to different physical implementations of the field bus (Profibus, Foundation Fieldbus), for displaying the diagnosis results (GUIAgent) or for archiving them (ArchiveAgent).

There are various development environments available for building and implementing multi-agent systems, which provide the developer with basic functions for deploying software agents. For the implementation of the proposed diagnostic concept the JADE development environment (Java Agent Development Framework) was selected [2, 3].

3 Diagnosis and Monitoring of Process Behaviour

When considering an integrated diagnosis of industrial processes, the field bus, being the central communications network of the system components, represents both the object of the diagnosis in terms of communications technology, and at the same time it is an ideal access point for the diagnosis of the physical process behaviour of the underlying automation process. To guarantee diagnostic functionality that is as transparent and robust as possible, the concept that has been developed splits up the diagnosis task into the diagnosis of the communication behaviour of the field bus and the diagnosis or monitoring of the physical system behaviour of the underlying automation process.

To diagnose the communication behaviour, relevant features are generated from the field bus messages, such as the response time of a field device. These features are evaluated by frequency-based neuro-fuzzy membership functions and an expert knowledge base implemented in a fuzzy-rulebase (*if ... then ...*) [4]. A detailed presentation of this neuro-fuzzy based design is given in [3].

In contrast to the diagnosis of the communications layer, in which only signal-based feature generation methods are used, the diagnosis of the underlying physical process behaviour relies on model-based methods. The basic principle of model-based diagnosis for technical systems is based on a quantitative mathematical model of the process to be monitored. Using a suitable distance measure, the measured process variables are compared with the calculated process variables of the process model. The greater the distance between measured and modelled variables, the greater the probability of a deviation from the normal behaviour of the process assumed in the modelling step [1, 5]. The difficulty in implementing such model-based diagnostic systems lies in generating a suitable process model, in particular if analytical modelling approaches are used. This means that, for example, in industrial process-engineering applications, setting up a robust model for feature generation often requires a comprehensive development effort, calling on detailed expert knowledge about the physical interactions involved in the process.

An alternative to the analytic model-based methods is offered by data-driven adaptive modelling approaches. In these approaches, the most prominent technique is that of artificial neural networks. On the basis of measured historical process variables (training data), neural networks can use a learning algorithm to acquire the static and/or dynamic transmission behaviour of the process [6, 7]. The advantage of this approach is that no analytically formulated process model is needed a priori, and thus the developer is not hampered by "unsafe" assumptions about the physical interactions of the process that are present when modelling. Frequently used methods in the so-called Black-Box Modelling approach are back-propagation networks. These neural structures are based on a supervised learning task, which requires a correct assignment of the available process variables to input and output variables of the process model.

In the context of the diagnosis of an unknown physical behaviour in field bus based automation systems, however, the following problem arises. While in principle all in- and output values of the physical sub-processes can be retrieved from the messages in the field bus data stream, the allocation in terms of input/output values itself requires a very detailed knowledge of the underlying system, which has usually to be obtained from the system documentation or expert knowledge. Based on these boundary conditions, the concept developed for the diagnosis of the unknown physical behaviour uses the properties of self-organizing maps for data-driven modelling of the process behaviour.

The so-called self-organising feature maps (SOM), a special neural network, are capable of generating a topology-preserving mapping of a high-dimensional feature space into an output space of lower dimensionality [8]. These neural models, also known as Kohonen maps, are capable of extracting and displaying unknown clusters in the database to be analysed (structure discovery), "unsupervised" without a priori information in terms of input/output assignment.

The basic components of SOMs are referred as neurons, but they differ from the neuronal model of the back-propagation networks in their basic function. The neurons do not function as processing units that respond to particular inputs with particular outputs, but assume the role of simple memory units, the content of which, known as stored pattern vectors or also prototypes, can be read out and rewritten with new data. The neurons of the SOMs, as shown in Fig. 2, are arranged in a specific topology, that is, the neurons are in a topological relationship to one another. This quality of the network is generally referred as neighbourhood. Self-organizing maps are based on an unsupervised, concurrent learning process – during the learning process a feature vector M from the learning task is presented to the network and its distance from the prototype vectors W stored in the neurons is calculated. The neuron with the smallest distance from the input vector, the so-called Best Matching Unit (BMU), is then modified according to the following equation:

$$W_j^{k+1} = W_j^k + \eta\mu(M - W_j^k)$$

Beside the "winner neuron," the Kohonen learning rule also includes neighbouring neurons in the learning process. The introduction of neighbourhood learning enables "similar" feature vectors to be projected onto topographically similar regions of the map. Using the neighbourhood coefficient n, this property can be made stronger or weaker during the learning process. A measure of how well the map represents a training dataset is provided by the so-called quantisation error, which is calculated using, for example, the Euclidean distance between input feature vector M and prototype vector W of the winner neuron.

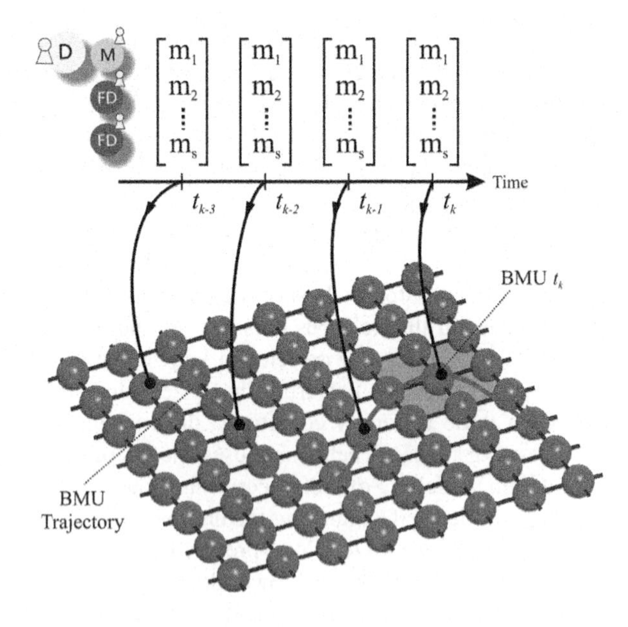

Fig. 2 Self-organizing map with a two-dimensional lattice topology

By applying the so-called UMatrix-representation of a self-organizing map, it is possible to perform a classification or a clustering of the feature space [7]. The UMatrix-transformation is based on the idea to add an additional dimension to the SOM lattice topology, which reflects the distance between the individual neuron to its surrounding neighbours. As an example of this, such a UMatrix representation of a 2D SOM is visualized in Fig. 3. The valleys in this UMatrix-plot (blue) represent

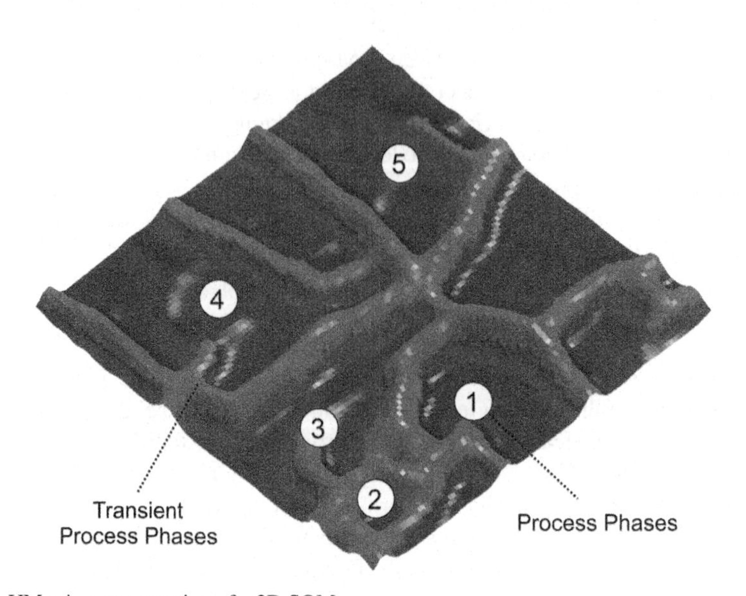

Fig. 3 UMatrix representation of a 2D SOM

regions in which the stored pattern vectors are very similar, the regions shown in red characterise a transition from one feature-space cluster to the next cluster.

In terms of the diagnosis and monitoring of a technical process, the valleys in the UMatrix-plot can be interpreted as stationary process phases, which are separated by so-called transient process phases. It should be noted here that the topology of the SOM corresponds to a toroid: opposite edges of the UMatrix-plot are joined together, that is, valleys of the UMatrix can extend across the boundaries of the map to the opposite side.

In principle, the segmentation or subdivision of a UMatrix into cluster regions can be performed manually by the observer. This method rapidly becomes impractical however with increasing size of the map or structural complexity of the UMatrix. For the automatic segmentation of the UMatrix, the diagnosis and monitoring concept we have developed uses the method of the Watershed transformation which is well-known from the image processing field [9]. The basic idea of the method can be made clear with the analogy of "drops of water" which fall on the UMatrix "mountains." Through the "mountain ridges" flooded regions are formed, which correspond to the clusters of the UMatrix. Figure 4 shows the result of such a Watershed transformation – the UMatrix has been segmented into six different clusters, corresponding to process phases.

As already indicated, in the data stream of the field bus, or alternatively in the messages of the field devices, all process variables are in principle retrievable and can be extracted by the diagnostic agents without detailed expert knowledge. Depending on the complexity of the system under consideration and the transmission speed of the field bus, the state variables of the process are summed over time and combined into an integrated state variable vector M (cf. Fig. 2). This

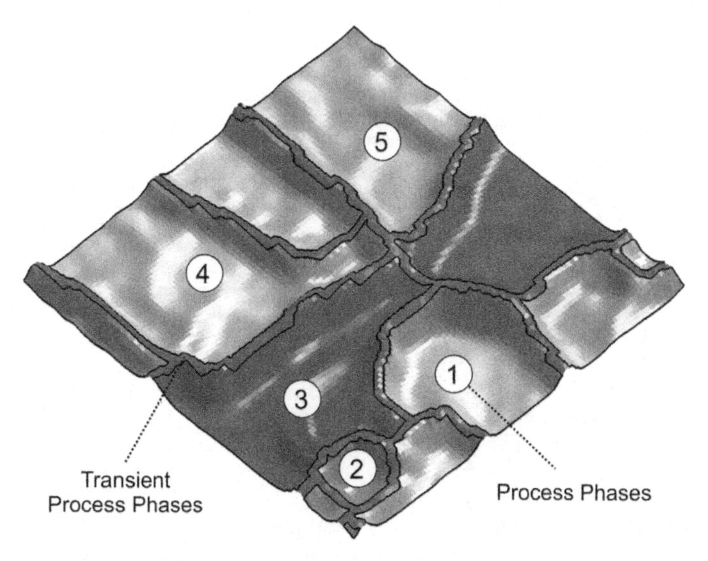

Fig. 4 Segmentation of a UMatrix into cluster regions by applying the watershed transformation

high-dimensional state variable vector describes the interaction of all field devices with the overall system and constitutes the basis of the model-based diagnosis. In the initial training phase of the diagnostic concept the data-driven model of the system is generated in the form of a SOM using the feature vector. The duration of the training process depends essentially on the complexity of the process (number of field devices), or in the case of cyclical process behaviour, on the duration of the batch. The success of the learning process can be evaluated with the aid of the resulting quantisation error of the map. It is clear that during this data-driven procedure, as close as possible to all operating states of the system should be acquired in order to obtain a robust diagnostic performance. Based on this trained map, an online diagnosis and monitoring of the process behaviour can then be performed. By analysing the quantisation error of the map, or the progress of the process phases respectively the BMU-trajectory, deviations from the normal behaviour of the system can be detected and thus traced back to possible errors in the behaviour of the system.

4 Testing the Diagnostic Concept

The experimental system, illustrated in Fig. 5, consists essentially of two containers between which liquid is pumped around in cycles at varying pumping powers and valve positions. It should be noted here that in the case of the demonstrator system

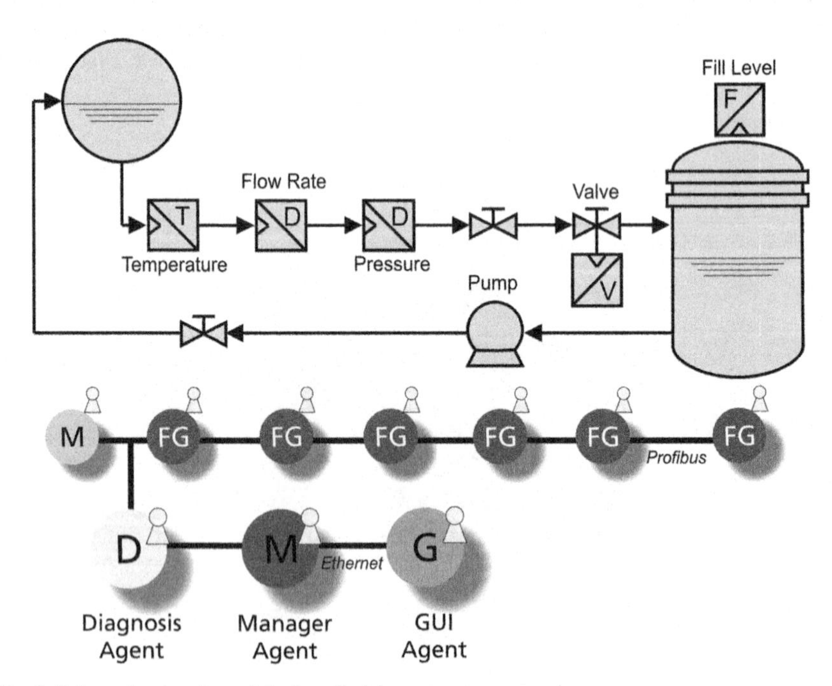

Fig. 5 Schematic overview of the installed demonstrator system

the nature of the field devices (measured or manipulated variable), the physical meaning of the process variables acquired (e.g., flow rate meter, temperature sensor), and in principle the processing behaviour implemented by the control system are all known. The diagnostic concept developed however does not rely on this additional information.

Figure 6 shows an example of the temporal course of the standardised process data obtained from the field bus data stream, or field bus messages as appropriate. From this curve the cyclically recurring behaviour of the process can be clearly recognised. On the basis of this training data, a data-driven model of the system in the form of a self-organizing map was generated. With the aid of the UMatrix transformation, the various process phases or operating states of the system (e.g., emptying containers or filling containers) were identified using the Watershed transformation. Based on this trained feature map an online-diagnosis of the process behaviour is then made. To do this, the IO-data of the field devices is filtered out online from the field bus data stream by the diagnosis agents and presented to the feature map. By analysis of the quantisation error of the map, deviations from the learned normal behaviour can be detected. As an example, Fig. 7 shows the behaviour of the quantisation error over the duration of the process during abnormal process behaviour. In error case 1 the venting of the system was reduced in order to disrupt the behaviour of the process, while in error case 2 the flow cross-section was reduced. These interventions in the process behaviour of the system are clearly revealed in the curve of the quantisation error of the map.

By observing the BMU trajectory in combination with the process phases found by watershed segmentation, it is also possible to analyse the progress of the process phases. Figure 7 shows, in addition to the quantisation error, the time course of the identified process phase – the modified progress of the process phases in error case 2 can be clearly seen.

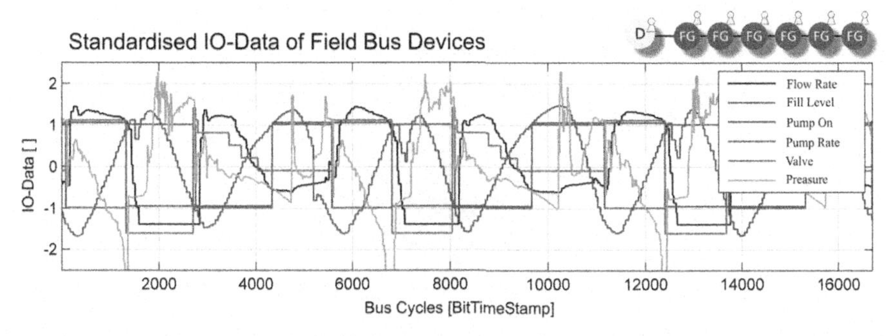

Fig. 6 Time course of the process state variables extracted from the field bus messages during normal operation

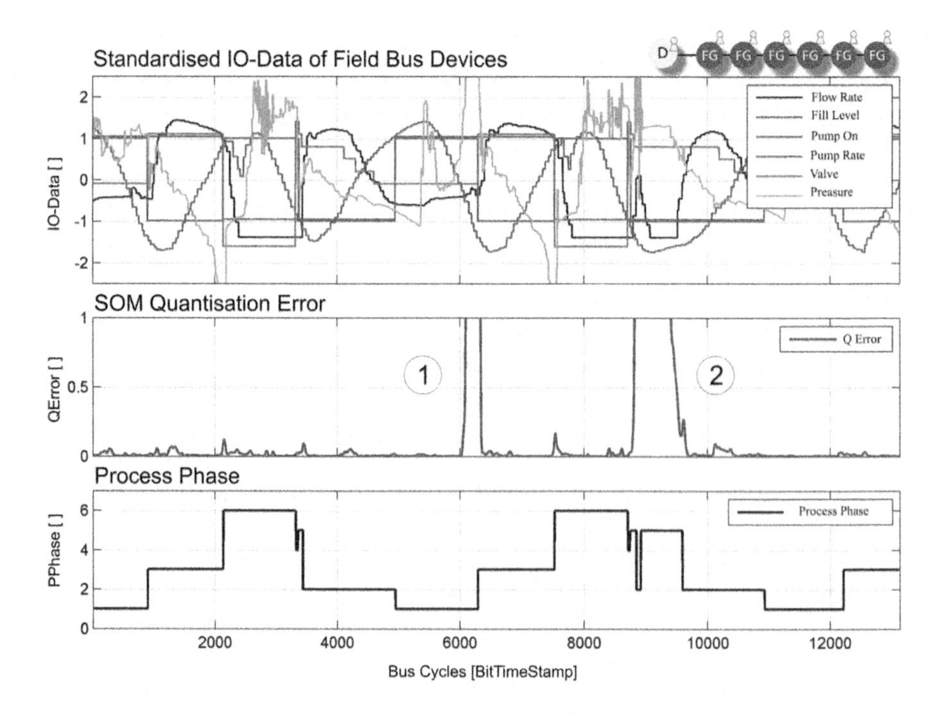

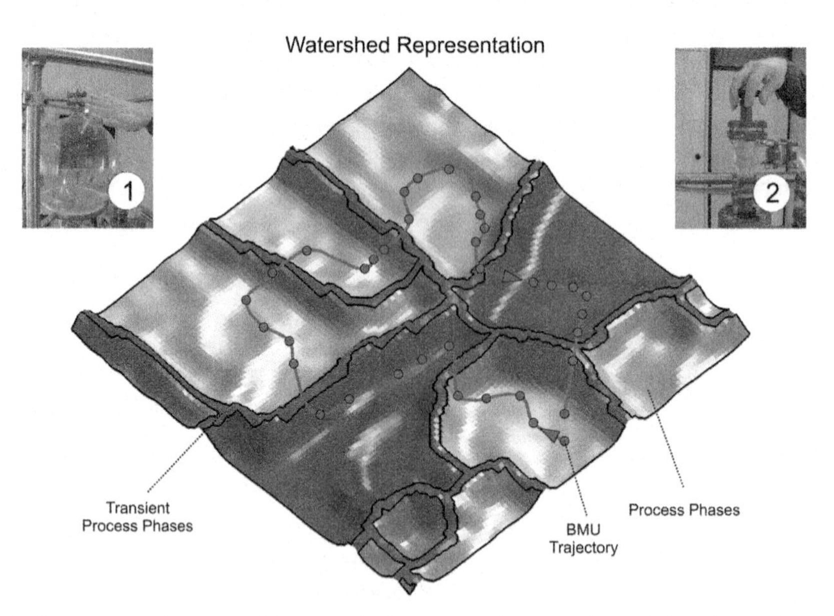

Fig. 7 Time course of state variables, quantisation error and process phase during abnormal process behaviour

5 Summary

The agent-based diagnostic concept that has been developed represents an integral component of the automated system, and enables a continuous integrated diagnosis and monitoring of the process under consideration. The present paper has outlined the architecture of this distributed diagnostic concept based on software agents and explained its embedded functionality for diagnosing the unknown process behaviour using self-organizing feature maps. The functionality and performance of the developed diagnostic concept was validated with the aid of a process-engineering demonstrator system. The concept is currently being transferred to a wide range of industrial process in the chemical industry.

References

1. Isermann, R.: Model-based fault-detection and diagnosis – status and applications. *Annual Reviews in Control*, 29(1), 71–85, 2005.
2. Bradshaw, J.M.: *Software Agents*. AAAI Press, Cambridge, MA, 1997.
3. Endl, H.: *FELDBUS – Erhöhung der Sicherheit und Zuverlässigkeit von eingebetteten, echtzeitfähigen Softwaresystemen durch die Entwicklung von Methoden und Werkzeugen zur agentenbasierten Diagnose von Feldbusnetzen in verteilten Automatisierungssystemen*. Schlussbericht BMBF Förderprojekt, Förderkennzeichen 01 IS C 16 A (Final Report of BMBF-funded project 01 IS C 16 A)
4. Nauck, D.; Klawonn, F.; Kruse, R.: *Neuronale Netze und Fuzzy Systeme*. Vieweg Verlag, Braunschweig/Wiesbaden, 1994.
5. Frey, C; Kuntze, H.-B.: *A neuro-fuzzy supervisory control System for industrial batch processes*. IEEE Transactions on Fuzzy Systems 9(4), S. 570–577, 2001.
6. Vermasvuori, M.; Enden, P.; Haavisto, S.; Jamsa-Jounela, S.-L.: The use of Kohonen self-organizing maps in process monitoring. *Intelligent Systems, 2002. Proceedings. 2002 First International IEEE Symposium*, vol. 3, 2002.
7. Haykin, S.: *Neural Networks a Comprehensive Foundation*. 2nd edition. Prentice Hall, Upper Saddle River, New Jersey, 1999.
8. Kohonen, T.: *Self-Organizing Maps*. 3rd Edition. Auflage Springer, Berlin, 1995.
9. Vincent, L.; Soille, P.: Watersheds in digital spaces: An efficient algorithm based on immersion simulation. *IEEE Transaction on Pattern Analysis and Machine Intelligence*, 13(6), June 1991.

Printed by Printforce, the Netherlands